# International Conference on River Regime

# International Conference on River Regime

*Edited by*
**Dr W.R. White**
Hydraulics Research Limited
Wallingford, Oxon OX10 8BA, UK

*Published on behalf of*
*HYDRAULICS RESEARCH LIMITED, WALLINGFORD*
by
JOHN WILEY & SONS
Chichester · New York · Brisbane · Toronto · Singapore

The selection and presentation of material and the opinions expressed in this publication are the sole responsibility of the authors concerned. The papers have been reproduced by offset printing from the authors' original typescript to minimise delay.



***Library of Congress Cataloging in Publication Data:***

International Conference on River Regime (1988 :
Wallingford, Oxfordshire)
International Conference on River Regime / edited by
W. R. White. p. cm.
Papers presented at the International Conference on River Regime held at Wallingford, England, 18–20 May, 1988.
Organised and sponsored by Hydraulics Research Limited and co-sponsored by the International Association for Hydraulic Resarch—P.
ISBN 0 471 91955 1
1. Rivers—Congresses. 2. River channels—Congresses.
I. White, W. R., Dr. II. Hydraulics Research (Firm)
III. Title.
GB1201.2.1575 1988
551.48'3—dc19

***British Library Cataloguing in Publication Data:***

International Conference on River Regime.
(1988 : Wallingford : England).
International Conference on River Regime
1. Rivers. Channels. Changes—Conference proceedings
I. Title II. White, W. R. III. Hydraulics Research
Limited
551.48'3

ISBN 0 471 91955 1

Printed and bound in Great Britain
by Anchor Brendon, Tiptree, Essex.

Papers presented at the International Conference on RIVER REGIME held at Wallingford, England : 18–20 May, 1988. Organised and sponsored by Hydraulics Research Limited and co-sponsored by the International Association for Hydraulics Research.

## Acknowledgements

The valuable assistance of the U.K. Organising Committee, the International Organising Committee and Panel of Referees is gratefully acknowledged.

## U.K. Organising Committee

| | |
|---|---|
| Dr. W.R. White (Chairman) | Hydraulics Research Ltd. |
| P. Ackers | Consultant |
| R. Angier | Consultant |
| Dr. R. Bettess | Hydraulics Research Ltd. |
| P.B. Clark | Binnie & Partners |
| T. Evans | Sir M. MacDonald & Partners |
| Dr. R.D. Hey | University of East Anglia |
| Professor P. Novak | Consultant |
| T.D. Pike | Overseas Development Administration |
| Professor B. Willetts | University of Aberdeen |
| R.M. Wingfield | Sir Alexander Gibb & Partners |
| Jacqueline Pounsford (Conference Organiser) | |

## International Organising Committee

| | |
|---|---|
| Professor M.S. Yalin (Chairman) | Queen's University at Kingston, Canada |
| Professor K. Ashida | Kyoto University, Japan |
| Professor W.H. Graf | Laboratoire d'Hydraulique, Lausanne, Switzerland |
| Professor J.F. Kennedy | University of Iowa, U.S.A. |
| Dr. L. Rakoczi | Research Centre for Water Resources, VITUKI, Hungary |
| Professor K.G. Ranga Raju | University of Roorkee, India |
| Professor H.W. Shen | University of California, U.S.A. |

International Conference on

# RIVER REGIME

18-20 May 1988

## CONTENTS

# Preface

Recently there have been major developments in the understanding of river regime which appear to be changing both the approach to the subject and its application. These are the concern of the present conference.

Regime theory was originally developed by Lacey and others. It was based on measurements of the geometry and slope of canals in the Indian sub-continent. These canals run at near constant discharges and generally carry a sediment load of sand and silt. The so called régime theory was entirely empirical and related the equilibrium dimensions and slope of the canals to discharge and sediment load.

Subsequently, regime concepts were applied to natural rivers with their variable flows and sediment loads. The concept of a significant or "dominant" discharge was introduced and the difficulties of defining and determining its value immediately became apparent. These problems remain with us today and it seems that further advances must rely on a better understanding of the physical processes of erosion and accretion and on the acquisition of field data from natural streams.

One of the most radical advances in regime theory that has occurred recently has been the development of an analytical or rational approach. This new approach relies on a theoretical description of the dominant fluvial processes rather than on a purely empirical approach. Whilst this represents a quantum leap we still have much to learn about river regime. Much remains to be done on the development of a theory which relates the channel size and shape to the fundamental physical processes involved. While progress has been made on combining equations describing sediment transport and alluvial friction, no approach to the problem of determining channel width has gained universal acceptance.

The range of application of regime theory has been steadily increasing — from canals to sand bed streams and rivers and, more recently, to gravel bed rivers — but these latter applications have indicated the shortcomings in our knowledge and have highlighted a lack of data. It should not be forgotten that regime theory started as an empirical subject and it is necessary that the subject remains firmly based on observations.

To take full advantage of these developments it is necessary that those working in these diverse areas remain in contact. It was therefore felt that there was a need for an international conference at which contacts could be developed and strengthened. That this conference should be held at Wallingford seems appropriate as HR has been closely involved with regime theory and its development since the time of its first Director, Sir Claude Inglis.

I would like to acknowledge the contribution made by the members of the National and International Organising Committees, particularly that of Professor Selim Yalin, Chairman of the International Committee. My gratitude also goes to Dr. Roger Bettess and other colleagues at Hydraulics Research who have worked so hard for the success of the conference. Finally, I would like to thank Jacqueline Pounsford, the Conference Organiser, for the efficient and enthusiastic way in which she has carried out her duties.

RODNEY WHITE
Hydraulics Research, Wallingford

International Conference on
**RIVER REGIME**
18-20 May, 1988



# DIMENSIONAL AND DIMENSIONLESS REGIME EQUATIONS FOR GRAVEL-BED RIVERS

R.D. Hey and G.L. Heritage
School of Environmental Sciences, University of East Anglia
Norwich, NR4 7TJ.

SUMMARY: Dimensional and non dimensional regime equations for gravel-bed rivers are compared and these show significant differences, even for the same data base. This is shown to be due to the omission of controlling variables and to statistical procedures. Aggregated data is used to derive more general regime equations.

## 1. INTRODUCTION

Several dimensional and dimensionless regime equations have been developed to predict the width, depth and slope of gravel-bed channels. However there are often major differences between equations, both within and between types, which can cause problems regarding choice of regime equations for practical engineering application.

This paper addresses this particular problem by reviewing existing dimensional and non dimensional regime equations, and those derived from published field data, to determine the nature and the reasons for differences between equations. Aggregated data is then used to derive general equations for engineering application.

## 2. DIMENSIONAL REGIME EQUATIONS

Traditional regime equations are dimensionally incorrect. Regression procedures are simply used to obtain a statistical relation between bankfull width (W), depth (d) and slope (S) and one or more parameters considered to control their values; for example, bankfull discharge (Q), median bed material size ($D_{50}$) and bed load transport ($Q_s$). These equations, being empirical, reflect the nature of the data base.

For gravel-bed rivers, equations developed by Bray (1) and those derived from Andrews' (2) and Hey and Thorne's data (3) are compared in table 1. Although Andrews' and Hey and Thorne's data distinguished between different bank vegetation densities and the latter included an estimate of bed load transport, this information is disregarded so that comparisons can be made between equations.

The depth equations compare very favourably with only slight variations in the values of the coefficients and exponents. This

International Conference on River Regime, edited by W.R.White.

| | Width (W) | | | Depth (d) | | | Slope (S) | | |
|---|---|---|---|---|---|---|---|---|---|
| | m | | | m | | | | | |
| | Exponent | | | Exponent | | | Exponent | | |
| | Coeff. | Q | $D_{50}$ | Coeff. | Q | $D_{50}$ | Coeff. | Q | $D_{50}$ |
| Bray (1) | | | | | | | | | |
| Dimensional | 3.86 | 0.53 | -0.07 | 0.25 | 0.33 | -0.03 | 0.06 | -0.33 | 0.59 |
| Non dimensional | 2.73 | 0.50 | -0.24 | 0.21 | 0.40 | 0.01 | 0.21 | -0.38 | 0.94 |
| Andrews (2) | | | | | | | | | |
| Dimensional | 3.53 | 0.53 | - | 0.24 | 0.37 | - | 0.22 | -0.36 | 0.87 |
| Non dimensional | 3.41 | 0.42 | -0.05 | 0.22 | 0.42 | 0.05 | 0.10 | -0.30 | 0.73 |
| Hey & Thorne (3) | | | | | | | | | |
| Dimensional | 3.67 | 0.45 | - | 0.22 | 0.36 | -0.12 | 0.09 | -0.29 | 0.69 |
| Non dimensional | 3.57 | 0.42 | -0.05 | 0.23 | 0.41 | -0.03 | 0.09 | -0.28 | 0.70 |
| Rationally based | | | | | | | | | |
| Dimensional | 4.33 | 0.39 | - | 0.15 | 0.49 | -0.14 | 0.11 | -0.32 | 0.48 |
| Non dimensional | 4.39 | 0.39 | 0.01 | 0.15 | 0.50 | -0.13 | 0.12 | -0.31 | 0.48 |
| Aggregated | | | | | | | | | |
| Dimensional | 2.55 | 0.53 | -0.06 | 0.24 | 0.37 | -0.04 | 0.06 | -0.43 | 0.33 |
| Non dimensional | 2.73 | 0.45 | -0.12 | 0.23 | 0.41 | -0.03 | 0.05 | 0.21 | 0.53 |

Table 1 Coefficients and exponents in power function regime equations (Dimensional and reduced non dimensional).

suggests that bankfull discharge and bed material size are the main factors controlling flow depth. Bank vegetation type and density and bed load transport would not appear to significantly affect its value (3). The exclusion of bed material size in Andrews' equation is probably due to small sample size.

In contrast there are significant differences between the various width equations. Previous work (2, 3, 4) has shown that bank vegetation has a major effect on channel width and, as the data could not be standardised to take it into account, this explains the variation in the values of the coefficients and exponents. If the sites in each data base were drawn from an unbiased sample of river sections, with regard to bank vegetation density, discharge and bed material size, then the discharge and bed material exponents would be independent of bank vegetation type and the coefficient would reflect the average bank vegetation condition and physical constants.

The slope equations also differ quite significantly, particularly the coefficients. Bank vegetation does not appear to be the cause of this variation and it probably results from the effect of bed load transport (3). The values of the coefficients and the exponents of discharge and bed material size being affected by covariance between the unmeasured bed load transport and the designated independent variables.

## 3. DIMENSIONLESS REGIME EQUATIONS

Dimensionless regime equations have previously been developed by Andrews (2) and Bray (1) based on different dimensionless groupings namely:-

$$W/D_{50},\ d/D_{50},\ Q/(\gamma_s/\gamma_w - 1)^{0.5}\ g^{0.5}\ D_{50}^{2.5} \qquad \text{(Andrews)}$$

and
$$\left[\frac{W\ g^{0.2}}{Q^{0.4}}\right],\ \left[\frac{V}{(gdS)^{0.5}}\right],\ D_{50}\ g^{0.2}/Q^{0.4} \qquad \text{(Bray)}$$

where $g$ is the acceleration due to gravity, $\gamma_s$ the specific weight of the sediment (normally 2.65 tonnes $m^{-3}$), $\gamma_w$ the specific weight of water and $V$ the bankfull average velocity.

It was anticipated that the dimensional and reduced dimensionless equations would show similar diversity; simply reflecting the values of the unmeasured controlling variables in each data base. Comparisons between Bray's and Andrews' dimensionless equations (table 1) indicate major differences between the width and slope equations which is surprising as the dimensionless groups are interchangeable. To illustrate this point consider Bray's dimensionless width as a function of dimensionless discharge.

$$\left[\frac{W\ g^{0.2}}{Q^{0.4}}\right] = f_1 \left[\frac{D_{50}\ g^{0.2}}{Q^{0.4}}\right]^{-x} \qquad (1)$$

Raising each side to the power of 2.5 and re-arranging gives

$$\left[\frac{W^{2.5}}{D_{50}^{2.5}}\right] \left[\frac{D_{50}^{2.5}\ g^{0.5}}{Q}\right] = f_2 \left[\frac{D_{50}^{2.5}\ g^{0.5}}{Q}\right]^{-x} \qquad (2)$$

$$\left[\frac{W}{D_{50}}\right] = f_3 \left[\frac{Q}{D_{50}^{2.5}\ g^{0.5}}\right]^{(1 + x)/2.5} \qquad (3)$$

which is identical to Andrews' dimensionless width and discharge assuming $(\gamma_s/\gamma_w - 1)^{0.5}$ is a constant.

The Hey and Thorne data were analysed using Andrews' dimensionless variables and this produced similar equations to those derived from Andrews' data (table 1). This suggests that there are either inconsistencies with Bray's data or problems with the regression model.

To assess these alternatives Bray's dimensionless width model was applied to Andrews' data with the following results

$$\left[\frac{W\ g^{0.2}}{Q^{0.4}}\right] = 2.76 \left[\frac{D_{50}\ g^{0.2}}{Q^{0.4}}\right]^{-0.282} \qquad (r^2 = 0.6) \qquad (4)$$

which reduces to

$$W = 1.54\ Q^{0.51}\ D_{50}^{-0.282} \qquad (5)$$

For comparison, Andrews' model gave

$$W/D_{50} = 6.12 \left[\frac{Q}{(\gamma_s/\gamma_w-1)^{0.5}\ g^{0.5}\ D_{50}^{2.5}}\right]^{0.42} \qquad (r^2 = 0.9) \qquad (6)$$

which reduces to

$$W = 3.41\ Q^{0.42}\ D_{50}^{-0.05} \qquad (7)$$

assuming constant values for $\gamma_s$, $\gamma_w$ and $g$.

Clearly the differences between equations (5) and (7) must result from the application of the regression model. This arises because:-

a) The Bray model produces relatively little variance in the dimensionless width variable, since Q increases as W increases. In these circumstances dimensionless discharge is unlikely to explain a large percentage of this variance, particularly when all the controlling variables are not specified and there is some measurement and sampling error in the field data. Hence correlation coefficients are low. In contrast, Andrews' model produces a large variance in dimensionless width, since $D_{50}$ tends to decrease as W increases, which tends to promote higher correlation coeffients with dimensionless discharge.

b) As a result of the low variance in Bray's dimensionless width, the exponent for the dimensionless discharge variable will approach zero, i.e. dimensionless width is virtually independent of the dimensionless discharge. Under these conditions the value of the dimensionless discharge exponent is very sensitive to the characteristics of the data base, particularly the value and distribution of any unmeasured controlling variables.

Ideally dimensionless groups need to be developed which produce exponents approaching unity as these are significantly different from both zero and infinity and are likely to have higher correlation coefficients and stable exponents. The Bray model suffers from this particular problem. As the discharge exponents in the traditional dimensional regime width equations lie in the range 0.45-0.53 (table 1), the dimensionless discharge exponent in equation (6) has to be low to accommodate the Q exponent of 0.4 in the dimensionless width term. In contrast the dimensionless groups in the Andrews model will produce a larger significant exponent for the dimensionless discharge.

Application of Bray's dimensionless width model to Hey and Thorne's data illustrates this point as the correlation coefficient is zero and the exponent is not significantly different from zero (eq. 8).

$$\left[\frac{W\ g^{0.2}}{Q^{0.4}}\right] = 2.63 \left[\frac{D_{50}\ g^{0.2}}{Q^{0.4}}\right]^{-0.045} \qquad (r^2 = 0) \qquad (8)$$

## 4. DIMENSIONAL V. DIMENSIONLESS EQUATIONS

Comparison between dimensional and dimensionless equations for each set of data indicate that Hey and Thorne's equations are very similar, while those of Andrews and Bray show some differences in coefficients and/or exponents (table 1). Bray's dimensionless equations are likely to be unstable (section 3) and this could account for the observed differences between his equations. The same argument does not apply to Andrews' equations and consequently it is necessary to determine whether the dimensional and dimensionless equations should give the same results.

To investigate this problem, 62 data sets were generated which were physically prescribed by continuity, flow resistance and sediment transport equations. The range of values in the data base were similar to Hey and Thorne's field data and were derived as follows. Values of width (W), depth (d), maximum depth ($d_m$), slope (S), bed material size ($D_{50}$ and $D_{84}$) were assumed according to the following rules. Bankfull width was less than 100m, width/depth ratios ranged between 5 and 15 and $\phi_{50}$ (the ratio of the Shields stress for the median diameter of the sub armour bed material to a Shields stress reference value of 0.0876) in the Parker, Klingeman and McClean (5) bed load transport equation was less than 1.65. Actual values were chosen to represent a range of river environments within the designated limits and values of discharge, sediment load and velocity were computed using the continuity, Hey flow resistance equation (6) and the Parker et al. (5) bed load transport equations. Generally the predicted velocity was less than 3m $s^{-1}$ and discharge and bed load transport were below 400 $m^3$ $s^{-1}$ and 3 kg $s^{-1}$ respectively which matched the UK field data.

Best fit dimensional and dimensionless (Andrews' model) equations were derived from this rationally based data (table 1). These are effectively identical, in spite of the problem of spurious correlation with the latter (7), indicating that this is a characteristic of a well ordered data set which has a rational physical basis. Any differences that occur between the Hey and Thorne equations and between the Andrews equations must be due to noise in the data base, or the operation of slightly different process equations in parts of the data base as, for example, when different flow resistance equations operate at low and high relative roughnesses.

The equations derived from the rationally based data differ somewhat from the Hey and Thorne equations. This results from the fact that width was prescribed within limits based on field data. Without the restrictions on W, W/d and $\phi_{50}$ prescribed above, there was even less conformance with the equations based on field data.

## 5. TOWARDS STANDARD REGIME EQUATIONS

Information from a variety of sources was screened to obtain standardised field data for a wide range of gravel-bed rivers. This included data from sites in Canada (1), USA (2, 8, 9, 10) and the UK (3, 4). In total 229 sets of data were assembled which met the following criteria.

a) The rivers should be self formed single thread gravel-bed channels flowing through alluvial material. They should be free from constraints such as bed rock outcrops or river training works.

b) The river should have a well defined flood plain.

c) The channel should be in regime and in natural equilibrium with the flow regime and sediment supply.

d) The reach average width and depth should be based on a minimum of four cross sections to define the average pool-riffle geometry. Width being defined as the water surface width and mean depth as the cross sectional area divided by water surface width.

e) Bankfull hydraulic geometry and discharge values were defined by the level of the valley flat.

f) Bankfull discharge values were based on gauged records.

g) Bed material size measurements relate to surface armour.

h) Bankfull water surface slope was obtained from river or bank edge longitudinal profiles over a distance of up to 30 channel widths.

Dimensional and dimensionless (Andrews' model) equations were derived from the data and the latter were reduced to conventional form for comparison purposes (table 1). There is not perfect agreement between the equations and this indicates a lack of uniformity in terms of data acquisition, noise in the data or data lying outside the limits of a particular governing equation.

These equations cannot be regarded as general equations but represent point estimates of width, depth and slope and are specific to a particular, and unknown, range of bank vegetation types and bed load transport rates. Specification of general regime equations, in which dimensional and dimensionless equations are identical awaits the collection of data which includes information on all the variables controlling channel morphology in particular bank vegetation type and bed load transport rates. Without this information the coefficients and exponents in the equations will reflect the values of the unmeasured controlling variables and not just physical constants. The collection of such data and the development of general equations remains a research priority.

## 6. CONCLUSIONS

Differences between dimensional and dimensionless regime equations can cause problems regarding choice of equation for practical engineering application.

Comparisons between the two approaches for different data sets indicates that the coefficients and the exponents of discharge and bed material size in the equations are very sensitive to the values of the unmeasured controlling variables. In addition the dimensionless equations are dependent on the form of the dimensionless groups used in the regression model.

Given a well ordered data base, one in which the measurement error is minimal and which is defined by a unique set of physical equations, albeit not necessary prescribed, the two approaches give identical results.

General regime equations for engineering application await the collection of data which includes all the controlling variables. The coefficients and exponents in the regression equation will then reflect the physical controls on channel morphology.

## 7. NOMENCLATURE

| | |
|---|---|
| $d$ | bankfull depth (m) |
| $d_m$ | bankfull maximum depth (m) |
| $D_{50}$ | median bed material size (surface armour) (m) |
| $D_{84}$ | bed material size, 84% finer (surface armour) |
| $g$ | acceleration due to gravity (m $s^{-2}$) |
| $Q$ | bankfull discharge ($m^3$ $s^{-1}$) |
| $Q_s$ | bed load transport rate (kg $s^{-1}$) |
| $S$ | bankfull water surface slope |
| $V$ | bankfull velocity (m $s^{-1}$) |
| $W$ | bankfull width (m) |
| $\gamma_s$ | specific weight of sediment |
| $\gamma_w$ | specific weight of water |
| $\phi_{50}$ | ratio of Shields stress for median diameter of sub armour material to Shields stress reference value of 0.0876 |

## 8. REFERENCES


1. Bray, D.I.: 'Regime equations for gravel-bed rivers'. In Gravel-Bed Rivers, eds. R.D. Hey, J.C. Bathurst and C.R. Thorne, Wiley, 1982, pp517-542.
2. Andrews, E.D.: 'Bed material entrainment and hydraulic geometry of gravel-bed rivers in Colorado'. Geol. Soc. Amer. Bull. 95, 1984, pp371-378.
3. Hey, R.D. & Thorne, C.R.: 'Stable channels with mobile gravel beds'. Jour. Hydraulic Engineering, Amer. Soc. Civ. Eng. 112 (8), 1986, pp671-689.
4. Charlton, F.G., Brown, P.M. & Benson, R.W.: 'The hydraulic geometry of some gravel bed rivers in Britain'. Report IT 180, Hydraulic Research Station, Wallingford, UK, 1978.
5. Parker, G., Klingeman, P.C. & McClean, D.G.: 'Bedload and size distribution in paved gravel-bed streams'. Journ. Hydraulic Division, Amer. Soc. Civ. Eng. 108 (HY4), 1982, 544-571.
6. Hey, R.D.: 'Flow resistance in gravel-bed rivers'. Journ. Hydraulics Division, Amer. Soc. Civ. Eng. 105 (HY4), 1979, pp365-379.
7. Benson, M.A.: 'Spurious correlation in hydraulics and hydrology'. Journ. Hydraulics Division, Amer. Soc. Civ. Eng. 91 (HY4), 1965, pp35-42.

8. Emmett, W.W.: 'The hydraulic geometry of some Alaskan streams south of the Yukon River'. U.S.G.S. Open File Report, 1972, 101pp.

9. Emmett, W.W.: 'The channels and waters of the Upper Salmon river area, Idaho'. U.S.G.S. Professional Paper 870-A, 1975, 116pp.

10. Williams, G.P.: 'Bankfull discharge of rivers'. Water Resources Research Vol. 14, No. 6, 1978, 1141-1154.

International Conference on
**RIVER REGIME**
18-20 May, 1988

PAPER A2

# DOWNSTREAM HYDRAULIC GEOMETRY OF NONCOHESIVE ALLUVIAL CHANNELS

Pierre Y. Julien
Engineering Research Center, Colorado State University
Fort Collins, Colorado 80523 U.S.A.

SUMMARY: Four fundamental relationships are combined to theoretically define the downstream hydraulic geometry of noncohesive alluvial channels. The flow depth, channel width, velocity and slope of similar channels depend primarily on discharge and sediment size.

## 1. INTRODUCTION

Alluvial channels are known to self adjust their slope, width, depth, and velocity. The shape of alluvial channels is modified during extreme events depending on the magnitude of water and sediment discharges. Downstream hydraulic geometry relationships describe the shape of alluvial channels as it progresses in the downstream direction. Early investigations on downstream hydraulic geometry relationships by Leopold and Maddock (Ref. 6) showed that the channel width, the flow depth, the mean flow velocity and the slope can be written as power functions of discharge. Similar relationships have been derived by Li et al. (Ref. 8).

## 2. FUNDAMENTAL RELATIONSHIPS

In this paper, hydraulic geometry relationships for the channel width $W$, mean flow depth $h$, mean flow velocity $\bar{U}$ and channel slope $S$ are theoretically derived after considering the similarities between noncohesive alluvial streams. The following four fundamental concepts are considered: (1) continuity; (2) flow resistance; (3) longitudinal mobility; and (4) radial mobility. The first three concepts are briefly summarized whereas the radial mobility is discussed more thoroughly.

### 2.1 Continuity

The continuity equation describes conservation of mass flux or discharge $Q$ in steady open-channels:

$$Q = Wh\bar{U} \tag{1}$$

The cross section area, $Wh$, is measured normal to the mean velocity vector.

## 2.2 Flow Resistance

The resistance relationship proposed by Einstein and Chien (Ref. 2) for turbulent flows over an hydraulically rough boundary reads:

$$\bar{U} \sim \sqrt{g} \left(\frac{h}{d_s}\right)^a h^{1/2} S^{1/2} \tag{2}$$

in which $g$ is the gravitational acceleration, $d_s$ is the sediment size, and $a$ is an exponent proportional to the relative roughness. For very small relative roughness, $a = 0$, and Eq. 2 reduces to the Chezy equation. The well-known Manning-Strickler relationship corresponds to $a = 1/6$. In gravel and cobble bed streams, the exponent $a$ becomes larger. For example, Kellerhals (Ref. 5) suggested $a = 0.25$; Leopold and Wolman (Ref. 7) found $a = 0.5$ when $0.7 < h/d_s < 10$; and Church (Ref. 1) showed that $0.43 < a < 3.35$.

## 2.3 Longitudinal Mobility

For steady uniform flow conditions in open channels, the longitudinal boundary shear stress $\tau_\theta$ is

$$\tau_\theta = \rho g\ h\ S \tag{3}$$

in which $\rho$ is the mass density of water. The stability of straight alluvial channels is described by the relative magnitude of the shear forces exerted on the boundary and the resistance to motion of individual noncohesive particles. The ratio of these two forces defines the longitudinal mobility factor or Shields number $\tau_\theta^*$:

$$\tau_\theta^* = \frac{\tau_\theta}{(\rho_s - \rho) g\ d_s} \tag{4}$$

in which $\rho_s$ is the density of particles. The Shields diagram defines the threshold conditions for sediment motion, and $\tau_\theta^*$ remains constant for turbulent flows over a rough boundary. It must also be recognized that the sediment load increases with $\tau_\theta^*$ beyond the critical value.

Considering constant values of $\rho_s$, $\rho$ and $g$ in Eqs. 3 and 4, the longitudinal mobility of sediments reduces to

$$\tau_\theta^*\ d_s \sim h\ S \tag{5}$$

In straight channels without secondary circulation, this relationship describes the cross-sectional geometry of an alluvial channel. Indeed, Lane (Ref. 3) analyzed the stability of particles on channel banks after considering the shear stress in the downstream direction and the weight of the particles. The ideal cross-sectional geometry of noncohesive alluvial channels has been theoretically defined (Refs. 9, 13) when the threshold condition for all sediment particles on the banks and the bed is satisfied.

## 2.4 Radial Mobility

In curved channels, the concept of radial mobility is introduced to describe the lateral stability of alluvial channel bends subjected to secondary flows. Secondary circulation is generated through a change in channel orientation in the downstream direction. The streamlines near the surface are deviated toward the outer bank whereas those near the boundary are deviated toward the inner bank. As a result, the velocity near the bed, the tangential bed shear stress and the drag on the bed particles are deviated toward the inner bank.

Flow in bends is analyzed in a cylindrical system of coordinates. The relative magnitude of the radial acceleration terms shows that the centrifugal acceleration is counterbalanced by a pressure gradient and radial shear stress, as suggested by Rozovskii (Ref. 10)

$$\frac{\bar{u}^2}{r} = g\, S_r - \frac{1}{\rho}\frac{\partial \tau_r}{\partial z} \tag{6}$$

in which $\bar{u}$ is the local downstream velocity, $r$ is the radius of curvature, $S_r$ is the radial water surface slope, $\tau_r$ is the radial boundary shear stress, and $z$ is the vertical coordinate. With reference to Fig. 1, the transverse boundary shear stress $\tau_{rR}$ at point R and the transverse water surface slope $S_{rA}$ at point A serve as scaling factors. These scaling factors are used to define the following dimensionless parameters for the channel width $w' = w/W$; the flow depth, $z' = z/h$; the radius of curvature, $r' = r/W$; the velocity $u' = \bar{u}/\bar{U}$; the radial shear stress, $\tau_r' = \tau_r/\tau_{rR}$; and the radial surface slope, $S_r' = S_r/S_{rA}$. The element of volume $d\forall = r d\phi dz dw$ of a given reach of length $r d\phi$ is reduced to a dimensionless volume $d\forall' = d\forall/W^2 h$.

The equation of motion (Eq. 6) is multiplied by $\rho$ and $d\forall$, reduced to a dimensionless form, then integrated over the dimensionless volume $\forall'$ of the reach. The resulting dimensionless momentum equation in the radial direction is

$$Wh\rho\bar{U}^2 \int_{\forall'} \frac{u'^2}{r'}\, d\forall' = \rho g W^2 h S_{rA} \int_{\forall'} S_r'\, d\forall' - W^2 \tau_{rR} \int_{\forall'} \frac{\partial \tau_r'}{\partial z'}\, d\forall' \tag{7}$$

The corresponding force diagram is sketched in Fig. 1 after denoting the centrifugal force on the left-hand side of the equation by $F_c$, the pressure force on the right-hand side by $F_p$ and the shear force exerted on the fluid by $F_s$. The pressure force balances the sum of the centrifugal force exerted at a distance $c$ above point A and the shear force exerted at a distance $d$ below point A. Moment equilibrium around point A gives:

$$\pi_r = \rho \frac{h\bar{U}^2}{W\,\tau_{rR}} = \frac{d}{c}\, \frac{\int_{\forall'} \frac{\partial \tau_r'}{\partial z'}\, d\forall'}{\int_{\forall'} \frac{u'^2}{r'}\, d\forall'} \; . \tag{8}$$

The dimensionless parameter $\pi_r$ denotes the ratio of the centrifugal force generating motion to the shear force abating the motion and dissipating energy. Owing to the similarities in velocity profiles and shear stress distributions in dimensionless form, the ratio d/c in Eq. 8 and the volume integrals remain constant. It is therefore assumed as a first approximation that a constant value of $\pi_r$ represents flow conditions in bends.

The mobility of sediment particles in a curved channel is now considered. In dimensionless form, the radial mobility factor or radial Shields parameter $\tau_r^*$ describes the ratio of the radial shear force to the submerged weight of a single particle.

$$\tau_r^* = \frac{\tau_{rR}}{(\rho_s - \rho) g d_s} \tag{9}$$

The stability of bed particles on sloping planes when the drag force deviates from the downstream direction has been analyzed by Stevens and Simons (Ref. 14). It is shown from this method that a downward shear stress component mobilizes the particles near the outer bank. On the other hand, a slight upward shear stress component stabilizes the particles and induces deposition. When this upward component becomes large, however, the particles are mobilized and scour occurs near the inner bank. Removal of bed material at the toe of the outer bank is initiated and at low values of the radial Shields parameter, deposition occurs near the river bank. At larger values of the radial Shields parameter, however, particles are eroded along the full width of the channel, which results in channel widening. The result of combining Eq. 8 with Eq. 9 while keeping $\pi_r$, $\rho_s$, $\rho$ and $g$ constant is:

$$\tau_r^* \sim \frac{h\bar{U}^2}{Wd_s} \tag{10}$$

Equation 10 relates radial mobility to hydraulic geometry and bed sediment size, the radial mobility factor being inversely proportional to the width-depth ratio at a given flow velocity. Therefore, widening of a narrow channel through bank erosion is expected to occur until $\tau_r^*$ is reduced sufficiently to stabilize the bed and bank material.

## 3. DOWNSTREAM HYDRAULIC GEOMETRY OF NONCOHESIVE ALLUVIAL CHANNELS

The above cited relationships are combined to algebraically derive the downstream hydraulic geometry of alluvial channels: continuity (Eq. 1); resistance to flow (Eq. 2); longitudinal mobility (Eq. 5); and radial mobility (Eq. 10). As a result of a similar derivation presented by Julien and Simons (Ref. 4), flow depth $h$, channel width $W$, velocity $\bar{U}$, and slope $S$ can be written as power functions of discharge $Q$, bed sediment size $d_s$, longitudinal and radial mobility factors $\tau_\theta^*$ and $\tau_r^*$:

$$h \sim Q^{\frac{1}{2+3a}}\, d_s^{\frac{6a-1}{4+6a}}\, \tau_\theta^{*\frac{-3}{4+6a}}\, \tau_r^{*\frac{1}{2+3a}} \tag{11}$$

$$W \sim Q^{\frac{1+2a}{2+3a}}\, d_s^{-\frac{1+4a}{4+6a}}\, \tau_\theta^{*\frac{1}{4+6a}}\, \tau_r^{*-\frac{1+a}{2+3a}} \tag{12}$$

$$\bar{U} \sim Q^{\frac{a}{2+3a}}\, d_s^{\frac{1-a}{2+3a}}\, \tau_\theta^{*\frac{1}{2+3a}}\, \tau_r^{*\frac{a}{2+3a}} \tag{13}$$

$$S \sim Q^{-\frac{1}{2+3a}}\, d_s^{\frac{5}{4+6a}}\, \tau_\theta^{*\frac{7+6a}{4+6a}}\, \tau_r^{*\frac{-1}{2+3a}} \tag{14}$$

These equations are essentially valid for similar channels in noncohesive alluvial material. Typical values of the exponent $a$ range from 0 to more than 0.4 depending on the relative roughness. For constant values of $\tau_\theta^*$ and $\tau_r^*$, the resulting relationships describing flow depth, channel width, velocity and slope reduce to simple power functions of discharge $Q$ and sediment size $d_s$.

## 4. DISCUSSION

### 4.1 Flow Depth, h (Eq. 11)

The exponent of water discharge varies slightly around 0.4 and the exponent of sediment size can be either positive or negative depending on the relative roughness. Interestingly, the flow depth becomes independent of sediment size when Manning's equation is applicable (a = 1/6). The negative values of the exponent of $\tau_\theta^*$ indicate that flow depth is inversely proportional to sediment mobility. Thus, flow depth is proportional to discharge and inversely proportional to sediment transport. These results are in agreement with the well-known (Refs. 11, 12) qualitative geomorphic principles for fluvial systems: $Q^+ \sim h^+$; $Q_s^+ \sim h^-$ (the signs + and - denote respectively an increase or decrease of the variable considered).

### 4.2 Channel Width, W (Eq. 12)

Channel width is proportional to discharge and sediment transport and inversely proportional to sediment size. The exponent of discharge remains close to 0.5 regardless of the relative roughness. This is in perfect agreement with Lacey's regime equation. Sediment mobility appears to have a slight influence on the channel width since the exponent of $\tau_\theta^*$ remains relatively small. An increase in sediment load results in a small increase in channel width. Here again, the qualitative concepts ($Q^+ \cong W^+$; $Q_s^+ \cong W^+$) proposed by Schumm (Ref. 11) find theoretical support in Eq. 12.

### 4.3 Average Velocity, $\bar{U}$ (Eq. 13)

The average velocity is proportional to discharge, sediment size, and sediment mobility. A significant reduction of the exponent of $d_s$ and a slight increase in the exponent of discharge result from increasing relative roughness.

### 4.4 Slope, S (Eq. 14)

The slope is proportional to sediment size and sediment mobility and inversely proportional to discharge. The discharge exponent increases while the sediment size exponent decreases with increasing relative roughness. Equation 14 supports qualitative geomorphic principles ($Q^+ \sim S^-$; $d_s^+ \sim S^+$; $Q_s^+ \sim S^+$) found by Simons et al. (Ref. 12) and Schumm (Ref. 11).

## 5. SUMMARY AND CONCLUSION

The downstream hydraulic geometry of alluvial channels described by flow depth, channel width, velocity, and slope is examined. Four fundamental relationships (continuity, flow resistance, longitudinal mobility, and radial mobility) are combined to derive theoretical hydraulic geometry relationships for turbulent flows over a rough boundary.

As a result, the hydraulic geometry can be written as a power function of discharge, bed sediment size, and two mobility factors. The theoretical results support very well the qualitative relationships of Simons and Schumm. For constant values of the mobility factors, the hydraulic geometry is then reduced to a power function of sediment size and discharge.

## 6. ACKNOWLEDGEMENTS

This study has been undertaken with the support of a NATO post-doctoral fellowship from the Natural Sciences and Engineering Research Council of Canada. The writer is indebted to Prof. D. B. Simons for his guidance during the course of this study.

## 7. REFERENCES


1. Church, M.: "Baffin Island sanders: A study of arctic fluvial process". Geological Survey of Canada Bulletin, 216, 1972, 208 pp.

2. Einstein, H. A. and Chien, N.: "Similarity of distorted river models with unstable bed". Proceedings of the ASCE, Vol. 80, No. 566, 1954, 21 pp.

3. Lane, E. W.: "Design of stable channels". Transactions of the ASCE, Vol. 120, Proc. Paper 2776, 1955, pp. 1234-1279.

4. Julien, P. Y. and Simons, D. B.: "Analysis of hydraulic geometry relationships in alluvial channels". Report No. CER83-84PYJ-DBS45, Civil Engineering Department, Colorado State University, Fort Collins, Colo., 1984, 47 pp.

5. Kellerhals, R.: "Stable channels with gravel-paved beds". Journal of the Waterways and Harbors Division, ASCE, Vol. 93, No. WW1, Proc. Paper 5091, Feb. 1967, pp. 63-84.

6. Leopold, L. B. and Maddock, T.: "The hydraulic geometry of stream channels and some physiographic implications". USGS Prof. Paper 252, 1953.

7. Leopold, L. B. and Wolman, M. G.: "River channel patterns: braided, meandering and straight". USGS Prof. Paper 282-B, 1957, pp. 38-85.

8. Li, R. M., Simons, D. B. and Stevens, M. A.: "Morphology of cobble streams in small watersheds". Journal of the Hydraulics Division, ASCE, Vol. 102, No. HY8, Aug. 1976, pp. 1101-1117.

9. Raudkivi, A. J.: "Loose Boundary Hydraulics" Pergamon Press, 2nd Ed., 1976, pp. 250-254.

10. Rozovskii, I. L.: "Flow of water in bends of open channels". Academy of Science of the Ukrainian SSR, Kiev, Translation by Y. Prushansky, Israel Program for Scientific Translations, S. Monson, Jerusalem, PST Cat. No. 363, 1957.

11. Schumm, S. A.: "The Fluvial System" Wiley Interscience, 1977, 338 pp.

12. Simons, D. B., Richardson, E. V. and Mahmood, K.: "One-dimensional modeling of alluvial rivers". Chapter 19 in Unsteady Flow in Open Channels, Water Resources Publications, Fort Collins, Colo., 1975, pp. 813-877.

13. Simons, D. B. and Senturk, F.: "Sediment transport technology" Water Resources Publications, Fort Collins, Colo., 1977, 807 pp.

14. Stevens, M. A. and Simons, D. B.: "Stability analysis for coarse granular material on slopes" Chapter 17 in River Mechanics," H. W. Shen, Ed., Water Resources Publications, Fort Collins, Colo., 1971, 27 pp.


## 8. NOTATION

The following symbols are used in this paper:

a exponent of the resistance equation

c distance between $F_c$ and $F_p$

d distance between $F_p$ and $F_s$

| | |
|---|---|
| $d_s$ | sediment size of bed material |
| $F_c$ | centrifugal force |
| $F_p$ | pressure force |
| $F_s$ | shear force |
| $g$ | gravitational acceleration |
| $h$ | average depth |
| $Q$ | discharge |
| $r$ | radius of curvature |
| $r'$ | dimensionless radius of curvature |
| $S$ | longitudinal slope |
| $S_r$ | radial water surface slope |
| $S_{rA}$ | radial water surface slope at point A |
| $S'_r$ | dimensionless radial water surface slope |
| $\bar{u}$ | local longitudinal velocity |
| $u'$ | dimensionless longitudinal velocity |
| $\bar{U}$ | average longitudinal velocity |
| $\forall$ | volume of fluid within a given reach |
| $\forall'$ | dimensionless volume of fluid |
| $w$ | channel width at the elevation z |
| $w'$ | dimensionless width |
| $W$ | channel width |
| $z$ | vertical coordinate |
| $z'$ | dimensionless depth |
| $\gamma$ | specific weight of water |
| $\phi$ | angle between the downstream direction and the down-valley direction |
| $\pi_r$ | dimensionless ratio of centrifugal to radial shear forces in bends |
| $\rho$ | mass density of water |
| $\rho_s$ | density of particles |
| $\tau_o$ | longitudinal boundary shear stress |

| | |
|---|---|
| $\tau_r$ | radial boundary shear stress |
| $\tau_{rR}$ | radial boundary shear stress at point R |
| $\tau'_r$ | dimensionless radial boundary shear stress |
| $\tau^*_\theta$ | longitudinal mobility factor |
| $\tau^*_r$ | radial mobility factor |

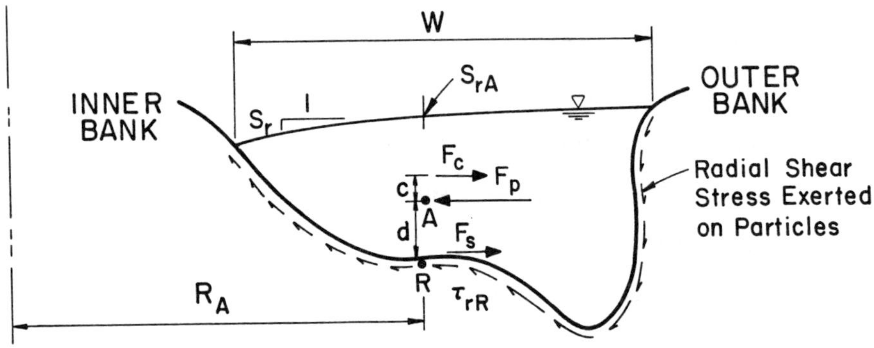


Figure 1. Force diagram in the radial direction.

International Conference on
RIVER REGIME
18-20 May, 1988

PAPER A3

# DYNAMIC ADJUSTMENTS OF CHANNEL WIDTH AND SLOPE

Chih Ted Yang
U.S. Bureau of Reclamation
Denver, Colorado, U.S.A.

Albert Molinas
Department of Civil Engineering, Colorado State University
Fort Collins, Colorado, U.S.A.

SUMMARY: The theory of minimum rate of energy dissipation and its simplified versions are applied to explain the variation of channel width and longitudinal profile adopted by an alluvial channel under different flow and sediment conditions. Field and computer generated data are used to verify and support the theory.

## 1. INTRODUCTION

Rivers adjust their channel geometry and longitudinal profile in order to be compatible with the changing hydrologic, hydraulic, geological, and man-made constraints. Different theories and hypotheses have been proposed in the past to explain these adjustments. The theory of minimum rate of energy dissipation (Yang and Song, 1986) and its simplified versions will be used in this paper to explain these adjustments at a given reach of river.

Theoretical analyses based on the theory of minimum rate of energy dissipation will be made to explain the impact of changing water discharge, channel width, and longitudinal profile on the rate of energy dissipation of an alluvial system. The effect of exchanging objective and constraint functions in the minimization process will be studied to explain some seemingly paradoxical phenomena. Data from a natural river and a stable canal will be used to verify and support the results of theoretical analyses.

## 2. THEORETICAL ANALYSES

The general theory of minimum rate of energy dissipation states that for a closed and dissipative system under dynamic equilibrium conditions, the system's total rate of energy dissipation is a minimum. This minimum value is dependent upon the constraints applied to the system. If the system is not at its equilibrium condition, its total rate of energy dissipation is not at its minimum value. However, the system will adjust itself in such a manner that its rate of energy dissipation can be reduced to reach a minimum value, which is compatible with the constraints, and regain equilibrium (Yang and Song, 1986). For an equilibrium river system carrying a discharge Q through a reach with a horizontal distance L and a vertical drop of $\Delta H$, the total rate of energy dissipation is

$$P = \gamma Q \Delta H = \gamma \iiint us\,dx\,dy\,dz \qquad (1)$$

International Conference on River Regime, edited by W.R.White.

where P is the total rate of energy dissipation; $\gamma$ is the specific weight of water; Q is the water discharge; $\Delta H$ is the vertical drop of the study reach with a reach length L; u is the local velocity; s is the local energy slope; x, y, z are the coordinates in the longitudinal, lateral, and vertical directions, respectively. According to the theory of minimum rate of energy dissipation,

$$P = \gamma Q \Delta H = \text{a minimum} \tag{2}$$

when the rate of energy dissipation due to sediment transport can be neglected. When an average energy slope S can be used for the study reach with a constant horizontal length L, equation (2) becomes

$$QS = \text{a minimum} \tag{3}$$

Equation (3) has been commonly referred to as the theory of minimum stream power. This simplified theory was applied by Chang and Hill (1977), Chang (1979a, 1979b, 1980), Yang and Song (1979), and Yang (1983) to the study of delta development, channel geometry, channel pattern, stable channel design, and the dynamic adjustments of alluvial channels. When the local velocity does not vary significantly across a channel of constant width and can be represented by an average velocity V, equations (1) and (2) can be reduced to

$$VS = \text{a minimum} \tag{4}$$

Equation (4) has been referred to as the theory of minimum unit stream power. The theory of minimum unit stream power has been used by Yang (1971a, 1971b, 1976) to explain the basic reason for river meandering, the formation of riffles and pools, and the prediction of the roughness coefficient of alluvial channels under different conditions of flow, sediment, and bed forms. Equations (2), (3), and (4) are special cases of the general theory of minimum rate of energy dissipation (Yang and Song, 1986) with varying degrees of simplification. These three equations will be used in the following sections to explain the dynamic adjustments of channel geometry and slope. The inclusion of the rate of energy dissipation due to sediment transport in the study of channel geometry and slope in accordance with the theory of minimum rate of energy dissipation, was done by Yang, et al. (1981).

The rate of energy dissipation increases with increasing water discharge regardless of whether VS, P, or QS is used in the minimization process. In other words, when Q is considered as a constraint in the minimization process, a higher discharge should be associated with a higher rate of energy dissipation.

Under high flows or channel forming discharge, minimization of P in the longitudinal direction requires that

$$\frac{dP}{dx} = L\gamma\left(Q\,\frac{dS}{dx} + S\,\frac{dQ}{dx}\right) = 0 \tag{5}$$

Equation (5) was used by Yang (1983) to explain the variation of S in the longitudinal direction. For a short reach without an increase in discharge in the downstream direction, $dQ/dx = 0$. This requires that $dS/dx = 0$ or S = a constant throughout the reach. Thus, at high flows, the water surface slope or energy slope should be fairly uniform throughout the reach. However, for a much longer reach, the discharge Q will gradually increase in the downstream direction due to inflows from ground water and tributaries. A positive value of $dQ/dx$ in equation (5) requires $dS/dx$ to be negative. This is in agreement with data from natural rivers, where the energy slopes decrease gradually in the downstream direction.

In the previous analysis, Q is considered as a constraint in the minimization process. At low flows, equation (1) can be rewritten as

$$P = \gamma Q \Delta H = \gamma Q(LS) = \gamma(\frac{AL}{\Delta t}) \Delta H \tag{6}$$

The drop in the energy gradient, $\Delta H$, over a given reach is predetermined by the overall longitudinal profile formed by the channel forming discharge. Thus, during low flows, $\Delta H$ acts as a constraint in the minimization process. According to equation (6), minimization of P with a given $\Delta H$ means minimization of Q or maximization of time-of-travel $\Delta t$ through the reach with a given horizontal length L. It has been shown by Yang (1971b) that a uniform slope will have a minimum time-of-travel $\Delta t$. According to the theory of minimum rate of energy dissipation, a river will adjust itself to increase $\Delta t$ by the formation of pools and riffles with uneven slopes throughout the reach at low flows. The effect of pool-riffle sequence on the rate of energy dissipation decreases with increasing water depth or discharge. Thus, the uniformity of water surface or energy slope increases with increasing discharge and the slope becomes uniform at high flows.

The requirement for a natural river to take a uniform slope at a high flow and a non-uniform slope at a low flow seems to be a paradox at first glance. These phenomena can be explained by the duality principle (Hadley, 1964) used in the minimization process. The duality principle states that for each minimization problem, there is a dual maximization problem by interchanging the objective function and constraint. Furthermore, the minimum of the original problem is equal to the maximum of the dual problem. At the channel forming high discharge, channel geometry and slope are being shaped by the given Q. In this case, Q is the constraint and the objective is to minimize P through the minimization of $\Delta H$ or S in the longitudinal direction, in accordance with equations (5) and (6). The result is a uniform slope when Q is a constant, or a concave longitudinal profile when Q increases in the downstream direction. In the case of low flows, the overall channel geometry and longitudinal profile are determined by the previous channel forming discharge. In this case, the total drop $\Delta H$ of a study reach is predetermined and acts as the constraint. The minimization of P is achieved by minimization of Q or maximization of $\Delta t$ in accordance with equation (6). The response of a river at low flow is the formation of riffles and pools to slow down the flow and increase its time-of-travel through a given reach. Not recognizing these dual phenomena could lead to misinterpretation and improper applications of the theory of minimum rate of energy dissipation and its simplified versions.

Song and Yang (1980) and Molinas (1984) pointed out the importance of having the upstream and downstream boundary conditions or channel controls in the stream power minimization for gradually varied open channel flows. The importance of having well-defined upstream and downstream controls and their impacts on P will be illustrated by the analyses of data.

## 3. ANALYSIS OF DATA

Stall and Yang (1972) measured the water surface profiles and their associated hydraulic and geometric characteristics of a reach of the Middle Fork Vermilion River near Oakwood, Illinois. Measurements were made at low, medium low, and high flow conditions. The water discharges were 0.533 cms (18.85 cfs), 1.235 cms (43.64 cfs), and 19.272 cms (681.00 cfs) with the associated frequency of occurrence of 0.8, 0.6, and 0.15, respectively. Figure 1 shows the water and bed surface profiles measured at these three discharges. The times-of-travel through the study reach were computed by Stall and Yang (1972). The mean width, depth, slope, and Manning's roughness coefficient of an equivalent channel with uniform width, depth, slope, and roughness through the study reach were computed by them. The time-of-travel through this hypothetical equivalent channel reach was then computed. It was assumed that the total drop $\Delta H$ through the study reach for a given Q remains the same for the actual and the equivalent uniform slope profiles. Table 1 summarizes the values of unit stream power under different flow conditions computed by Stall and Yang (1972). It is apparent that the formation of riffles and pools at a low flow with uneven slopes through the study reach can reduce the unit stream power. Similar conclusions can be reached for P and QS. Results in table 1 also indicate the effectiveness of the riffle-pool sequence in reducing the rate of energy dissipation decreases with increasing discharge. At a high flow near the channel forming discharge, the water surface slope or energy slope becomes uniform through the study reach and the riffle-pool phenomenon disappears.

The characteristics of channel geometry and longitudinal slope at channel forming discharge can be determined by the analysis of either a natural river at high flows or a stable regime channel which was designed to carry a certain discharge. In this case, discharge Q becomes the constraint and $\Delta H$ becomes the objective function to be minimized. This minimization process was used by Molinas (1984) to study a stable reach of the Rangpur Canal in the Punjab region of Pakistan (Harza Engineering, 1963). Some of the computed results are shown in figures 2 to 9. The channel cross sections are trapezoidal in shape and the longitudinal bed profile is fairly uniform, as shown in figure 2. The original or the existing case in figure 2 is compared with three hypothetical pool-riffle cases by raising station 13000 artificially 0.061 m (0.2 ft), 0.183 m (0.6 ft), and 0.366 m (1.2 ft), respectively. A water surface profile computer model (Molinas, 1983) was used to compute water surface profiles with various discharges and bed elevation adjustments at station 13000. Figure 3 shows the computed water surface profiles with the bed elevation at station 13000 raised by 0.366 m (1.2 ft). The total rate of energy dissipation P through the study reach was computed for each case and compared with the original uniform bed profile case. The total stream power ratio shown in figure 4 is the ratio of total stream power between a hypothetical pool-riffle profile and the existing smooth longitudinal bed profile throughout the study reach. Figure 4 indicates that the total stream power ratio decreases with increasing discharge. Figure 4 also shows that when Q is considered as a given constraint, the existing smooth profile is the one with minimum P value. This can be explained by the results shown in figures 2 and 3. For Q = 14.15 cms (500 cfs) the total head loss between stations 3000 and 15000 is

$$\Delta H_1 = 147.66 \text{ m} - 147.20 \text{ m} = 0.46 \text{ m } (1.5 \text{ ft})$$

for the existing smooth bed profile in figure 2. For the same discharge, the total head loss in figure 3 is

$$\Delta H_4 = 147.77 \text{ m} - 147.22 \text{ m} = 0.55 \text{ m } (1.8 \text{ ft})$$

which is larger than $\Delta H_1$. Thus, when Q is acting as a given constraint, minimization of P means minimization of $\Delta H$ to form a uniform longitudinal profile in accordance with theory of minimum rate of energy dissipation. For a given discharge Q, increase of channel width will reduce P and QS. Figure 5 shows an example of the variation of P with respect to W at station 11000 of the Rangpur Canal. It should be pointed out that sediment transport was not considered in the results shown in figure 5. An alluvial channel in equilibrium must be able to carry the sediment supplied to the system. Further increase of channel width beyond the equilibrium width will cause sediment deposition and channel narrowing. Thus, sediment load should also be considered as a constraint in the minimization process for the determination of optimum channel width.

In the preceding minimization exercise, no upstream or downstream constraints were attached to the width, depth, or energy slope. It was pointed out by Song and Yang (1980) and Molinas (1984) that the upstream and downstream channel controls have to be specified in the minimization process to reach realistic and meaningful solutions. The effects of these constraints will be illustrated with some numerical examples. Figure 6 shows the effect of downstream control on water surface profile for a discharge of 55.47 cms (1960 cfs) in the Rangpur Canal. For a given bed thalweg profile, a fixed downstream water surface elevation and a water discharge, increase of channel width W will reduce stream power QS or slope S. The increase of W is permissible only if the water surface elevation at the upstream end is allowed to be lowered. Once the hydraulic parameters are computed, the corresponding sediment load in the Rangpur Canal was computed by Yang's (1973) sediment transport equation. The relationship between stream power, channel width, and bed material load of the Rangpur Canal with a constant downstream controlled water surface elevation was computed by Molinas (1984) and is shown in figure 7. Figure 7 shows that stream power decreases with increasing channel width. However, increase of channel width will reduce the canal's sediment-carrying capacity. Channel width cannot be increased indefinitely without causing sediment deposition and reverse the process. Thus, the existing equilibrium width is the one with the minimum rate of energy dissipation under the constraint that a given amount of bed material load has to be transported through the study reach. Similar results can be obtained by having the control at the upstream end and allowing the downstream water surface elevation to vary.

The upstream and downstream water surface elevations can both be fixed for a given discharge and the bed surface profile can be raised or lowered parallel to the existing profile. In this case, QS is a constant for a given discharge, as shown in figure 8. The increase of channel width is accompanied by raising the bed surface profile, or vice versa, as shown in figure 8. However, figure 9 indicates that even if Q, S, and QS are given, sediment transport capacity decreases with increasing channel width. Thus, the maximum width a channel can have in order to minimize its rate of energy dissipation is still limited by the constraint of sediment load or a channel's sediment carrying capacity.

## 4. SUMMARY AND CONCLUSION

The theory of minimum rate of energy dissipation and its simplified versions have been applied to explain the adjustments of channel width and slope. The importance of having proper constraints in the minimization process and the effect of exchanging the objective and constraint functions have been discussed. Field data and hypothetical situations generated by a computer model are used to verify the theoretical results. These studies reached the following conclusions:

1. The rate of energy dissipation of an alluvial system increases with increasing water discharge within a relatively short study reach. The overall channel slope should decrease with increasing discharge in the downstream direction.

2. For a given water discharge, the rate of energy dissipation decreases with increasing channel width.

3. With low flows in a channel carved by its channel forming discharge, the formation of pool-riffle sequences can increase the time-of-travel through a given reach and reduce the rate of energy dissipation. The effect of pool-riffle sequence on the reduction of the rate of energy dissipation decreases with increasing discharge.

4. At high flows or channel forming discharge, water discharge becomes a constraint in the minimization process. Under this condition, the longitudinal profile associated with minimum rate of energy dissipation is a uniform profile.

5. The upstream and downstream boundary conditions or controls have to be clearly defined in the application of the theory of minimum rate of energy dissipation to obtain realistic channel width and profile.

6. Sediment load is a constraint which should not be overlooked in the minimization process. Channel width cannot be increased indefinitely in reducing the rate of energy dissipation without causing sediment deposition and reversing the process.

## 5. NOMENCLATURE

$A$ = Channel cross-sectional area;
$\Delta H$ = Vertical drop of a study reach;
$L$ = Horizontal distance of a study reach;
$P$ = Total rate of energy dissipation;
$Q$ = Water discharge;
$QS$ = Stream power;
$s$ = Local energy or water surface slope;
$S$ = Average energy or water surface slope;
$\Delta t$ = Time of travel;
$u$ = Local velocity;
$V$ = Average velocity;
$VS$ = Unit stream power;
$x, y, z$ = Coordinates in the longitudinal, lateral, and vertical directions, respectively; and
$\gamma$ = Specific weight of water.

## 6. ACKNOWLEDGEMENT

The authors appreciate the financial support provided by the U.S. Bureau of Reclamation which made this study possible.

## 7. REFERENCES

1. Chang, H. H., "Geometry of Rivers in Regime," Journal of the Hydraulics Division, ASCE, Vol. 105, No. HY6, 1979a, pp. 691-706.

2. Chang, H. H., "Minimum Stream Power and River Channel Patterns,"Journal of Hydrology, Vol. 41, 1979b, pp. 303-327.

3. Chang, H. H., "Stable Alluvial Canal Design," Journal of the Hydraulics Division, ASCE, Vol. 106, No. HY5, 1980, pp. 873-891.

4. Chang, H. H., and J. C. Hill, "Minimum Stream Power for Rivers and Deltas," Journal of the Hydraulics Division, ASCE, Vol. 103, No. HY12, 1977, pp. 1375-1389.

5. Hadley, G., "Nonlinear and Dynamic Programming," Addison-Wesley Publishing Company, 1964, pp. 36, 204, 238, and 467.

6. Harza Engineering Company, "Canal and Headworks Data Observation Programme, 1962 Data Tabulation," Report prepared for West Pakistan Water and Power Development Authority, April 1963.

7. Molinas, A., "Stream Power Minimization to Determine Equilibrium Channel Geometry in Alluvial Rivers and Canals," Department of Civil Engineering, Colorado State University, Report No. CER84-85AM24, December 1984.

8. Molinas, A., "Mathematical Simulation of Rivers by the Use of Streamtubes," Department of Civil Engineering, Colorado State University, Report No. CER83-84AM11, September 1983.

9. Song, C. C. S., and C. T. Yang, "Minimum Stream Power: Theory," Journal of the Hydraulics Division, ASCE, Vol. 106, No. HY9, 1980, pp. 1477-1487.

10. Stall, J. B., and C. T. Yang, "Hydraulic Geometry and Low Streamflow Regimen," University of Illinois Water Resources Center Research Report No. 54, Urbana, Illinois, July 1972, pp. 11-16.

11. Yang, C. T., "On River Meanders," Journal of Hydrology, Vol. 13, 1971a, pp. 213-253.

12. Yang, C. T., "Formation of Riffles and Pools," Water Resources Research, Vol. 7, No. 6, 1971b, pp. 1567-1574.

13. Yang, C. T., "Incipient Motion and Sediment Transport," Journal of the Hydraulics Division, ASCE, Vol. 99, No. HY10, 1973, pp. 1679-1704.

14. Yang, C. T., "Minimum Rate of Energy Dissipation and River Morphology," D. B. Simons Symposium on Erosion and Sedimentation, Colorado State University, Fort Collins, Colorado, 1983, pp. 3.2-3.19.

15. Yang, C. T., "Minimum Unit Stream Power and Fluvial Hydraulics," Journal of the Hydraulics Division, ASCE, Vol. 102, No. HY7, 1976, pp. 919-934.

16. Yang, C. T., and C. C. S. Song, "Dynamic Adjustments of Alluvial Channels," Adjustments of the Fluvial System, edited by D. D. Rhodes and G. P. Williams, Kendall/Hunt Publishing, Dubuque, Iowa, 1979, pp. 56-67.

17. Yang, C. T., and C. C. S. Song, "Theory of Minimum Energy and Energy Dissipation Rate," Encyclopedia of Fluid Mechanics, Vol. 1, Chapter 11, Gulf Publishing Company, 1986, pp. 353-399.

18. Yang, C. T., C. C. S. Song, and M. J. Woldenberg, "Hydraulic Geometry and Minimum Rate of Energy Dissipation," Water Resources Research, Vol. 17, No. 4, 1981, pp. 1014-1018.

Table 1. - Unit stream power reduction for three flow conditions, Middle Fork Vermilion River near Oakwood, Illinois (from Stall and Yang, 1972)

| Flow Q | | Unit stream power (watts/kg) | | |
|---|---|---|---|---|
| Amount Q (cms) | Frequency F (% of days) | Through pools and riffles | Through uniform equivalent channel | Reduction % |
| 0.53 | 0.80 | 0.00084 | 0.00114 | 26.2 |
| 1.23 | 0.60 | 0.00135 | 0.00176 | 23.3 |
| 19.27 | 0.15 | 0.00754 | 0.00754 | 0 |

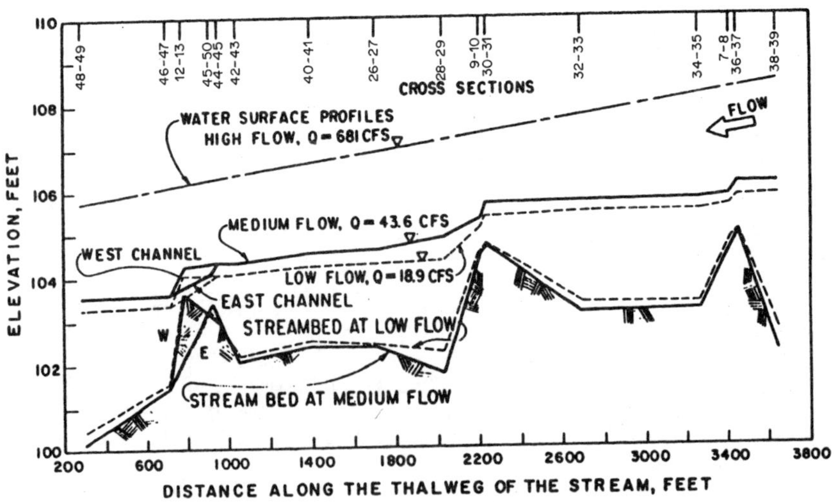


Figure 1. Bed and water surface profile for the test reach of the Middle Fork Vermilion River near Oakwood, Illinois, for the three flow conditions (from Stall and Yang, 1972).

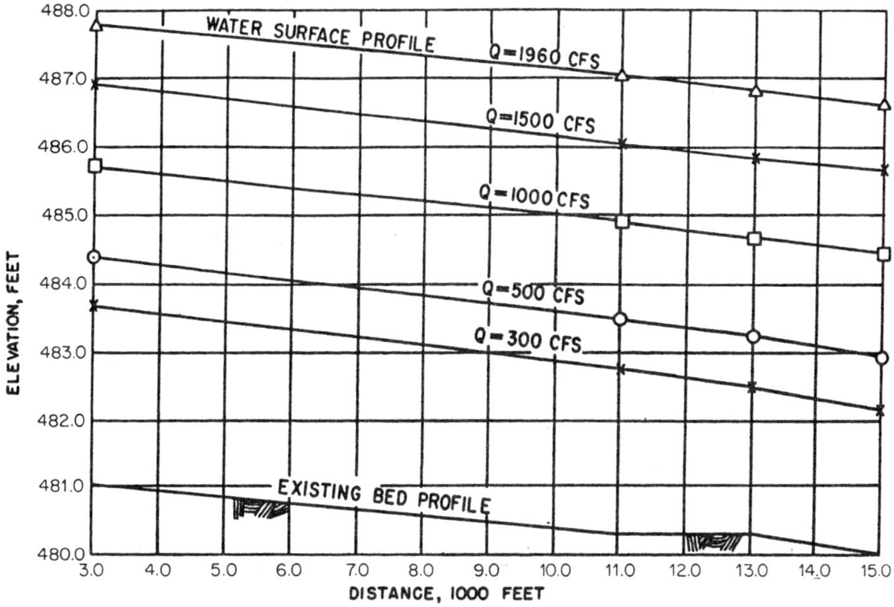


Figure 2. Water surface profiles for case No. 1, existing smooth bed profile.

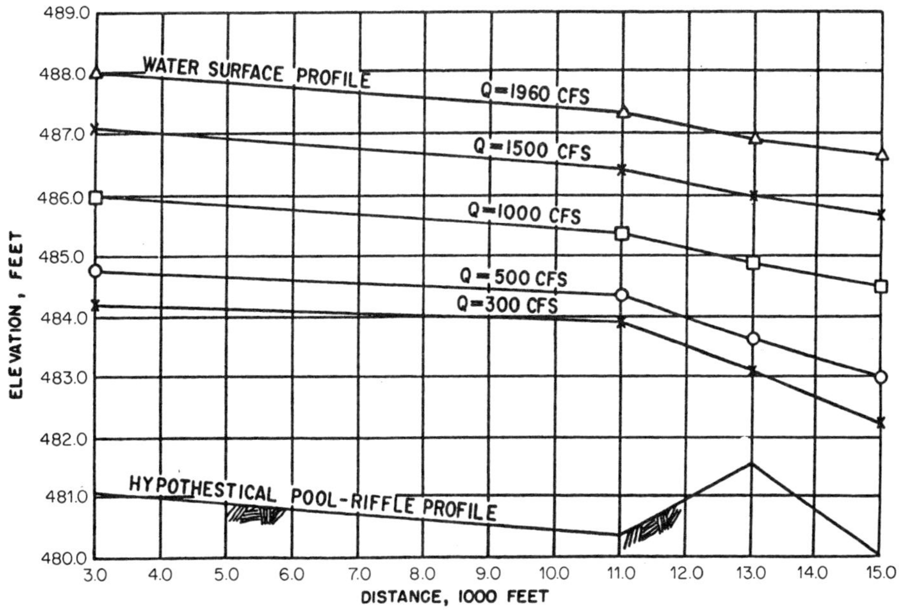


Figure 3. Water surface profiles for case No. 4, raising station 13000 by 0.366 m (1.2 ft).

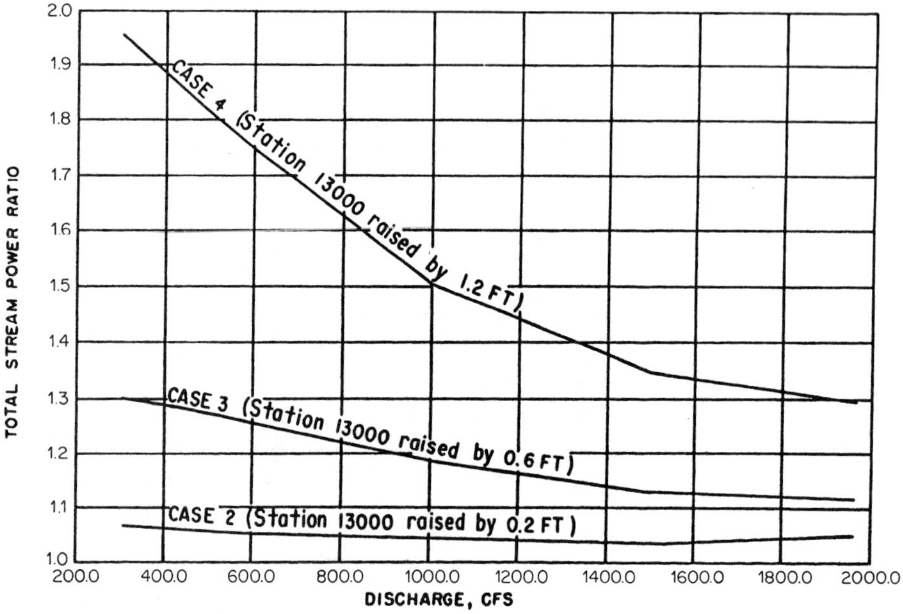


Figure 4. Stream power ratio versus discharge for different cases.

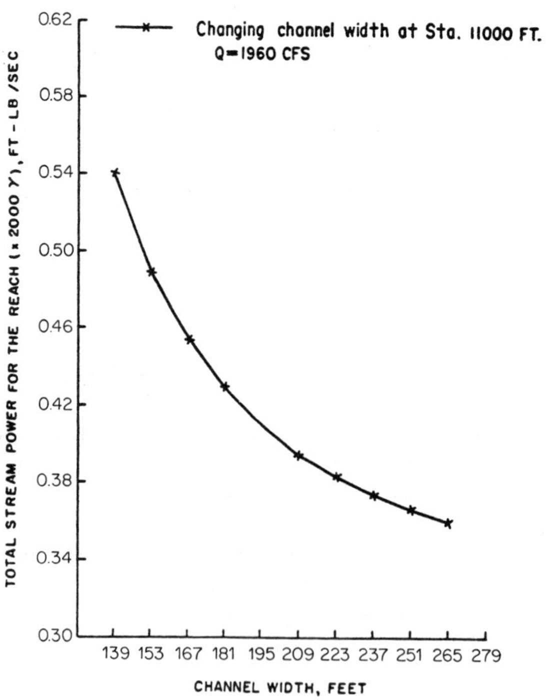


Figure 5. Channel width versus computed total stream power at station 11000.

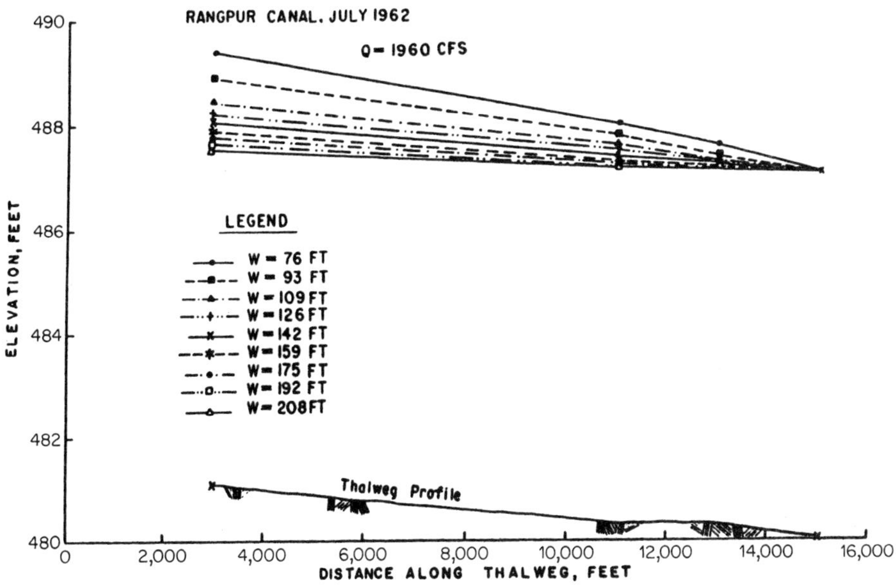


Figure 6. Water surface profiles for a constant downstream control elevation.

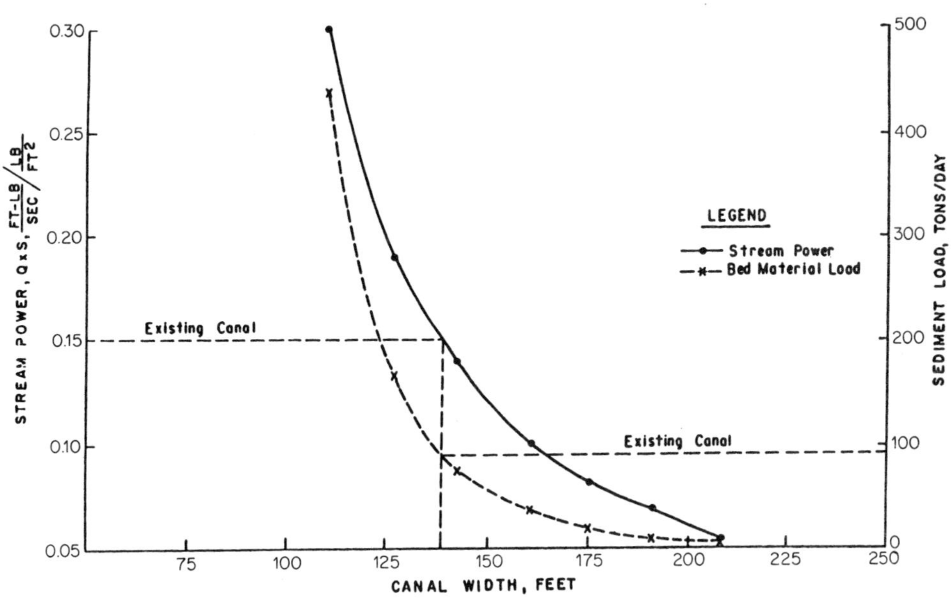


Figure 7. Stream power and sediment load versus width for constant downstream water surface control elevation.

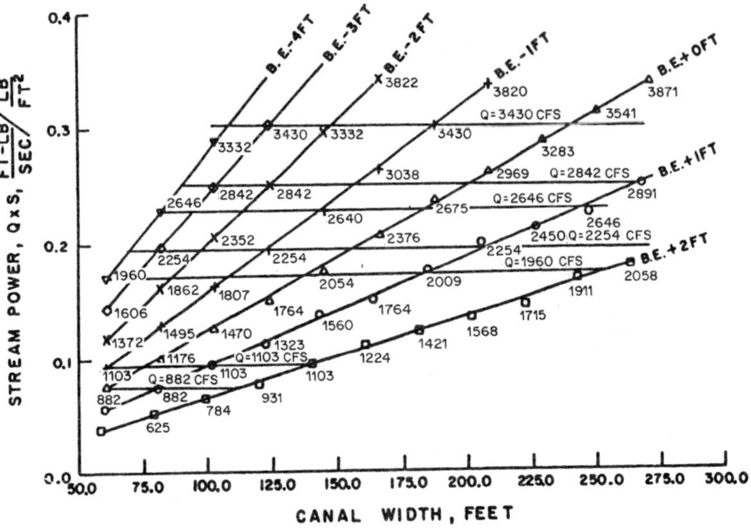


Figure 8. Stream power versus canal width for a family of constant discharge curves with fixed upstream and downstream water surface elevations.

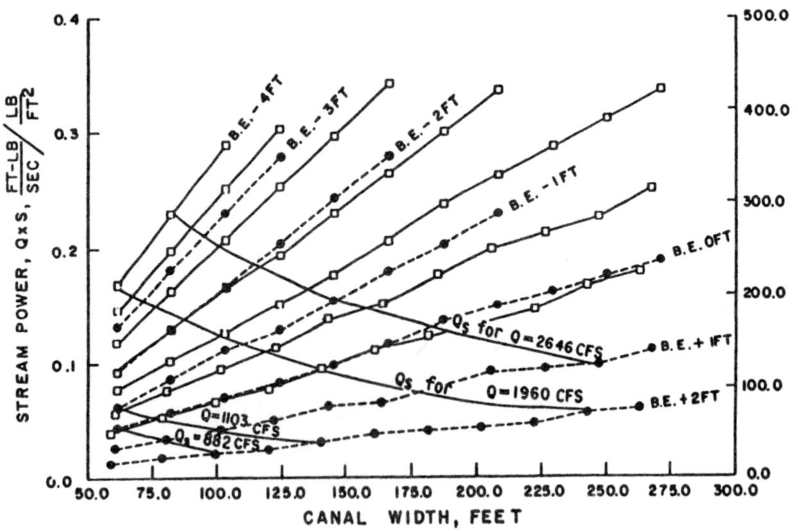


Figure 9. Sediment transport capacity curves for constant discharge with fixed upstream and downstream water surface elevations.

International Conference on
**RIVER REGIME**
18-20 May, 1988

PAPER A4

# PREDICTION OF HYDRAULIC GEOMETRY OF GRAVEL-BED STREAMS USING THE MINIMUM STREAM POWER CONCEPT

C.R. Thorne, Queen Mary College, Univ. London, London E1 4NS, UK
H.H. Chang, San Diego State University, San Diego, Ca 92182, USA
R.D. Hey, University of East Anglia, Norwich NR4 7TJ, UK

SUMMARY: The hydraulic geometry of gravel-bed streams can be predicted using the minimum stream power concept. Application of the theory to data for British rivers shows that although the predictions are reasonable, considerable scatter exists. This scatter can be reduced by taking into account the strength of bank vegetation at the sites and a modification that greatly improves the accuracy of the predictions is suggested.

## 1. INTRODUCTION

Although the hydraulic geometry of gravel-bed rivers is significantly different to that of sand-bed rivers, relatively little information is available on their characteristics and few predictive methods specifically designed for gravel-bed rivers are available. Two fundamental approaches to the prediction of hydraulic geometry have been developed. The first and most widely adopted approach is based on empirical regime equations. The second is based on solution of the governing equations of flow.

In a recent paper Hey and Thorne (1) presented regime equations for gravel-bed rivers with mobile bed material. In this paper the data collected by Hey and Thorne are used to test, evaluate, and modify the theoretical approach developed by Chang (2) based on simultaneous solution of equations for flow resistance, bed load transport and minimum stream power. Chang's method was first published in 1980, and included testing against available data for American and Canadian rivers. However, to the best of our knowledge, it has not been subsequently tested against other reliable data sets.

In this paper we first present the governing equations in the minimum stream power analysis. Unlike those in the original paper, these are expressed in SI units. Next we describe the field data for British gravel-bed rivers that is used to test the prediction method. Calculated and measured values of the dependent variables of width, depth and velocity are presented graphically. The results are then critically examined with particular emphasis being placed on the role of vegetation in affecting the geometry. Empirical modifications to the calculated widths and depths to account for bank vegetation are suggested and it is shown that they greatly improve the accuracy of the predictions. Finally, the major findings are listed as a series of conclusions.

## 2. THEORETICAL MODEL

### 2.1 Rational Model

In the model, the channel is assumed to be straight, with a roughly trapezoidal cross-section consisting of a mobile central region and immobile banks. Analytical determination of the hydraulic geometry can be accomplished if the applicable physical relations are sufficient to describe the unknowns or

International Conference on River Regime, edited by W.R.White.

degrees of freedom in the channel morphology. If the bank slope z is selected according to the physical properties of the bank material, then it is necessary to determine the width and depth in order to define the hydraulic geometry. The velocity is not treated as an unknown as it is defined by continuity for a given combination of discharge, width and depth.

The hydraulic geometry of alluvial channels is self-formed. Over a long period the hydraulic geometry is adjusted so that the channel is able to carry the discharge of water and sediment input from upstream, through the materials forming the channel boundary. Therefore, in the long-term, the independent variables are water and sediment discharge, and the bed and bank materials, while the dependent variables are width, depth and channel slope. The discharge is highly variable through time and so it is necessary to select a dominant discharge to represent the effects of the range of flows experienced by the channel. There is a considerable volume of field and flume data to support the use of bankfull discharge as the dominant discharge (3). The sediment discharge used in the analysis is that associated with bankfull flow.

Physical relations useful for the analysis of hydraulic geometry are grouped into four categories pertaining to fluid mechanics, sediment transport, bank stability and dynamic equilibrium. Although the role of each category of physical relations can be separately identified, they are nevertheless closely interrelated in the fluvial processes. Under dynamic equilibrium, the dynamics of two-phase (liquid-solid) flow and the channel form have adapted to a state of mutual compatibility. The physical relationships for steady, uniform flow are:

1. Relationships related to fluid dynamics. The steady forms of the flow continuity equation and the flow resistance equation are included in this category. Because it is a stage-discharge predictor, the flow resistance equation is useful for depth determination.

2. Bed load relationships. The mobile bed as molded by the inter-action of flow and bed load, must be compatible with sediment transport mechanics. Therefore, appropriate bed load equations are useful in determining the hydraulic geometry. For a prescribed cross-section, discharge and load, the channel slope can be computed using the bed load equation.

3. Criteria for bank stability. The submerged bank slope is related to the bank stability, which varies with the bank material. For noncohesive (sand or gravel) banks, the bank slope is limited by the angle of repose of the material. For cohesive (silt and clay) banks, the bank slope can be steeper, generally in direct relation to the silt-clay content. Although bank stability will certainly affect the hydraulic geometry, it does not constrain the development of the stable channel configuration in the long run. In other words, bank stability is a necessary factor for width formation, but it is not the sufficient factor.

4. Relationships for the state of dynamic equilibrium. An alluvial channel in dynamic equilibrium must comply with certain physical relationships governing the equilibrium state. In the analytical approach, the concept of minimum stream power is used to provide the condition of dynamic equilibrium. This concept states that for an alluvial channel, the necessary and sufficient condition of equilibrium occurs when the stream power per unit channel length, $\gamma QS$, is a minimum subject to given constraints. Hence, an alluvial channel with water discharge Q and sediment load $Q_s$ as independent variables tends to establish its width, depth and slope such that QS is a minimum. Since Q is a given parameter, minimum $\gamma QS$ also means minimum channel slope S.

In establishing the relationship for hydraulic geometry, physical relations employed include a flow resistance relation, a bed load equation and the concept of minimum stream power. The bank slope is assumed in the analysis for average conditions of bank stability. Only a brief outline of the basis of the analysis is given here. A complete description may be found in Chang (2).

2.2 Flow Resistance Equation

In gravel-bed rivers flow resistance is primarily due to grain roughness and can be predicted using an equation based on the relative roughness (4). Usually equations are either of logarithmic or power form. In this analysis the equation of Bray (5) is used as this has been found to give reliable results, especially for large rivers. Bray's equation is:

$$1/f^{0.5} = 1.36\ (D/d_{50})^{0.281} \tag{1}$$

with f = Darcy-Weisbach friction factor, D = mean flow depth, and $d_{50}$ = median bed material size.

2.3 Bed Load Equation

For rivers with coarse bed material the transport of bed material load is predominantly as bed load. The Parker bed load equation (6,7) was selected for use in the analysis because it is specifically intended for use with coarse sediments. The bed load is calculated from:

$$\Phi = 11.2\ (1/\Psi - 0.03)^{4.5}\ \Psi^{3} \tag{2}$$

$$\Phi = q_s\ /\{g[(\rho_s - \rho)/\rho]\ d_{50}^{3}\ \}^{0.5} \tag{3}$$

$$1/\Psi = T\ /\ \{\ g\ (\rho_s - \rho)\ d_{50}\ \}\ = T^{*} \tag{4}$$

with $q_s$ = unit bed load, T = shear stress, $T^*$ = dimensionless or Shields shear stress, $\rho$ = water density, $\rho_s$ = sediment density, $d_{50}$ = median sediment size, g = gravitational acceleration.

Eqn. 2 has its lower bound at $T^* = 0.03$, which is the threshold Shields stress for entrainment of gravel. The bed load equation was intended for use in the range $0.03 < T^* < 0.1$, which covers all the data used by Parker but does not extend to the steepest slopes used in this analysis. For steep slopes we use the equation:

$$\Phi = 6.62\ \{(1/\Psi) - 0.03\}^{5}\ \Psi^{3.9} \tag{5}$$

to extend the range of Parker's equation, based on Meyer-Peter's data (8).

2.4 Bank Slope Equation

The bank slope z must be selected as part of the input. As discussed previously, the bank slope for gravel banks will always be less than the angle of repose but it is also affected by other physical conditions. In general, small streams in confined valleys have steeper bank slopes than large streams in broad flood plains. In the analysis, the z value is selected to be 2.5 for bankfull discharge of 0.3 cumecs, 3 for 3 cumecs, 3.5 for 30 cumecs, 4 for 300 cumecs, and 4.5 for 3000 cumecs. This selection of z values is taken to represent the general trend for gravel-bed rivers. It was noted in Chang's original paper that other values of z were not precluded, but at that time insufficient information was available to allow the z factor to be adjusted to account for bank material properties or vegetation.

2.5 Rational Regime Relations

The three independent conditions of flow resistance, sediment transport and minimum stream power are solved using a computer program. For given values of discharge, sediment load, bed material size and bank slope, the width and depth at a unique minimum slope are obtained by successive trials in the computation. Detailed accounts of the program may be found in references (9,10).

The computer output was used to develop regime style equations for width and depth. Relations were first obtained for $d_{50}$ = 50 mm. Other gravel sizes yield similar relations, differing only in their slope. If the slope is multiplied by a factor $(50/d_{50})^{1.15}$ all the relations can combined and these are plotted in Fig.1.

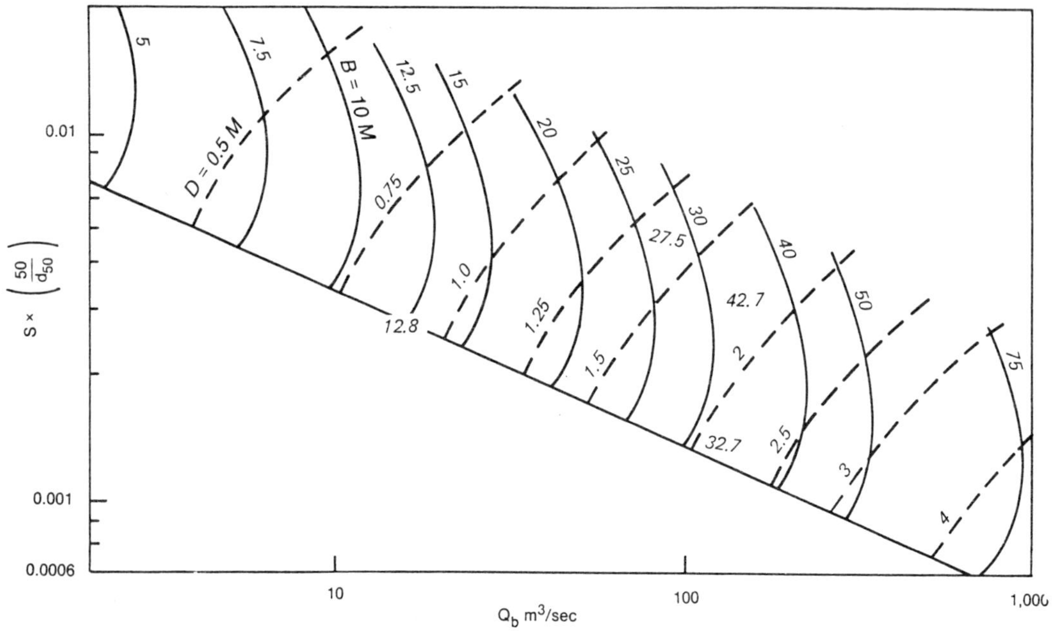


Figure 1. Analytical Curves for the Hydraulic Geometry of Gravel-Bed Streams

The equations are:

$$B = \{3.101 + 0.4053\ (\ln 0.0002383 d^{1.15}/SQ^{0.42})^2\}\ Q^{0.47} \qquad (6)$$

$$D = \{0.2829 + 0.0569\ \ln(0.0000989 d^{1.15}/SQ^{0.42})\}\ Q^{0.42} \qquad (7)$$

with B = water surface width, S = channel slope and Q = bankfull discharge. The equations and graphical solutions are intended for channels with mobile bed material at bankfull flow. The threshold condition for bed material movement is defined by:

$$S_{cr}' = 0.00009892\ d^{1.15}\ Q^{-0.42} \qquad (8)$$

with $S_{cr}$ = threshold slope for incipient motion of bed material. At the lower boundary of vanishingly small bed load, the theory reduces to that of Lane's threshold channel design (11).

For mobile bed conditions Fig.1 shows that width varies with both discharge and slope. Along lines parallel to $S_{cr}$ width varies only with discharge according to:

$$B = a\ Q^{0.47} \qquad (9)$$

which is similar to regime equations developed elswhere (1). For a constant discharge, width varies with the logarithm of slope in a parabolic relation so that at steep slopes well above the threshold condition width increases rapidly, tending toward a braided channel.

For a given sediment size, the depth varies with discharge and slope. Along lines parallel to $S_{cr}$, depth varies with discharge according to:

$$D = b\,Q^{0.42} \tag{10}$$

which is similar to the result found in regime analyses (1). At a constant discharge, depth varies inversely with slope, decreasing rapidly as the braided condition is approached at steep slopes.

The rational equation for velocity is simply the continuity equation for steady, uniform flow:

$$V = Q/BD \tag{11}$$

The velocity is directly proportional to discharge and slope, except for steep slopes where velocity actually decreases with slope due to reduced hydraulic efficiency of the channel resulting from rapid increase in width and decrease in depth.

## 3. COMPARISON TO FIELD DATA

### 3.1 Data Collection

Field data were collected from 62 sites adjacent to gauging stations on single thread, gravel-bed rivers in Great Britain. Full details of the site selection criteria, data collection procedures, and the complete data base may be found in Hey and Thorne (1) and only a brief summary is given here.

At each location five cross-sections were surveyed, covering the basic geomorphic unit of three riffles and two intermediate pools. These sections represented the maximum local variability for the reach. Long profiles of the channel bed, water surface and flood plain edge were also surveyed, extending from the study reach to the stage board at the gauging station. Surface bed material was sampled at each cross-section using the Wolman technique. All the sites had armoured beds and estimates of substrate size distribution were made for all the sites on the basis of comparison of surface and substrate sizes observed at ten sites. The banks were mostly composite in form, consisting of fine overbank deposits of silt and sand, underlain by coarse in-channel deposits of sandy gravel. Upper bank material was sampled by coring and strength tests were carried out on the undisturbed cores. Planform sinuosity was determined from large scale-maps. Finally, additional information on the geomorphic characteristics of the site was recorded on a data sheet with particular attention being paid to the type and density of bank vegetation.

### 3.2 Data Analysis

The field data were analysed by computer to produce values for the average bankfull hydraulic geometry for the four downstream sections. Bankfull elevation at each cross-section was identified using the minimum width/depth criterion for active flood plain height as suggested by Wolman (12). Bankfull discharge was obtained from the stage-discharge relation for the gauging station corresponding to bankfull flow in the reach, as suggested by Williams (13). No sediment transport measurements corresponding to bankfull flow were available. Sediment load associated with bankfull discharge was calculated using the Parker, Klingeman and McClean equation (14) with the data for the upstream riffle which was excluded from development of the hydraulic geometry values. This produced an independent estimate for sediment load.

The bank strength data were found to be very uniform between the sites and as a result this parameter had no significant affect on the variation in hydraulic geometry. Bank vegetation did vary between sites however and to incorperate this into the analysis sites were grouped into four categories on the basis of the type and density of vegetation present. The classes are:

| | |
|---|---|
| Type I | Grassy banks with no trees or shrubs; |
| Type II | 1 to 5% tree and shrub cover; |
| Type III | 5 to 50% tree and shrub cover; |
| Type IV | Greater than 50% tree and shrub cover, or incised into flood plain. |

3.3 Results

The theoretical analysis was applied to the data for British gravel-bed rivers and the calculated and measured values of width, depth and velocity are compared in Fig.2 a, b and c respectively.

The results all show a fair degree of general agreement between calculated and measured values, with the points in each graph scattered fairly evenly about the line of perfect agreement. However, the level of scatter is too high to allow recomendation of the 1980 method as the basis for the design of stable

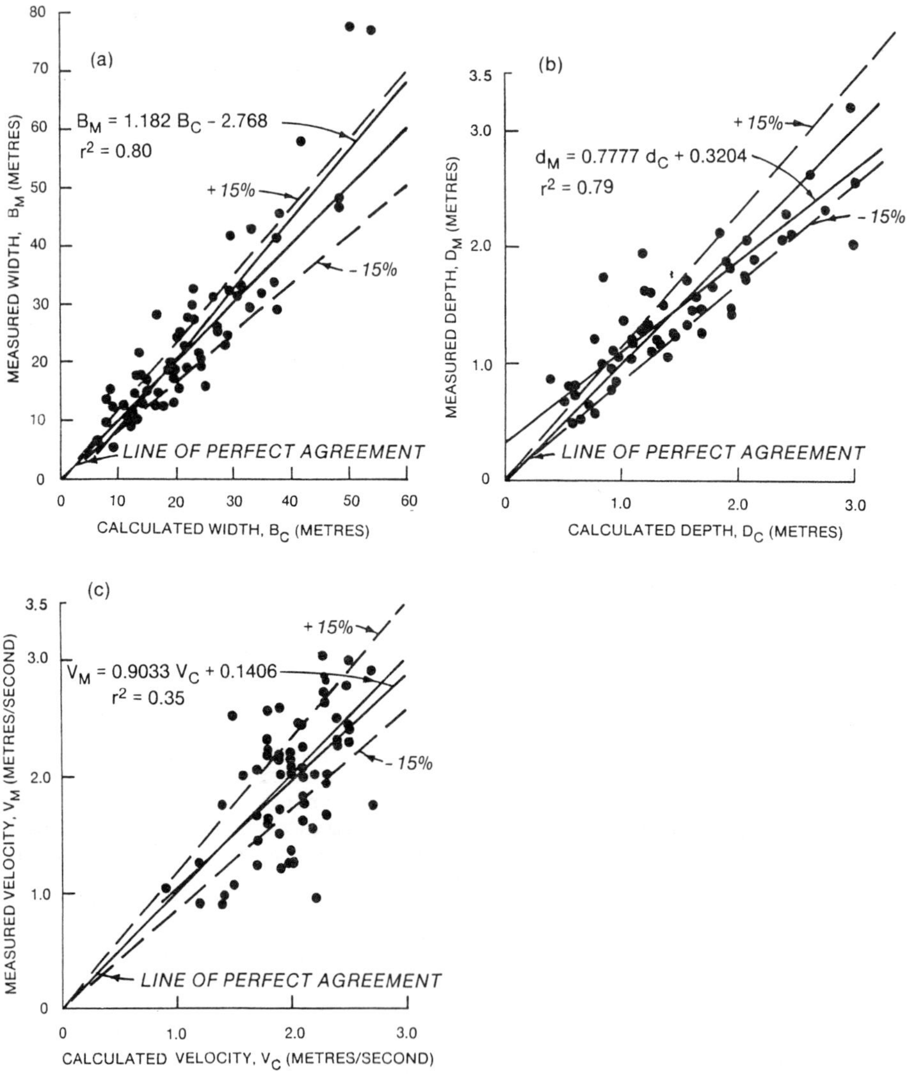


**Figure 2. Calculated versus Measured Values of a) Width, b) Depth, c) Velocity**

channels, without further enhancement. For example, taking an error margin of plus or minus 15% as being acceptable for design purposes, for width only 29 out of 62 sites (47%) fall into this range. For depth, the result is better at 39 cases (63%), but in only 17 cases (27%) does the method predict *both* width and depth to within 15% of their actual values. The analytical method does not do well in predicting velocity, with only 47% of predictions being within 15% of the measured values.

Fitting regression lines to the data highlights systematic errors in the analysis. A number of regression equations were tried, and linear relations were found to give the highest coefficients of determination. The $r^2$ values for width and depth were both about 0.8, but that for velocity was only 0.35. All $r^2$ values were significant at the 95% level of confidence.

For width, calculated widths tend to over-estimate true values for small channels and under-estimate width for large channels. This tendency was also noted by Chang in his tests (2). Conversely, depths are under-estimated in small streams and over-estimated in large streams. The limited range of velocity data, together with the fact that accurate velocity prediction depends on accuracy in the prediction of both width and depth are responsible for the low coefficient of determination in the velocity regression. The best fit line shows the analysis to over-estimate velocity in high velocity streams slightly, because the over-estimation of depth and under-estimation of width combine to produce unrealistically high hydraulic efficiency.

Initially, it was suspected that these discrepancies might be due to the fact that the analytical procedure was developed for straight channels, while only half the rivers used to test the method are either straight or are of very low sinuosity ($p<1.1$). The remainder are sinuous, with sinuosities as high as 2.5. To test if this might be the case, errors in width and depth were plotted and regressed against sinuosity using a wide variety of linear and nonlinear regression curves. There was no correlation between the errors and sinuosity, the highest $r^2$ values for both width and depth errors being less than 0.1. Also, removing the sinuous streams and dealing only with the straight rivers did not improve accuracy. It was therefore concluded that the errors were not due to the sinuous nature of the channels.

### 3.4 Consideration of Bank Vegetation

When the points in the graphs of calculated versus measured width and depth were coded by vegetation type, a pattern emerged. Specifically, channels with weak bank vegetation are generally wider and shallower than predicted, while those with strong bank vegetation tend to be narrower and deeper. In Tables 1, 2 and 3 the differences between calculated and measured parameters are classified by bank vegetation type. In the Tables the error is defined by:

$$\text{Error} = \{(\text{measured value} - \text{calculated value})/\ \text{measured value}\}\ 100\% \qquad (12)$$

Table 1. Distribution of Width Prediction Errors by Bank Vegetation Type

| Bank Veg. Type | Error >+15% (Wider) | Error < 15% (Similar) | Error >-15% (Narrower) | Row Total |
|---|---|---|---|---|
| I | 11 (4) | 2 (6) | 0 (3) | 13 |
| II | 8 (5) | 8 (7) | 0 (4) | 16 |
| III | 1 (4) | 19 (6) | 2 (3) | 13 |
| IV | 0 (6) | 9 (9) | 11 (5) | 20 |
| Column Totals | 20 | 29 | 13 | 62 |

Expected values are in brackets following observed values.
Calculated Chi-squared = 44 Critical Chi-squared = 7.815 (95%)

Table 1 shows that positive width errors, that is cases where the analytical method under-predicts the width, are associated almost exclusively with weak vegetation catagories (Types I and II). Negative error, that is over-prediction, occurs only for Type III and IV vegetation.

Table 2. Distribution of Depth Prediction Errors by Bank Vegetation Type

| Bank Veg. Type | Error >+15% (Deeper) | Error < 15% (Similar) | Error >-15% (Shallower) | Row Total |
|---|---|---|---|---|
| I | 1 (3) | 7 (8) | 5 (2) | 13 |
| II | 2 (4) | 11 (10) | 3 (2) | 16 |
| III | 4 (3) | 8 (8) | 1 (2) | 13 |
| IV | 7 (5) | 13 (13) | 0 (3) | 20 |
| Column Totals | 14 | 39 | 9 | 62 |

Expected values are in brackets following observed values.
Calculated Chi-squared = 12.7 Critical Chi-squared = 7.815 (95%)

Depth errors are less clear cut, but still indicate a relationship between the type of error and the vegetation type. Channels with weak vegetation tend to be shallower than predicted, while those with stronger vegetation tend to be deeper.

Table 3. Distribution of Velocity prediction errors by Bank Vegetation Type.

| Bank Veg. Type | Error >+15% (Faster) | Error < 15% (Similar) | Error >-15% (Slower) | Row Total |
|---|---|---|---|---|
| I | 0 (3) | 4 (6) | 9 (4) | 13 |
| II | 1 (4) | 11 (7) | 4 (5) | 16 |
| III | 3 (3) | 7 (6) | 3 (4) | 13 |
| IV | 11 (5) | 7 (9) | 2 (6) | 20 |
| Column Totals | 15 | 29 | 18 | 62 |

Expected Values are in brackets folowing observed values.
Calculated Chi-squared = 26 Critical Chi-squared = 7.815 (95%)

The impact of bank vegetation on flow velocitiy can also be demonstrated. The results in Table 3 show that the lower hydraulic efficiency of the shallow, wide channels associated with weak bank vegetation results in lower velocities than predicted. Conversely, the deep, narrow channels associated with strongly vegetated banks have higher velocities than predicted, because of their greater hydraulic efficiency.

These results indicate that bank vegetation strength influences channel hydraulic geometry in ways not accounted for in the analytical procedure.

3.5 Statistical Analysis of Calculated versus Measured Results

This working hypothesis may be tested by the significance of the groupings shown in Tables 1, 2 and 3 using Chi-squared tests. The null hypothesis in each case is that there is no significant difference between the error groups (columns) attributable to the categories of vegetation type (rows), and that the observed distribution is the product of random chance. The alternative hypothesis is that the error groupings of the vegetation classes are significantly different.

The expected values in the Chi-squared test are shown in brackets in the Tables. A slight problem arises because ideally none of the expected values should be less than 5 and not all of the expected values meet this criterion. However, as the calculated values of Chi-squared greatly exceed the critical values in all cases, the low values of some of the expected frequencies does not affect the outcome of the test. As calculated Chi-squares exceed critical values, the null hypotheses must be rejected and the alternative hypotheses accepted.

It is concluded ,therefore that the bank vegetation categories are significant and do help to explain the errors in the calculated widths, depths and velocities.

### 3.6 Adjustment of Calculated Parameters to Account for Bank Vegetation Strength

The next step in the analysis was to calculate regression curves for each of the bank vegetation categories separately. The curves and associated equations are shown in Fig.3.

| BANK VEG. TYPE | SYMBOL | REGRESSION EQUATION | $r^2$ |
|---|---|---|---|
| I | ○ | $B_M = 1.46\ B_C - 0.2535$ | 0.99 |
| II | ● | $B_M = 1.306\ B_C - 2.661$ | 0.95 |
| III | ▽ | $B_M = 1.161\ B_C - 5.130$ | 0.95 |
| IV | ▼ | $B_M = 0.9656\ B_C - 3.234$ | 0.93 |

| BANK VEG. TYPE | SYMBOL | REGRESSION EQUATION | $r^2$ |
|---|---|---|---|
| I | ○ | $D_M = 0.8815\ D_C + 0.0642$ | 0.87 |
| II | ● | $D_M = 0.5026\ D_C + 0.5350$ | 0.78 |
| III | ▽ | $D_M = 0.5413\ D_C + 0.8278$ | 0.73 |
| IV | ▼ | $D_M = 0.7648\ D_C + 0.4436$ | 0.80 |

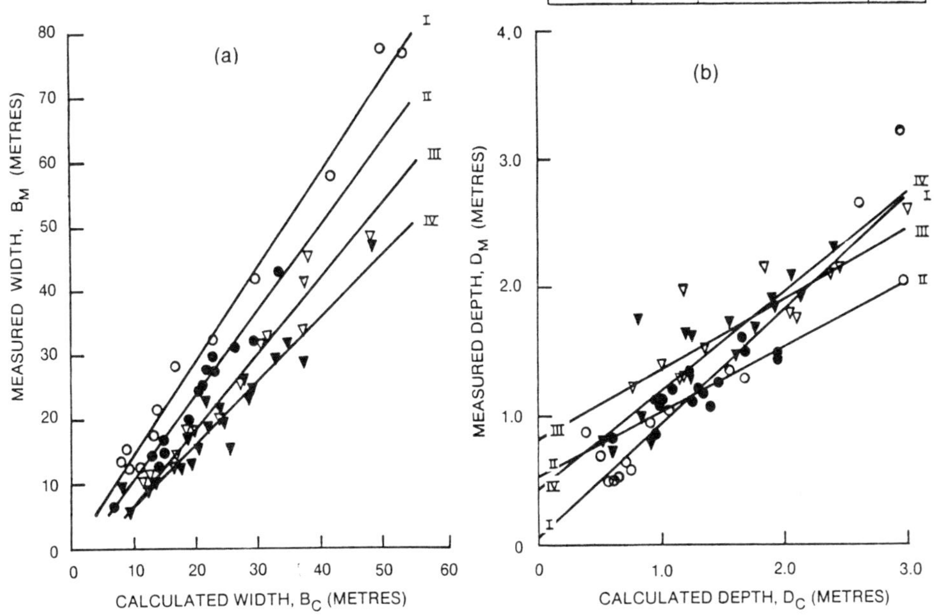


Figure 3. Calculated versus Measured values of a) Width by Vegetation Type, and b) Depth by Vegetation Type

The r² values are all significant at the 95% level of confidence, and show improvement over the values for the lumped data, suggesting that more reliable predictions of hydraulic geometry can be made if the calculated values of width and depth are be adjusted to account for the strength of bank vegetation in the equilibrium channel. Better velocity predictions should follow if the modification is valid, as accurate velocity prediction depends on width and depth predictions in eqn. 11. To check this, the widths and depths were adjusted using the regression equations in Fig.3 and the new values were used to modify the calculated velocities. The modified values are shown in Fig.4. The errors associated with these adjusted parameters are listed in Tables 4, 5 and 6. It should be noted that as we are using the regression equations to adjust the data from which they were developed, this is essentially a "goodness of fit test" rather than an independent verification of the equations. Verification would require another data set and that, unfortunately, was not availble.

Table 4. Modified Width Prediction Errors

| Bank Veg. Type | Error >+15% (Wider) | Error < 15% (Similar) | Error >-15% (Narrower) | Row Total |
|---|---|---|---|---|
| I | 2 (1) | 9 (10) | 2 (2) | 13 |
| II | 1 (2) | 13 (12) | 2 (2) | 16 |
| III | 1 (1) | 11 (10) | 1 (2) | 13 |
| IV | 2 (2) | 15 (15) | 3 (3) | 20 |
| Column Totals | 6 | 48 | 8 | 62 |

Calculated Chi-squared = 1.1 Critical Chi-squared = 7.815 (95%)

Table 5. Modified Depth Prediction Errors

| Bank Veg. Type | Error >+15% (Deeper) | Error < 15% (Similar) | Error >-15% (Shallower) | Row Total |
|---|---|---|---|---|
| I | 2 (1) | 6 (9) | 5 (3) | 13 |
| II | 0 (1) | 13 (12) | 3 (3) | 16 |
| III | 1 (1) | 11 (9) | 1 (1) | 13 |
| IV | 2 (2) | 15 (14) | 3 (4) | 20 |
| Column Totals | 5 | 45 | 12 | 62 |

Calculated Chi-square = 7.35 Critical Chi-square = 7.815 (95%)

Table 6. Modified Velocity Prediction Errors

| Bank Veg. Type | Error >+15% (Faster) | Error < 15% (Similar) | Error >-15% (Slower) | Row Total |
|---|---|---|---|---|
| I | 4 (2) | 5 (9) | 4 (2) | 13 |
| II | 1 (2) | 13 (11) | 3 (3) | 16 |
| III | 1 (2) | 9 (9) | 3 (2) | 13 |
| IV | 4 (3) | 14 (13) | 2 (4) | 20 |
| Column Totals | 10 | 41 | 11 | 62 |

Calculated Chi-square = 7.53 Critical Chi-Square = 7.815 (95%)

There is a marked improvement in accuracy. The results for width show the greatest improvement, with the width at 48 sites now being predicted to within plus or minus 15%. This represents a success rate of 77%, up from only 47% for the unadjusted figures. The distribution of errors by bank vegetation type in Table 4 shows no significance when tested by Chi-square, and so it may be concluded that the remaining errors are the product of random chance. For depth, 45 cases (73%) now fall within the acceptable range. This is up from 63% prior to adjustment. Again, the Chi-square value is not significant, indicating a random distribution of remaining errors.

The modified analysis now predicts both width and depth to within 15% in 36 cases or 58% of the time. This is more than double the pre-adjustment success rate. Also, in only 6 cases is either the width or the depth prediction off by more than 25% so that the results can be viewed with some confidence.

Velocity prediction shows marked improvement. The regression line in Fig.4 is very close to the line of perfect agreement and the $r^2$ is more than doubled to 0.72. 41 cases (66%) now fall within the acceptable range and at 32 sites (52%) the width, depth and velocity are predicted to within plus or minus 15%. The distributon of remaining errors in Table 6 is indicated as not being significant by the Chi-square value.

The modified, calculated values of width, depth and velocity could form the basis for prediction of the hydraulic geometry of stable gravel-bed streams.

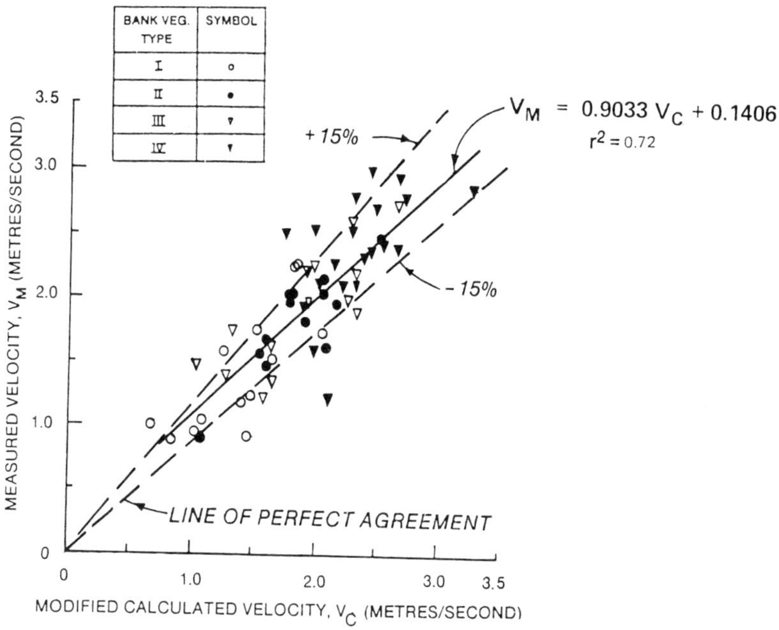


Figure 4. Adjusted Calculated Velocities versus Measured Velocities

## 4. CONLUSIONS

1. The hydraulic geometry of gravel-bed rivers can be reasonably well predicted using Chang's (2) analytical procedure based on the concept of minimum stream power, together with suitable equations for flow resistance and bed load transport. Linear regression of calculated versus measured widths and depths produces $r^2$ values of about 0.8 for both parameters. The method is not a good predictor of velocity. Calculated versus measured regression results in an $r^2$ of only 0.35, which although statistically significant, is unacceptably low for confident application of the method for design purposes.
2. Considerable scatter exists when predicted values are plotted against measured values. Width predictions are within 15% of measured values in about half of the cases and about two thirds of the cases for depth. Velocity is predicted to within 15% in half the cases.
3. The analysis tends to systematically under-estimate width and over-estimate depth with increasing channel size. This leads to under-estimation of velocity in larger channels.
4. Bank material properties in the streams studied were rather uniform, but bank vegetation density was observed to vary from grass to dense trees and shrubs. Classifcation of errors in the prediction of width and depth at each site according to the strength of bank vegetation, produced a statistically significant distribution.
5. Channels with strongly vegetated banks tend to be narrower and deeper than predicted, while those with weak vegetation tend to be wider and shallower. Velocities in the channels with strong bank vegetation tend to be higher than in those with weakly vegetated banks because of the greater hydraulic efficiency of the narrower, deeper channels.
6. The accuracy of predictions of hydraulic geometry can be markedly improved by adjusting the calculated widths and depths to account for bank vegetation effects. Adjusted widths were within 15% of measured values for 77% of the cases. The success rates for depth and velocity were 73% and 66%, respectively.
7. It should be possible to incorperate the effects of bank vegetation into the

analytical procedure by increasing the sophistication of the method of calculating the bank slope index z. However as there is no simple relationship between vegetation, bank stability and channel geometry, this will require a considerable research effort.

8. It is the intention of the authors to pursue this line of research with the aim of developing a process-based approach to incorporating bank properties, including vegetation, into the prediction of bank slope in the procedure. On the evidence of the results reported here, there is scope for considerable improvement in the accuracy of the analysis through this initiative.

## 5 ACKNOWEDGEMENTS

This paper was written while the first author was a visiting scientist at the US Army Waterways Experiment Station, Vicksburg, Mississippi, USA. The support of Phil Combs, David Biedenharn, Bobby Brown and John Ingram at Vicksburg is gratefully acknowledged. The field data used in this paper were collected under a grant from the Natural Environment Research Council (UK).

## 6 REFERENCES

1. Hey, R.D. and Thorne, C.R."Stable channels with mobile gravel beds" Jour. Hydrau. Eng., ASCE, 112(No.8),1986,pp 671-689.
2. Chang, H.H."Geometry of gravel streams" J. Hyd.Div., ASCE, 106(HY9),1980, pp 1443-1456.
3. Hey R.D. "Design equations for mobile gravel-bed rivers" In, "Gravel-Bed Rivers" J. Wiley and Sons, Chichester, UK, 1982, pp 553-574.
4. Bathurst, J.C. "Theoretical aspects of flow resistance" In, " Gravel-Bed Rivers" J. Wiley and Sons, Chichester, UK, 1982, pp 83-108.
5. Bray, D.I. "Estimating average velocity in gravel-bed rivers" Jour. Hydrau. Div., ASCE, 105(HY9), 1979, pp 1103-1122.
6. Parker, G. "Self-formed rivers with equilibrium banks and mobile bed: Part II, the gravel river" Jour. Fluid Mechanics, 89(1), 1978, pp127-148.
7. Parker, G. "Hydraulic gecmetry of active gravel rivers" Jour. Hydrau. Div., ASCE , 105(9), 1979, pp1185-1201.
8. Meyer-Peter,E. "Favre: Neuere Versuchsresulttate uber den Geschiebetrieb" SBZ, Zurich, Switzerland,1934.
9. Chang, H.H. "Minimum stream power and river channel patterns" Jour. Hydrology, 41, 1979, pp 303-327.
10. Chang, H.H. "Stable alluvial canal design" Jour. Hydrau. Div., ASCE, 106(5), 1980, 873-891.
11. Lane, E.W. "Design of stable channels" Trans. ASCE, 120, 1955, pp 1234-79
12. Wolman, M.G. and Leopold, L.B. "River flood plains: Some observations on their formation" US Geol. Surv. Prof. Paper 282-C,1957.
13. Williams, G.P. "Bankfull discharge of rivers" Water Res. Res.,14(6),1978, pp 1141-54.
14. Parker, G., Klingeman, P.C. and McClean, D.G. "Bedload and size distribu ticn in paved gravel-bed streams" J.Hyd.Div. ASCE, 103(4),1982,pp 544-571.

## 7 NOMENCLATURE

| | | | |
|---|---|---|---|
| B | Water surface width | S | Channel slope |
| D | Mean flow depth | $S_{cr}$ | Threshold channel slope |
| $d_{50}$ | Median bed material size | T | Shear stress |
| f | Darcy-Weisbach friction factor | T* | Dimensionless or Shields shear stress |
| g | Gravitational acceleration | z | Bank slope index |
| p | Sinuosity | $\rho$ | Water density |
| Q | Water discharge | $\rho_s$ | Sediment density |
| $Q_s$ | Sediment discharge | $\Phi$ | Bed load intensity |
| $q_s$ | Unit sediment discharge | $\Psi$ | Flcw intensity |

International Conference on
# RIVER REGIME
18-20 May, 1988

PAPER B1

## REGIME RELATIONS BASED ON DISPERSIVE-FORCE CONCENTRATION PROFILES

By Kenneth C. Wilson, Professor, Department of Civil Engineering, Queen's University at Kingston, Ontario, Canada K7L 3N6

SUMMARY

Regime equations pertain to series of erodible channels of different sizes, usually involving solids transport both as bed load and by turbulent suspension. Analysis of this combination by a new dispersive-force concentration model gave "derived" regime equations which compare well with the usual "field-based" ones.

## INTRODUCTION - THE REGIME EQUATIONS

The regime approach to channel design originated in the palmy days when vast irrigation works were planned and executed in areas under British rule or influence, and empirical relations were necessary to standardise the design of a multitude of unlined canals. The equations used, described in detail by Lacey (Ref. 1), later attracted the attention of many authors, the most noteworthy of whom is Blench (Ref. 2).

Although sometimes referred to as a theory, the regime equations did not arise from theoretical derivation from accepted principles of fluid mechanics. Instead, they comprise a series of design rules which are, in effect, scaling relations for networks of unlined canals excavated in the the erodible soil of alluvial plains. For each canal of the network the width, B, depth, h, and slope, S, must be determined on the basis of the design discharge, Q. Using power-law functions, the following basic proportionalities are obtained

$$B \propto Q^{\alpha} \tag{1}$$

$$h \propto Q^{\beta} \tag{2}$$

These two expressions (or equivalent ones with the wetted periphery and hydraulic radius of the channel substituted for B and h) are central to all versions of the regime equations. The proportionalities (1) and (2) are often stated as equations, but the required "constants" of proportionality depend on the size and grading of the soil particles, and other factors. However, the powers $\alpha$ and $\beta$, as evaluated from field data, show remarkably little variation. Blench (Ref. 2) gives $\alpha = 0.500$ and $\beta = 0.333$ whereas Simons and Albertson (Ref. 3), while in basic agreement with Blench regarding $\alpha$, generally use a $\beta$ value of 0.36.

International Conference on River Regime, edited by W.R.White.

The slope of regime channels also depends on discharge, but in an inverse sense, indicating a relation similar in form to (1) and (2) but with a negative power. This power may be taken as simply a further quantity to be obtained empirically (Ref. 3); but it can better be considered (as in Ref. 2) as defined in terms of $\alpha$ and $\beta$, thus

$$S \propto Q^{-(\alpha-\beta)} \tag{3}$$

The reason for this formulation is that each channel of a set (sharing the same soil properties but differing in discharge) can be regarded as a scale model of every other channel. The fact that there is a different scaling of horizontal and vertical dimensions, as given by (1) and (2), is required by considerations of fluid friction. This result implies that the scaling relations reflect a vertically - distorted modelling system. For such a system the slopes scale according to the ratio of vertical to horizontal lengths, in accord with (3) above.

However they may be arrived at, the regime equations can only constitute a successful basis for design if they are in accord with the priciples of fluid mechanics. These include the scaling of friction, mentioned above, by which larger channels have smaller fraction factors and thus, <u>cæteris paribus</u>, require smaller longitudinal slopes. Consideration of continuity is also required, not only for the flowing water, but also for the solids which it transports. Thus, if a single irrigation channel is divided into several smaller channels, not only must the outflow of water from the junction equal the inflow, but the same must apply to the solids, so that no net erosion or deposition occurs. This condition can be expressed by stating that the delivered concentration of solids must be the same in each channel of the network. With $C_d$ denoting the delivered concentration, this relation is written

$$Q_s = C_d \, Q \tag{4}$$

where $Q_s$ is the volumetric discharge of solids through any given channel.

## IMPLICATIONS OF CONSTANT-CONCENTRATION CONDITION

In exploring the implications of the condition of constant delivered solids concentration, it is appropriate to begin by considering a unit breadth of flow, for which the discharge is q. This unit discharge equals $Q/B$ or $h\,V$, where V is the mean velocity. The general case will be outlined before dealing specifically with regime relations.

In general terms q (or V) must be obtained by integrating the local velocity, which, for given solids, depends only on the boundary shear stress, $\tau$, and the distance above the boundary, y. On integration, it is found that the average velocity V depends only on $\tau$ and the flow depth h (again, for given solids). This relation may be written

$$V = V(\tau, h) \tag{5}$$

The discharge of solids per unit breadth, $q_s$, may include both bed-load and suspended-load components. For given solid particles, the bed-load component of $q_s$ depends only on the boundary shear stress, $\tau$, (see Yalin (Ref. 4), Chapter 5). The suspended-load component can be determined by integrating the product of local concentration (which depends on the ratio of fall velocity to shear velocity, and on position in the flow) and local velocity. As with the velocity integral mentioned above, it is found that the suspended load component, and hence all of $q_s$, depends only on $\tau$ and h (for given solids). This general relation may be written

$$q_s = q_s(\tau,h) \tag{6}$$

On combining (5) and (6) with the relation for constant delivered concentration, $C_d$, one obtains

$$q_s(\tau,h) = C_d \, h \, V(\tau,h) \tag{7}$$

This expression shows that, for given $C_d$ and solids properties, there must be a functional relationship between $\tau$ and h. Moreover as $\tau$ itself depends only on the product hS ($\tau = \rho ghS$, with $\rho g$ constant), the relationship must be equivalent to a function linking h and S.

Before investigating this linkage in general terms, it is appropriate to consider an example involving certain simplifications. Specifically, suppose that there is no suspension of particles by fluid turbulence -- a condition which occurs for large particles and for smaller particles at shear stresses not far above the threshold of motion. This condition implies that the function of expression (6) depends only on $\tau$, not on h *per se*. The problem to be solved here is to obtain the form of this function which is compatible with regime relations (1) to (3).

From expressions (7) and (1) it is seen that

$$q_s(\tau) \propto hV \propto Q/B \propto Q^{1-\alpha} \tag{8}$$

Moreover $\tau$ (i.e. the product hS) is proportional to $Q^{\beta} \, Q^{-(\alpha-\beta)}$ or $Q^{2\beta-\alpha}$, giving

$$q_s(Q^{2\beta-\alpha}) \propto Q^{1-\alpha} \tag{9}$$

The solution to (9) is a power-law expression whereby $q_s$ is proportional to $\tau^{\varepsilon}$, with $\varepsilon$ given by

$$\varepsilon = (1-\alpha) \;/\; (2\beta-\alpha) \tag{10}$$

Substituting typical values of $\alpha$ and $\beta$ of 0.50 and 0.33 respectively, $\varepsilon$ is found to be 3.0.

This result is interesting, and gives a good indication of what may be expected from the method. The prediction is that, for bed load transport to correspond to the constant-$C_d$ condition and the regime equations, the unit discharge of solids would have to depend on $\tau^3$. In other words, a log-log plot of $q_s$ *versus* $\tau$ would show a straight line with a slope of 3. In fact, such a log-log plot (or its equivalent in terms of the dimensionless co-ordinates $\phi$ and $\psi$) is distinctly curved (see, for sample, Yalin's text (Ref. 4), p. 140). Figure 1 shows a plot on logarithmic co-ordinates of relative $q_s$ *versus* the ratio $\tau/\tau_c$, where $\tau_c$ is the critical shear stress, at which particle motion just begins. The solid line represents a bed load transport equation of the type proposed by Meyer-Peter and Mueller (Ref. 5), i.e.

$$q_s = \frac{K_m}{g(s-1)} \left[\frac{\tau - \tau_c}{\rho}\right]^{1.5} \tag{11}$$

where $\rho$ is fluid density, s is relative density of solids, g is gravitational acceleration and $K_m$ is a constant, equivalent to 8 in the original formulation of Ref. 5. More recent evidence indicates a rather larger value of $K_m$ (Ref. 6). However, a change in $K_m$ merely displaces the curve on the logarithmic co-ordinates of Fig. 1, without changing its shape, and this feature is eliminated on the figure by plotting relative $q_s$, based on dividing $q_s$ by the value obtained at $\tau = 10\ \tau_c$.

The dashed line on Fig. 1 represents the power-law derived from the regime relations. Its slope of 3.0 corresponds to that of the curve at $\tau/\tau_c = 2.0$, i.e. a shear stress twice that required to set the first grains in motion. For shear stresses in the vicinity of this value the simple power law gives an adequate approximation to the curve, and it may be argued that this is the range for which the regime relations would be applied to cases for which turbulent suspension of solids is not significant. Nevertheless, the self-adjustment of channel shape, on which the regime relations depend, is favoured by flows with suspended sediment, in which settling particles can fill in areas of low fluid velocity.

As soon as suspended load is taken into account, the procedure used above in obtaining expression (10) can no longer be followed, as the relation for $q_s$, i.e. (6), now involves both $\tau$ and h. What can be done instead is to use known relationships for calculating V and $q_s$ and, working in the opposite sense to that of the previous example, to determine the values of the powers $\alpha$ and $\beta$ which are compatible with these relationships.

In this regard it should be recalled that the functional relations discussed above indicated that h depends only on S once the particle properties and delivered concentration have been specified. In practice it is necessary to analyse a large number of unit-breadth flow for various combinations of h and S. For given solid particles each combination defines the boundary shear stress, and hence determines the mean velocity, solids transport rate and delivered concentration. Lines of constant $C_d$ can then be plotted on the S-h plane to give the required link between h and S. As will be shown in a subsequent section, if the lines of constant $C_d$ are straight on a log-log S-h plot, a set of power-law "regime" relations will be produced.

Before dealing with the derivation of regime relations in this way, it is necessary to consider briefly the mathematical model by which the delivered solids concentration will be determined for each S-h combination. This is the subject of the following section.

## DISPERSIVE-FORCE MODEL FOR SOLIDS CONCENTRATION

As remarked above, the general case of flow in channels excavated into erodible material requires calculation of both bed-load and suspended-load transport rates. The calculation of the bed-load rate presents no computational difficulties. Many formulations are available and the more widely accepted versions are in substantial agreement in the intermediate shear-stress range needed for the present application. Likewise, there is general agreement as to how the concentration of suspended solids varies with height, provided that the concentration at any one height is known. However, the link between the bed load and the lower end of the suspended-load profile requires detailed consideration of conditions very near the bottom of the flow. Following the classical work of Bagnold (Ref. 7), it has become widely accepted that at substantial values of bed shear stress the particles of bed load will be travelling in a sheared layer which may be many particle diameters in thickness. As will be seen below, experimental evidence shows that within this layer the concentration of bed-load solids diminishes with height. From this viewpoint the problem of calculating suspended-load concentration reduces to one of linking the bed-load and suspended-load concentration profiles in the solids-rich zone just above the stationary bed.

In this zone both shear stress and solids concentration are often very large, and observation can be remarkably difficult under open-channel conditions. The same problem occurs in enclosed conducts such as slurry pipelines, where modelling of particulate motion is also required. Fortunately, measurements at high solids concentration are somewhat easier to carry out for the closed-conduct flows, and some excellent profiles have been obtained recently, especially at the laboratories of the Saskatchewan Research Council. Remarkably interesting results were obtained for coarse gravel particles 13 mm in diameter which were the subject of tests by Shook et al. (Ref. 8) at in situ volumetric concentration 0.09 in a pipe of 263 mm internal diameter. The data from this source are shown on Fig. 2 which plots local concentration C against relative height (compared with pipe diameter) y/D. The average velocity V is included as parameter, but clearly has virtually no effect on the concentration profile. The upper portion of this profile shows, as expected, that there is virtually no turbulent suspension of these large, heavy particles. The lower portion shows that within the zone where bed-load particles can move relative to each other, their concentration decreases with height in an essentially linear fashion.

This linear distribution of bed-load concentration has been combined with the stress analysis of Bagnold (Ref. 7) to derive formulations for frictional resistance and bed-load transport rate (Refs. 9, 10) It can also be used in obtaining the link between bed load and the bottom of the suspended-load concentration profile.

Both types of profile appear on Fig. 3, based on data from Shook et al. (Ref. 8) in the same pipe used for the coarse-particle experiments of Fig. 2. In addition to coarse (13 mm) particles, the experiments illustrated on Fig. 3 included fine particles 0.17 mm in size. The coarse particles, distributed linearly, are highly concentrated near the bottom of the flow, where they dominate the concentration profile. In the upper portion the coarse bed-load particles are virtually absent (as seen previously on Fig. 2) and the profile, dominated by fines, has the gently curving change of concentration with height to be expected for turbulent suspension.

If particles of a single intermediate size are substituted for the bimodal distribution of Fig. 3, the shift in profile type will be less clear-cut, but intergranular contact will still dominate in the high-concentration region near the bottom of the flow, with turbulent suspension dominant at a higher level (with the proviso that suspension will occur at all, i.e. that the particle fall velocity is less than the velocity of the turbulent eddies). At an intermediate level both bed load and suspended load will be present, with some particles supported by intergranular collisions and others by turbulent eddies. This type of behaviour has been studied by Pugh (Ref. 11), who set up a vertical force balance for the solids which included both the intergranular normal stress introduced by Bagnold (Ref. 7) and a dispersive force obtained from the interaction of the turbulent fluid flow with the solid grains. Numerical integration then led to total load (i.e. both bed- and suspended-load) profiles, which were found to compare well with experimental results, principally those obtained by Gillies et al. (Ref. 12). As this dispersive-force model for calculating solids concentration has already been described not only in Pugh's thesis (Ref. 11), but in other papers submitted for publication, it will not be discussed further here.

## SYNTHESIS

It is now possible to return to the earlier proposal of analysing a large number of combinations of slope and depth to determine (for given solids properties) the unit discharge of solids $q_s$, as well as the total unit discharge q. These quantities can readily be evaluated by computer; first using the dispersive-force model for the layer at the bottom of the flow in which bed-load transport takes place, and then applying the values of velocity and concentration obtained at the top of this layer as boundary conditions for a conventional mixing-length eddy-viscosity model of the remaining portion of the flow. Once q and $q_s$ have been calculated on this basis, the mean velocity V and delivered solids concentration $C_d$ are readily determined.

Typical results are shown on Fig. 4, a logarithmic S-h plot based on solids of relative density 2.65 and particle diameter 0.2 mm. The lines on the figure define conditions for constant values of mean velocity and of delivered concentration. As expected, all lines of constant mean velocity show depth deminishing with increasing channel slope at essentially the same rate, approximating a set of parallel straight lines on the logarithmic co-ordinates of Fig. 4. As these lines have virtually equal offsets for each doubling of velocity, it can be seen that the modelled velocities are in accord with the usual engineering approximations for the effects of friction, which link velocity depth and slope in a power-law relation.

Two lines of constant delivered concentration are shown on Fig. 4. The value of $C_d$ = 0.001 may be considered typical of regime flow, while the value of 0.01 is abnormally large but is intended primarily to give a check on the effects of differences in concentration. Like the constant-V lines, the line of C = 0.001 is virtually straight on the logarithmic co-ordinates of the figure, and the line for $C_d$ = 0.01, while showing some curvature at low flow depth, can also be approximated well by a straight line. Figures similar to Fig. 4 were prepared for a number of other particle diameters. It was found that the displacement of the constant-V lines by different particle size can hardly be seen at the scale of Fig. 4; whereas, because of effect of fall velocity, constant-$C_d$ lines are displaced to the right with increasing particle size, and the spread between the two constant-$C_d$ lines also increases with d.

Consider next a given line of constant $C_d$, which is effectively straight on logarithmic co-ordinates (for example the line for $C_d$ = 0.001 on Fig. 4). Clearly such a line can be represented by a relation of the type

$$h \propto S^{-m} \tag{12}$$

(For the example mentioned above, m = 2.55).

The constant-$C_d$ line intersects various lines of constant V, and these intercepts can also be plotted against log S (i.e. with log V replacing log h). Such plots were found to be virtually linear, and thus are represented by

$$V \propto S^{-n} \tag{13}$$

(where n = 0.83 for the example case).

As S is proportional to h/B for all vertically - distorted models, the breadth, i.e. h/S, must be proportional to $S^{-(1+m)}$. Hence Q, the product of B h and V, is proportional to $S^{-(1+2m+n)}$ or, equivalently

$$S \propto Q^{-1/(1+2m+n)} \tag{14}$$

(Here the power has the value -0.144 for the example.)

Substituting Eq. (14) into the immediately preceeding relations for breadth and depth gives two further power-law expressions in which the powers must equal α and β of Eq. (1) and Eq. (2). These derived expressions for α and β are

$$\alpha = (1+m)/(1+2m+n) \tag{15}$$

and

$$\beta = m/(1+2m+n) \tag{16}$$

(For the example, the numerical values are 0.512 and 0.368, respectively.)

Computer simulations were run for a series of particle diameters between 0.05 mm and 2.0 mm, and used to plot figures like Fig. 4 (and the derived plots of log V against log S). The slopes measured from such plots then were substituted into Eq. (15) and Eq. (16) to obtain derived values of α and β. Fig. 5 is a plot of these derived values against the logarithm of particle diameter. For the typical concentration of 0.001, the derived values show remarkably little variation with d, giving α close to the commonly accepted regime value of 0.50 and β equal to approximately 0.37 (compared with the previously mentioned values of 0.333 (Ref. 2) and 0.36 (Ref. 3)).

The dotted lines on Fig. 5 show similar results for the high delivered concentration of 0.01, indicating a somewhat greater variation of α and β with d, and a tendency for greater separation in the values of the two powers. Even for this atypically high concentration, however, the derived powers are close to being "constants".

## CONCLUSION

Regime equations pertain to series of unlined canals situated in the erodible soils of alluvial plains. These equations are based on power-law design relations for channel breadth and depth in terms of discharge, and can be considered as a method of representing each channel of the series as a vertically-distorted model of all the others.

Equations of this type must be compatible with continuity conditions for both water and sediment, implying that each channel in the series has the same (time-averaged) delivered concentration of solids. If all of the solids are transported as bed load, it has been shown that "traditional" regime equations can only be applied for a range of bed shear stress near twice the critical stress for the beginning of motion. In fact, the major application of the regime approach has been to channels in rather fine-grained alluvium, where solids transport involves a combination of bed-load motion and turbulent suspension. Analysis of this combination requires a link between suspended-load and bed-load profiles -- to be sought in a small zone just above the stationary bed, where shear stress and solids concentration are often large, and observation is remarkably difficult. The same problem occurs in modelling flow of particulate slurries in pipelines; but closed-conduit flows are somewhat easier to measure, as shown by solids-concentration profiles obtained recently in Saskatchewan.

These were analysed using a new dispersive-force model in which the submerged weight of solid particles is balanced by a combination of intergranular normal stress (first studied by Bagnold) and a dispersive force caused by fluid turbulence. The resulting profiles give delivered concentration, and for given solids the lines of constant concentration are approximated by power-law relations. Manipulation then produces "derived" regime relations for any desired combination of particle size and concentration, which compare well with the usual "field-based" equations.

## NOMENCLATURE

| | |
|---|---|
| $B$ | breadth of channel |
| $C_d$ | delivered volumetric concentration |
| $D$ | internal pipe diameter |
| $d$ | particle diameter |
| $g$ | gravitational acceleration |
| $h$ | depth of channel |
| $K_m$ | coefficient of Eq. (11) |
| $m$ | power of Eq. (12) |
| $n$ | power of Eq. (13) |
| $Q$ | discharge |
| $Q_s$ | discharge of solids |
| $q$ | discharge per unit breadth |
| $q_s$ | discharge of solids per unit breadth |
| $S$ | longitudinal slope of channel |
| $s$ | relative density of solids |
| $V$ | mean velocity in channel section |
| $y$ | vertical ordinate |
| $\alpha$ | power in Eq. (1), (15) |
| $\beta$ | power in Eq. (2), (16) |
| $\varepsilon$ | power in Eq. (10) |
| $\rho$ | fluid density |
| $\tau$ | boundary shear stress |
| $\tau_c$ | value of $\tau$ for beginning of sediment motion |

## REFERENCES


1. Lacey, G.: "Stable Channels in Alluvium", Proc. Inst. Civil Engrs, 229, pp 259-384 (1929).

2. Blench, T.: Regime Behaviour of Canals and Rivers, Butterworths, London (1957).

3. Simons, D.B., and Albertson, M.L.: "Uniform Water Conveyance Channels in Alluvial Material", Trans. Am. Soc. Civil Engrs, 128/I, pp 65-167 (1963).

4. Yalin, M.S.: Mechanics of Sediment Transport, Pergamon, Oxford (1972).

5. Meyer-Peter, E., and Müller, R.: "Formulas for Bed-Load Transport ", Proc. 2nd Congr. Internat. Assoc. for Hydr. Structures Research, Stockholm (1948).

6. Wilson, K.C.: "Bed-Load Transport at High Shear Stress", Jour. Hydr. Div., ASCE, 92, HY6, pp 49-59 (1966).

7. Bagnold, R.A:: "The Flow of Cohesionless Grains in Fluids", Phil. Trans. Roy. Soc. London, Ser.A, 249, pp 235-297 (1956).

8. Shook, C.A., Gillies, R., Haas, D.B., Husband, W.H.W., and Small, M.: "Flow of Coarse and Fine Sand Slurries in Pipelines", Journal of Pipelines, 3, pp 13-21 (1982).

9. Wilson, K.C.: "Analysis of Contact-Load Distribution and Application to Deposition Limit in Horizontal Pipes", Journal of Pipelines, 4, pp. 171-176 (1984).

10. Wilson, K.C.: "Analysis of Bed-Load Motion at High Shear Stress", Jour. of Hydr. Engrg., 113, 1, pp. 97-103 (1987).

11. Pugh, F.J.: Effects of Single-Size particles on Turbulent Slurry Flow, M.Sc. Thesis, Dept. Civil Engrg, Queen's University, Kingston, Canada (1986).

12. Gillies, R.G., Husband, H.W.H., and Small, M.H.: Study of Flow Conditions Arising in Horizontal Coarse Slurry Short Distance Pipelining Practice Phase 1: Sand Slurry Tests in a 250 mm Pipeline, Publ. No. R-833-2-C-85, Saskatchewan Research Council, Saskatoon, Sask., Canada (1985).

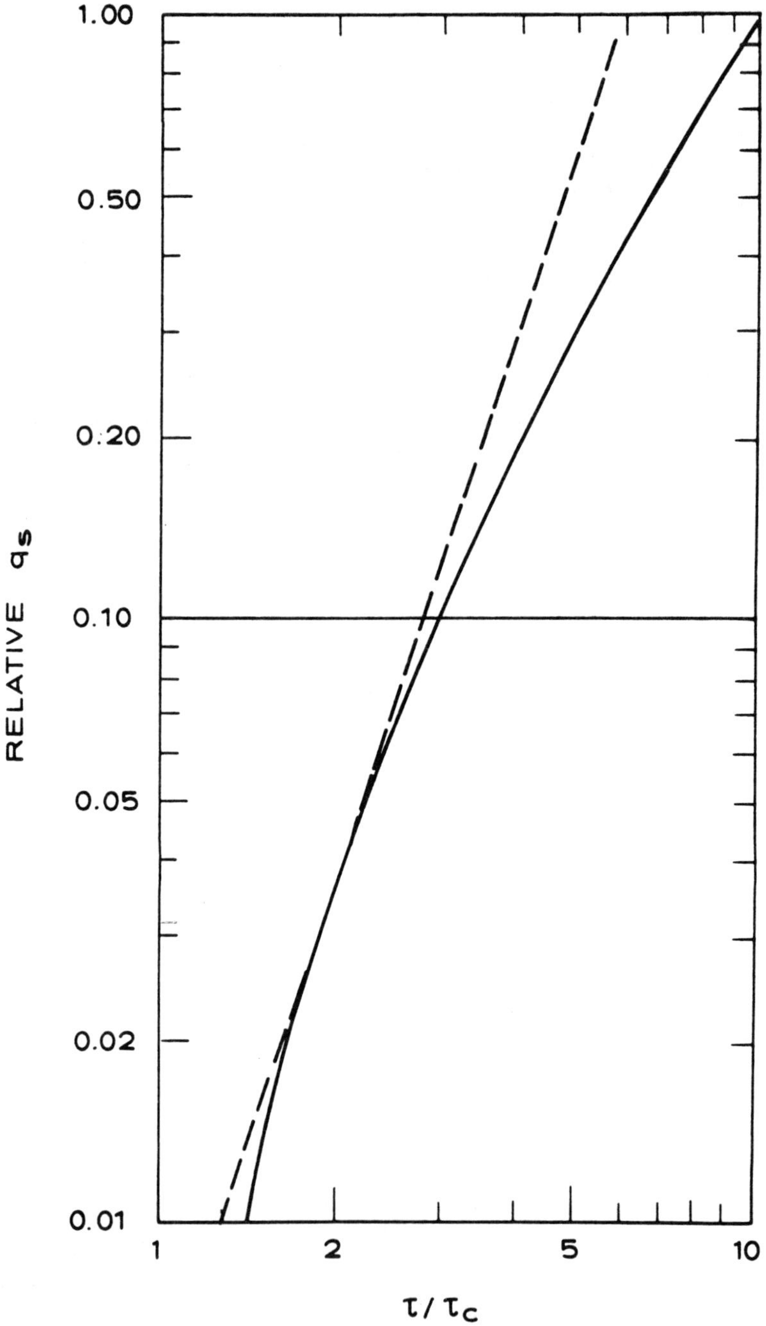


Fig. 1 Logarithmic Plot of Relative $q_s$ versus $\tau/\tau_c$

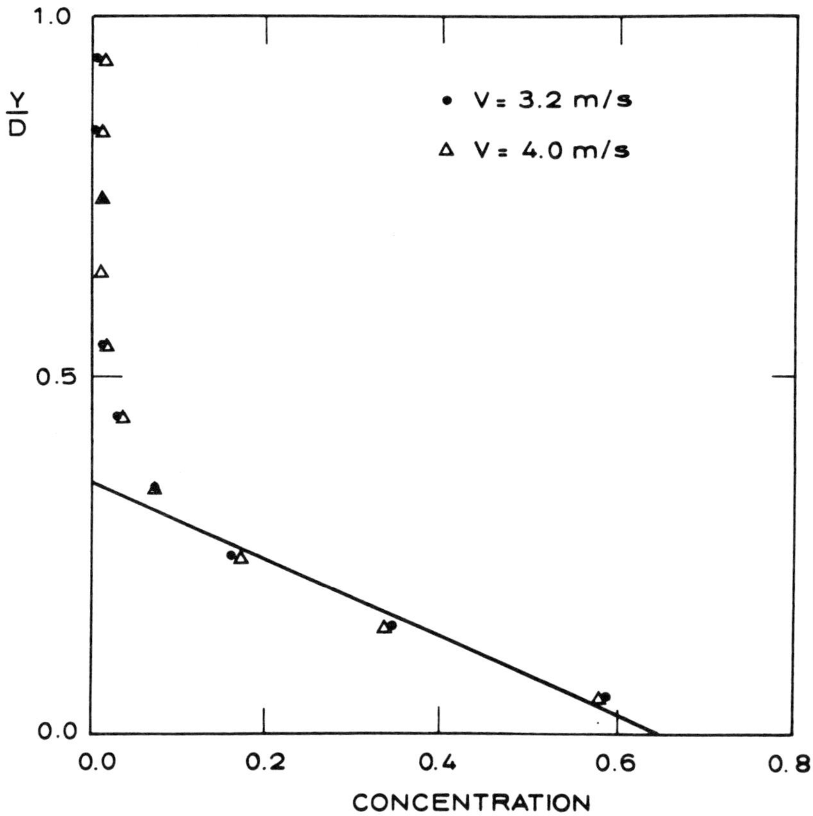


Fig. 2 Concentration Profile for Coarse Particles (after Shook et al., 1982)

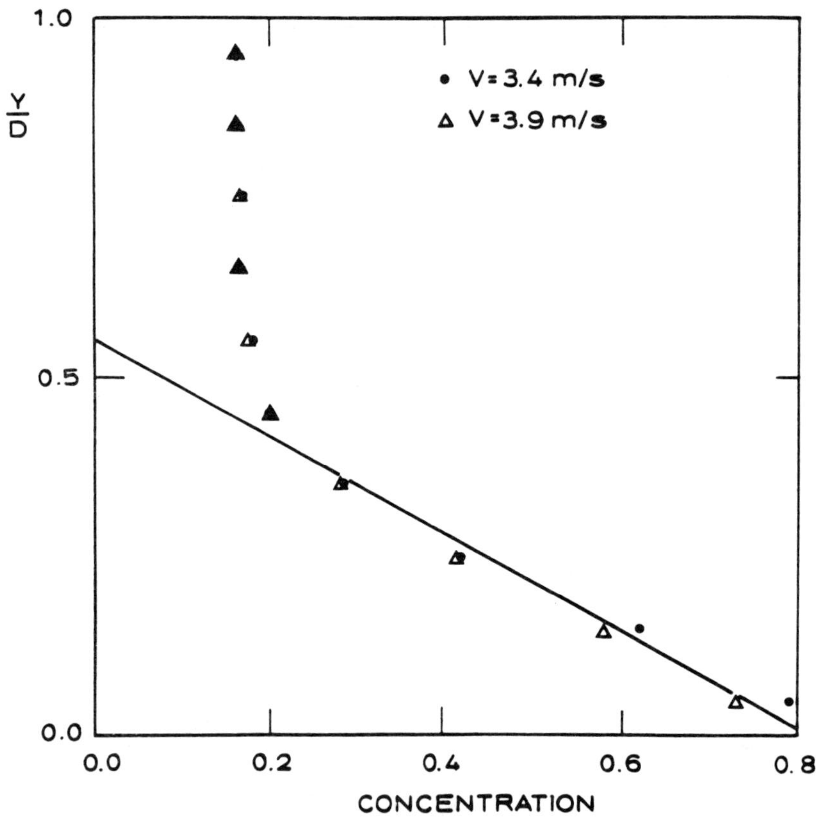


Fig. 3 Concentration Profile for Mixture of Coarse and Fine Particles (after Shook et al., 1982)

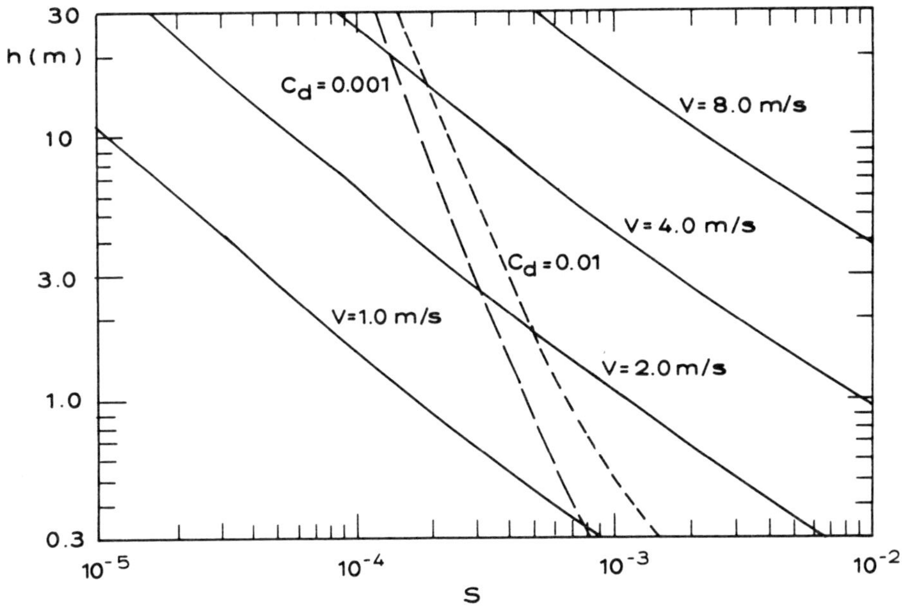


Fig. 4 Calculated Unit Flow Characteristics for 0.2 mm Sand

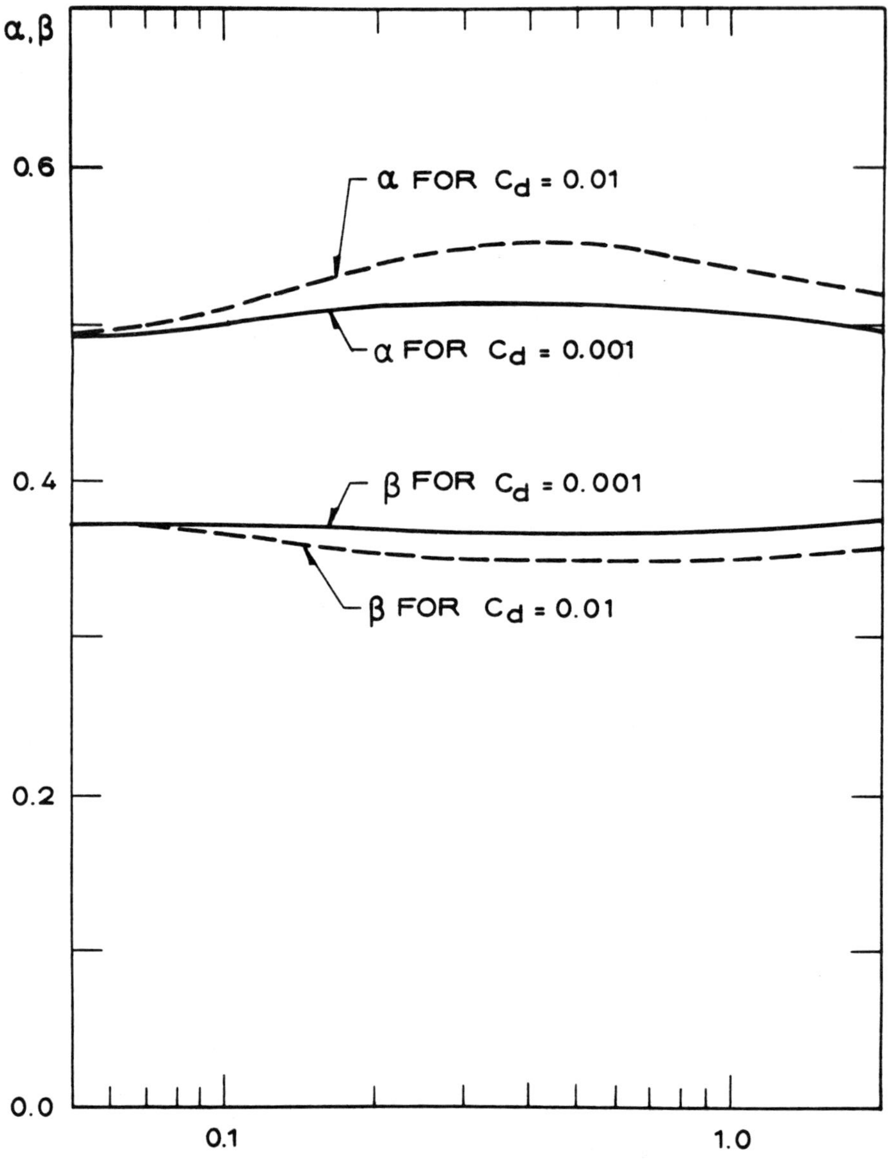


Fig. 5 Derived Powers for Regime Relations

International Conference on
RIVER REGIME
18-20 May, 1988

PAPER B2

# REGIME CHANNELS OF AN EXTREMELY ARID ZONE

M. Nouh
Associate Professor, Department of Civil Engineering
King Saud University, P.O.Box 70178, Riyadh-Diriyah 11567, Saudi Arabia

**SUMMARY :** Field observations were collected from 37 ephemeral channels of an extremely arid zone in Saudi Arabia, and used to develop regime formulae relating the channel dimensions and pattern to the characteristics of flash flood and sediment flow in the channels.

## 1. INTRODUCTION

A channel in regime is defined as the channel that has acquired a long-term average steady state in boundaries of its transported sediments (Ref 1). The properties of such a channel has attracted the attention of many investigators. Kennedy (Ref 2) was the first to analyse data from stable canals in the Upper Bari Doab, Punjab, India, and to develop an empirical relationship between the allowable critical mean velocity and the depth of flow. Subsequent investigations reported by Lindley (Ref 3) related both the critical mean velocity and the depth of flow to the channel bottom width. He also proposed a relation between the width and the depth of a channel in regime. Lacey (Refs 4 to 9) introduced concepts and correction factors to the regime equations in order to explain variations in stable channel behaviour. He was able to establish equations describing the relationships among the various variables; including the mean velocity, the hydraulic radius, the wetted parameter, the flow cross-sectional area, the slope, the Manning's coefficient, and the discharge of a stable channel. Inglis (Ref 10) extended Lacey regime equations to include cases when sediment charge is existing. Blench (Refs 1, 11, and 12) expanded the regime equations to include cases where the soil on the channel side is different from that on the channel bed, and developed relationships for the width, the depth, and the slope as a function of the discharge of a channel in regime. His contribution is reviewed together with the contributions on this subject of other investigators and reported in many literatures (Refs 13 to 16).

While the above excellent research works, among others, have succeeded to provide valuable information to describe the behaviour of regime channels under varying soil and flow conditions in humid areas it must be emphasized that there is a lack of similar information on the behaviour of regime ephemeral channels ( which carry water during storms only) in arid areas. To the author's knowledge, only limited data collected from ephemeral streams in a semiarid area were analysed and reported by Leopold et al. (Refs 17 and 18). The present study is directed to fill this gap of knowledge. It intends to provide information on the characteristics of ephemeral channel dimensions, pattern, and bed armouring under varying flash flood and sediment flow conditions in an extremely arid zone. Data collected from a number of stable ephemeral channels located in the south and in the south-west regions of Saudi Arabia were used in the study. General description of

International Conference on River Regime, edited by W.R.White.

of the study zone and its channels is given hereafter.

## 2. THE STUDY ZONE

As it has been mentioned above, the study zone covers the south and the southwest regions of Saudi Arabia. This particular zone was selected for the study because it has the relatively most significant and reliable flood flow observations in the country. The climate of this zone is classified as arid (Ref 19). It is affected, mainly, by two contrasting and conflicting systems of climate; the cyclonic system during winter and the monsoon in summer. The maximum heat index is about 86 and the range of daily temperature variation is 32°C. The annual relative humidity has an average value of 62%.

Altitude is the major factor controlling the quantity and pattern of rainfall in the zone (Ref 20), the amount of rain increasing progressively with elevation. Rainfall at the Sea Level has an annual average of 100 mm. In lands of altitude of about 2000 m the mean annual rainfall is 850 mm. Runoff coefficients fluctuate between 0.133 and 0.185 resulting in an overall average of 0.158. Runoffs are characterized by spates which generally last on average 12 hours from start to finish. Flash flood hydrographs are characterized by steep rise and rapid recession. Details of the characteristics of these hydrographs and the factors affecting their shape and pattern is outside the scope of this paper, but can be found elsewhere (Ref 21).

The ephemeral channels in the zone carry water only during each flash flood resulting from an effective rainstorm. The dry period between two successive flash floods varys from 25 days to 4 months, on average, depending on the season and the altitude of channel. This period is longer at low altitudes than at high altitude channels, and is longer during summer (from May to November) than during winter (from December to April). Slope of the channels also varys, according to altitude, from 18-20 cm/km at high altitudes to 4-8 cm/km at low altitudes. Soils vary from sand to sandy clay. Flood flows are characterized by carrying large amount of fine suspended sediment compared with the transported amount of bed load sediment (Ref 22).

A number of stable ephemeral channel cross sections in the above zone were identified. The observed values of each of the channel bed slope, the channel width, and the channel depth at each cross section, while fluctuate from year to year around a mean value, indicate no trend and vary insignificantly with time during the observations' period, which is of at least 18 years. At each of these cross sections flow hydrograph and concentration of suspended sediments were measured during several flood events. The suspended sediment and water discharge measurements were taken using a depth integrating sampler for the sediment and a Wardar type cableway for the flow measurements. The main characteristics of the channel cross sections are given in Table 1.

## 3. REGIME EQUATIONS FOR WIDTH, DEPTH, AND SLOPE

Several equations have been proposed by investigators to predict the width, the depth, and the slope of a channel in regime. The general form of these equations is to express each of the width, the depth, and the slope as a function of the discharge of channel (Refs 13 to 16). The applicability of such form of equations in ephemeral channels of a semiarid area was studied by Leopold et al. (Refs 17 and 18).Their results indicated that in ephemeral streams, generally, velocity increases downstream at a faster rate than in perennial streams. In addition, they recommended the use of mean annual discharge instead of average discharge in the regime equations in order to explain the variation of hydraulic characteristics in a downstream direction of an ephemeral channel.

Moreover, previous investigations (Ref 19) indicated that in ephemeral channels of extremely arid areas, such as in Saudi Arabia, the correlation with average discharge of each of the width, the depth, and the slope of channel is not suitable to study the variation of hydraulic characteristics in a downstream direction. There are two reasons for such finding; the high variability of discharge even from one hour to another during a flash flood event, and the large amount of suspended sediments that are carried by some (not all) streams during the floods in the study zone of Saudi Arabia.

To describe the variation of water and sediment flow conditions in an ephemeral channel in the study zone variables identifying the flow hydrograph were selected. These are peak flowrate, mean flowrate, time to peak flowrate, and duration of hydrograph. Previous study (Ref 23) made on the characteristics of flash flood flow hydrograph in the study zone showed that the ratio of a hydrograph duration to the hydrograph time to peak flowrate is almost constant, and that the hydrograph peak flowrate can be expressed as a linear function of the time to peak flowrate of the hydrograph. Based on these results, two parameters of hydrograph may be selected to describe the characteristics of flow hydrograph in the study zone in a regression regime type formula. These parameters are peak flowrate and mean flowrate of hydrograph.

Further, different definitions for the peak flowrate and the mean flowrate of hydrograph in the study zone were investigated to best fit the regime equations (Ref 19). It has been found that the ratio of annual peak flood flowrate for a return period of 50 years, $Q_{p_{50}}$ (in $m^3/s$), to annual mean flood flowrate, Q (in $m^3/s$), of hydrograph provides the best correlation with the width, the depth, and the slope of ephemeral channel in the study zone.

Thus, for the purpose of developing a general empirical relation for the regime width, depth, and slope of ephemeral channel in the study zone, the following form is proposed

$$B, D, \text{ or } S = \xi_1 (Q_{p_{50}}/Q)^{\alpha} + \xi_2 (1+d)^{\beta} C^{\gamma} \qquad (1)$$

in which B is the channel width (in meters); D is the channel mean depth (in meters); S is the channel bed slope (in cm/km); d is the median size diameter of suspended sediment (in millimeters); C is the mean concentration of suspended sediment (in kg/m ); $\xi_1$, $\xi_2$, $\alpha$, $\beta$, and $\gamma$ are constants to be determined through regression; and $Q_{p_{50}}$ and Q are as defined before.

The first and second terms on the right side of Eq 1 reflect the effect of flash flood flow variability and concentration of suspended sediments, respectively, on the predicted values of B, D, and S.

It is realized that the use of both $Q_{p_{50}}/Q$ and C as independent variables in a regression equation may not be proper if there is a high correlation between them. Previous study (Ref 22) examined the correlation between C and flowrate for various recurrence intervals, and between C and ratios of flowrate of various recurrence intervals using observations from 118 streams in the study zone. It has been found that the correlation coefficient is not significant for C and a ratio of flowrates, but significant for C and a given flowrate of a certain recurrence interval. In particular, the correlation coefficient between C and $Q_{p_{50}}/Q$ was, on average, equal to 0.341, meaning these two independent variables in Eq 1 are not highly correlated in the study zone.

Based on the above argument, the available records at 37 ephemeral channel cross sections were used to develop the following regression relationships

$$B = 28.30 \, (Q_{p_{50}}/Q)^{0.83} + 0.018 \, (1+d)^{0.93} \, C^{1.25} \qquad (2)$$

$$D = 1.29\ (Q_{p_{50}}/Q)^{0.65} - 0.01\ (1+d)^{0.98}\ C^{0.46} \quad (3)$$

$$S = 18.25\ (Q_{p_{50}}/Q)^{-0.35} - 0.88\ (1+d)^{1.13}\ C^{0.36} \quad (4)$$

in which $Q_{p_{50}}/Q$ is in hundreds of $m^3/s$.

Initially, the regression was performed excluding the variable (1+d)C from Eq 1. The correlation coefficient between the measured and predicted log B values was 0.638, between the measured and predicted log D values was 0.781, and between the measured and predicted log S values was 0.715. When the variable (1+d)C is included in the regression general form, Eq 1, the correlation coefficients for B, D, and S improved to be 0.861, 0.913, and 0.906, respectively. Fig 1 shows the developed relationships (Eqs 2 to 4) with the observations in case of excluding the variable (1+d)C resulting in predicted width $B_w$, depth $D_w$, and slope $S_w$, and in case of including the variable (1+d)C resulting in predicted width $B_m$, depth $D_m$, and slope $S_m$. The figure indicates that the variable (1+d)C is important and should be included in the regime equations in the study zone.

## 4. CHANNEL PATTERN

Natural streams tend to form meanders. Three dimensions are generally used by investigators to describe such meanders. These are the length $M_L$, the width $M_B$, and the radius of curvature $M_R$ of a meander. Jefferson (Ref 24) suggested a relationship between the width B and the meandering width $M_B$ of a channel. Based on his approach Inglis (Ref 25) was able to relate both the meandering length $M_L$ and the meandering width $M_B$ with the channel width B of a meander. Same type of relationships were proposed by Leopold et al. (Ref 26), Zeller (Ref 27), among others. On the other hand, analysis of actual observations led Inglis (Ref 28) to conclude that both meandering length and meandering width vary with the discharge. This conclusion was admitted by Blench (in Ref 15). The present study intends to develop relationships by which $M_L$, $M_B$, and $M_R$ in ephemeral channels of extremely arid areas can be reasonably predicted. Observations from the 37 ephemeral channels of Saudi Arabia were used in the investigations.

First, the approach of relating $M_L$ and $M_B$ to B was examined. The relationships proposed by Inglis (Ref 25), Leopold et al. (Ref 26), and Zeller (Ref 27) are plotted together with observations collected from the ephemeral channels and shown in Fig 2. These relationships are listed in the appendix. Inspection of Fig 2 indicates that none of the relationships produces satisfactory prediction in the ephemeral channels. However, the relationship of Inglis (Ref 25) for $M_L$ and those of Leopold et al. (Ref 26) and Zeller (Ref 27) for $M_B$ were the relatively best predictive relationships among the investigated ones.

To improve the prediction of $M_L$ and $M_B$ in an ephemeral channel another approach was utilised. The approach is based on the argument described earlier regarding the characteristics of flash flood flow in the study zone. Both $M_L$ and $M_B$ were assumed to vary with $Q_{p_{50}}/Q$ according to the following equation

$$M_L \text{ or } M_B = \xi_3\ (Q_{p_{50}}/Q)^{\eta}\ \varepsilon \quad (5)$$

where $(Q_{p50}/Q)$ is as defined earlier in hundreds of $m^3/s$; $\xi_3$, and $\eta$ are regression parameters to be determined; and $\varepsilon$ is the error term.

Using the available observations, the following regression relationships are developed

$$M_L = 45.9\ (Q_{p_{50}}/Q)^{1.47} \tag{6}$$

$$M_B = 70.4\ (Q_{p_{50}}/Q)^{1.34} \tag{7}$$

From Eq 6 and Eq 7

$$M_B/M_L = 1.533\ (Q_{p_{50}}/Q)^{-0.13} \tag{8}$$

In addition, the plot of the observed $M_L$ and $M_R$ (Fig 2) suggests the following relationship

$$M_L = 6.60\ M_R \tag{9}$$

The correlation coefficient between the observed and predicted log $M_L$ values was 0.924, and that between the observed and predicted log $M_B$ values was 0.906. The plot of Eqs 6 through 9 together with the observations is shown in Fig 2. It can be seen that the developed relationships (Eqs 6 to 9) produce satisfactory prediction for $M_L$, $M_B$, and $M_R$ in the study zone. In addition, the ratio $M_B/M_L$ is almost constant and varies insignificantly with $(Q_{p_{50}}/Q)$. It is approximately equal to 1.30.Similar finding was reported by Inglis (Ref 28) who found the ratio of $M_B/M_L$ in humid areas is about 2.50.

## 5. BED ARMOURING

Several procedures have been proposed to predict the composition of the armoured layer of an alluvial channel. These included in the pioneer studies of Gessler (Refs 29 and 30), Little and Mayer (Refs 31 and 32), Williman (Ref 33), and Shen and Lu (Ref 34), among others. While these studies have contributed significantly towards better understanding of the process of selective erosion in alluvial channels it must be emphasized that they are based on limited data collected from flume experiments under steady flow conditions. They are not applicable in natural ephemeral channels where flow is significantly unsteady (Ref 19). The present study uses field observations in a trial to develop a procedure by which the composition of the armoured layer in an ephemeral channel in the extremely arid zone of Saudi Arabia can be reasonably predicted.

Generally, the grain size distribution of the armoured layer in an alluvial channel can be considered as functions of the original size distribution of alluvial sediments and the flow conditions in the channel. Based on the assumption that the original size distribution is close to the lognormal distribution, one can describe it with two parameters: median size, $d_{mo}$, and standard deviation, $\sigma_0$ (Ref 34). The flow conditions in an ephemeral channel in the study zone, due to its high variability, can be considered as functions of peak flood flowrate and mean flood flowrate of hydrograph. Based on the above argument, one may select the following equation form for regression analysis

$$d_{ma}/d_{mo} = \psi\ (Q/Q_p)^{\phi_1} (\sigma_0)^{\phi_2}\ \varepsilon \tag{10}$$

in which $d_m$ is the size of sediment for which m% of the sediment by weight is finer, subscripts o and a refer to the armoured layer and the original sediment, respectively; $\psi$ , $\phi_1$, and $\phi_2$ are regression parameters to be determined; $\varepsilon$ is the error term; $Q_p$ is the annual peak flood flowrate for a given return period (in $m^3/s$); and Q is as defined earlier (in $m^3/s$).

Field observations were collected at the center of several ephemeral channel cross sections in the study zone. For each cross section, the original size distribution of sediments was determined through the analysis of at least 5 samples collected from the channel cross section at its berm level. Samples of the bed sediments were taken after each flash flood event; during the dry period. The size distribution of the samples that did not show significant variation with time during the observations' period of two years (12 flash flood events, on average) was considered as the stable size distribution of the armoured layer at the center of the cross section.

It was possible to identify in the study zone 50 cross sections at which the size distribution of the original sediments and that of the armoured layers as well as the flash flood flow hydrographs were observed.

The identified observations were split into two sets. The first set of observations (referred to as the calibration observations), contains the observations of 37 ephemeral channel cross sections, whereas the second set of observations (referred to as the verification observations), contains the observations of the remaining 13 cross sections. The calibration set of observations was used in the regression analysis of the proposed model form (Eq 10), whereas the verification set of observations was used in the verification process of the model. The main characteristics of the cross sections and their observations are given in Table 1.

Several recurrence intervals for $Q_p$ in Eq 10 were investigated, and the regression parameters were estimated using a forward stepwise algorithm. It was found that $Q_{p10}$, the peak flood flowrate for a return period of 10 years, is the best to be used in Eq 10 to have a high correlation between the measured and predicted log $d_{ma}/d_{mo}$. Based on this finding, the following regression equations were determined

$$d_{30}(\text{armoured})/d_{30}(\text{original}) = 0.326\ (Q/Q_{p_{10}})^{0.21}\ (\sigma_0)^{1.46} \tag{11}$$

$$d_{50}(\text{armoured})/d_{50}(\text{original}) = 0.410\ (Q/Q_{p_{10}})^{0.53}\ (\sigma_0)^{1.21} \tag{12}$$

$$d_{84}(\text{armoured})/d_{84}(\text{original}) = 0.932\ (Q/Q_{p_{10}})^{0.72}\ (\sigma_0)^{0.33} \tag{13}$$

Inspection of Eqs 11 through 13, it can be seen that the exponent values of $Q/Q_{p10}$ increased from $d_{30}$, $d_{50}$, to $d_{84}$, whereas the exponent values of $\sigma_0$ decreased from $d_{30}$, $d_{50}$, to $d_{84}$. This indicates that the variable log $(Q/Q_{p10})$ is more important in predicting $d_{84}$, and the variable log $(\sigma_0)$ is more important in predicting $d_{30}$. Since the coarser fractions are more important in the armouring process than are finer fractions, it may be concluded that the variable $\log(Q/Q_{p10})$ is more important than the variable log $(\sigma_0)$ in describing the armouring process in ephemeral channels.

Fig 3 shows the relationship between different sizes of the armoured layer of the calibration observations and the corresponding sizes predicted by Eqs 11 to 13. It can be seen that the equations produce satisfactory prediction for the armoured layer in the investigated ephemeral channels.

In addition, the merit of the above equations was examined through applications in another 13 ephemeral channels; i.e. the verification set of observations. Fig 4 shows a comparison between the predicted sizes of the armoured layer and those measured in the channels. It is apparent that the prediction is generally of reasonable accuracy, in spite of the fact that coarser fractions are slightly overestimated. This encourages the use of the equations in ephemeral channels.

## 6. CONCLUSIONS

The characteristics of flash flood and suspended sediment flow conditions should be given consideration in formulae describing the regime of ephemeral channels of arid areas. A ratio of a peak flood flowrate for a given return period to annual mean flood flowrate is proposed to describe the flow conditions in these channels. The ratio is used successfully in regression type formulae to predict the dimensions and pattern as well as the bed armouring characteristics of a regime ephemeral channel in an extremely arid zone.

## 7. NOMENDATURE

The following symbols are used in this study

| | |
|---|---|
| B | channel width (m) |
| C | mean concentration of suspended sediment ($kg/m^3$) |
| D | channel mean depth (m) |
| d | median size of suspended sediment (mm) |
| $d_{ma}$ | size of sediment for which m% of the sediment by weight is finer in the armoured layer (mm) |
| $d_{mo}$ | size of sediment for which m% of the sediment by weight is finer in the original alluvial sediment (mm) |
| $M_B$ | meandering width of channel (m) |
| $M_L$ | meandering length of channel (m) |
| $M_R$ | meandering radius of curvature (m) |
| Q | annual mean flood flowrate ($m^3/s$) |
| $Q_{p_t}$ | annual peak flood flowrate for a return period of t years ($m^3/s$) |
| S | bed slope of channel (cm/km) |
| $\alpha$ | parameter |
| $\beta$ | parameter |
| $\gamma$ | parameter |
| $\eta$ | parameter |
| $\xi_1$ | coefficient |
| $\xi_2$ | coefficient |
| $\xi_3$ | coefficient |
| $\varepsilon$ | error term |
| $\sigma_0$ | standard deviation of original alluvial sediment (mm) |

## 8. ACKNOWLEDGEMENTS

This study was supported by King Abdul-Aziz City for Science and Technology, KACST, Saudi Arabian National Water Research Project Number AR-5-62. The author appreciates the assistance of the Deputy Minister of Agriculture and Water in Riyadh, Saudi Arabia, for supplying data used in the study.

## 9. REFERENCES

1. Blench, T: "Mechanics of plains rivers" Edmonton-Alberta, Canada, University of Alberta Press, 1986.

2. Kennedy, R G: "The prevention of silting in irrigation canals" Paper No. 2826, Minutes of Proceedings of the Institution of Civil Engineers, Vol. CXIX, 1895, pp 281-290.

3. Lindley E S: "Regime channels", Punjab Engineering Congress, Lahore, Government of India, 1919, pp 63-74.

4. Lacey G: "Stable channels in alluvium" Paper No. 4736, Minutes of Proceedings of the Institution of Civil Engineers, Vol. 229, 1930, pp 259-284.

5. Lacey G: "Uniform flow in alluvial rivers and canals". Paper No. 4893, Minutes of the Proceedings of Civil Engineers, Vol. 237, 1935, pp 421-453.

6. Lacey G: "Regime flow in incoherent alluvium". Publication No. 20, Central Board of Irrigation, India, 18 July, 1939.

7. Lacey G: "Shock in regime channels". Punjab Engineering Congress, Government of India, 1940.

8. Lacey G: "A general theory of flow in alluvium". Paper No. 5518, Journal of the Institution of Civil Engineers, Vol. 27, 1946,pp 16-47.

9. Lacey G: "Flow in alluvial channels with sandy mobile beds". Proceedings of the Institution of Civil Engineers, Vol. 9, 1958, pp 145-164.

10. Inglis C C: "Historical note on empirical equations, developed by engineers in India for flow of water and sand in alluvial channels", 2nd Meeting of the International Association for Hydraulic Research, Stockholm, 1948.

11. Blench T: "Canals, channels and rivers", London, England, Civil Engineering Reference Book, Butterworth, 1951.

12. Blench T: "Regime behaviour of canals and rivers", London, England, Butterworth, 1957.

13. Chitale S V: "Hydraulics of stable channels", New Delhi, India, Ministry of Irrigation of the Government of India, 1966, pp 276-323.

14. Graf W H: "Hydraulics of sediment transport", New York, USA, McGraw Hill Book Company, 1971, pp 243-272.

15. Bogardi J: "Sediment transport in alluvial streams", Budapest, Akademiai Kiado, 1974, pp 630-653.

16. Mahmood K and Shen H W: "Regime concept of sediment-transporting canals and rivers", Chapter 30, Vol. 2 in "River mechanics" edited and published by Shen H W, P.O.Box 606, Fort Collins, Colorado, USA 80521, 1971, pp 30/1-30/39.

17. Leopold L B and Maddock T: "The hydraulic geometry of stream channels and some physiographic implications". Professional Paper No. 252, U.S. Geological Survey, 1953.

18. Leopold L B and Miller J: "Ephemeral streams-hydraulic factors and their relation to the drainage net". Professional Paper No. 282-A, U.S.Geological Survey, 1956.

19. Nouh M: "Flood damages to constructions in wadi Ad-dilah of Saudi Arabia",Technical Report Prepared for King Abdul-Aziz City for Science and Technology (KACST), Riyadh, Saudi Arabia, 1986.

20. Nouh M: "Analysis of rainfall in the southwest region of Saudi Arabia". Proceedings Part 2, The Institution of Civil Engineers, Vol. 83, March 1987, pp 339-349.

21. Nouh M: "On the prediction of flood frequency in Saudi Arabia". Proceedings Part 2 (In Press), The Institution of Civil Engineers, 1987.

22. Nouh M and Jamjoom T: "Sediment transport in wadis in Saudi Arabia", Technical Report Prepared for King Abdul-Aziz City for Science and Technology (KACST), Riyadh, Saudi Arabia, 1981.

23. Nouh M: "A rainfall-runoff model for Saudi Arabia". In: Proc. 5th Congress of Asian and Pacific Regional Division, IAHR (Seoul, Korea: August 18-20, 1986) International Association for Hydraulic Research, 1986, 121 pp.

24. Jefferson M: "The limiting widths of meander belts". Natl. Mag. Oct. 1902.

25. Inglis C C: "Meanders and their bearing on river training". Maritime and Waterways Engineering Division, The Institution of Civil Engineers, Jan. 1947.

26. Leopold L B and Wolman M G: "River meanders". Bull. Geol. Soc. Am., Vol. 71, 1960.

27. Zeller J: "Meandering channels in Switzerland". In: Proc. Intern. Assoc. Sci. Hydr. Symp. on River Morphology (Bern, Switzerland: May 2-4, 1967) IAHR, 1967, 137 pp.

28. Inglis C C: "Relationship between meander belts ...". Tech. Note No. 12, Central Irrigation and Hydrodynamic Research Station, Poona, India, 1938.

29. Gessler J: "The process of selective erosion in alluvial channels". Lect. Note, Colorado State University, Fort Collins, Colorado, USA, 1967.

30. Gessler J: "Self stabilizing tendencies of alluvial channels". Journal of Waterways and Harbours Division, ASCE, Vol. 96, No. 2, 1970, pp. 235-249.

31. Little W C and Mayer P G: "The role of sediment gradation on channel armouring". Technical Report No. ERC-0672, Georgia Institute of Technology, USA, 1972.

32. Little W C and Mayer P G: "Stability of channel beds by armouring". Journal of Hydraulics Division, ASCE, Vol. 102, No.HY11, 1976, pp. 1647-1661.

33. Williman E B: "The development of armoured surfaces in streambed channels". M.Eng. Thesis, University of Canterbury, New Zealand, 1975.

34. Shen H W and Lu J: "Development and prediction of bed armouring". Journal of Hydraulic Engineering, ASCE, Vol. 109, No. 4, 1983, pp. 611-629.

## 10. APPENDIX

Relatioships proposed by Inglis (Ref 25)

$$M_L = 6.6\ B^{0.99}$$
$$M_B = 18.6\ B^{0.99}$$

Relatioships proposed by Leopold et al. (Ref 26)

$$M_L = 10.9\ B^{1.01}$$
$$M_B = 2.7\ B^{1.1}$$
$$M_L = 4.7\ M_R^{0.98}$$

Relatioships proposed by Zeller (Ref 27)

$$M_L = 10.0\ B^{1.025}$$
$$M_B = 4.5\ B$$

Table 1. Main characteristics of channel cross sections and records utilised for study.

| Name of channel | Length (km) | Number of cross sections | Distance from mouth (km) | Total period of records (yr) |
|---|---|---|---|---|
| Ashran | 613 | 9 | 18.6*, 25.8*, 33* 78, 109*, 258, 401 458, 527* | 18 |
| Saliman | 914 | 15 | 103*, 198, 314*, 367, 411* 541, 645*, 698, 742, 799* 813, 845*, 866, 873*, 890 | 22 |
| Wadi Sabha | 1498 | 26 | 60*, and every 50 km | 21 |

* used for verification.

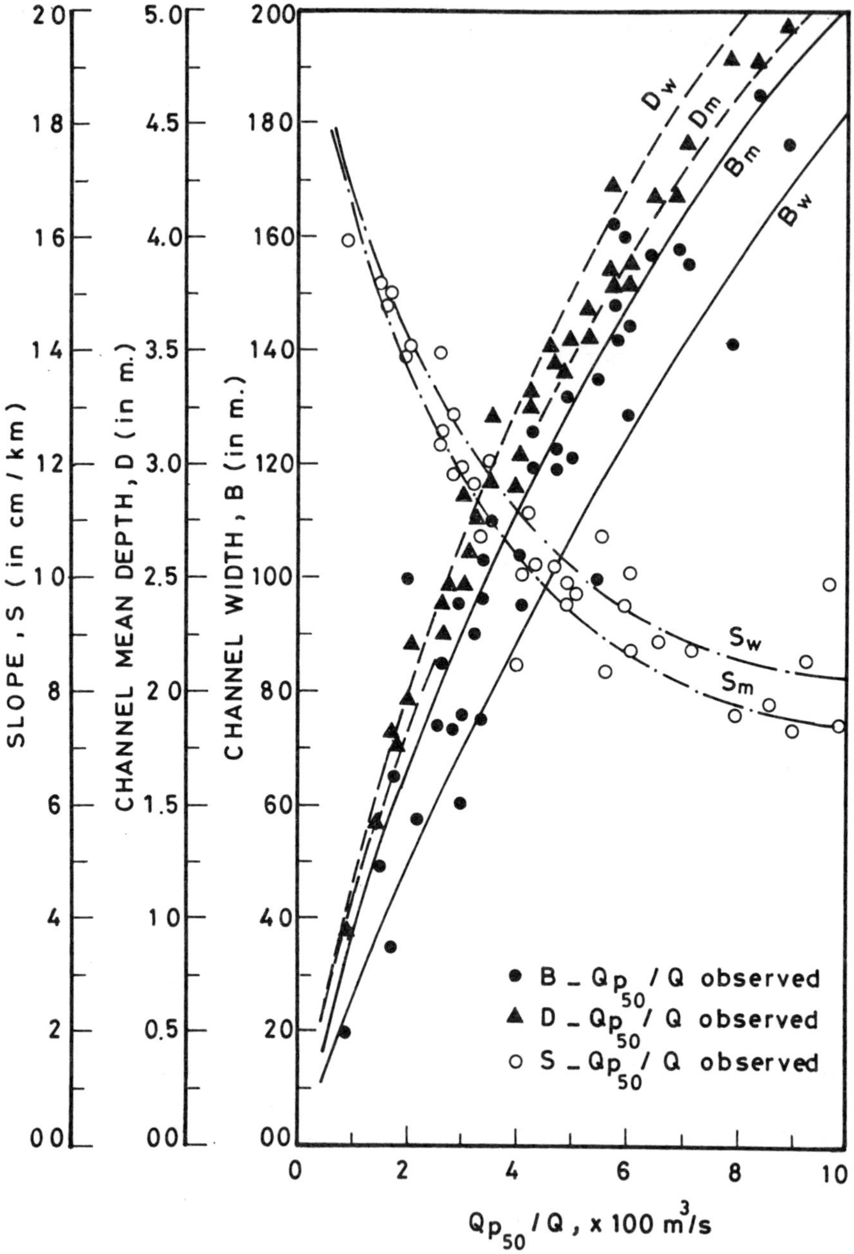


Fig 1. Observed and predicted regime width, depth, and slope of channel.

1 $M_L$-B relationship: Leopold et al. (Ref 26) and Zeller (Ref 27)
2 $M_L$-B relationship: Inglis (Ref 25)
3 $M_B$-B relationship: Leopold et al. (Ref 26)
4 $M_B$-B relationship; Zeller (Ref 27)
5 $M_B$-B relationship: Inglis (Ref 25)
6 $M_L$-$M_R$ relationship; Leopold et al. (Ref 26)

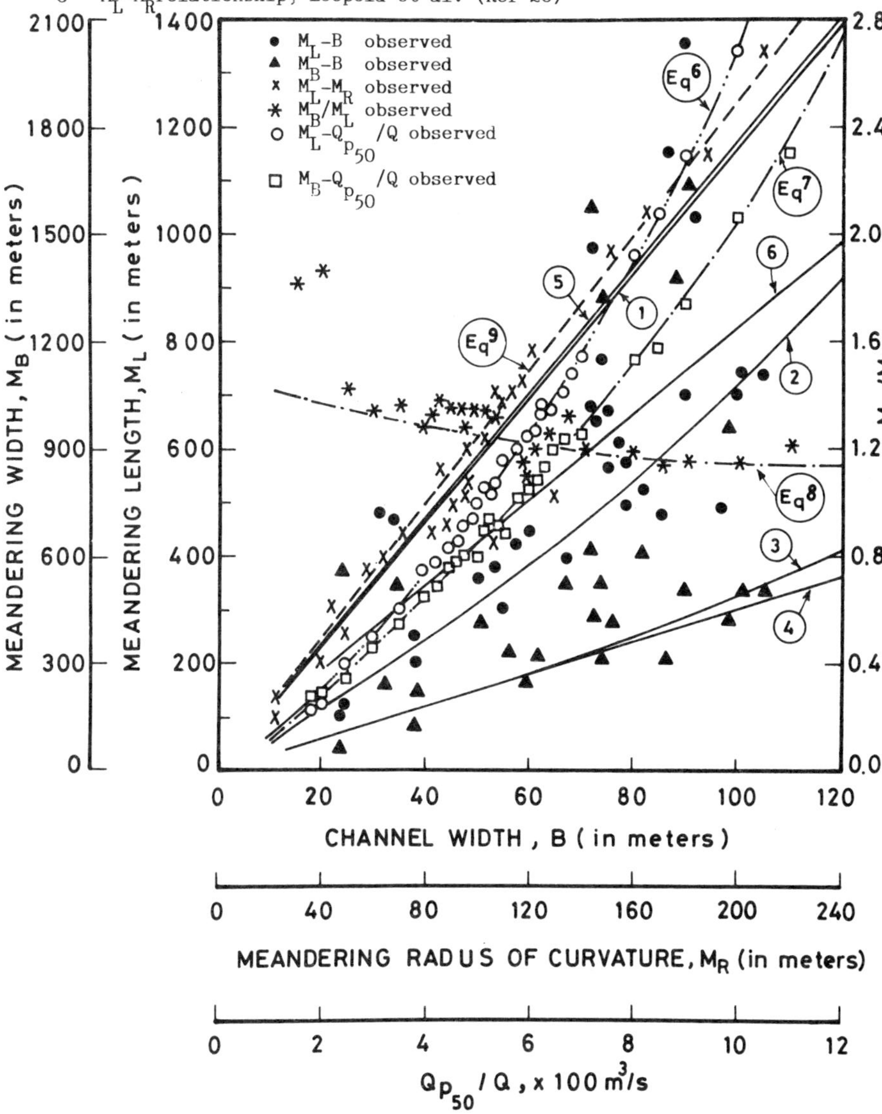


Fig 2. Observed and predicted geometry of meanders.

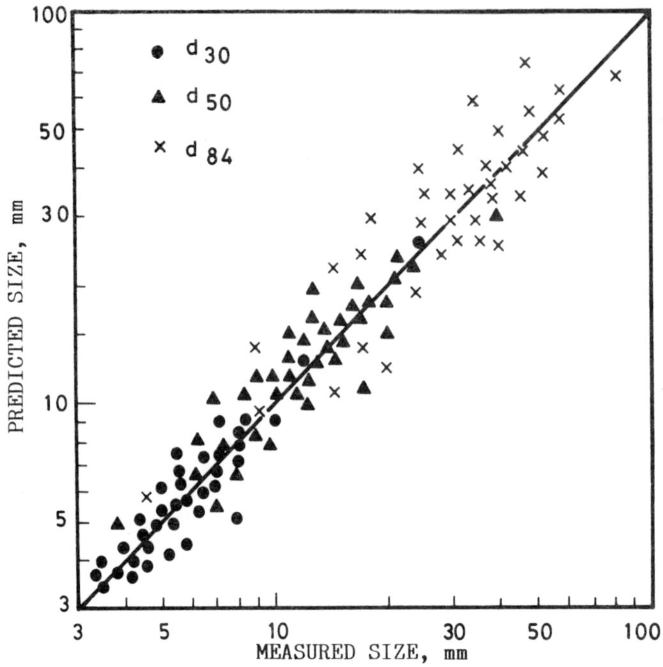


Fig 3. Sizes predicted and measured in the calibration process.

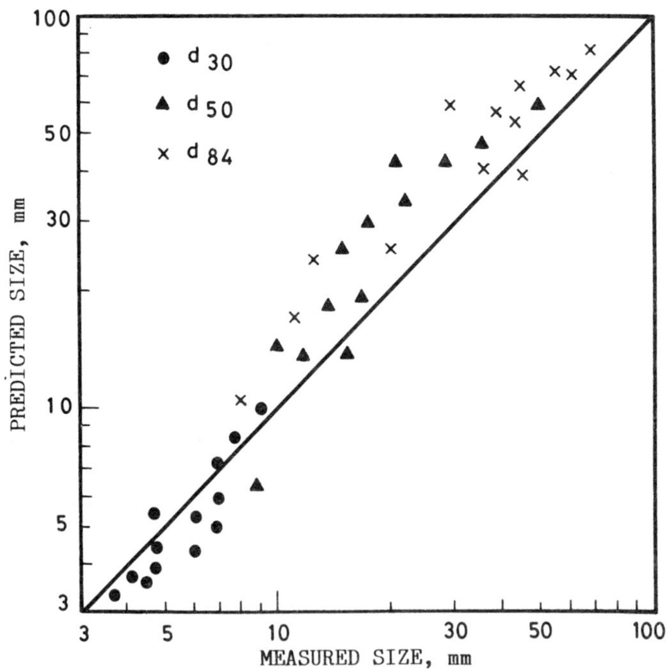


Fig 4. Sizes predicted and measured in the verification process.

International Conference on
RIVER REGIME
18-20 May, 1988 PAPER B3

# THE APPLICATION OF REGIME THEORY TO WADI CHANNELS IN DESERT CONDITIONS

P B Clark and S M A Davies,
Binnie & Partners
Grosvenor House, Redhill, Surrey, UK

SUMMARY: In the arid conditions of PDR Yemen, wadis experience only short duration spate flows. Channels are steep and large movements of bed material occur. Channel widths in particular are subject to considerable change. The concept of a 'regime' in such conditions is discussed and the applicability of some of the relationships developed for other conditions considered.

## 1. INTRODUCTION

In March 1982 a very intense storm moved northwest through South Yemen (The Peoples Democratic Republic of Yemen). Devestating floods occurred throughout the south western part of the country. Rainfall and flow records of the event are sparse but in many catchments the floods were estimated to have a return period in excess of 100 years, and in the worst affected regions perhaps approaching 500 or even 1000 years. In one small isolated valley at the heart of the storm the estimated flood was of comparable magnitude to the Probable Maximum Flood (PMF).

The area is arid and much of the country is desert. The watercouses are ephemeral wadis, normally dry or with a small, spring-fed baseflow. Most of the rural population of the country lives along these wadis, farming the narrow alluvial terraces either side of the wadi channel, traditionally using spate irrigation methods although a substantial area now uses groundwater. The effects of the floods were devastating in many areas with houses and villages destroyed and much irrigable land lost. The damage was caused in most cases not by the wadis overtopping their banks but by the widening of their channels, eroding the alluvial terraces. In places the channel widths increased by 100% and more, which represents an enormous movement of material in the relatively short duration floods.

This paper describes some of the hydraulic studies carried out on two projects, the general aims of which were similar - to advise on rehabilitation measures and the river training and bank protection works necessary in the wake of the flood damage of 1982. However, the emphases of the two projects were somewhat different. The earlier study was concerned with a reach of about 35 km of Wadi Bana, an agriculturally important watercourse running to the Indian Ocean some 50 km north east of Aden. It therefore allowed a fairly detailed look at the wadi regime and resulted in detailed recommendations for the layout of river training works, the reconstruction of barrages and other control structures, and the requirements for bank protection.

International Conference on River Regime, edited by W.R.White.

The second study involved some 600 km of watercourses on 30 wadis in various locations throughout South Yemen. It was not feasible within the limits of the study to carry out design of all the possible works. Therefore, the approach was to try and establish general hydraulic parameters for each reach which would allow design of individual training and bank protection works at a later stage. Such parameters included optimum channel widths to be left by any training or reclamation works; maximum depths and energy levels of flow for the design floods to enable determination of the height of protection works; and maximum velocities, to provide guidance on potential scour depths.

The approaches in both studies have attempted to establish the regime of each wadi reach. It was recognised, however, that the characteristics of the watercourses are significantly different from those rivers and channels for which regime relationships have been established. The equations which are available have therefore been used with some caution particularly in the light of the uncertainties of the data. Where more than one relationship has been proposed they have been compared and contrasted and, in establishing the design parameters listed above, generally it has been necessary to adopt fairly conservative values.

## 2. DESCRIPTION OF THE WADIS

Though the wadis studied were spread over an area about 800 km by 200 km, the general characteristics of the watercourses are similar. The catchments are all in barren, mountainous regions with rapid run-off responses. Wadi flows tend to consist therefore of short duration spates. Despite the aridity of the country - some of the wadis are on the fringe of the main central desert region of the Arabian peninsular - such spates in the upper wadis are not infrequent. In two of the larger wadis in the southern part of the country perhaps 50 floods of varying magnitude may occur each year.

Infiltration and diversion for irrigation reduce the flows as they pass down the wadis, offsetting any effect of increased catchment areas commanding the lower reaches. The magnitude of any flood thus increases in the upper reaches to a peak, but tails away in the lower reaches with only major floods extending to the lowest extents of the definable channels. Thus not only the magnitudes of the peak flows but also the locations where they occur, depend on the severity of the storm events. This is illustrated in Figure 1.

Throughout their length the wadis are steep, particularly so in their upper reaches where gradients of 1% or greater are not uncommon. Bed material is coarse with boulders and cobbles predominating. Lateral movement of the channels is largely controlled by rock outcrops but there are terraces of finer alluvial material in sheltered embayments. Lower down, in the middle reaches, there is some slackening of the gradients and correspondingly a reduction in size of bed material. The channels typically are fairly deeply incised between terraces of fine alluvial material.

In the lower reaches gradients have reduced to about 0.3% and the bed material is tending to sand. The watercourses are wide and ill-defined with multiple channels. These may be a metre or more deep and clearly move their positions considerably during any major flood. It is unlikely, except possibly in very major events, that the full width of the watercourse is inundated.

With such generally steep channels movement of sediment is considerable, however, the nature of the spate flows means that material is being moved in a series of waves down the wadi. At the points where the flow magnitude starts to decrease, floods begin to deposit their loads. Thus there is considerable mixing of material of different sizes within the overall gradation from boulders to sand down each wadi. Although local bed levels and gradients may vary significantly from one spate to the next the overall longitudinal slopes are relatively stable.

## 3. AVAILABLE DATA

In countries such as PDR Yemen the available data was limited and the time and money available in the project budget for data collection extremely restricted. The major uncertainties were in the hydrological studies and the estimates of floods of various return periods for the many wadis involved. Most of the wadis were ungauged and a regional flood study approach was therefore adopted. It is not the purpose of this paper to detail the hydrological studies (Ref. 1) but the general approach was to apply multiple regression analyses to the mean annual flood (MAF) and catchment characteristics for 5 gauging stations in PDRY with records of between 5 and 25 years duration and 8 stations in Saudi Arabia with records of less than 10 years duration. The most satisfactory relationship was found to be:-

$$MAF = 3.27\ CA^{1.163}\ MSL^{-0.935} \qquad (1)$$

(multiple correlation coefficient of 0.988)

where CA is the catchment area, and MSL is the mean stream length. The other catchment parameters (Mean annual rainfall, MAR, and main channel slope, S) did not add significantly to the unexplained variance.

In addition growth factors were established to relate longer return-period floods to the MAF. Various approaches were applied resulting in the following adopted values for the growth factors ($Q_T$/MAF - where $Q_T$ is the peak flow of the flood with return period of T years) applied to flood peak:-

| | MAF | T=5 | 10 | 20 | 50 years |
|---|---|---|---|---|---|
| growth factor | 1.0 | 1.5 | 2.2 | 3.1 | 4.6 |

It must be stressed that these estimates can be approximate only, reflecting the small available data base. However, the first, detailed study, was on Wadi Bana, one of the gauged watercourses, and more reliable estimates of the T-period floods were possible.

It was also possible to obtain some checks on the estimates, for a number of the ungauged wadis, by comparing them with discharges assessed by other consultants during studies on those wadis. Such discharges were generally estimated by visual assessment of maximum water levels and converting these to flows by some form of slope-area approximation. The comparisons are shown in Table 1. Given the approximations inherent in both sets of estimates and the short period of observations on most of these wadis, there is reasonable agreement. The only major discrepancies are the apparently low estimates of the peak flows of the floods of March 1982, produced by Strojexport (Ref. 2) for the two wadis they studied, Markhah and Khawrah in the Nissab region. Their estimates, however, appear incompatible with the extent of damage caused by the floods in these two wadis.

Infiltration, as discussed above, has a significant effect on flood discharges. The gauged flows are net discharges with infiltration in the upstream channel already accounted for. Thus the T-period flood estimates using equation (1) and the growth factors allow for typical bed infiltration losses. However, in many of the lower reaches of the wadis, infiltration becomes so significant that only exceptional flows reach downstream areas. For these reaches additional allowances for such losses were included. Adopted values for these infiltration rates were of the order of 50 to 150 mm/hr.

Topographical information was of better quality. In most cases some information, albeit on small scale mapping, was available for the period before the 1982 flood. More recent aerial photography was available but, with the exception of Wadi Bana, mapping had not been produced from this. In some cases, where other studies were in progress, post-1982 mapping was available. It was possible to carry out some check surveys to provide corroboration of widths and slopes assessed from the other data sources.

In the Wadi Bana study, resources were made available for sampling the bed material. The results are shown in Figure 2. There is obviously a wide scatter of the data, which shows the difficulty of assessing a representative size of bed material in a particular location. It does, however, illustrate the general reduction of particle size along the wadi. Lack of resources precluded such sampling in any of the other wadis. In the circumstances all that was possible was a visual inspection and estimate of the representative size of the surface material at frequent points along each wadi. It is appreciated that there are inherent inaccuracies in such an approach but, in practice, ranges of values were taken and, in establishing design criteria, conservative estimates adopted.

## 4. DOMINANT DISCHARGE AND CHANNEL REGIME

It was considered appropriate to try and establish general criteria for the design of training works in terms of the channel regime. It is not essential that the defined watercourse should be 'in regime' provided the training works are designed to resist the appropriate water levels, velocities and scour patterns and depths. However, the regime channel can be considered as an optimum both hydraulically and economically. A wider channel would encourage the formation of sub-channels with transverse flow patterns allowing oblique attack on the training and bank protection works. A narrower channel would constrict the flows causing greater scour depths and possibly affecting in the longer term the longitudinal slope. The 'regime' width thus allows reclamation, where required, but provides a reasonable limit to the narrowing of any channel.

There have been various definitions of the regime theory as applied to channels and natural watercourses. Generally, however, it can be considered as an empirically-based method of describing the geometrical properties of the stable channel in terms of the parameters which influence its formation. In the case of the wadis studied and within the limitations of the available data, it was possible to assign values to some of these parameters for each reach; slope, S; a representative bed material size, D; and the width, B, and the depth, Y, of the channel in each case. However, considerable caution had to be exercised in applying the latter two values and particularly in assigning a value to the representative water discharge, Q.

Inglis (Ref 3) first introduced the concept of a dominant discharge as being the representative flow which can be considered as the channel-forming discharge. As Hey (Ref 4) puts it, by using dominant discharge concepts it is possible to avoid the problems of modelling channel response to all flows that can transport sediments, and simply focus attention on the average response of the channel over time periods in excess of ten years. The dominant discharge has been defined in various ways (e.g. Inglis (Ref 3), Blench (Ref 5), Terrel (Ref 6)) but is now generally considered as being the bankfull discharge, although some researchers define it terms of the flow with a particular return period, such as 1.5 or 2 years.

The wadis studied are subject to major changes of width in single flood events,and the channel sections cannot be considered as stable. Moreover the concept of bankfull discharge as being the 'dominant discharge' is misleading for these wadis. In a stable natural watercourse the level of the terraces or flood plains is a function of the pattern of discharge. As the terraces are built up so the frequency of flooding and the amount of material deposited reduces until an equilibrium is reached and the depth of channel can be considered as stable. In South Yemen, however, the terraces are frequently irrigated and have been so for many centuries. With spate irrigation methods large quantities of fine sediments are carried onto the terraces. As the terraces build up so the diversion canals are extended upstream to maintain command of the irrigated areas. Thus the terraces are raised well above the levels which would be formed with a natural channel, and moreover tend to be formed of finer, more easily erodible material. 'Bankfull' discharge in such circumstances is thus a very rare event if indeed it ever occurs.

The pattern of changes to the channel width follow a cycle in which major rare events cause large increases in width by erosion of the fine material of the terraces. This is followed by years of retrenchment with smaller floods tending to form narrower channels within the wider watercourses. This process is again aided by man as bunds and training works are extended into the wadi to reclaim land - there are obvious advantages in owning land close to the main channel with better access to water.

The 'regime' channel section in such cases thus represents a long-term averaged section. With this definition the concept of a channel-forming, dominant discharge is open to question. However the regime section can be related to a representative flow using suitable regime equations. This flow can be considered as the dominant discharge for definition of the other channel parameters. It will not generally represent a bankfull flow.

Particularly in the lower reaches of some of the wadis, flows of any size may occur only occasionally and only rare floods may have a duration and intensity that would be significant in terms of the channel-forming processes. Higher up, the frequency of spate flows increases but the significant changes to the channel sections still occur only in relatively rare, major floods. It appears probable, therefore that the return period of the representative flow in any reach is a function of the frequency of spates, with the longer return periods corresponding to the less frequent occurrence of spate flows.

## 5. REGIME WIDTHS

It is probably true to say that each channel has its own unique regime relationship, although close similarities are found for rivers in the same region with similar bed and bank materials and similar sediment loads. However, the authors are not aware of any set of regime equations developed for channels in desert conditions similar to those in South Yemen. The resources of the project did not allow the carrying out of the detailed studies necessary to establish such relationships, particularly in view of the added complication over the definition of the dominant discharge. The authors were therefore faced with the difficulty of how to establish the 'regime' width for each channel based on such limited data.

One approach would have been merely to accept the best estimate based on the topographical evidence available in each case. This would have resulted, however in widely differing estimates for the widths of wadis of similar character and catchment size. It is reasonable to assume that there should be a general similarity of regime for all the wadis being studied. The approach adopted was to utilise such regime equations as are available to impose an overall pattern on the selection of regime widths.

The relationships that are available were developed for channels in very different environments to those in South Yemen and the authors do not claim that they are directly applicable to the wadis studied. Indeed, with the dominant discharge defined as the 1.5- or 2- year flow, any correlations with the data are very poor. However, as discussed above, there are strong indications that the return period of the dominant discharge is a function of the frequency of spate flows. If the return period of the dominant discharge is treated as an additional variable in this way, then it is possible to obtain better correlations between the available regime equations and the data. The selection of regime widths was thus based on the use of an appropriate regime equation with the return of periods of the dominant discharges modified give a consistent pattern both along each wadi and throughout the project area.

Of the many regime-type equations that have been proposed, those relating channel width, B, to the dominant discharge have generally been found to be the most robust in application to rivers other than those for which they were developed. A number of such equations have been proposed by various researchers. These are mainly of the form:-

$$B = KQ^{m} \tag{2}$$

with Q defined either as the bankfull discharge or as the flow with 2 year return period. Values assigned to K and m for gravel bed rivers are listed in Table 2 (S.I. units).

**Table 2 : Coefficients in Regime Width Equation**

| Reference | K | m |
|---|---|---|
| Nixon (Ref 7) | 2.99 | 0.50 |
| Simons and Albertson (Ref 8) | 2.85 | 0.50 |
| Kellerhals (Ref 9) | 3.26 | 0.50 |
| Charlton, Brown & Benson (Ref 10) | 3.74 | 0.45 |
| Bray (Ref 11) | 4.75 | 0.527 |
| Hey and Thorne (Ref 12) | 3.67 | 0.45 |

Bray extended his analysis to include the effect of the bed material and suggested the relationship:-

$$B = 3.83\, Q_2^{0.528}\, D_{50}^{-0.07} \qquad (3)$$

Where $D_{50}$ is the size of bed material such that 50% by weight of the total is smaller. Hey (Ref 13) has suggested other relationships involving the sediment discharge, $Q_S$. The lack of data on sediment movement has precluded their consideration.

A comparison of the various forms of equation (2) and equation (3) is illustrated in Figure 3, although, since Q is not defined the same way in each case, strictly the equations are not comparable in this simple way. The relationships proposed by Bray give significantly greater channel widths than those of any of the other researchers. It is worth noting, however, that the widths given by equation (3) for the smallest gravel material are not totally incompatible with those given by Simons and Albertson (Ref 8) for sand-bed rivers, defined in terms of $Q_B$, the bankfull discharge:-

$$B = 6.4\, Q_B^{0.5} \qquad (4)$$

Although not demonstrated quantitatively by any other researchers it is perhaps reasonable to expect a gradual variation of width from a coarse gravel or cobble-bed reach to a sand-bed section. Certainly this was the general case in the wadis studied here although other factors such as flow variation also have an influence.

At the time the initial studies were carried out, the paper by Hey and Thorne (Ref 12) was not available and the relationship of Bray (equation 3) was used for cobble and gravel-bed reaches while equation (4) was used for sand-bed reaches. The former requires estimates of the representative bed-material sizes which, as discussed earlier, were, in this study, based in most cases only on a visual inspection. However, as illustrated by Figure 3, even an order of magnitude error in $D_{50}$ would produce only about a 20% error in predicted B. It is not thought that this level of inaccuracy was involved in the assessment of the size of the bed material.

Hey and Thorne's results do not support the equation derived by Bray and indeed when they introduce bank vegetation type into the analyses the equations derived are similar to those of Nixon, Simons and Albertson, and Kellerhals with m = 0.50 and K varying between 2.34 and 4.33. However, in none of the wadis study is there bank vegetation of any significance and a 'K' value at least at the highest end of that range would be expected. Indeed, in view of the general erodibility of the bank material it might be expected to be higher still. Using K = 5.0 and m = 0.5 would make equation (2) not incompatible with equation (3) over the range of flows of interest (generally between 50 $m^3/s$ and 1000 $m^3/s$). The use of Bray's equation is thus not considered inappropriate even in the light of Hey and Thorne's results.

The approach adopted in practice was to utilise the available mapping and survey data to make a first estimate of the regime width for each reach of each wadi. (Reaches were defined primarily on the basis of longitudinal slope and size of bed material.) Using equation (3), values of the representative discharge were then calculated and, using equation (1) and the growth factors listed in section 3 with allowance for additional seepage losses where appropriate, return periods were calculated for these flows. As might be expected, there was considerable variation in these return periods, values ranging from about 2 years to over 50. The overall distribution of these return periods was then reviewed and individual values adjusted to give a consistent pattern for all the wadis studied. Inevitably there was a degree of subjective judgement involved in making these adjustments but there was some guidance available. Firstly, it appeared, as discussed above, that the return periods were a function of the frequency of spates and that the return periods increased in the lower reaches of the wadis. This fitted the general pattern of return periods as calculated for the individual reaches. Secondly, the study of Wadi Bana had suggested that the return periods of the dominant discharges in the upper reaches under consideration were about 10 years. This study had had the benefit of reasonably accurate mapping and surveys from before the 1982 flood and hence the measured channel widths were more representative of the 'regime' widths. The upper reaches of Wadi Bana were subject to spates about fifty times a year - a relatively high frequency compared to most of the wadis. This return period could be considered, with reasonable confidence, to be typical of the return periods of dominant discharges in the upper reaches of the wadis with the most frequent flows. Thirdly, the increases of width in the upper wadi reaches, caused by the 1982 floods, were generally less than those lower down the systems. Thus there was less variation in the estimated return periods of the dominant discharges in these reaches and again greater confidence in the adopted values.

When the return periods had been modified to a consistent pattern, revised dominant discharges could be calculated working back through equation (1) and the growth factors. Similarly modified regime widths could be calculated from equation (3) utilising these dominant discharges. Table 3 gives details of the measured and adopted values.

The return periods of the adopted dominant discharges varied typically from 5 to 10 years in the upper reaches of wadis to 20 years in the lower reaches. The authors do not suggest, however, that too much attention should be paid to the precise numbers. The use of any of the other available equations would have yielded somewhat different values. Although equation (3) appears to give reasonable results when used in the way outlined above, the data was not of sufficient accuracy to confirm its use in preference to any of the other relationships. What is perhaps of greater interest is the relative variation of the return periods of the dominant discharges down the wadis.

## 6. OTHER REGIME RELATIONSHIPS

It had been hoped that it might prove possible to crosscheck the choice of regime width and corresponding dominant discharge by applying regime relationships involving other measurable parameters. A number of equations have been proposed relating, for example, longitudinal slope and the bankfull depth to the dominant discharge and representative bed-material size. As discussed earlier, however, bankfull discharge is not a meaningful concept with regard to the wadis studied and a comparison of the estimated dominant discharges with those obtained using the 'slope' equations proposed by Simons and Albertson (Ref 8), Kellerhals (Ref 9) and Bray (Ref 11) for the gravel bed reaches showed very little correlation.

The measured slopes and estimated dominant discharges are plotted in Figure 4. Multiple regression analyses gave the relationship (in S.I. units):

$$S = 0.028\, Q_D^{-0.18}\, D_{50}^{0.09} \tag{5}$$

or, with the sediment size parameter eliminated:-

$$S = 0.026\, Q_D^{-0.23} \tag{6}$$

These compare with the equations proposed by the authors quoted above, which are respectively:-

$$S = 0.0062\, Q_B^{-0.24} \qquad \text{(Ref 8)} \tag{7}$$

$$S = 0.086\, Q_B^{-0.4}\, D_{90}^{0.92} \qquad \text{(Ref 9)} \tag{8}$$

$$S = 0.059\, Q_2^{-0.33}\, D_{50}^{0.59} \qquad \text{(Ref 11)} \tag{9}$$

The differences in the way the representative flow and bed-material size have been defined, make it difficult to draw direct conclusions from a comparison of these equations. However, all three of the latter equations give slopes for comparable values of Q about an order of magnitude less than those given by equations (5) and (6), as shown in Figure 5. Obviously part of this difference is explained by the differing definitions of Q. However, even if $Q_D$ in equations (5) and (6) were replaced by the Mean Annual Flood values, which are between about 1.5 and 3.0 times lower and much closer to the $Q_2$ used in equation (9), there would still be a major discrepancy.

Compared to the slopes of the channels, from which the authors quoted above obtained their data, the slopes found on the wadis in South Yemen are very steep for the magnitude of flows experienced. In the long term the gradients may slacken but in terms of the durations relevent to any training works they appear relatively stable. All the channels, as would be expected, have slopes which are well above the boundary between braided and meandering channels as given by Leopold and Wolman (Ref 14):-

$$S = 0.012\, Q^{-0.44} \tag{10}$$

The Kellerhals equation (8) appears to apply to conditions close to this limit and the other two, (7) and (9), apply to slopes significantly steeper than this only at relatively high discharges.

At low discharges, the channels show patterns of braiding but at high flows, comparable to the estimated dominant discharges, they revert to a single channel.. These main channels, in many cases, exhibit a sinuous pattern, in plan form, of fairly regular wavelength. Analysis of the measured wave lengths, and estimated dominant discharges gave the relationship

$$\lambda = 254\, Q_D^{0.29} \tag{11}$$

Where $\lambda$ is the wavelength (S.I. units)

The authors know of no other published work on such sinuosity in conditions similar to this, but it appears to be a measurable parameter that might provide a useful cross-check on the estimated dominant discharge.

## 7. FRICTION/DEPTH ESTIMATES

For design of the training and bank protection works estimates had to be made of the depths and velocities of flow for different flood conditions. Generally works protecting villages and towns were designed for floods of 50-year return period; works protecting agricultural land were designed for floods with 10-year return period.

Application of any of the available regime 'depth' equations (i.e. resistance-type relationships) gave very high values for depths and the corresponding roughnesses. In view of the steepness of the wadi channels, compared to those for which the equations were derived, this is not surprising. Recourse was therefore made to Manning's equation for the gravel-bed reaches with the coefficient defined by the Limerinos equation (Ref 15):-

$$n = \frac{0.133Y^{0.167}}{1.16 + 2.0 \log (Y/D_{84})} \tag{12}$$

where Y is the depth of flow and $D_{84}$ is the size of bed material such that 84% by weight of the total is smaller. This was considered by Bray (Ref 16), in his review of flow resistance equations applicable to gravel-bed rivers, to be one of the 'most acceptable' - the other being the Lacey equation (Ref 17):-

$$V = 10.8\, Y^{2/3} S^{1/3} \tag{13}$$

where V is the average flow velocity. This equation, however, strictly relates only to bankfull conditions and is not therefore appropriate to the conditions applying in this study.

Equation (12) requires the estimate of $D_{84}$ - again, in this study, a subjective decision, with inherent uncertainties. Some guidance on the $D_{84}/D_{50}$ ratios was available from the Wadi Bana data (Figure 2). In practice a range of values was taken with the highest resulting 'n' value taken for assessment of the height of the training works and the lowest value taken for determining the maximum velocities to which the works might be subjected.

To complete the picture, for the sand-bed, lower reaches, flow conditions were estimated for each of the two possible flow conditions. For the sub-critical flow regime where bed forms may exist and depths of flow are greatest, the method proposed by White, Paris and Bettess (Ref 18) was used. For the upper regime, when bed forms are assumed to be swept clear and the flow is likely to be supercritical, a single value of n = 0.017 was used in conjunction with Manning's equation to establish the maximum likely velocities of flow. This value for n was considered to be the lowest practicable value equivalent to a uniform smooth sand bed, and was used only to give an upper limit to the velocities for design of the works.

## 8. CONCLUSIONS

This paper has been written to demonstrate some of the practical difficulties the practising engineer has in applying regime concepts in conditions for which the available equations were not developed. In South Yemen slopes are extremely steep, channel sections cannot be considered stable and the ephemeral nature of flows and their infrequency raises problems as to the nature of any regime condition and even to the concept itself. Moreover the activities of man have had a significant effect on the channel section.

However, the evidence, such as it is, suggests that Bray's regime relationship for channel width (equation 3) gives reasonable values when used with representative flows of greater rarity than are usually considered applicable. Such flows have been estimated to have return periods of between about 5 and 20 years and there is some indication that the return period is a function of the frequency of spate flows.

Other available regime relationships relating depth of flow and slope to a representative flow appear not to be applicable for these very steep channels. Equations for the slope/discharge relationships have been developed from the data set but, particularly in view of the difficulty in defining the dominant discharge, these must be viewed as tentative. A relationship between the channel wavelength and dominant discharge has also been developed as this wavelength appears to be a measurable parameter that might provide an independent check on the dominant discharge for any further work on this type of channel.

### NOMENCLATURE

| | |
|---|---|
| B | Channel width (m) |
| CA | Catchment Area ($km^2$) |
| D | Equivalent diameter of bed material (m) |
| $D_x$ | Diameter of bed material such that x% by weight of the total is smaller (m) |
| MAF | Mean Annual Flood ($m^3/s$) |
| MAR | Mean Annual Rainfall (mm/year) |
| MSL | Mean Stream Length (km) |
| n | Coefficient in Manning's (Strickler's) equation |
| $Q_B$ | Bankfull discharge ($m^3/s$) |
| $Q_S$ | Sediment discharge (t/s) |
| $Q_T$ | Wadi discharge with a return period of T years ($m^3/s$) |
| S | Channel slope |
| V | Mean flow velocity (m/s) |
| Y | Depth of flow (m) |
| $\lambda$ | Wavelength of sinuosity (m) |


### ACKNOWLEDGEMENTS

The authors acknowledge with gratitude the work of Peter Ackers and Stephen Priestley both formerly of Binnie and Partners, on the Wadi Bana and Abyan Delta Project, which was carried out by Binnie and Partners in association with W S Atkins. The authors were not directly concerned in that project, and, although some of the concepts put forward in this paper were based on the ideas formulated in that study, the authors have only themselves to blame for the views expressed here.

## REFERENCES


1. Binnie & Partners, "Wadi Hadramaut Agricultural Development Project (Phase II): Feasibility study for long-term wadi rehabilitation", Report to Ministry of Agriculture and Agrarian Reform, PDR Yemen, 1987.

2. Strojexport, "Hydrogeological survey of Nissab region: Final Report" Report to Ministry of Agriculture and Agrarian Reform, PDR Yemen, 1982.

3. Inglis, C.C., "Meanders and their bearing on river training", Proc. ICE, Maritime Paper No 7, 1947.

4. Hey, R.D., "River dynamics, flow regime and sediment transport", in "Sediment Transport in Gravel Bed Rivers" (ed C R Thorne, J C Bathurst and R D Hey) pub. by John Wiley and Sons, 1987, pp 17-40.

5. Blench, T., "Regime behaviour of canals and rivers", pub. by Butterworths Scientific Publications, 1956, p 87.

6. Terrell, P.W., and Borland W.M., "Design of stable canals and channels in erodible material", Transactions ASCE, Vol 123, 1958.

7. Nixon, M., "A study of bankfull discharges of the rivers in England and Wales", Proceedings of Institution of Civil Engineers, Vol 12 1959, pp 157-174.

8. Simons, D.B., and Albertson, M.L. "Uniform water conveyance channels in alluvial material", Transactions ASCE, Vol 128, Pt1 No 3399, 1963, pp 65-167.

9. Kellerhals, R., "Stable channels with gravel paved beds" Journal of Waterways and Harbours Division, ASCE, Vol 102, No HY7, July 1976, pp 813-829.

10. Charlton, F.G., Brown, P.H., and Benson, R.W., "The hydraulic geometry of some gravel bed rivers in Britain", Report IT 180, Hydraulics Research Station, Wallingford, England, 1978.

11. Bray, D.I., "Regime equations for gravel-bed rivers", in "Gravel Bed Rivers" (ed R D Hey, J C Bathurst and C R Thorne), John Wiley and Sons, 1982.

12. Hey, R.D., and Thorne, C.R., "Stable channels with mobile gravel beds", Journal of Hydraulic Engineering, ASCE, Vol 112, No HY8 August 1986, pp 671-689.

13. Hey, R.D., "Regime equations for mobile gravel-bed rivers", in "Gravel Bed Rivers" (ed R D Hey, J C Bathurst and C R Thorne), John Wiley and Sons, 1982.

14. Leopold, L.B., and Wolman, M.G., "River channel patterns - braided, meandering and straight". US Geological Survey Professional Paper No 282B, 1957 pp 39-85.

15. Limerinos, J.T., "Determination of the Manning coefficient from measured bed roughness in natural channels", Water Supply Paper 1898-B, UK Geological Survey, Washington DC, 1970.

16. Bray, D.I., "Flow resistance in gravel-bed rivers", in "Gravel Bed Rivers" (ed R D Hey, J C Bathurst and C R Thorne) John Wiley & Sons, 1982.

17. Lacey G., "A theory of flow in alluvium", Journal of the Institution of Civil Engineers Vol 27, Paper no 5518, 1946-47 pp 16-47.

18. White, W.R., Paris, E., and Bettess, R., "A new general method for predicting the frictional characteristics of alluvial streams", Report IT 187 Hydraulics Research Station, Wallingford, England, 1979.

TABLE 1 Comparison of Regional Flood Estimates and Observed Peak Discharges

| Area | Wadi | Location | CA | Regional estimates | | Max. observed peak | Period of observation |
|---|---|---|---|---|---|---|---|
| | | | | MAF | 50 yr flood | | |
| | | | ($km^2$) | ($m^3/s$) | | ($m^3/s$) | (years) |
| Tuban | Tuban | Dukeim | | 656 | 3020 | 2640 | 20 |
| Dhala | Azariq | Ad Darb | 48 | 29 | 133 | 38 | 0.25 |
| Hassan | Hassan | Al Mijuh | 3190 | 492 | 2260 | 2000 | -(1) |
| Beihan | Khier | Qurayn/ Nahr confluence | 530 | 93 | 430 | 700 | -(2) |
| Beihan | Nahr | Beihan Ford/ Khirr confluence | 2830 | 417 | 1920 | 270 | -(3) |
| Beihan | Beihan | Nuqub weir | 3630 | 380 | 2100 | 2600 | -(1) |
| Nissab | Dura | Mukhail | 850 | 182 | 840 | 2070 | 1.75(1) |
| Nissab | Abadan | Husn al Wasr | 740 | 211 | 970 | 1310 | 1.75(1) |
| Nissab | Markhah | Hajar | 1450 | 366 | 1680 | 300 | 1.75(1) |
| Nissab | Khawrah | Gerashah | 1360 | 240 | 1100 | 255 | 1.75(1) |
| Hadramaut | Am'd/Duan | Qaudah | 6280 | 970 | 4460 | 985 | 3 |
| Hadramaut | Ain | u/s Ajlaniyah | 1500 | 200 | 920 | 500 | 3(5) |
| Hadramaut | Sarr | Otfa | 2160 | 410 | 1890 | 272 | 3 |
| Hadramaut | Idim | Sah | 3370 | 550 | 2530 | 775 | -(5) |
| Hadramaut | Idim | Ghuraf | 5420 | 650 | 2990 | 1314 | 3 |
| Hadramaut | Masilah | Qassam | - | 690 | 3170 | 2100 | 3 |

Notes:
(1) Estimate of March 1982 flood
(2) Estimate of 1975 flood (biggest in memory)
(3) Estimate of 1977 flood but known to have been exceeded in 1982
(4) Estimate of 1975 flood known to be exceptional.
(5) Estimate of October 1977 flood which is not included in the 3 years' observation period

Table 3 ; Measured and 'regime' widths for selected wadis

| (1) Area | (2) Wadi | (3) Reach length (km) | (4) Slope | (5) Bed material ($D_{50}$)(1) (m) | (6) Width(2) est. from maps etc. (m) | (7) Discharge(3) ($m^3/s$) | (8) Return period (yrs) | (9) Modified values: Return period (yrs) | (10) Modified values: Dominant discharge (m3/s) | (11) Modified values: Regime width | (12) Spate frequencies (4) |
|---|---|---|---|---|---|---|---|---|---|---|---|
| Abyan delta | Bana(5) | 20 | 0.0058 | 0.05 | 250 | 1850 | 9 | 10 | 1950 | 260 | H |
| | | 21 | 0.0026 | sand | 360 | 3160 | 35 | | | | M |
| Tuban | Tuban | 4 | 0.009 | 0.15 | 290 | 2700 | 40 | 10 | 1400 | 210 | H |
| | | 4 | 0.009 | 0.10 | 370 | 4000 | >50 | 10 | 1500 | 215 | M/H |
| | | 3 | 0.007 | 0.10 | 330 | 3300 | 50 | 15 | 1800 | 240 | M/H |
| | Khadir(6) | 8 | 0.008 | 0.05 | 190 | 1100 | 18 | 15 | 1000 | 180 | M |
| | | 11 | 0.0075 | 0.02 | 183 | 900 | 15 | 15 | 750 | 165 | M |
| | | 6 | 0.005 | 0.05 | 319 | 2200 | >50 | 15 | 750 | 160 | M |
| | Saghir(6) | 20 | 0.008 | 0.05 | 116 | 450 | 3 | 10 | 850 | 165 | M |
| | | 10 | 0.005 | 0.01 | 88 | 210 | 3 | 12 | 720 | 170 | M |
| Nissab | Markhah | 15 | 0.003 | 0.05 | 240 | 1700 | 30 | 20 | 1200 | 200 | L |
| | | 9 | 0.003 | 0.01 | 420 | 3800 | >50 | 20 | 1400 | 240 | L |
| | | 15 | 0.003 | sand | 550 | 8000 | ≫50 | 20 | 1300 | 240 | L |
| | Khawrah | 6 | 0.007 | 0.15 | 150 | 800 | 20 | 10 | 520 | 120 | M/L |
| | | 18 | 0.006 | 0.05 | 215 | 1400 | >50 | 15 | 630 | 140 | L |
| | | 11 | 0.004 | 0.01 | 280 | 1900 | >50 | 20 | 680 | 160 | L |
| | Dura | 6 | 0.004 | 0.02 | 195 | 1000 | >50 | 10 | 400 | 120 | M |
| | | 8 | 0.004 | sand | 160 | 630 | 40 | 20 | 500 | 140 | M/L |
| | Abadan | 9 | 0.006 | 0.05 | 190 | 1100 | 50 | 20 | 700 | 150 | M/L |
| | | 11 | 0.004 | sand | 450 | 5500 | ≫50 | 25 | 700 | 170 | L |
| | Hamman | 8 | 0.004 | sand | 350 | 3000 | >50 | 20 | 1000 | 200 | L |
| | | 10 | 0.004 | sand | - | - | - | 30 | 1000 | 200 | L |
| Mudia | Azzan | 8 | 0.008 | 0.05 | 114 | 430 | 50 | 10 | 155 | 70 | M/H |
| | Wajr | 7 | 0.010 | 0.10 | 35 | 52 | 10 | 10 | 52 | 35 | M/H |
| Hassan | Hassan | 12 | 0.007 | 0.05 | 265 | 2050 | 40 | 13 | 1200 | 200 | M/L |
| | | 5 | 0.005 | 0.02 | 260 | 1750 | 30 | 15 | 1350 | 225 | M/L |
| Beihan | Nahr | 26 | 0.004 | 0.15 | 90 | 300 | 2 | 5 | 600 | 130 | H |
| | | 11 | 0.004 | 0.05 | 150 | 700 | 6 | 6 | 700 | 150 | M |
| | Khirr | 17 | 0.007 | 0.20 | 110 | 470 | 50 | 10 | 200 | 70 | M |
| | Beihan | 14 | 0.003 | 0.01 | 250 | 2000 | 35 | 15 | 1300 | 200 | M |
| | | 14 | 0.0025 | sand | 220 | 1200 | 15 | 15 | 1200 | 220 | L |

<u>Notes</u> :
(1) Average bed-material size in reach estimated by visual observation only.
(2) Widths (col.6) represent best estimate of regime width as assessed from mapping, aerial photographs etc. In some cases they represent post-1982 values only.
(3) Discharge (col.7) calculated from width (col.6) using eqn.(3) or (4) as appropriate.
(4) Spate frequencies classed as High (H) where more than about 25 floods occur annually; Medium (M) where between about 5 and 25 floods occur; and Low (L) where less than about 5 occur. Frequencies assessed in most cases by questioning local farmers.
(5) Values for Wadi Bana are average values for the gravel- and sand-bed lengths of the wadi.
(6) Wadis Saghir and Khabir in the Tuban area are the two downstream channels of Wadi Tuban. Before the construction of a dividing control weir at the bifurcation, W.Khabir received the greater part of the flow from W.Tuban. The weir now distributes the flow approximately equally. W.Saghir is thus presently narrower than the regime width.

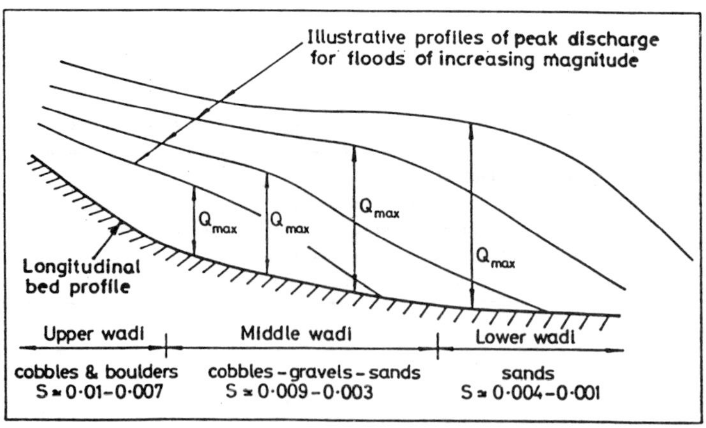


Figure 1 : Dragrammatic representation of wadi longitudiual profile with typical bed materials and ranges of slopes (s).

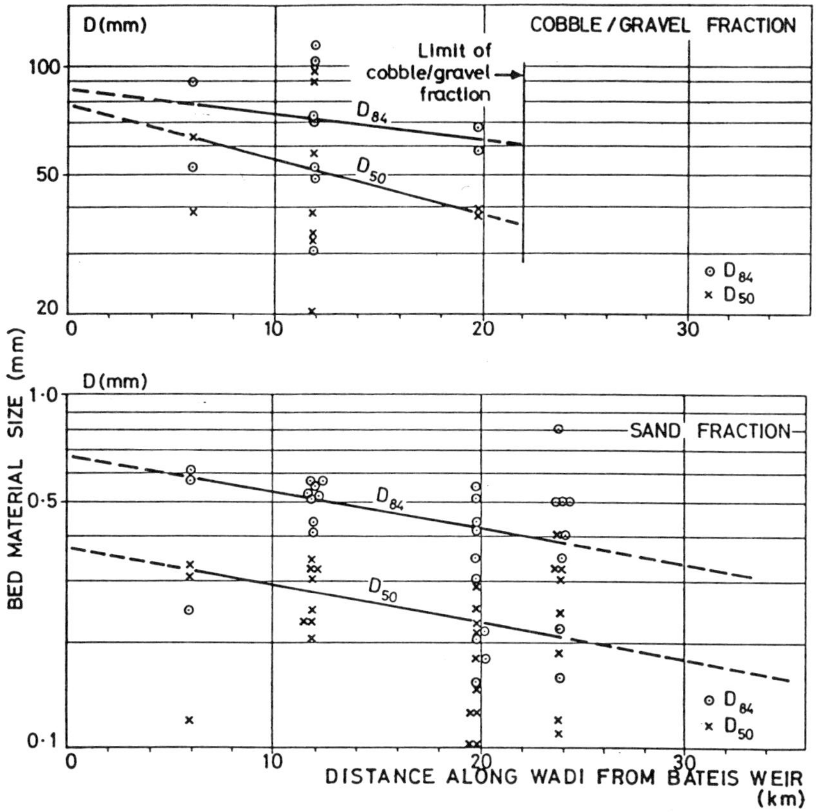


Figure 2 : Bed material sampling from Wadi Bana

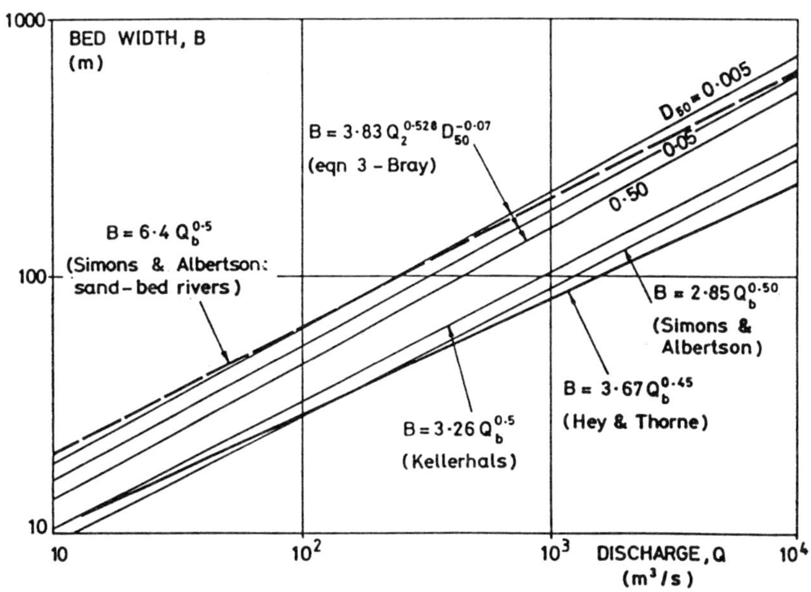


Figure 3 : Comparison of regime width relationships for Gravel-bed Rivers

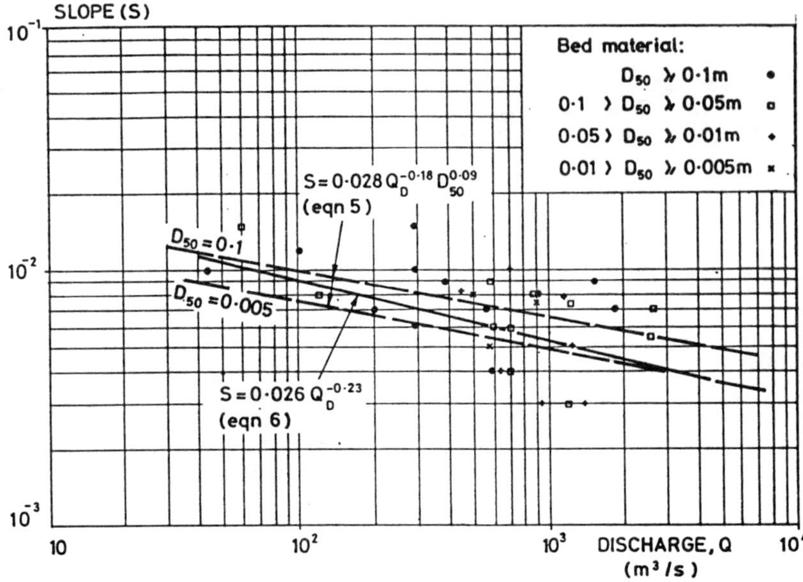


Figure 4 : Slope/Dominant discharge relationship based on South Yemen data

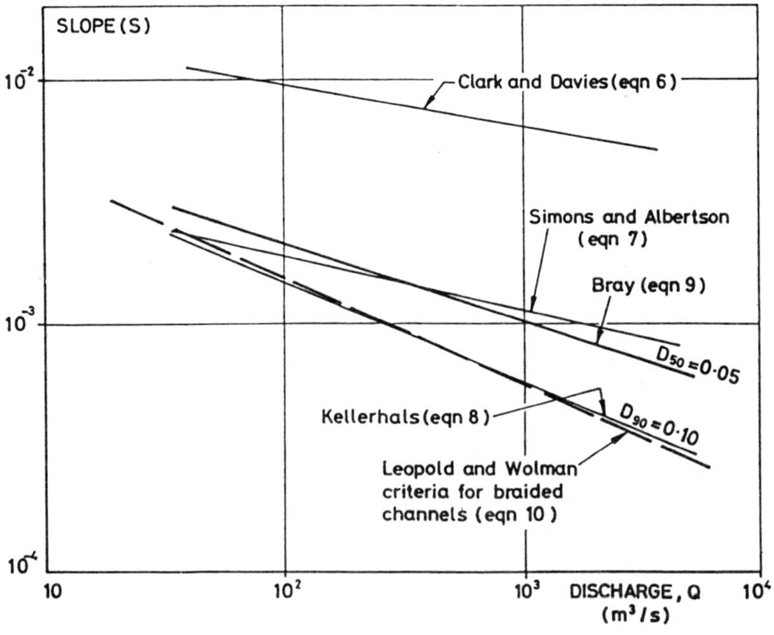


Figure 5 : Comparison of slope/discharge relationships for gravel-bed rivers with the relationship for data of this study

International Conference on
RIVER REGIME
18-20 May, 1988

PAPER C1

# ON THE CAUSE OF RIVER MEANDERING

Howard H. Chang
Professor of Civil Engineering
San Diego State University

**SUMMARY:** Previous theses on the cause of river meandering are reviewed and criticized on the basis whether the cause and effect are identified in interpreting the fluvial processes. An explanation for the origin of meandering is presented, which elucidates why the channel slope of meandering rivers is flatter than the alluvially-formed valley slope.

## 1 INTRODUCTION

The cause of river meandering has been a subject of persistant debate in scientific and engineering literature. This subject, for which an explanation has been pursued throughout human history, remains an obscurity wanting a complete answer. Previous explanations of what causes river meandering have been extensively reviewed by Yang (Ref 33), Chitale (Ref 9), Callander (Ref 3) and Hickin (Ref 15), among others. Proposed causes have included the most probable path (minimum variance), secondary currents, dynamic instability, and silt-and-clay content.

The present discussion focuses on the cause from which meandering follows as an effect. Since meandering is a feature of river channel formation, determining the cause of meandering requires identifying the cause of river channel formation. In technical terms, the cause refers to the independent or controlling variables which are imposed upon the river channel. The effect which results from the cause is described by the dependent variables, which include the channel geometry and meandering pattern. Determination of the independent and dependent variables, i.e., cause and effect, is subject to the time scale under consideration, as detailed in a following section. It is clear that any explanation of the cause of meandering must be derived from the independent variables and their interactions with the dependent variables. This is the concept upon which criticism of previous theses on meandering is based. This is also the key to the new explanation presented herein. Failure to identify the cause and effect usually leads to misinterpretation of the fluvial processes. Future progress in river research, as pointed out by Kennedy (Ref 17), will rest in part on the generally adequate understanding of riverflow processes that has evolved in recent years, and in particular on the interdependency of the variables and their roles as dependent variables at some times and independent variables at other times.

Any thesis on the origin of meandering must explain why the channel slope must be flatter than the alluvially-formed valley slope. This slope

International Conference on River Regime, edited by W.R.White.

relationship is the necessary and sufficient condition for the occurrence of meanders. The ratio of valley slope to channel slope is defined as meander sinuosity. A previous study by Leopold and Wolman (Ref 21) established from physical evidence that meandering is not necessarily related to a river's flow and sediment characteristics because rivers of similar characteristics can be more or less straight or meandering with a wide range of sinuosities. Any explanation of the cause of meandering should be consistent with these findings. Before presenting the new explanation, it is necessary to review the variables of river-flow processes and to discuss, relative to these variables, why other theses are inadequate to explain the cause of meandering.

## 2 VARIABLES OF RIVER-FLOW PROCESSES

Variables of river analysis are classified into independent and dependent variables, i.e., cause and effect. Those which are imposed upon the river by any source are independent or controlling variables, and those which result are dependent variables (Refs 18,23,28). In selecting variables for non-tidal rivers, it is necessary to consider the time scale, which is differentiated as geologic time, graded time, and steady time following Schumm's (Ref 28) classification. The absolute length of these time spans is not important, though it should vary directly with the size of the drainage basin. What is significant is the concept that a variable can be a cause or an effect depending upon the time scale.

For the geologic time, measured in perhaps millions of years climate geology are independent variables and the valley slope is a dependent variable. Valley-floor gradients are generally established in this very long time span. During the graded time, which generally lasts a few hundred years, the valley slope does not change, so it is considered an independent variable. Also, dominant discharges of water and sediment that develop within the drainage basin are independent variables for the river. Thus, in the graded time span, the inflow quantities of water and sediment together with the valley slope constitute the cause from which river channel formation (channel morphology) follows as an effect. The concept of grade as a condition of equilibrium (or regime) in streams as agents of transportation is the basis for the understanding of river meandering.

During the short-term steady time span, measured perhaps in days, sediment discharge is a dependent variable which may be expressed as a function of the velocity. Such a functional relationship implies that velocity is the cause and sediment discharge is the effect. This is usually true for the short-term situation, but not so for the long-term (graded time) equilibrium under which a river must establish its velocity, slope, and other characteristics to maintain a balance between its transport capacity and the inflow quantities imposed. Therefore, the velocity and channel characteristics (width, depth, slope) so established are dependent variables for the graded time. The velocity in a channel as an agent of transportation is determined by, or adjusted to, the load (sediment inflow rate). Velocity is an effect and load is the cause: this relationship is not transposable (Ref 23).

## 3 REVIEW AND CRITICISM OF PREVIOUS THESES

Different theories explaining the origin of meandering for regime rivers are reviewed and criticized herein. The criterion of criticism is whether a theory considers the cause or simply describes the meandering phenomenon as an effect. Three major theories: the most probable path, dynamic instability, and secondary currents are included in the following discussion, but different theories such as the Coriolis force, bank erosion, and statistical analysis, etc., that have been reviewed by others are not.

3.1 *Most Probable Path* - The cause of meandering was attributed to the river's tendency to follow the most probable path between fixed points. With a fixed number of steps, Von Schelling (Ref 32) assumed the normal probability distribution for the change of direction at the end of each step. For a path that is a continuous curve, Von Schelling showed that a general condition for the most probable path is for the variance to be a minimum, i.e.,

$$\sum \frac{\Delta s}{r^2} = \text{a minimum} \tag{1}$$

where $\Delta s$ = increment in s; s = curvilinear coordinate along meander path; and r = radius of curvature.

The theory of minimum variance was offered by Langbein and Leopold (Ref 20) as a cause of meandering because meanders which achieve the minimum variance are more stable than a straight alignment. From the most probable path obtained by Von Schelling, Langbein and Leopold derived the sine-generated curve for the meander path as

$$\phi = \omega \sin (2\pi s/M) \tag{2}$$

where $\phi$ = angle that a meander path at a given point makes with mean downpath direction; $\omega$ = maximum angle a path makes with mean downpath direction; and M = meander arc length. This equation contains a scale parameter M and a shape parameter $\omega$, both of which need to be specified in application. With properly selected values of M and $\omega$, the sine-generated curve fits the shape of river meanders quite well.

The sine-generated curve has contributed to the understanding of meandering as a phenomenon but not the cause. Because the controlling variables of water and sediment inflows and valley slope were not included in the derivation of this equation, it does not explain the cause of meandering and because the shape parameter needs to be prescribed, the sine-generated curve does not predict the channel slope nor sinuosity of the meander path.

3.2 *Stability Theory* - A group of researchers, including Callander (Ref 2), Sukegawa (Ref 30), Engelund and Skovgaard (Ref 12), Hayashi (Ref 14), Parker (Ref 25) and many others, believe that meanders result from an unstable response of the bed to a small perturbation. The stability theory states that the mobile bed of a channel with straight banks is unstable because sediment transport at the bed intensifies the perturbation by moving bed material to one bank where a point bar is initiated. This process grows as it accentuates the thalweg oscillation until true meandering develops.

The stability analysis generally employs a two-dimensional model with two space coordinates along the longitudinal and transverse directions of flow. Dependent variables are the two components of depth-averaged velocity, depth of flow, water-surface elevation, bed shear stress and bed load discharge.

The maximum instability of a linear stability analysis of meandering is generally used to estimate the wave length of the finite amplitude meanders realized in nature. Thus, the stability theory enables the analytical determination of this important meander feature. By bringing in physical relations he flow in bends, Ikeda, et al, (Ref 16) and Parker, et al, (Ref 26) obtained a bend equation describing the meander path. This effort provided the mechanistic foundation of the sine-generated curve.

The existence of perturbations in river flow is beyond any doubt because there are unlimited opportunities for perturbations in nature. Stability analysis has been advanced with impressive successes; it has been applied to reproduce many meander features, climaxing in the bend equation. However, the approach is that of fluid mechanics for the steady-time span. It deals with the dependent variables of flow characteristics, but the controlling variables - discharge, load and valley slope - which cause equilibrium channel morphology are not included in the analysis as independent variables. The bend equation may be used to describe the geometry of the channel centerline (meander path) only if the shape parameter (arc angle or sinuosity) is specified. In other words, the bend equation provides the meander shape if the shape parameter is given; therefore, it does not predict, nor explain, the origin of the meander shape.

3.3 *Secondary Currents* - Development of secondary currents has been proposed as the possible cause of meandering. This thesis was derived from the finding that secondary currents can occur in a straight channel with or without sediment motion, as exemplified by the studies of Einstein and Li (Ref 10), Tanner (Ref 31), Einstein and Shen (Ref 11), and Shen and Komura (Ref 29).

However, there is more evidence that secondary currents are actually induced by channel curvature or channel roughness. In fact, secondary currents reflect the presence of meanders but not their cause. Sinuosity of the river is the cause and the resulting spiral motion is the effect: this relation is not transposable.

Secondary currents undoubtedly play a very important role in the morphology of river meanders because many meander features, except sinuosity, can be modeled based upon secondary currents. This is exemplified by the meander path model (Ref 7), which assumes that, for the equilibrium channel geometry, the river flow and river channel formation have adapted to a state of mutual compatibility. Therefore the meander path can be determined based upon the mechanics of flow governing the streamwise variation in spiral motion. The variation arises from the interaction of the centrifugal force and the turbulent shear that the flow has to overcome in transforming from the spiral pattern to parallel flow and vice versa. The smoothly changing channel curvature reflects the pattern of gradual helical flow development. This model provides the scale parameter of meanders because meander arc length can be determined by the channel length required for the growth and decay of helical flow. In using this model, the shape parameter (sinuosity) has to be prescribed as an independent variable. In other words, the sinuosity or channel slope is not determined from secondary currents; the sinuosity of the river defines the helical strength instead.

## 4 THE CAUSE OF RIVER MEANDERING

The cause of river meandering is explained herein by the fluvial processes involved in the formation of the alluvial valley and the river channel, and the interrelation of their gradients. The explanation is based upon the independent variables of river channel formation, from which meandering follows as a dependent variable, i.e., an effect. It is described in the following four stages: (1) formation of the channel slope, (2) relation between channel slope and valley slope, (3) formation of the valley slope, and (4) formation of river meanders. This explanation elucidates why certain rivers have a distinctly meandering pattern with a channel slope that is significantly flatter than the valley slope.

4.1 *Formation of the Channel Slope* - For the graded-time span, the channel gradient is under the control of the water and sediment inflow rates and their characteristics. From the relationship developed by Lane (Ref 19), the channel slope S can be written as:

$$S \propto \frac{Q_s^x}{Q^y} d^z \tag{3}$$

or

$$S = fn\ (Q_s,\ Q,\ d) \tag{4}$$

where $Q_s$ = bed-material discharge; Q = water discharge; d = representative sediment size; and x,y,z = exponents. This functional relationship implies that the slope of the river channel is the effect while $Q_s$, Q, and d constitute the cause. With these independent variables given or predetermined, the slope in the functional relationship is indeterminate because the unknowns for the channel (width, depth, and slope) outnumber the applicable physical relationships (flow resistance and sediment transport). This indeterminacy was resolved by the minimum stream power concept (Ref 4) which states that the channel geometry and slope are so formed to transport the inflow quantities of water and sediment at the minimum power expenditure per unit channel length, i.e., minimum $\gamma QS$, where $\gamma$ is the unit weight of the water-sediment mixture. Since Q is a given variable, minimum $\gamma QS$ is equivalent to minimum channel slope. Under this definition, the channel geometry of streams as agents of transportation develops to maximize channel efficiency in transporting sediment while fulfilling the laws of sediment transport and flow resistance. This concept initially proposed as a hypothesis was later proved by Yang and Song (Ref 34).

Following this concept, the regime channel slope for sand-bed rivers was established as (Ref 8)

$$\frac{S}{\sqrt{d}} = 0.28\,\frac{Q_s^{0.58}}{Q^{0.75}}\ , \quad \text{for} \quad 0.00704\ Q^{-0.55} \geq \frac{S}{\sqrt{d}} \geq \frac{S_c}{\sqrt{d}} \tag{5}$$

$$\text{and} \qquad \frac{S}{\sqrt{d}} = 2.82\left(\frac{Q_s}{Q}\right)^{0.87}, \quad \text{for} \quad \frac{S}{\sqrt{d}} \geq 0.00763\ Q^{-0.51} \qquad (6)$$

in which $S_c$ = slope for bed-load threshold; d is in mm; and Q and $Q_s$ are in $m^3$/sec. The discontinuity between these two equations is due to the threshold in river morphology. Eqs. 5 and 6 which express S in terms of $Q_s$, Q, and d, represent the quantification of Lane's functional relationship. A similar relation to Eqs. 5 and 6 was obtained for gravel-bed rivers (Ref 5).

4.2 *Relation between Channel Slope and Valley Slope* - Because the values of Q and $Q_s$ are determined in the drainage basin, the channel slope produced by these inflow quantities is usually different from the valley slope, which is another independent variable for the river during the graded-time span. With consideration given to all possibilities, the channel slope could be steeper than, equal to, or flatter than the valley slope. The first case can be eliminated because the channel slope cannot physically be steeper than the valley slope under dynamic equilibrium or regime. It may therefore be concluded that for rivers in regime, the channel slope must be equal to or flatter than the valley slope. It follows that regime rivers may be straight or meandering in channel pattern. Since the channel slope is primarily determined by the values of Q, $Q_s$, and d as Eqs. 5 and 6 show, it is possible for rivers of identical slopes to have different sinuosities depending upon each valley slope. This concurs with the findings of Leopold and Wolman (Ref 21) in their study of channel patterns because they found rivers with similar flow and sediment characteristics can be either straight or meandering. Of course, the question why certain rivers are meandering still needs further elucidation. The following discussion of fluvial processes illustrates that the channel slope can be quite flatter than the valley slope.

4.3 *Formation of the Valley Slope* - Alluvial valleys are usually formed under different hydrologic or geologic regimes during the very long-term geologic time span, and are affected by changes in climate, surface vegetation, paleohydrology, catastrophic floods, tectonic events, forest fires, channel piracy, over grazing, man-made factors, etc. Under different regimes represented by separate combinations of the inflow quantities of water and sediment, the river channel develops different slopes. If the channel slope, during any episode in the geologic time, is steeper than the valley slope, then deposition occurs to steepen the valley slope. The valley slope may also be reduced depending upon the water-sediment mixture supplied. Therefore, the valley slope, for the geologic time span, is a dependent variable while the climate (or paleohydrology) and geology of the basin are independent variables. Examples of valley formation include the paleo channels of the Murrumbidgee River reported by Schumm (Ref 27), effects of a prehistoric flood on the Snake River discussed by Malde (Ref 24), and geomorphological response to an extreme flood in the Honda Valley, Spain reported by Harvey (Ref 13).

4.4 *Formation of Meanders* - Meandering rivers with channel gradients significantly flatter than valley gradients are explained by the different fluvial processes involved in forming a river channel and an alluvial valley. While each undergoes periods of erosion and deposition because of changes in water discharge and sediment load, the geomorphological responses of channels and valleys to erosion and deposition are different. During deposition, which may be caused by overloading of sediment, stream adjustment is usually characterized by widening and braiding, while the alluvial valley is built up uniformly. Most active depositional valley formation occurs during major historical hydrologic events; but an alluvial valley may also be filled over a longer time span through evolutionary adjustments. Whether it is through active braiding or gradual shifting of the stream course, deposition is usually rather evenly distributed on the valley floor. Therefore, deposition favors alluvial plain formation.

During episodes of erosion which result from a deficit in sediment supply, the stream channel seeks to develop a flatter slope. In sharp contrast to deposition, morphological response to erosion is characterized by incision of the stream into the alluvial plain. Therefore, erosion formation of a more confined river channel. Incision or degradation of a river channel is inhibited by coarsening of the bed material due to hydraulic sorting; it may also be prevented when an armored layer develops. Reduction of channel slope through incision would require tremendous degradation. For these reasons, a river channel usually adjusts by developing a flatter slope through meandering.

Because of the different geomorphological responses to deposition and erosion, an alluvial valley may be continually built up during episodes of deposition. The deposited alluvium is not as easily removed during episodes of erosion because of sediment sorting and development of meanders. Continued rise of alluvial valleys is generally a rule rather than an exception.

River channel formation is associated with the dominant, or bankfull, discharge, which has a return period of no more than a few years (Ref 35). Given the length of the geologic time span and its many hydrologic and geologic regimes, the valley slope usually results from historical conditions different from those forming the existing channel. Because the valley slope is steepened during any episode in geologic time when the channel slope is steeper than the valley slope, the valley slope can be quite steeper than the current channel slope. This explains why certain rivers have a distinctly meandering pattern.

A river in a broad alluvial plain has sufficient freedom to develop a slope much flatter than the valley slope by the combination of channel meanders and valley meanders as illustrated in Fig. 1. The smooth serpentine meander path is compatible with the gradual flow development of secondary currents. For highly sinuous channels, the curvature at an apex, defined by the radius of curvature to channel width ratio, has an average value of about three (Ref 22). It represents the maximum curvature for which a river does the least work in turning while maintaining streamwise uniformities in sediment transport and energy expenditure (Ref 6). Sharper turning angles which cause greater energy loss, or a steeper energy gradient, are normally eliminated in the process of meander formation.

## 5 ILLUSTRATIVE EXAMPLE

In the foregoing explanation for river meandering, consideration was given to the alluvial-valley formation in geologic time. Meandering represents subsequent changes in slope and pattern reflecting the river's adjustment to altered discharge and load. It is interesting to note that deformation of the valley floor by active tectonism produces the same types of pattern changes, as reported by Burnett and Schumm (Ref 1). They collected information on channel pattern, gradient, depth, and channel behavior for five rivers crossing the Monroe uplift in Louisiana and three rivers crossing the Wiggins uplift in Mississippi. Uplift rates of 5 mm/year have been indicated by repeat surveys. Big Colewa Creek provides a good example of the nature of channel response to uplift in these areas.

Above the axis of the uplift, the reduction of valley slope has caused a decrease in channel sinuosity and an anastomosing pattern has developed with frequent overbank flooding. The straight channel pattern indicates approximately equal channel slope and valley slope. During any further decrease in valley slope, the river must make adjustment in channel morphology since its slope cannot physically be steeper than the valley-floor gradient. As a result, the anastomosing pattern develops, which has a lower energy gradient because of the lower velocity and the larger overall channel width. With reduced transport capacity for sediment, this braided channel pattern is generally associated with deposition and steepening of the valley floor.

Downstream of the uplift axis, the valley slope is increasing. There is bank erosion and the channel is meandering. In this case, the river maintains its gradient with slowly steepening valley slope by an increase in sinuosity. It exemplifies that meandering should be explained with consideration given to the formation of the valley-floor gradient.

## 6 CONCLUSION

Previous theses on the cause of river meandering have been review. These analyses, whether probabilistic or deterministic, treated meandering as a phenomenon without giving the cause. The meander path based upon each thesis is a curve with an undetermined shape parameter (sinuosity); the origin of the sinuosity is not explained. In this paper, the cause and effect, i.e., independent and dependent variables, are identified in interpreting the fluvial processes. In the graded time span for river channel formation, the regime channel slope is a dependent variable controlled by the water and sediment inflows from the drainage basin, and it is the minimum slope needed to transport such inflow quantities. The valley slope, which is formed during the much longer geologic time span, is an independent variable for the river during the graded time span. The river slope cannot physically be steeper than the valley slope for a river in regime. It may either be equal to or flatter than the valley slope. Therefore, both straight and meandering rivers may be and have been found in nature. For distinctly meandering rivers, the channel slope is significantly flatter than the valley slope. This is because the alluvial valley is usually formed during the very long-term time span under various hydrologic and geologic conditions different from those shaping the current channel. In the geologic time span, the valley slope is built up during episodes of deposition whenever the channel slope exceeds the valley slope. But the valley slope is not easily reduced during periods of erosion. Because degradation is inhibited by sediment sorting, the morphological response to sediment deficit is to form a meandering channel which in turn reduces erosion. For these reasons, the present channel can be quite flatter than the alluvially-formed valley.

## ACKNOWLEDGEMENTS

This research was supported by the National Science Foundation under Grant No. CEE-8209029. Comments on the paper from Bruce Westermo are greatly appreciated.

## REFERENCES

1. Burnett, A W and Schumm, S A: "Alluvial river response to the neotectonic deformation in Louisiana and Mississippi". Science, Vol 222, Oct 1983, pp 49-50.

2. Callander, R A: "Instability and river meanders". Journal of Fluid Mechanics, Vol 36, 1969, pp 465-480.

3. Callander, R A: "River Meandering". Annual Review of Fluid Mechanics, Vol 10, 1978, pp 129-158.

4. Chang, H H: "Geometry of rivers in regime". Journal of the Hydraulics Division, ASCE, Vol 105, No HY6, June 1979, pp 691-706.

5. Chang, H H: "Geometry of gravel streams". Journal of the Hydraulics Division, ASCE, Vol 106, No HY9, Sept 1980, pp 1443-1459.

6. Chang, H H: "Analysis of river meanders". Journal of Hydraulic Engineering, ASCE, Vol 110, No 1, Jan 1984, pp 37-50.

7. Chang, H H: "Regular meandering path model". Journal of Hydraulic Engineering, ASCE, Vol 110, No 10, Oct 1984, pp 1398-1411.

8. Chang, H H: "River channel changes: adjustments of equilibrium". Journal of Hydraulic Engineering, ASCE, Vol 112, No 1, January 1986, pp 43-55.

9. Chitale, S L: "Theory and relationship of river channel patterns". Journal of Hydrology, Vol 19, 1973, pp 283-308.

10. Einstein, H A and Li, H: "Secondary currents in straight channels". Trans. Am. Geoph. Union, Vol 39, No 6, 1958, pp 1085-1088.

11. Einstein, H A and Shen, H W: "A study of meandering in straight alluvial channels". Journal of Geoph. Research, Vol 69, No 34, December 1964, pp 5239-5247.

12. Engelund, F and Skovgaard, O: "On the origin of meandering and braiding in alluvial streams". Journal of Fluid Mechanics, Vol 57, Part 2, 1973, pp 289-302.

13. Harvey, A M: "Geomorphological response to an extreme flood: a case from southeast Spain". Earth Surfaces Processes and Landforms, Vol 9, 1984, pp 267-279.

14. Hayashi, T: "The formation of meanders in rivers". Trans. Japan Soc. of Civil Engineers, Vol 180, 1970.

15. Hickin, E J: "River channel changes", in Modern and Ancient Fluvial Systems, J D Collins and J Lewin, Eds, Int. Asso. Sediment. Spec. Pub 6, Blackwell Scientific Pub., Oxford, 1983, pp 61-83.

16. Ikeda, S. Parker, G and Sawai, K: "Bend theory of river meanders. Part I. Linear Development". Journal of Fluid Mechanics, Vol 112, 1981, pp 363-377.

17. Kennedy, J F: "Reflections on rivers, research, and Rouse". Journal of Hydraulic Engineering, Vol 109, No. 10, Oct 1983, pp 1253-1271.

18. Kennedy, J F and Brooks, N H: "Laboratory study of alluvial streams at constant discharge". Proc. Federal Inter-Agency Sedimentation Conference, Misc. Pub. No 970, Agr. Res. Ser., 1963, pp 320-330.

19. Lane, E W: "The importance of fluvial morphology in hydraulic engineering", Proceedings, ASCE, Vol 81, Paper 745, 1955, pp 1-17.

20. Langbein, W B and Leopold, L B: "River meanders-theory of minimum variance". U. S. Geological Survey Professional Paper 422-H, 1966, 15 pages.

21. Leopold, L B and Wolman, M G: "River channel patterns: braided, meandering and straight". U. S. Geological Survey Professional Paper 282-B, 1957, 85 pages.

22. Leopold, L B and Wolman, M G: "River meanders". Geological Society of America Bulletin, Vol 71, 1960, pp 769-794.

23. Mackin, J H: "Concept of the graded river". Geological Society of America Bulletin, Vol 59, 1948, pp 463-512.

24. Malde, H E: "The catastrophic late pleistocene Bonneville flood in the Snake River Plain, Idaho". U. S. Geological Survey Professional Paper 596, 1968, 52 pages.

25. Parker, G: "On the cause and characteristic scales of meandering and braiding in rivers". Journal of Fluid Mechanics, Vol 76, Part 3, 1976, pp 457-480.

26. Parker, G, Sawai, K and Ikeda, S: "Bend theory of river meanders. 2. nonlinear deformation of finite-amplitude bends". Journal of Fluid Mechanics, Vol 115, 1982, pp 303-314.

27. Schumm, S A: "River adjustment to altered hydrologic Regimen-Murrumbidgee River and paleochannels, Australia". U. S. Geological Survey Professional Paper 598, 1968, 65 pages.

28. Schumm, S A: "Fluvial geomorphology: the historical perspective". River Mechanics, I, edited by H W Shen, PO Box 606, Ft. Collins, CO, 1971, pp 4.1-4.30.

29. Shen, H W and Komura, S: "Meandering tendencies in straight alluvial channels". Journal of the Hydraulics Division, ASCE, Vol 94, No HY4, April 1968, pp 893-908.

30. Sukegawa, N: "Condition for the formation of alternate bars in straight alluvial channels". J. Faculty Eng. Tokyo Univ, Vol 30, Part B, 1970, pp 289-306.

31. Tanner, W F: "Helical flow, a possible cause of meandering". Journal of Geophysical Research, Vol 65, 1960, pp 993-995.

32. Von Schelling, H: "Most frequent particle paths in a plane". Trans. Am. Geophys. Union, Vol 32, 1951, pp 222-226.

33. Yang, C T: "On river meanders". Journal of Hydrology, Vol 13, 1971, pp 231-253.

34. Yang, C T and Song, C S: "Theory of minimum rate of energy dissipation". Journal of the Hydraulics Division, ASCE, Vol 105, No HY7, July 1979, pp 769-784.

35. Williams, G P: "Bank-full discharge of rivers". Water Resources Research, Vol 14, No 6, 1978, pp 1141-1154.

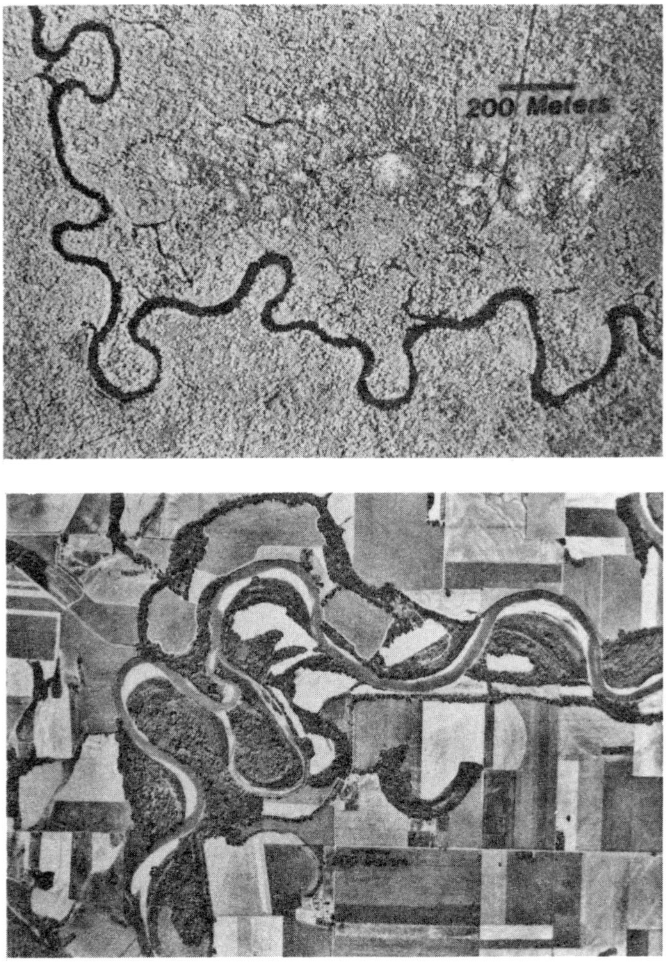


Figure 1. Examples of meandering rivers: (top) Lumber River near Boardman, N. Carolina; and (bottom) West Fork of White River near Newberry, Indiana (courtesy of James C. Brice).

International Conference on

RIVER REGIME

18-20 May, 1988

PAPER C2

# STABILITY OF INITIALLY STRAIGHT RIVERS

Panayiotis Diplas, John F. Kennedy, A. Jacob Odgaard
Institute of Hydraulic Research
The University of Iowa, Iowa City, Iowa 52242-1585 USA

**SUMMARY:** During the last two to three decades stability analysis based on small perturbation theory has emerged as the methodology of choice for predicting meandering or braiding tendencies of initially straight streams. The state of knowledge of this topic is examined here. Several stability analyses are reviewed. Emphasis is directed toward formulation and understanding of, and criteria for, channel stability.

## 1. Introduction

The inherent tendency of streams to meander or braid has been well documented both in the field and laboratory (e.g. Refs 1, 2). Anna Commena, Princess of Byzantium and historian of her era, gives what might be the first account of river instability leading to braiding (Ref. 3). She sets forth her observations regarding the rivers succeptible to, as well as the cause of, this type of instability. It was around 1150 AD when she wrote: "It is the case with all the greatest rivers that when a pretty considerable level of earth has been built up in the form of a deposit, they flow on to low-lying ground as if deserting their first beds and leave the old course empty of water and dry; the new bed they fill with abundant streams." It was several centuries later when Lord Kelvin (Ref. 4) published his work on the origin of "windings of rivers", which can be considered the first rational analysis of river meandering.

More recently, the study of river planform geometry has attracted the attention of numerous engineers and scientists. The main emphasis had been given to placing fluvial morphology on a more analytical footing, with the goal of enabling the prediction of river-channel evolution. Gravity waves (Ref. 5), secondary flow (e.g. Ref. 6), transverse oscillations (Ref. 7) are some of the mechanisms considered to be responsible for the deviation of channels from the straight configuration. Stability-based analyses have received considerable attention during the last two decades (e.g. Ref. 8). They include alternate-bar type and bend-stability analyses. Two- and three-dimensional flow models have been used.

The stability characteristics of straight streams are of interest in the present study. Thus, the pertinent type of instability is that of the alternate-bar. Some physical considerations regarding the stability of straight streams are introduced first. Then, a linear stability analysis is formulated and several relevant theories are reviewed.

## 2. Physical Consideration on the Conditions of Alternate Bar Formation

The remarkable variety of bed topography and planform geometry observed in natural streams, and partially reproduced in laboratory flumes, is the result of the complex interaction of the fluid flow with the channel boundaries and sediment transport. The concept of linear stability has been proven to be a powerfull tool for the prediction of several channel configurations and for improving our understanding of the processes leading too their occurrence.

International Conference on River Regime, edited by W.R.White.

Detailed laboratory experiments (Ref. 21) and field observations (Ref. 19) have demonstrated that bed deformation is the fundamental cause of river meandering. The wave-like pattern of scour and deposition of sediment particles on the bed can result in slight irregularities of the initially flat bed of a straight channel, which in turn disturb the flow. The irregularities further disturb the pattern of sediment transport over the bed such that the initial disturbances are either dampened or amplified. The bed irregularities can be resolved mathematically into a number of doubly periodic bed wave components. Due to the infinitesimal amplitude of these waves their interaction with the disturbed flow can be considered linear. Then, the interaction of each wave component with the flow and spatial and temporal distributions of sediment discharge can be examined separately. The temporal amplification of one wave component is sufficient to render the channel unstable. Because all possible wave componentes can be excited by small irregularities, for the channel to be stable it should remain unchanged for all slight disturbances.

A three-dimensional bed perturbation (double-sinuousoidal) on a straight channel with an otherwise flat bed and the response of the flow to this perturbation are shown schematically in Figure 1. It has been suggested (Ref. 14) that the phase shift $\delta$ between the bed shear stress and the local longitudinal bedload discharge is not important for the case of the three-dimensional bed disturbance; this phase shift is essential for the amplification of two-dimensional bed disturbances (dunes and antidunes) (Ref. 22). Consequently, it is only the lateral bedload transport, in the absence of suspended load, that will determine the fate of a three-dimensional bed disturbance. The doubly sinuousoidal bed perturbation forces the thalweg to follow a sinuousoidal path. The result of the continuously varying curvature of this path is a secondary flow and net fluid mass flux in the lateral direction. For subcritical flow, the cross-flow depth-averaged velocity component lags behind the disturbed bed profile by a phase angle ranging from 0 and $\pi/2$ (Ref. 23). The first limit occurs when friction forces dominate the flow, while the latter occurs when the inertial forces are predominant. The lateral velocity component tends to amplify the bed disturbance and contributes to its downstream propagation, and thus the amplification is intensified as the friction forces become more important. The secondary-flow component also contributes to the growth of the bed perturbation. Lateral bed slope, opposes the growth of the bed disturbance. These three mechanisms and their relative importance for a given flow will determine whether the channel will remain stable or not. For an unstable channel, the bed perturbation that grows fastest is expected to prevail and dictates whether the channel will meander or braid. If the effect of the lateral bed slope is ignored, all channels become unstable with infinite braiding being the preferred channel configuration.

Dimensional considerations are very useful for relating the stability characteristics of a straight channel to the gross channel-flow parameters. For a longitudinal wavelength of the dominant doubly-sinuousoidal bed wave, $\ell$, the following functional relation is considered:

$$\ell = f(\rho, \mu, D_o, B, \tau_o, d, \rho_s, g) \tag{1}$$

where $\rho$ and $\mu$ = density and viscosity of the fluid; $D_o$ and B = depth and width of the undisturbed flow; $\tau_o$ = boundary shear stress of the undisturbed flow; d and $\rho_s$ = representative grain size and density of the bed material; and g = the acceleration of gravity. In dimensionless terms (1) becomes

$$\frac{\ell}{B} = f(\tau^*, R_p, R, \frac{D_o}{d}, \frac{B}{D_o}) \tag{2}$$

where $\tau^* = \tau_o/\rho gRd$ is the Shields term, $R = \rho u_* d/\mu$ = particle Reynolds number; $R = \rho_s/\rho - 1$ = submerged specific gravity of sediment, and $u_* = \sqrt{\tau_o/\rho}$ is the shear velocity. For fluvial sediments $R \simeq 1.65$; for fully rough flow conditions the influence of $R_p$ is not as important compared to the rest of the parameters in (2). Then (2) becomes

$$\frac{\ell}{B} = f(\tau^*, \frac{D_o}{d}, \frac{B}{D_o}) \tag{3}$$

If the mean flow velocity is considered instead of $\tau_o$ in the list of independent variables in (1), the Froude number will replace $\tau^*$ in (3). The independent variables in (3) also determine whether the channel will remain stable, meander or braid. The role of the relative roughness, $D_0/d$, becomes less important for large values of $D_0/d$. It has been suggested (Ref. 24) that this is the case for $D_0/d \gtrsim 100$, while for $D_0/d \gtrsim 1000$ it can be disregarded from (3) (Ref. 13). Typically, $3.8 < D_0/d < 422$ (Ref. 25) which suggests that the deletion of $D_0/d$ from (3) is not warranted.

The parameter $B/D_0$ is probably the most important for determining the mode chosen by a channel. The experiments conducted by Ikeda (Ref. 26) and Stebbings (Ref. 27) lend support to this statement. In the first case, gradual increase of the flow rate, while keeping the channel width, slope and bed material the same, systematically changed the bedform pattern from braided to alternate bar to antidune. Stebbings examined the shape of stable channels. A straight channel with continuously curved cross section was initially formed. No sediment was fed or recirculated in the channel. To accomodate the increasing bedload discharge in the downstream direction the channel increased its width along its length. This resulted in a straight stable channel in the upstream end and incipient braiding along the downstream end of the channel. After the bedload transport ceased along the entire length of the channel the flow returned to a single thread channel. The latter experiments indicate also that sediment transport is essential for the occurrence of fluvial instability.

The flow boundary also plays an important role in channel stability. When the stream banks are close together, they tend to constrain the development of a disturbance and render the channel more stable. Most channels are stable when $B/D_0 \lesssim 8$. However, they can also promote flow instability when strong shear is generated at the banks which produces turbulence that is diffused outwards. This was illustrated in the experiments conducted by Shen and Komura (Ref. 6). A channel with a width-depth ratio of ten developed an alternate bar pattern only after its walls were made artificially very rough.

## 3. Formulation of the Linear Stability Theory

Fluid flow in a natural alluvial stream is far too complicated to lend itself to complete rigorous analysis. Simplifying the flow description, while still retaining those features of the flow that influence the fluvial stability phenomena, is an art that has been used effectively by several scientists.

Kinoshita (Ref. 20), was the first to record that alternate bars and meandering flow form readily in a channel with straight, inerodible walls, as long as the bed is erodible. The experiments of Chang, Simons and Woolhiser (Ref. 28) corroborated this observation. Parker (Ref. 12) observed the development of meandering in an initially straight channel with erodible banks in the laboratory. An alternate bar pattern developed first while the channel was still straight. Latter, the growth of the bars caused bank erosion and channel meandering without altering the wavelength of the bars. Thus, bar development appears to be a precursor of incipient meandering and/or braiding.

The model river used in a typical stability analysis consists of a shallow flow in a channel with straight, nonerodible, vertical banks and erodible bed. The shallow flow assumption is based on observations from natural streams and laboratory flumes which indicate that the width-depth ratio of meandering and braiding streams is large. Two-dimensional flow is considered; the vertical velocity component is disregarded, and depth-averaged values are used for the horizontal velocity components. This flow model cannot account for the effect of the secondary currents. The undisturbed flow is taken to be steady and uniform. The wavelength of the bed disturbances is assumed to be longer than the wavelength of ripples, dunes, or antidunes, so that the latter can be considered only as roughness elements. Because of the very small propagation speed of the bed disturbances, the flow is assumed to be quasi-steady (Ref. 11). The equations of motion and continuity for hydrostatic pressure distribution in the vertical are:

$$u \frac{\partial u}{\partial x} + v \frac{\partial u}{\partial y} = - g \frac{\partial h}{\partial x} - \frac{\tau_x}{\rho D} \tag{4}$$

$$u \frac{\partial v}{\partial x} + v \frac{\partial v}{\partial y} = - g \frac{\partial h}{\partial y} - \frac{\tau_y}{\rho D} \tag{5}$$

$$\frac{\partial}{\partial x}(UD) + \frac{\partial}{\partial y}(VD) = 0 \tag{6}$$

$$\frac{\partial}{\partial t}(h-D) + \frac{1}{1-p}\left(\frac{\partial q_x}{\partial x} + \frac{\partial q_y}{\partial y}\right) = 0 \tag{7}$$

with B.C.

$$V = q_y = 0 \quad \text{at} \quad y = 0,\ B \tag{8}$$

where u and v = depth-averaged flow velocity in x- and y-directions; $q_x$ and $q_y$ = x- and y-components of the volumetric bedload transport rate; $\tau_x$ and $\tau_y$ = respective bed shear stresses; h = height of the water surface above some reference level; and $\lambda$ = porosity of the bed material. Several of the symbols, as well as the model channel used, are shown in Fig. 1.

Small perturbations about the undisturbed flow are introduced

$$u = U_o(1+u')\ , \qquad v = U_o v'$$
$$D = D_o(1+D')\ , \qquad h = D_o(h^*{}_o - s_o x^* + h'),\quad h^*{}_o = \frac{h_o}{D_o} \tag{9}$$
$$\tau_x = \tau_o(1+\tau_x')\ , \qquad \tau_y = \tau_o\tau_y',\ q_x = q_o(1+q_x')\ ,\quad q_y = q_o q_y'$$

where primed variables refer to small perturbation. Substituting these expressions into easy (4) to (8) and using $x^* = x/D_o$, $y^* = y/D_o$, $t^* = U_o/D_o$ lead, after appropriate linearization, to the following equations for the perturbed quantities:

$$F_r^2\frac{\partial u'}{\partial x^*} + \frac{\partial h'}{\partial x^*} + S_o(\tau_x' - D') = 0 \tag{10}$$

$$F_r^2\frac{\partial v'}{\partial x^*} + \frac{\partial h'}{\partial y^*} + S_o\tau_y' = 0 \tag{11}$$

$$\frac{\partial u'}{\partial x^*} + \frac{\partial v'}{\partial y^*} + \frac{\partial D'}{\partial x^*} = 0 \tag{12}$$

$$\frac{\partial}{\partial t^*}(h'-D') + \beta\left(\frac{\partial q_x'}{\partial x^*} + \frac{\partial q_y'}{\partial y^*}\right) = 0 \tag{13}$$

$$v' = q'_y = 0 \quad \text{at } y^* = 0,\ \frac{B}{D_o} \tag{14}$$

where $S_o$ = water surface slope of the undisturbed flow; $F_r = U_o/\sqrt{gD_o}$ = Froude number of the undisturbed flow; and $\beta = q_o/(1-p)U_oD_o$.

Constitutive relations for $\tau'$, $\tau'$, $q_x'$, and $q_y'$ in terms of flow parameters are necessary for solving the system of eqs. (10) to (14). Various researchers have used different expressions for the bottom shear stress and sediment transport. These expressions usually are linearized by means of series expansions about the unperturbed flow parameters (Ref. 12, 13, 16). Substitution of the linearized expressions into eqs (10) to (13) results in a system of four linear partial differential equations with constant coefficients and four unknowns: u', v', h', D'. This system accepts solutions that are harmonic waves. Due to the finite width of the channel, standing waves should be considered in the cross-stream direction; traveling waves are appropriate for the streamwise direction. To satisfy the boundary condition (14), v' should have a node at each bank, while the other variables have antinodes at the banks. A stability analysis then is carried out by introducing the following doubly-sinusoidal perturbations into the balance equations:

$$u' = \hat{u}\cos(mk_2y^*)\exp i(k_1x^*-\phi t^*) \qquad v' = \hat{v}\sin(mk_2y^*)\exp i(k_1x^*-\phi t^*)$$
$$h' = \hat{h}\cos(mk_2y^*)\exp i(k_1x^*-\phi t^*) \qquad D' = \hat{D}\cos(mk_2y^*)\exp i(k_1x^*-\phi t^*) \tag{15}$$

where $k_1 = 2\pi D_o/\ell$ = downstream wavenumber of the perturbation; $\ell$ = wavelength; $\phi = \phi_r + i\phi_i$ = complex celerity of the wave, with $c = \phi_r/k_1$ indicating the propagation speed of the disturbance, and $\phi_i$ = temporal exponential growth rate; $mk_2 = m\pi D_o/B$ = cross-stream wavenumber with m indicating the mode of perturbation to (e.g. m = 1 corresponds to alternate bar type of perturbation, and m>1 to braiding); and $\hat{u}$, $\hat{v}$, $\hat{h}$, and $\hat{D}$ = amplitudes of the respective perturbations. For a perturbation to grow (instability) $\phi_i$ should be greater than zero.

The resulting linear and homogeneous algebraic system of four equations with unknowns, $\hat{u}$, $\hat{v}$, $\hat{h}$, $\hat{D}$ constitutes an eigenvalue problem that possesses nontrivial solutions if and only if there is a relation, called the dispersion relation, among $k_1$, $mk_2$, $\phi_r$, and $\phi_2$ for given flow conditions. Such a relation reads (Ref. 12, 23):

$$f(k_1, mk_2, \phi; \tau^*, \frac{D_o}{d}) = 0 \tag{16}$$

and determines whether a doubly sinusoidal bed perturbation characterized by $k_1$ and $mk_2$ will grow or diminish for given flow conditions. Lord Rayleigh (Ref. 29) advanced the thesis that the prevailing perturbation will be the one that grows faster. Raichlen and Kennedy (Ref. 30) provided experimental support for this thesis for the case of ripples and dunes. Although the linear stability analysis is valid only for infinitesimal perturbations there is sufficient evidence to indicate that the so obtained dominant perturbation compares favorably with the characteristics of the fully developed bed configurations (e.g. Ref. 30). Nevertheless, the wavelengths of meandering streams in nature exhibit considerable scatter even under the same flow conditions. Data from reaches of the East Nishnabotna and Des Moines Rivers demonstrate this clearly (Fig. 2). This scatter may be partially attributed to variations of the formative discharge along the same river. It also indicates the complexity of the problem examined here, and suggests that the simplified approach presented here is at best expected to detect the mean trend.

## 4. Review of Stability Theories

Since the original contributions of Hansen (Ref. 9) and Callander (Ref. 10) in late 1960's, there has been a proliferation of papers that treat the meandering or braiding tendencies of straight channels as a stability problem. Representative works from this period are examined here.

Hansen (Ref. 9), Callander (Ref. 10), and Parker (Ref. 12) used the same two-dimensional flow model. All three ignored the effect of the lateral bed slope and therefore, are unable to predict any stable channels.

Hansen obtained the expression $\ell/D_o = 7F_r^2\, S_o^{-1}$ for the dominant wavelength which, because of its simplicity, is often quoted in the literature. However, this simple expression was obtained at the expense of severe approximations, most notably that the downstream wavelength of the perturbation is much larger than the wavelength in the cross-stream direction, while in reality both usually are of the same order of magnitude. However, a more serious flaw in Hansen's analysis is the consideration of traveling waves for the perturbation in both the downstream and cross-stream directions. Such a perturbation cannot satisfy (8) and also requires that the channel continuously shift laterally, with the same speed that the perturbation propagates downstream. However, this is contrary to observations in the field and laboratory, and consequently represents a serious shortcoming of Hansen's approach.

Callander used simplified expressions for the bed shear stress and bedload transport. He assumed that they depend only on the flow velocity. Parker generalized these expressions and utilized perturbation techniques and physical reasoning to obtain criteria for separating meandering from braided streams. No criterion was obtained for stable streams. Neither Parker nor Callander accounted for the effect of secondary currents. Their formulation suggests that the existence of sediment transport is necessary for instability to occur. Ikeda (Ref. 26) compared the expression for the dominant meandering wavelength obtained by Parker with laboratory and field data and concluded that it is a valid approximation of observed wavelengths for flows with $F_r < 0.8$.

The seminal paper of Engelund and Skovgaard (Ref. 11) introduced several new ideas, including the quasi-steadiness of the perturbed flow and the important of the lateral bed slope and its effect on the bedload transport. The former was mentioned in the preceeding section. If the latter is not included in the analysis, the maximum amplification rate of a perturbation is proportional to the lateral wavenumber, $mk_2$, of the perturbation. Thus, infinite braiding is the most unstable bed configuration (Ref. 11). The inclusion of the lateral bed slope resolves this paradoxical result. Its stabilizing effect is proportional to $(mk_2)^2$. Then for small values of $B/D_o$ the stream is stable, for intermediate values of $B/D_o$ meandering is the preferred mode, and for large values of $B/D_o$ braiding occurs. Furthermore, the growth rate of the instability becomes higher for a meandering stream as the width-depth ratio increases. These results agree well with observations (Ref. 1, 31). Careful examination of Fig. 4 of Ref. 11 also reveals that for given flow conditions and bed material, the longitudinal wavelength of the dominant perturbation is proportional to its lateral wavelength in accordance with the regime theory (Ref. 1, 26). However, the constant of proportionality is dependent on the flow condition and the bed material (Ref. 13). Engelund and Skovgaard used a three-dimensional flow model which accounts for helicity. The calculated vertical velocity component is very much smaller than the lateral velocity component. The major drawback of their analysis is the assumption of constant eddy viscosity which results in a finite velocity at the channel bed.

Fredsoe (Ref. 13) extended the work of Engelund and Skovgaard (Ref. 11), although he used a two-dimensional flow model, by accounting separately for the suspended load in addition to the bedload. This was motivated by the fact that suspended load responds with a certain space lag to bed stress changes, while such a lag is insignificant for the bedload; in addition, the transverse channel bed slope does not affect the suspended load. Fredsoe developed stability diagrams and concluded that the prevailing mode of instaiblity depends only on the Shields stress and the width-depth ratio. The former is not as important in establishing the limits of straight-meandering and meandering-braiding regimes; however, it influences considerably the longitudinal wavelength, $\ell/D_o$, selected by the dominant bed disturbance which increases as the Shields stress does, though $\ell/D_o$ approaches a constant value for $\tau^* \gtrsim 1.2$. Nevertheless, his suggestion that bed friction is not important is in disagreement with his own calculations, which indicate that a dune-covered bed becomes unstable for lower values of the width-depth ratio compared to the case of a flat bed (Figs 4 and 6 of Ref. 13). While the channel stability is rather insensitive to changes in $D_o/d$ for values around 1000 (used in Ref. 13), it is expected that the relative roughness, $D_o/d$, would be an important parameter when more realistic $D_o/d$ values are considered (Ref. 26).

Fredsoe also pointed out that the suspended load plays an important role in the amplification of a bed perturbation, especially for high Shields stress values. More specifically, he suggested that in the absence of suspended load and $\tau^* > 1$, all streams should remain straight. This criterion implies that a gravel bed stream ($d>0.2$cm), in the absence of suspendable material ($1.0 < \tau^* < 4.0$) will be invariably stable regardless of the depth-width ratio. By employing a Manning-Strickler resistance relation and $D_o/d = 1000$ it is found that $\tau^* = 1.0$ is equivalent to $F_r \simeq 1.0$. Thus, the fundamental cause of instability for supercritical flows is the suspended load. This statement is in disagreement with observations of several researchers (Refs 25, 26, 31), who have reported meandering and braided stream supercritical flows in the absence of suspended load. Consequently, it seems that the role of suspended load on channel stability is overstated in Fredsoe's analysis.

Data reported by Parker (Ref. 12) are presented in Fig. 3 to examine the stability limits suggested by Fredsoe (Ref. 13) for the case of a dune-covered bed. Finally, Fredsoe (Ref. 13) recommends the following stability limits: for $B/D_o < 8$ the stream will remain straight and for $B/D_o > 60$ it will braid. He also suggests that for meandering strems $5 < \ell/B < 15$, which is supported by field data.

The most recent contributions to channel stability are due to Blondeaux and Seminara, and Kuroki and Kishi (Refs 15, 16, 14). Both employ a two-dimensional flow model and consider only bedload transport. Blondeaux and Seminara used Parker's formulation, and extend it by including the effect of the lateral bed slope on bedload transport and the influence of the secondary flow on the direction of the bed shear stress. The expression obtained for the amplification rate was similar to that suggested by Engelund and Skovgaard (Ref. 11). That is it consists of a destabilizing component due to the lateral bedload transport, which is proportional to the number of braids in (Ref.

23); and a stabilizing component due to the effect of the lateral bed slope on the transverse bedload transport, which is proportional to $m^2$ (Ref. 16). The integer value of m that maximizes, the growth rate of the disturbance determines the dominant mode of instability. An important conclusion of this work is that the stability of a stream or the preferred mode of instability depends on three parameters, in agreement with the dimensional analysis presented earlier. $\tau^*$, $D_o/d$, $D_o/B$, or another equivalent set, and not on two as was suggested by Fredsoe (Ref. 13). Also, increased bed friction rendered the channel unstable for smaller values of the width-depth ratio, in accordance with field and laboratory experience. The limits for straight-meandering and meandering-braiding streams are reproduced here from Ref. 15, in Fig. 4. The data used for comparison are those compiled by Parker (Ref. 12). The agreement between predicted and observed mode of instability is good. The shaded areas in the same figure indicate the dependence of the stability mode on the third parameter, which here is taken as the particle Froude number, $F_d$.

Kuroki and Kishi (Ref. 14) followed a parallel path and their conclusions are similar to those of Blondeaux and Seminara. However, they chose as independent parameters $\tau^*$, $S_o$, $B/D_o$, and suggested, based on data, that they can be reduced to the following two: $\tau^*$ and $B/D_o\ S_o^{0.2}$. The limits for straight-meandering and meandering-braiding based on this theory are reproduced here from Ref. 14 in Fig. 5. Their theory compares favorably with data obtained from the field and laboratory in Japan.

Kuroki and Kishi, and Fredsoe used similar coordinate parameters in their stability diagrams. However, the former used in addition $S_o$ as a free parameter. A comparison of these diagrams for the use of a channel with flat bed (Ref. 13, Fig. 6; Ref. 14, Fig. 5) indicates that they are in agreement for the meandering-braiding limits; for the straight-meandering limits both diagrams give similar results for $\tau^* < 0.2$ and $\tau^* > 0.6$. Because Fredsoe used only two parameters to determine the stability diagram, his limit expresses the mean trend of the limits obtained by Kuroki and Kishi for various water surface slopes, $S_o$. The agreement for large Shields stress values is rather surprising because Kuroki and Kishi did not consider suspended load. Thus, it might be inferred that even in the absence of suspended load and for $\tau^* > 1.0$, channels instability still occurs.

## Conclusions

The characteristics of several linear stability analyses of straight channels, were examined here. The most recent theories (Refs. 11, 13, 14, 15, and 16) give a better understanding of the mechanisms responsible for meandering and braiding. Reasonably good predictions of the stability limits are also obtained (Refs. 14, 15). The width-depth ratio has the dominant influence on the type of river pattern that develops and on the stability limits. However, two other parameters, $\tau^*$ and $D_o/d$, are essential to obtain complete stability diagrams. Also, consideration of the roles of bedload and friction are necessary to obtain accurate predictions of instability.

The role of suspended load should be examined more carefully. Sediment grading effect should be also included in future treatments. Three-dimensional flow models deserve more attention, and supercritical flow instability should be investigated more closely.

## Nomenclature

| | |
|---|---|
| B | channel width |
| $C = \phi_r/k_1$ | propagation speed of a disturbance |
| D | flow depth |
| d | representative grain size of bed material |
| $F_r = U_o/\sqrt{gD_o}$ | Froude number of uniform flow |
| $F_d = U_o/\sqrt{gd(\rho_s-\rho/\rho)}$ | particle Froude number |
| h | elevation of the free surface above a horizontal plane |
| $k_1 = 2\pi\ D_o/\ell$ | downstream wavenumber of a double-sinusoidal wave |
| $k_2 = 2\pi\ D_o/B$ | cross-stream wavenumber of an alternate bar perturbation |

| | |
|---|---|
| $\ell$ | downstream wavelength of a double-sinusoidal wave |
| m | number of braids |
| q | volumetric bedload transport rate |
| $R_p = \rho u_* d/\mu$ | particle Reynolds number |
| $S_o$ | water surface slope |
| u | downstream velocity component |
| $U_o$ | velocity of uniform flow |
| $\tau$ | bed shear stress |
| $\tau^*$ | Shields stress |
| $\mu$ | viscosity of fluid |
| $\rho$ | density of fluid |
| $\rho_s$ | density of bed material |
| $\phi$ | complex celerity of a wave |
| $\phi_r$ | temporal exporential growth rate of a disturbance |
| Subscript | |
| o | refers to uniform flow conditions |
| Superscripts | |
| ' | refers to small perturbations |
| ^ | refers to the amplitude of the small perturbations |

**Acknowledgement**

This material is based upon work supported by the U.S. Army Corps of Engineers, Contract No. DACW39-86-K-0013.

**References**

1. Leopold, L.B. and Wolman, M.G.: "River channel patterns: braided, meandering, and straight." U.S.G.S. Prof. Paper, no. 282-B, 1957.
2. Kinoshita, R.: "Formation of dunes on river beds - an observation on the condition of river meandering." Proc. Japan Soc. Civil Engrs. no. 42, 1957. (in Japanese)
3. Sewter, E.R.A.: "The Alexiad of Anna Comnena." Translated from the Greek, Penguin Books, 1969.
4. Thomson, J.: On the origin of windings of rivers in alluvial plains, with remarks on the flow of water round bends in pipes." Proc. Royal Society, 25, May 1876, pp. 5-8.
5. Werner, P.W.: "On the origin of river meanders." Trans. Am. Geophys. Union, 32, 1951, pp. 898-902.
6. Shen, H.W. and Komura, S.: "Meandering tendencies in straight alluvial channels." J. Hydraul. Div., ASCE, 94, 1968, pp. 997-1016.
7. Anderson, A.G.: "On the development of stream meanders." In: Proc. 12th Cong. IAHR, Colorado USA, 1967, vol. 1, pp. 370-378.
8. Callander, R.A.: "River Meandering." Annual Review of Fluid Mechanics, 10, 1978, pp. 129-158.
9. Hansen, E.: "The formation of meanders as a stability problem." Basic Res. Prog. Rep. 13, Hydraul. Lab., Tech. Univ. Denmark, 1967, pp. 9-13.
10. Callander, R.A.: "Instability and river channels." J. Fluid Mech., 36, 1969, pp. 465-480.
11. Engelund, F. and Skovgaard, O.: "On the origin of meandering and braiding in alluvial streams." J. Fluid Mech., 57, 1973, pp. 289-302.
12. Parker, G.: "On the course and characteristic scales of meandering and braiding in rivers." J. Fluid Mech., 76, 1976, pp. 457-480.
13. Fredsoe, J.: "Meandering and braiding of rivers." J. of Fluid Mech., 84, 1978, pp. 609-624.
14. Kuroki, M. and Kishi, T.: "Regime criteria on bars and braids." Hydraulic Papers, The Research Lab. of Civil and Encironmental Eng., Hokkaido University, Sapporo, Japan, July 1985.
15. Blondeaux, P. and Seminara, G.: "Bed topography and instabilities in sinuous channels." In: Proc. ASCE Rivers '83, New Orleans, USA, 1983, pp. 747-758.
16. Blondeaux, P. and Seminara, G.: "A unified bar-bend theory of river meanders." J. Fluid Mech., 157, 1985, pp. 449-470.

17. Ikeda, S., Parker, G., and Sawai, K.: "Bend theory of river meanders. Part 1. Linear development." J. Fluid Mech., 112, 1981, pp. 363-377.
18. Kitanidis, P.K. and Kennedy, J.F.: "Secondary current and river-meander formation." J. Fluid Mech., 144, 1984, pp. 217-229.
19. Lewin, J.: "Initiation of bed forms and meanders in coarse-grained sediment." Geol. Soc. Am. Bull., 87, 1976, pp. 281-85.
20. Kinoshita, R.: "Investigation of the channel migration of the Ishikari River." (in Japanese). Science and Technology Agency, Bureau of Resources, Memorandum No. 36, Nov., 138p. Appendix to Memorandum No. 36, Dec. 1961.
21. Charlton, F.G. and Benson R.W.: "Effect of discharge and sediment charge on meandering of small streams in alluvium." Hydraulic Research Station Wallingford, Jan. 1966.
22. Kennedy, J.F.: "The formation of sediment ripples, dunes and antidunes." Ann. Rev. Fluid Mech., 1, 1969, pp. 147-168.
23. Colombini, M., Seminara, G., and Tubino, M.: "Finite amplitude alternate bars." J. Fluid Mech., 181, pp. 213-232.
24. Ashida, K.: "Medium-size river bedforms in three dimensional structures of river flood and channel process." Edited by Ashida (1982) (in Japanese).
25. Ikeda, S.: "Prediction of alternate bar wavelength and height." J. of Hydraulic Engineering, ASCE, 110, no. 4, 1982, pp. 371-386.
26. Ikeda, H.: "Experiments on bedload transport, bed forms, and sedimentary structures using fine gravel in the 4-meter-wide flume." Environmental Research Center, The University of Tsukuba, Japan, 1983, 78p.
27. Stebbings, J.: "The shapes of self-formed model alluvial channels." Proc. Inst. of Civ. Engrs., London, England, 25, 1963, pp. 485-510.
28. Chang, H.Y., Simons, D.B. and Woolhiser, D.A.: "Flume experiments on alternate bar formation." J. Waterways, Harbors Coastal Engng Div., ASCE, 97, no. 1, 1971, pp. 155-165.
29. Rayleigh, Lord: The theory of sound. 1945, New York; Dover.
30. Raichlen, F. and Kennedy, J.F.: "The growth of sediment bed forms from an intially flattened bed." Proc. XI Congr. Intern. Assoc. Hydraulic Res., 3, 1965.
31. Fujita, Y. and Muramoto, Y.: "Studies on the process of development of alternate bars." Bulletin of the Disaster Prevention Research Institute, Kyoto University, Japan, 35, 1985, pp. 55-86.

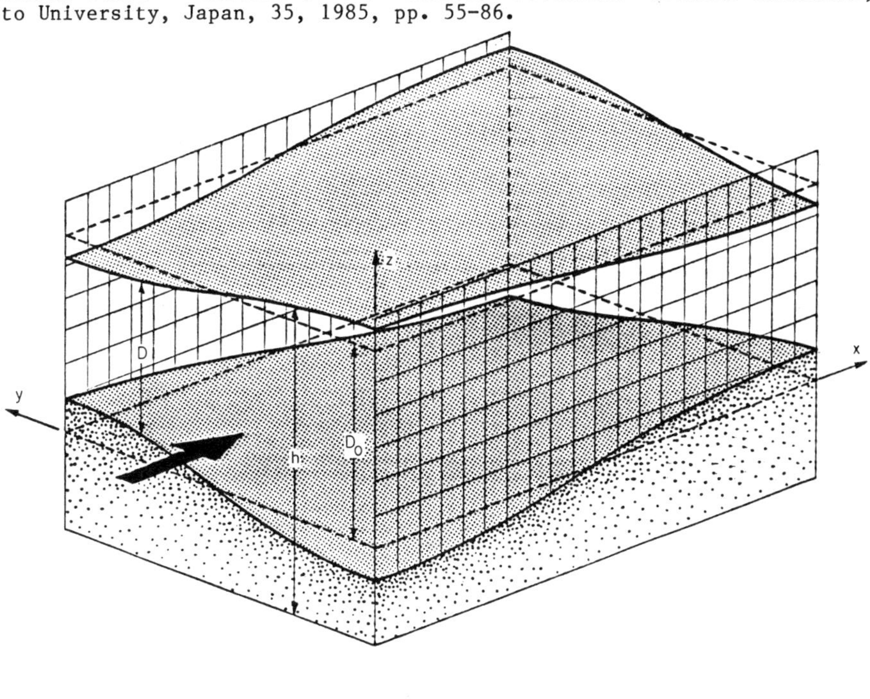


Figure 1. Definition sketch of a doubly-sinusoidal disturbance.

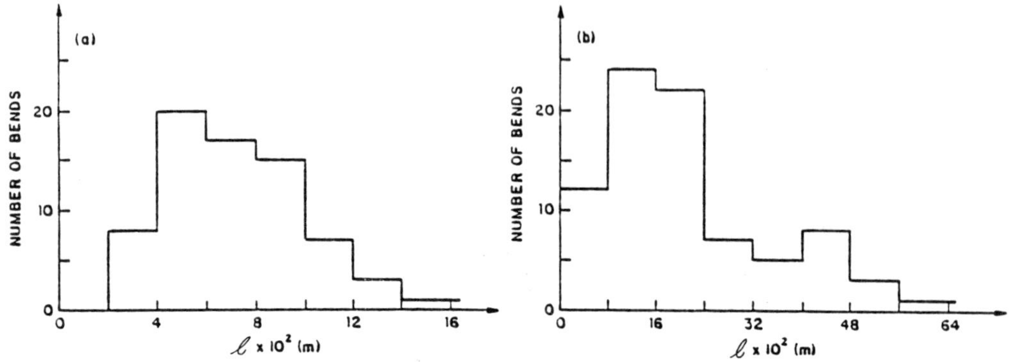


Figure 2. Histogram of meander wavelengths for
(a) East Nishnabotna and
(b) Des Moines Rivers, Iowa.

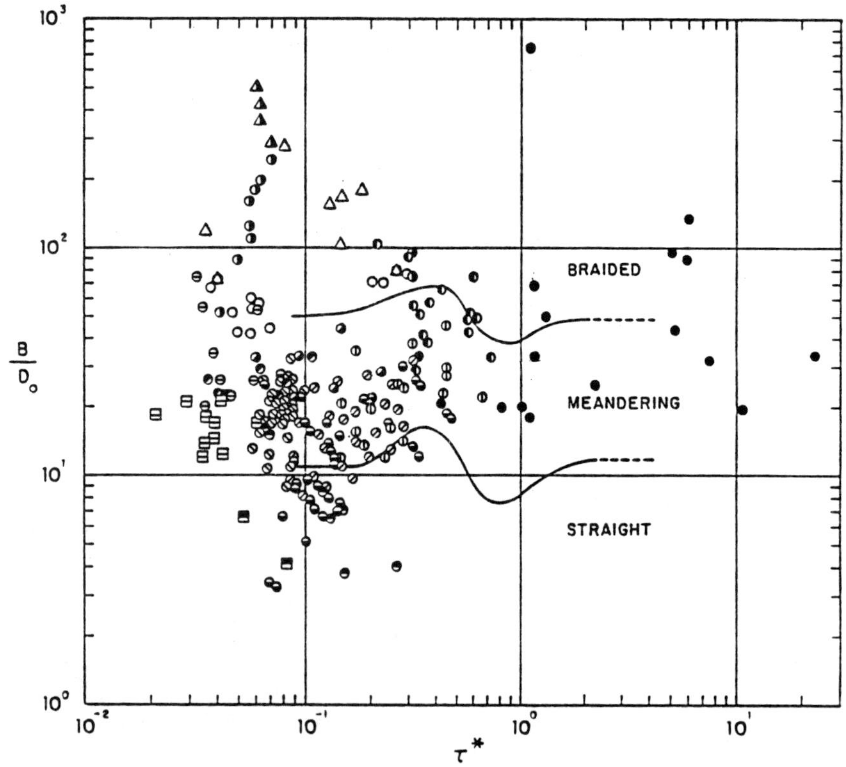


Figure 3. Straight-Meander-Braid stability diagram obtained from Ref. 13. Triangles mark braided river reaches; circles meandering reaches; and squares straight reaches. △○ Anderson et al. (1975); ⊖ ⊟ Wolman and Brush (1961; ⦶ Chang et al. (1971); ◐ Ackers and Charlton (1970); ▲◑ Schumm and Khan (1972); ● Chitale (1970); (see Ref. 12). ◓ ⬒ Ashida and Narai (1969); ⦸ Ikeda (1984); ⊘ Kinoshita (1961); ◕ Fujita (1980); ⬤ Iguchi (1980); ◒ Yoshino (1967); (Ref. 25).

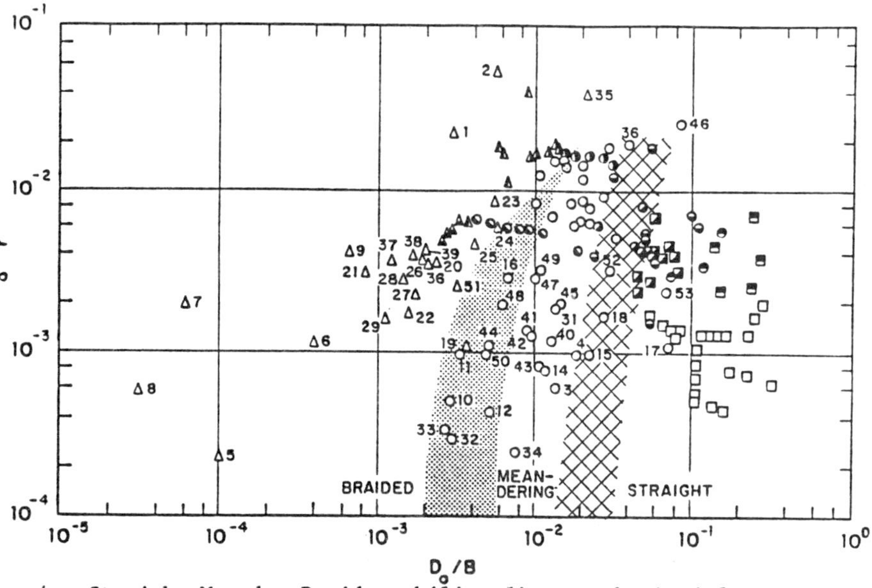


Figure 4. Straight-Meander-Braid stability diagram obtained from Ref. 15. Triangles mark braided river reaches; circles meandering reaches; and squares straight reaches. The data points are from Ref. 12.

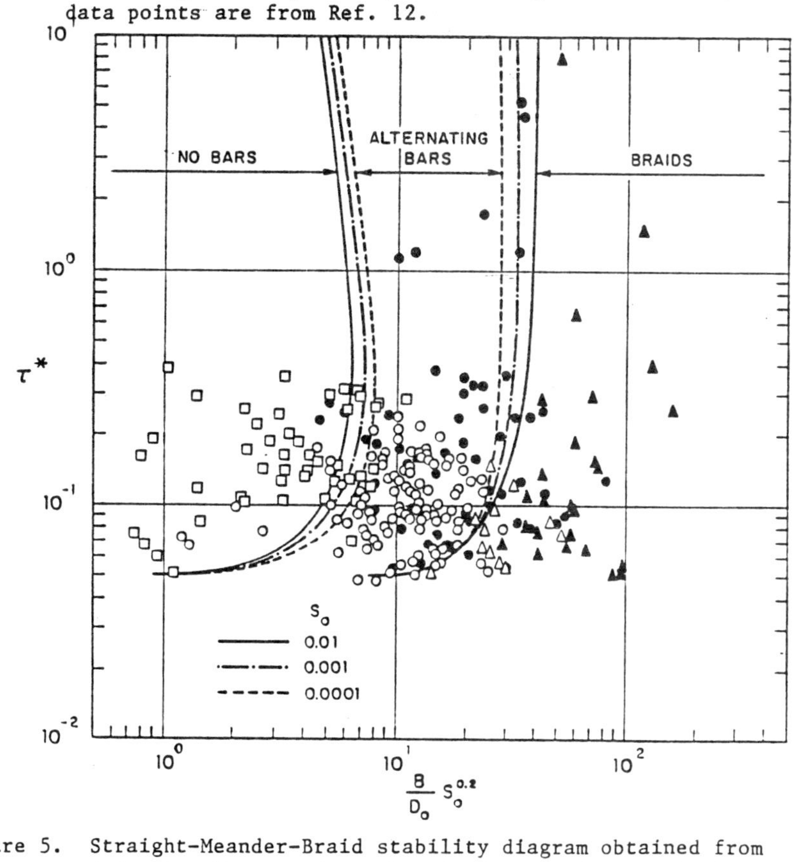


Figure 5. Straight-Meander-Braid stability diagram obtained from Ref. 14. Squares describe ripples and dunes; circles alternate bars, and triangles double-row bars. Solid symbols represent field data and open symbols represent laboratory data.

International Conference on
RIVER REGIME
18-20 May, 1988

PAPER C3

# ON THE THRESHOLD BETWEEN MEANDERING AND BRAIDING

by

N. Struiksma and G.J. Klaassen

DELFT HYDRAULICS, P.O.Box 152,
8300 AD Emmeloord, The Netherlands

Summary

A prediction method for planform classification is developed by extending on the work by Struiksma et al (1985) on bed deformation in curved alluvial channels. By comparing the results of the classification method with natural rivers and laboratory channels it is shown that the method yields fair results. The relation with the classic Leopold and Wolman (1957) classification method is shown by introducing Lacey's regime equations into the proposed criterion. Several implications of the proposed method are discussed.

## 1. Introduction

Traditionally, channel patterns of alluvial rivers have been classified as straight, meandering or braided (Leopold and Wolman, 1957), although such a classification is slightly obscured by changes in channel characteristics over the years (see e.g. Klaassen and Vermeer (1988) data for the braided Jamuna River, Bangladesh). Initially empirical classification graphs were derived, the first one by Leopold and Wolman (1957) who plotted bankfull discharge against channel slope. A separations between meandering and braiding could be made (see Figure 1). The derived equation for the separation line reads:

$$i = 0.0116\ Q^{-0.44} \qquad (1.1)$$

where Q is the bankfull discharge in $m^3/s$ and i the channel slope. Later studies provided similar separation criteria, although the scatter involved "seems to make the distinction less satisfactory" (Bettess and White, 1983). Recently, Ferguson (1984) proposed to include the bed material size of the alluvial channels as additional parameter to reduce the scatter in the predictions.

From a theoretical point of view the problem of the transition from meandering to braiding has been studied by various authors using stability theory. Bettess and White (1983) provide a short review of this approach, but according to them "none of the theories adequately reproduce the sort of behaviour observed by Leopold and Wolman (1957) and others".

International Conference on River Regime, edited by W.R.White.

In the present paper the recent theoretical and experimental work of Struiksma et al (1985) on the bed deformation in river bends is extended to a predictive method to distinguish between meandering and braiding. It is shown that this approach results in a classification diagram that allows to make such a distinction, provided that the channel dimensions are known in advance. By introducing Lacey's regime equations a threshold criterion similar to Leopold and Wolman (1957) is obtained. Instead of plotting i versus Q it is proposed to use the bed material size D as an additional parameter either via $i/\sqrt{D}$ or via $i/\sqrt[3]{D}$ (conform the suggestions by Ferguson, 1984; see also Figure 2). Furthermore it is suggested that these parameters also govern the number of parallel braids (conform the empirical expressions derived by Howard et al, 1970).

## 2. Development of a criterion for planform classification

### 2.1 General

To arrive at a predictive discrimination method between meandering and braiding use will be made of the bed topography analysis presented by Struiksma et al (1985). They studied the equilibrium bed topography in curved channels and developed criteria for the spatial amplification of steady bed level perturbations. These results are used to derive a predictor for the threshold between meandering and braiding. In this chapter first a summary of the analysis carried out by Struiksma et al (1985) is given (Section 2.2) and subsequently the results are extended to arrive at the prediction method (Section 2.3).

### 2.2 Linear perturbation analysis

For their study of the bed deformation in curved alluvial channels Struiksma et al (1985) developed a mathematical model under the assumption that time-averaged flow velocities and overall sediment transport relations can be used. In addition to running a comprehensive numerical model of the relevant phenomena they carried out a linear perturbation analysis of the equilibrium bed topography according to this model. The main difference between this linear analysis and previous stability analyses (see Bettess and White (1983) for a review) is that, a non-uniform but steady perturbation is considered, whereas a uniform but unsteady perturbation is considered in the stability analyses. In view of the relatively large propagation velocity of the bed perturbations, as predicted by the latter and the generally low erodibility of the banks, the stability analyses cannot provide an adequate explanation for the development of meanders but only an explanation for the development of alternate bars (see Olesen, 1983 and Blondeaux and Seminara, 1985).

The steady perturbation analysis formally applies to mildly curved channels. Although in strongly curved channels this mild curvature approximation may give rise to some quantitative influence on the results, it is postulated that the general character of the solutions remains unaltered. Under a number of other simplifications the complete non-linear mathematical model is linearised. The analysis considers harmonic bed level perturbations of the form:

$$h' = \hat{h} \exp i(kx + k_w y) \tag{2.1}$$

where $\hat{h}$ is a complex number in which the absolute value gives the amplitude and the modulus the phase of the perturbation, $i = \sqrt{-1}$, $k_w$ is the wave number in transverse y-direction and k is a complex wave number in downstream x-direction. The imaginary part of k describes the development of the amplitude

of the perturbation in the flow direction. The solution of the remaining variables is in a similar form.
Impermeable side walls put constraints onto the transverse wave number, namely:

$$k_w = m\frac{\pi}{B}, \quad m = 1, 2, 3... \tag{2.2}$$

For a meandering channel m = 1. Higher values of m pertain to braiding channels. Introducing Equation (2.1) into the linearized model yields non-trivial results if k satisfies a certain condition, in the form of a complex polynomial equation. The solution of this quation is generally dominated by two roots which have mostly the form:

$$k = \pm k_r + i\,k_i = \pm\frac{2\pi}{L_P} + i\frac{1}{L_D} \tag{2.3}$$

where: $L_P$ = wave lenght and $L_D$ = damping length.

The roots mainly depend on the ratio of the adaptation length $\lambda_s$ of the bed topography development and the adaptation length $\lambda_w$ of the main flow. These adaptation lengths are defined as (Struiksma, 1983; de Vriend and Struiksma, 1983):

$$\lambda_s = \frac{1}{\pi^2}\left(\frac{B}{h}\right)^2 f(\Theta)\,h \tag{2.4}$$

$$\lambda_w = \frac{C^2}{2g}\,h \tag{2.5}$$

where B = channel width, h = channel depth, C = Chezy coefficient, g = acceleration of gravity and $f(\Theta)$ = weighing function of influence of sloping bed, $\Theta$ = Shields parameter defined via $\Theta = hi/\Delta D$, $\Delta$ = relative density defined via $\Delta = (\rho_s-\rho)/\rho$, $\rho_s$ = specific density of sediment, $\rho$ = specific density of water. The expression for the ratio $\lambda_s/\lambda_w$ reads:

$$\frac{\lambda_s}{\lambda_w} = \frac{2}{\pi^2}\cdot\frac{g}{C^2}\cdot\left(\frac{B}{h}\right)^2\cdot f(\Theta) \tag{2.6}$$

In Struiksma et al (1985) for $f(\Theta)$ the following expression is used:

$$f(\Theta) = 1.5\,\Theta \tag{2.7}$$

but there is evidence (van Mierlo, 1985; Odgaard, 1981) that for natural channels an improved expression reads:

$$f(\Theta) = 0.85\,\Theta^{0.5} \tag{2.8}$$

The dependence of $L_D$ and $L_P$ in Equation (2.3) on the ratio $\lambda_s/\lambda_w$ is shown in Figure 3. This figure presents the results of a strongly simplified model suggested by Struiksma (1983), which very well represents the behaviour of the roots according to the complete linearised model. Figure 3 also shows that the roots depend also on the value of a coefficient b. This coefficient b relates the sediment transport s per unit width to the velocity u via:

$$s :: u^b \tag{2.9}$$

It is assumed that b is a constant. This implies that the sediment transport formula is approximated by a power law of the flow velocity. The parameter b depends on the Shields parameter. If the value of the Shields parameter lies close to the threshold value the parameter b is large and strongly variable. For larger transport rates, lower values for b are found. Figure 4 provides the approximate relationship between b and the Shield parameter $\Theta'$ (corrected for form roughness) as derived from the Meyer-Peter and Müller transport formula. The applied expression for $\Theta'$ reads:

$$\Theta' = \frac{u^2}{C_{90}^2 \cdot \Delta \cdot D} \tag{2.10}$$

where $C_{90} = 18 \log(12h/D_{90})$. Equation (2.10) allows to estimate b for given u, $C_{90}$, $\Delta$ and D. Another sediment transport formule results in a different relationship between b and $\Theta'$ (e.g. for the Engelund-Hansen formula b = 5).

## 2.3 Criterion for planform classification

The foregoing summary of the linear perturbation analysis shows that the damping length $L_D$ (Figure 3(a)) and wave length $L_P$ (Figure 3(b)) are related to the ratio $\lambda_s/\lambda_w$. Now Figure 3 is considered in some more detail to derive a criterion for planform classification. According to Figure 3(a) the perturbation are damped in the downstream direction for $\lambda_w/L_D > 0$. If $\lambda_w/L_D < 0$, the perturbations will grow in downstream direction. Figure 3(b) shows that only for a restricted range of $\lambda_s/\lambda_w$ (harmonic range) the wave number $2\pi\lambda_w/L_P$ can be defined. Beyond that range the present approach based on m = 1 (one single channel) is not appropriate any more. Probably braiding will occur, more channels will develop and the value of $\lambda_s/\lambda_w$ is relatively large. Thus tentatively the following criterion can be formulated for the transition from meandering to braiding:

$\lambda_w/L_D > 0$ meandering

within harmonic range for m = 1

$\lambda_w/L_D < 0$ transition

within harmonic range for m = 1

$\lambda_w/L_D < 0$ braiding

beyond harmonic range for m = 1.

For the time being the value of $\lambda_s/\lambda_w$ where $\lambda_w/L_D = 0$ is indicated as $\lambda_s/\lambda_w{}^*$ and where $2\pi\ \lambda_w/L_P = 0$ as $\lambda_s/\lambda_w{}^{**}$ (upper limit of harmonic range). Thus the above criterion can also be formulated as:

$\lambda_s/\lambda_w < \lambda_s/\lambda_w^*$ meandering

$\lambda_s/\lambda_w^* < \lambda_s/\lambda_w < \lambda_s/\lambda_w^{**}$ transition

$\lambda_s/\lambda_w^{**} < \lambda_s/\lambda_w$ braided

The following remarks are made:

(i) The values of $\lambda_s/\lambda_w^*$ and $\lambda_s/\lambda_w^{**}$ depend on the value of b (see Figure 3), according to the following relationship:

$$\lambda_s/\lambda_w^* = 2/(b - 3) \tag{2.11}$$

$$\lambda_s/\lambda_w^{**} = \{\tfrac{1}{2}(b + 1) - \tfrac{1}{2}[(b + 1)^2 - 2(b - 3)]^{0.5}\}^{-1} \tag{2.12}$$

(ii) It can be shown that always $\lambda_s/\lambda_w^{**} > \lambda_s/\lambda_w^*$.

## 3. Verification of proposed criterion

For the verification of the proposed criterion data are required from alluvial channels, of which both the planform and channel characteristics should be available. The required channel characteristics are B, u (velocity), i, h and D, $D_{90}$. For the verification data from both natural rivers and laboratory channels are available. Only data readily available in the literature are used here; no special tests or measurements were done.

Data was selected from the following sources:

- natural rivers (see Table 1):
  - Leopold and Wolman (1957)
  - Chitale (1970, 1973)
  - Mosley (1976)
  - RPT/Nedeco/BCL (1987)
- Laboratory channels (see Table 2):
  - Fujita and Muramoto (1981) (Items II-1 ... II-7)
  - Hong and Davies (1979) (Items III-1)
  - Schumm, and Khan (1972) (Items IV-1 ... IV-3)
  - Leopold and Wolman (1957) (Items V-1)
  - Ashmore (1980) (Items VIII-1)

The following remarks are made regarded the selected data:

(i) Other river and channel data included in the above literature was not used because of one of the following reasons:
- the data was not complete,
- the (test)conditions did not deviate substantially from other sets of data already included.

(ii) Information from other sources (Ashmore and Parker, 1983; Parker and Anderson, 1975; Khan, 1978; Southard et al, 1981 could not be used for the same reasons.

From the basic data obtained from the literature the following parameters were derived (see Tables 1 and 2): $\lambda_s/\lambda_w$, $C_{90}$ (whereby it was assumed that $D_{90}$ = 2.5 $D_{50}$, in case $D_{90}$ was not known), $\Theta'$, $\lambda_s/\lambda_w^*$ and $\lambda_s/\lambda_w^{**}$. For the transformation of $\Theta'$ to b Figure 4 was used. By comparing the computed value of $\lambda_s/\lambda_w$ with the threshold values $\lambda_s/\lambda_w^*$ and $\lambda_s/\lambda_w^{**}$, the proposed criterion for planform classification can be evaluated. The results are given in the last columms of the Tables 1 and 2. From an inspection of these results it follows that the proposed criterion results in a fair prediction of channel pattern, although in the case of laboratory channels the predictions are less good than for natural rivers. The latter may be due to e.g. the limited accuracy in measuring the average water depth in small laboratory channels the applied relation between $\Theta'$ and b, etc.

The proposed criterion merely consists of computing the values of $\lambda_s/\lambda_w$ and b comparing these with the appropriate values of $\lambda_s/\lambda_w^*$ and $\lambda_s/\lambda_w^{**}$. The latter can also be done graphically as is shown in Figure 5. In this Figure points have been introduced that correspond to the data sets listed in the Tables 1 and 2. The index B and M indicate whether the channel pattern is braided or meandering.

Considering the results of the verification as presented in the Tables 1 and 2 and Figure 5 it may be concluded that starting from the results of the linear analysis by Struiksma et al (1985) a criterion was developed that results in a fair prediction of the channel pattern. Still, however, certain channels are not correctly classified, which may be caused by e.g. less correct data or a less distinct threshold boundary than assumed here, but also because the present theory is based on a linear analysis that does not necessarily reflect all characteristics of the complete phenomenon.

## 4. Discussion

A major disadvantage of the proposed classification criterion is that it requires à priori knowledge on channel characteristics like widths and average depths during bankfull discharge. One would prefer a method that starts from more basic parameters like the bankfull discharge and the valley slope (or alternatively the yearly amount of sediment the river has to carry). Only by combining the present method with a method that predicts the channel characteristics a more universal classification method can be obtained.

It is of interest to combine the method proposed here with Lacey's regime equations. From Equation (2.6), the following relationship should hold for the transition from meandering to braiding:

$$\frac{2}{\pi^2} \cdot \frac{g}{C^2} \cdot \left(\frac{B}{h}\right)^2 \cdot f(\Theta) = \text{constant} \tag{4.1}$$

or:

$$\left(\frac{B}{h}\right)^2 \cdot f(\Theta) \cdot hi :: u^2 \tag{4.2}$$

Lacey's regime equations read:

$$h :: Q^{1/3} \tag{4.3}$$

$$B :: Q^{1/2} \tag{4.4}$$

Introducing these expressions into Equation (4.2) and using Q = B.h.u it is possible to find a relationship between Q and i. This relationship depends on the assumption made as to how $f(\Theta)$ reads. If $f(\Theta) :: \Theta$, then one obtains:

$$i :: Q^{-1/3} (\Delta D)^{1/2} \tag{4.5a}$$

For $f(\Theta) :: \Theta^{0.5}$ a slightly different expression is found, notably:

$$i :: Q^{-1/3} (\Delta D)^{1/3} \tag{4.5b}$$

The exponent of Q in both expressions should be compared to e.g. Equation (1.1), proposed by Leopold and Wolman (1957). Although the exponent -1/3 is closer to -0.44 than the value of -1/6 found by Ferguson (1984) by combining Parker's (1976) prediction method with Lacey's regime equation, there is still a considerable discrepancy. Tentatively, however, it can be stated that the classification method proposed here can be shown to be similar to some extent to the empirical method proposed by Leopold and Wolman (1957).

Next the influence of the bed material size is considered. From Equation (4.5a) and (4.5b) it follows that the threshold is also affected by $\Delta$ and by D. Figure 2, taken from Ferguson (1984), provides empirical evidence for this. From the data plotted in this figure it may be concluded which of the two expressions is more appropriate. Until now, the authors have not yet carried out an analysis to verify this. Anyhow it seems more appropriate to plot $i/(\Delta D)^{1/2}$ or $i/(\Delta D)^{1/3}$ against the bankfull discharge Q than only i versus Q.

Finally, one more implication of the proposed classification method is discussed here. It was proposed that if $\lambda_s/\lambda_w$ exceeds $\lambda_s/\lambda_w^{**}$ that in that case braiding would occur resulting in multiple channels. To study this in some more detail it is for the time being assumed that when the total bankfull discharge $Q_t$ exceeds the threshold value for that particular valley slope a number of parallel channels will develop. All these individual channels should have such channel dimensions that $\lambda_s/\lambda_w$ is smaller than $\lambda_s/\lambda_w^{*}$ and anyhow smaller than $\lambda_s/\lambda_w^{**}$. Equation (4.5) can be written as:

$$i :: \left(\frac{Q_t}{n}\right)^{-1/3} (\Delta D)^{\alpha} \qquad (4.6)$$

where n = number of parallel braids and $\alpha$ = exponent depending as which expression for $f(\Theta)$ is used. Equation (4.6) can also be written as:

$$i :: Q_t^{-1/3}\, n^{1/3}\, (\Delta D)^{\alpha} \qquad (4.7)$$

The only study known to the authors that provides information on the number of braids is the work by Howard et al (1970). Based on observations on natural rivers they derived the following relationship:

$$n = 0.58\, R_q^{-0.19}\, i^{0.41}\, Q^{0.25} \qquad (4.8)$$

where $R_q$ = ratio of mean annual flood to the mean annual discharge. Neglecting the influence of $R_q$ for the time being, Equation (4.8) can be written as:

$$i :: n^{2.4}\, Q^{-0.60} \qquad (4.9)$$

The exponent of n is quite different for the Equations (4.7) and (4.9), but the predicted behaviour is to some extent similar: an increase of the slope results is an increase of the number of braids. In fact this behaviour was predicted by Bettess and White (1983) and it was used recently for the scaling of a model of a braided river (Klaassen, 1987). The large deviation between the exponent of n in the Equations (4.7) and (4.9) could be due to the limited data set used by Howard et al (1970). It is worthwile to study the number of braids as a function of i and Q in more detail in the future.

## 5. Conclusions

Based on the study presented here the following conclusions can be drawn:

(i) It is possible to extend the work of Struiksma et al (1985) on bed deformation in curved alluvial channels to arrive at a classification method for channel patterns. The proposed method predicts observed channel patterns fairly well.

(ii) It can be shown that the proposed classification method in combination with Lacey's regime equations results in a more universal method that is comparably to the classic Leopold and Wolman (1957) classification diagram.

(iii) It is preferable to plot $i/\sqrt{(\Delta D)^{1/2}}$ or $i/\sqrt{(\Delta D)^{1/3}}$ against bankful discharge Q than only i versus Q as was done by Leopold and Wolman (1957).

(iv) Increase of slope or of discharge will probably result in an increase of the number of braids in a braiding river system.

(v) More research is needed to substantiate in particular the conclusions (iii) and (iv).

## 6. Acknowledgements

The present research work was carried out as part of the internally funded basic research programme of DELFT HYDRAULICS. It was initiated by and data were used from the Jamuna Bridge Appraisal Study. This study is being funded by the United Nations Development Programme (UNDP) through the World Bank. The Jamuna Multi-purpose Bridge Authority (JMBA) is furthering the studies for and trhe construction of the bridge. The Jamuna Bridge Appraisal Study is carried out by Rendel, Palmer and Tritton (U.K.), Nedeco and Bangladesh Consultants Ltd. The second author is the senior morphologist in this study. The present research work extends on research work carried out within the Applied Research Program (TOW) on Rivers, funded by Rijkswaterstaat (Netherlands Governmental Water Control and Public Works Department). The collection of river and laboratory data and the verification of the proposed method was carried out by Mr. P.J. de Boer, trainee from the Polytechnical College, Leeuwarden. Finally Dr. H.J. de Vriend is acknowledged because of his review of the paper.

## 7. REFERENCES

ASHMORE, P. (1982), Laboratory modelling of gravel braided stream morphology, Earth Surface Processes, Vol. 7, pp. 201-225.

ASHMORE, P. and PARKER, G. (1983), Confluence scour in coarse braided streams, Water Resources Research, Vol. 19, no. 2, pp. 392-402.

BEGIN, Z.B. (1981), The relationship between flow-shear stress and stream pattern, Journ. Hydrology, Vol. 52, pp. 307-319.

BETTESS, R. and WHITE, W.R. (1983), Meandering and braiding of alluvial channels, Proc. Instn. Civ. Engrs., Part 2, Vol. 75, pp. 525-538.

BLONDEAUX, P. and SEMINARA, G. (1985), A unified bar-bend theory of river meanders. J. Fluid Mech. Vol. 157, pp. 449-470.

CHITALE, S.V. (1970), River channel patterns, Journ. Hydr. Div. ASCE, Vol. 96, no. HY1, pp. 201-221.

CHITALE, S.V. (1973), Theories and relationships of river channel patterns, Journ. Hydrology, Vol. 19, pp. 285-308.

FERGUSON, R.T. (1984), The threshold between meandering and braiding, Proc. 1st Intern. Conf. on "Channels and Channel Control Structures", Southampton, U.K.

FUJITA, Y. and MURAMOTO, Y. (1982), Experimental study on stream channel processes in alluvial rivers, Bull. Disaster Proc. Res. Inst., Kyoto Univ., Vol. 32, part 1, no. 288, pp. 49-96.

REFERENCES (continued)


HONG, L.B. and DAVIES, T.R.H. (1979), A study of stream braiding: Summary, Bull. Geol. Soc. Am., Vol. 90, pp. 1094-1095.

HOWARD, A.D., KEETCH, M.E. and VINCENT, C.L. (1970), Topological and geometrical properties of braided streams, Water Resources Research, Vol. 6, no. 6, pp. 1974-1988.

KHAN, H.R. (1978), Sediment discharge and channel patterns, Int. Conf. on Water Resources Enginnering, Bangkok, Thailand.

KLAASSEN, G.J. (1987), On the scaling of braided sand-bed rivers, NATO Workshop on Movable Bed Physical Models, DELFT HYDRAULICS.

KLAASSEN, G.J. and. VERMEER, K. (1988), Channel characteristics of the braiding Jamuna River, Bangladesh, Intern. Conf. on River Regime, Wallingford, England.

LEOPOLD, L. and WOLMAN, G. (1957), River channel patterns: braided, meandering and straight, U.S. Geol. Survey, Prof. Paper 282-B.

MIERLO, M.C.L.M. van, Influence of sloping bed on the sediment transport direction. DELFT HYDRAULICS, TOW report R657-XXIX/Q186.

MOSLEY, P. (1976), An experimental study of channel confluences, Journ. of Geology, Vol. 84, pp. 535-562.

ODGAARD, A.J. (1981), Transverse bed slope in alluvial channel bends, Journal of the Hydraulic Division, ASCE, Vol. 107, No. 12, pp. 1677.

OLESEN, K.W. (1983), Alternate bars in and meandering of alluvial rivers, Proc. ASCE Specialty Conference "River Meandering", New Orleans.

PARKER, G. (1976), On the cause and characteristic scales of meandering and braiding rivers, Journal Fluid Mech. Vol. 76, pp 457-480.

PARKER, G. and ANDERSON, A.G. (1975), Modelling of meandering in rivers, ASCE Symposium on Modelling, San Francisco.

SCHUMM, S.A. and KHAN, H.R. (1972), Experimental study of channel patterns, Geol. Soc. Am. Bull., Vol. 83, pp. 1755-1770.

SOUTHARD, J.P., SMITH, N.D., DRAKE, T.G. and KUHNLE, R.A. (1981), Field and laboratory studies of braiding in shallow gravel- bed straems, Proc. 2nd Intern. Symp. on Fluvial Sedimentology, Keele, England.

STRUIKSMA, N. (1983), Point bar initiation of alluvial rivers with dominant bed load transport. DELFT HYDRAULICS, TOW Report R657-XVII/W308-part II.

STRUIKSMA, N., OLESEN, K.W., FLOKSTRA, C. and VRIEND, H.J. de (1985), Bed deformation in curved alluvial channels, Journal of Hydraulic Research, Vol. 23, no. 1, pp. 57-79.

VRIEND, H.J. de, and STRUIKSMA, N. (1983), Flow and bed deformation in river bends, ASCE Speciality Conference "River Meandering", New Orleans.

| Name of river | | B(m) | h(m) | i (-) | u (m/s) | D (m) | $\frac{\lambda s}{\lambda w}$ (-) | $C_{90}$ (-) | θ' (-) | $\frac{\lambda s}{\lambda w}$* (-) | $\frac{\lambda s}{\lambda w}$** (-) | channel pattern predicted | channel pattern observed | +/- |
|---|---|---|---|---|---|---|---|---|---|---|---|---|---|---|
| 1. Jamuna | (Bangladesh) | 5000 | 6.0 | $6x10^{-5}$ | 1.33 | $0.2x10^{-3}$ | 251 | 86.0 | 0.725 | 10 | 42 | B | B | + |
| 2. Brahmaputra | (India) | 9449 | 1.52 | $2.52x10^{-4}$ | 1.73 | $0.3x10^{-3}$ | 7396 | 72.1 | 1.16 | 13 | 55 | B | B | + |
| 3. Saraswati | (India) | 320 | 2.54 | $12.5x10^{-4}$ | 2.50 | $0.3x10^{-3}$ | 34 | 76.3 | 2.17 | 13 | 55 | B/M | M | + |
| 4. Ranoli | (India) | 287 | 0.89 | $7.8x10^{-4}$ | 1.57 | $0.11x10^{-3}$ | 97 | 75.8 | 2.36 | 13 | 55 | B | B | + |
| 5. Yamuna | (India) | 366 | 3.46 | $3.1x10^{-4}$ | 1.38 | $0.32x10^{-3}$ | 14 | 79.1 | 0.576 | 10 | 42 | B/M | M | + |
| 6. Ganga | (India) | 4023 | 5.18 | $1.6x10^{-4}$ | 2.53 | $0.43x10^{-3}$ | 144 | 78.9 | 1.45 | 13 | 55 | B | M | - |
| 7. Yashwantpur | (India) | 122 | 1.68 | $4.3x10^{-4}$ | 1.40 | $2.05x10^{-3}$ | 48 | 57.9 | 0.273 | 5 | 22 | B | B | + |
| 8. Indus | (Pakistan) | 983 | 4.88 | $1x10^{-4}$ | 1.49 | $0.18x10^{-3}$ | 19 | 85.2 | 1.03 | 13 | 55 | B/M | M | + |
| 9. Tonle Sap | (Cambodja) | 1402 | 6.10 | $0.167x10^{-4}$ | 0.84 | $0.01x10^{-3}$ | 32 | 109.6 | 3.56 | 13 | 55 | B/M | B | + |
| 10. Savannah | (U.S.A.) | 107 | 5.18 | $1.1x10^{-4}$ | 1.55 | $0.8x10^{-3}$ | 0.175 | 74.0 | 0.332 | 10 | 42 | M | M | + |
| 11. Mississippi | (U.S.A.) | 1381 | 20.12 | $0.473x10^{-4}$ | 1.54 | $0.5x10^{-3}$ | 3.5 | 88.3 | 0.37 | 10 | 42 | M | M | + |
| 12. Mississippi | (U.S.A.) | 2144 | 16.76 | $0.815x10^{-4}$ | 1.19 | $0.3x10^{-3}$ | 45 | 90.9 | 0.346 | 10 | 42 | B | M | - |
| 13. Beaver Creek | (U.S.A.) | 1280 | 0.27 | $6.6x10^{-3}$ | 0.26 | $2.96x10^{-2}$ | 61 | 22.7 | $2.69x10^{-3}$ | 0.28 | 3 | B | B | + |
| 14. Cottonwood Creek | (U.S.A.) | 1310 | 0.24 | $5x10^{-5}$ | 0.55 | $1.8x10^{-2}$ | 2.9 | 25.7 | $1.55x10^{-2}$ | 0.28 | 3 | B/M | B | + |
| 15. Yellowstone River | (U.S.A.) | 1280 | 1.40 | $1.52x10^{-3}$ | 0.48 | $8.23x10^{-2}$ | 3.9 | 27.6 | $2.2x10^{-3}$ | 0.28 | 3 | B | B | + |
| 16. Watts Branch | (U.S.A.) | 399 | 0.1518 | $3.1x10^{-3}$ | 0.189 | $2.50x10^{-2}$ | 1.6 | 19.8 | $2.2x10^{-3}$ | 0.28 | 3 | B/M | B | + |
| 17. Middle Piney Creek | (U.S.A.) | 698 | 0.21 | $9.1x10^{-3}$ | 0.61 | $9.4x10^{-2}$ | 1.1 | 18.5 | $7.0x10^{-3}$ | 0.28 | 3 | B/M | M | + |
| 18. Beaver | (Canada) | 55 | 2.74 | $2.4x10^{-4}$ | 0.95 | $0.5x10^{-3}$ | 0.45 | 72.7 | 0.207 | 4 | 28 | M | M | + |
| 19. Ohau River | (Nieuw Zeeland) | 450 | 0.25 | $6.5x10^{-3}$ | 0.4 | $20x10^{-3}$ | 12414 | 25.2 | $7.66x10^{-3}$ | 0.28 | 3 | B | B | + |
| 20. Ohua River | (Nieuw Zeeland) | 450 | 0.56 | $6.5x10^{-3}$ | 1.5 | $20x10^{-3}$ | 590 | 31.5 | $9.60x10^{-3}$ | 0.28 | 3 | B | B | + |
| 21. Rio Parana | (Argentinië) | 1600 | 10.0 | $5x10^{-5}$ | 1.0 | $0.3x10^{-3}$ | 22 | 86.8 | 0.268 | 5 | 22 | B | B | + |

Table 1 Comparison of predicted and observed channel pattern of natural rivers

| Test number | B (m) | h (m) | i (-) | u (m/s) | D (m) | $\frac{\lambda s}{\lambda w}$ (-) | $C_{90}$ (-) | $\theta'$ (-) | $\frac{\lambda s}{\lambda w}^*$ (-) | $\frac{\lambda s}{\lambda w}^{**}$ (-) | channel pattern | | +/- |
|---|---|---|---|---|---|---|---|---|---|---|---|---|---|
| | | | | | | | | | | | predicted | observed | |
| II-1 | 1.55 | $2.13 \times 10^{-2}$ | 0.005 | 0.458 | $0.88 \times 10^{-3}$ | 1.24 | 33.2 | $1.89 \times 10^{-1}$ | 3.3 | 15 | M | B | - |
| -2 | 2.18 | $1.90 \times 10^{-2}$ | 0.005 | 0.374 | $0.88 \times 10^{-3}$ | 3.74 | 32.3 | $1.37 \times 10^{-1}$ | 2 | 4.8 | B/M | B | + |
| -3 | 3.10 | $1.34 \times 10^{-2}$ | 0.005 | 0.388 | $0.88 \times 10^{-3}$ | 8.70 | 29.6 | $1.71 \times 10^{-1}$ | 2.85 | 13 | B/M | B | + |
| -4 | 2.131 | $2.69 \times 10^{-2}$ | 0.005 | 0.286 | $0.88 \times 10^{-3}$ | 1.39 | 35.0 | $6.44 \times 10^{-2}$ | 0.4 | 2.5 | B/M | M | + |
| -5 | 2.77 | $1.97 \times 10^{-2}$ | 0.002 | 0.292 | $0.88 \times 10^{-3}$ | 2.56 | 32.6 | $7.47 \times 10^{-2}$ | 0.83 | 5.2 | B/M | M | + |
| -6 | 1.8725 | $2.19 \times 10^{-2}$ | 0.005 | 0.380 | $0.88 \times 10^{-3}$ | 2.59 | 33.4 | $1.29 \times 10^{-1}$ | 1.8 | 4.1 | B/M | M | + |
| -7 | 2.25 | $1.54 \times 10^{-2}$ | 0.005 | 0.380 | $0.88 \times 10^{-3}$ | 7.15 | 30.7 | $9.51 \times 10^{-2}$ | 1.1 | 6.3 | B | M | + |
| III-1 | 0.1 | $2.00 \times 10^{-2}$ | 0.07 | 0.052 | $0.17 \times 10^{-3}$ | 4.82 | 42.7 | $5.39 \times 10^{-3}$ | 0.28 | 3.0 | B | B | + |
| IV-1 | 1.55 | $1.8 \times 10^{-2}$ | 0.015 | 0.52 | $0.70 \times 10^{-3}$ | 6.09 | 30.8 | $2.47 \times 10^{-1}$ | 4 | 18.0 | B/M | B | + |
| -2 | 0.56 | $4.42 \times 10^{-2}$ | 0.0064 | 0.387 | $0.70 \times 10^{-3}$ | 0.26 | 37.8 | $9.08 \times 10^{-2}$ | 1 | 5.8 | B/M | M | + |
| -3 | 0.427 | $4.57 \times 10^{-2}$ | 0.0026 | 0.296 | $0.70 \times 10^{-3}$ | 0.065 | 38.1 | $5.22 \times 10^{-2}$ | 0.28 | 3.0 | M | M | + |
| V-1 | 0.49 | $3.6 \times 10^{-2}$ | 0.013 | 0.38 | $0.84 \times 10^{-3}$ | 0.62 | 34.8 | $8.62 \times 10^{-2}$ | 1 | 5.8 | M | B | - |
| VII-1 | 0.175 | $3.5 \times 10^{-2}$ | 0.013 | 0.163 | $1.2 \times 10^{-3}$ | 0.35 | 31.8 | $8.15 \times 10^{-2}$ | 0.83 | 5.2 | M | B | - |

Table 2 Comparison of predicted and observed channel pattern of laboratory channels

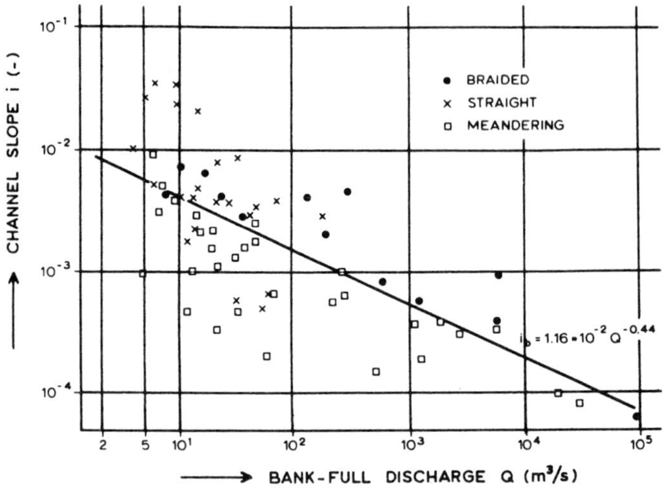


Figure 1 Classification diagram proposed by Leopold and Wolman.

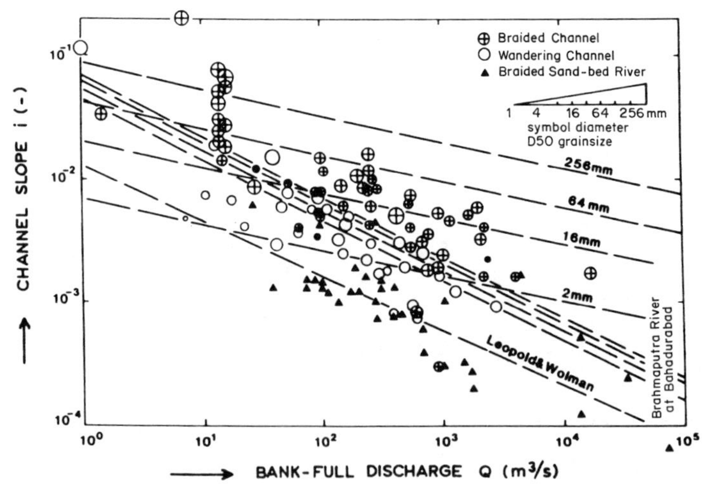


Figure 2 Discharge-gradient-grainsize plot of braided rivers (after Ferguson, 1984).
Note: the drawn lines are not relevant here.

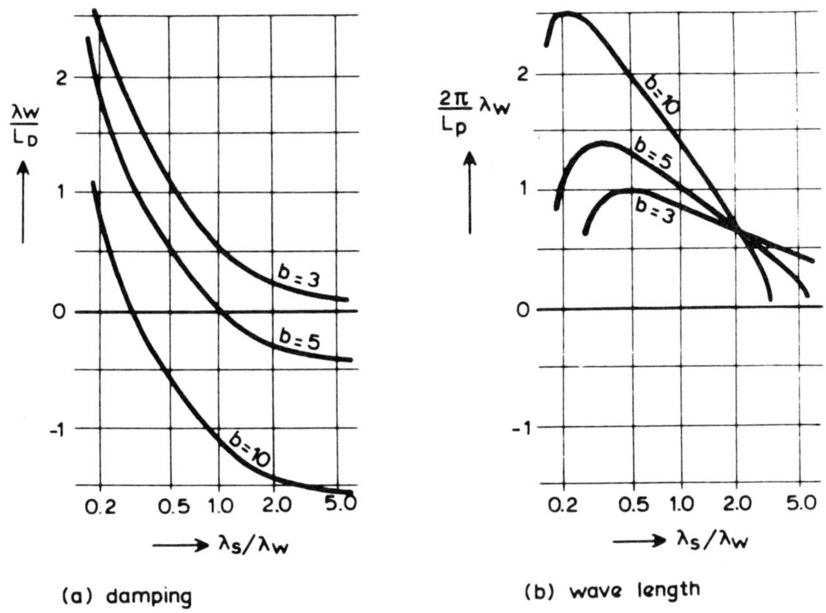


Figure 3 Damping and wavelength of perturbations in the linear analysis versus $\lambda_s/\lambda_w$.

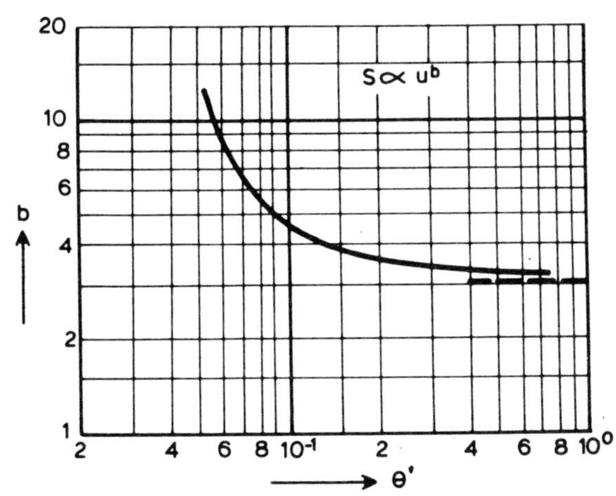


Figure 4 Approximate relationship between b and θ'.

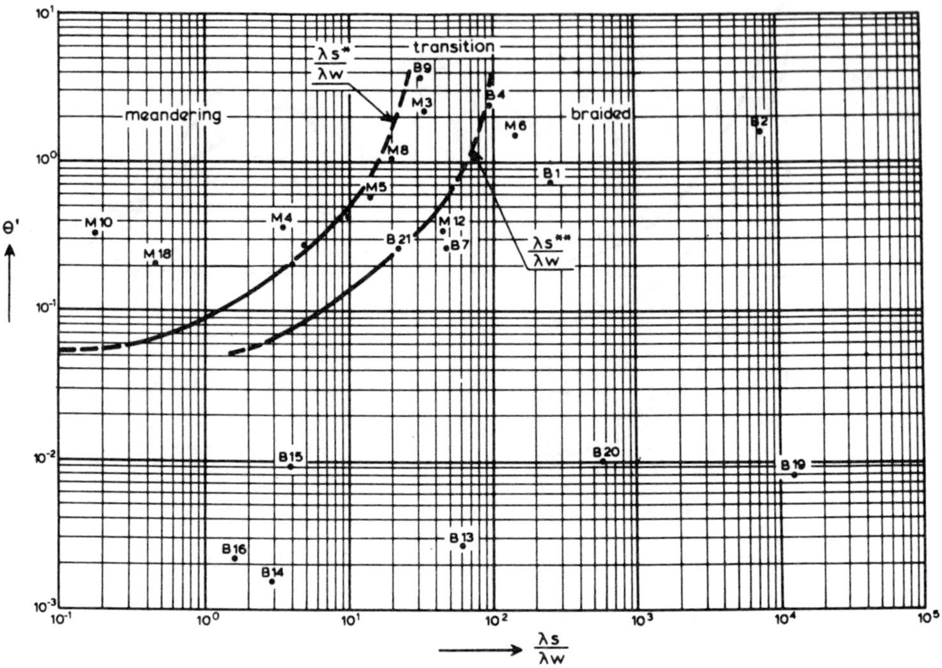


Figure 5 Classification diagram based on θ' and $\lambda_s/\lambda_w$.

International Conference on
RIVER REGIME
18-20 May, 1988

PAPER C4

# ABOUT EXTREMAL HYPOTHESES AND RIVER REGIME

by Alberto Lamberti
University of Bologna,
viale del Risorgimento 2, 40137 Bologna, ITALY

Summary: A lemma is proposed, useful to prove the equivalence of different variational formulations of related problems. Extremal hypotheses about regime alluvial channels are then examined and an equivalence class between them is pointed out, containing the minimum stream power and the maximum transport principles. Evidence is produced of contraddictions between state- ments derived from this class of hypotheses and well established opinions about regime channel. The internal consistency of the theory of minimum dissipation rate is examined and it is shown that it can't be derived from the principles of mechanics. The physical basis of extremum hypotheses about alluvial channels is analyzed and a functional is proposed whose minimum should be equivalent to transversal equilibrium of the channel section.

## 1. Introduction

Since it became clear that the classical (1-D) hydraulic or fluid-mechanic approach to alluvial channel dynamics description gave rise to a theory that needed some closure hypothesis beyond the resistance and sediment transport relations, many researchers, in order to avoid the use of the empirical relations named as 'regime theory' of alluvial channels, were tempted to close the theory with an extremal hypothesis. The temptation was made stronger by the fact the regime theory was formulated as a complete set of relations, among which it was not clear how to separate two relations that could be thought as a substitute of the resistance and transport relations and one that could be taken as an empirical closure of the hydraulic theory.

Various extremal hypotheses were proposed, some of which, as the Minimum Energy Dissipation Rate (MEDR), were supported as general laws of mechanics, similar or related to the enthropy law. But, as poor attention was paid to prove the equivalence of the different hypotheses or to define the set of the associated conditions necessary to produce their equivalence, and as the supporters were looking for favourable cases and passing over the existence of any counterexample, the supposed rational approach was in effect far away from his goal.

The trend begun to change about '80. I wish to mention in this sense the two papers of Parker (1978), which posed the mechanical problem of the tranverse equilibrium of the alluvial section, showing all the main physical interacting processes: gravity pull towards the channel center, pick-up, sedimentation and lateral dispersion associated to motion. White & Bettes & Paris (1982) posed and solved the problem of equivalence between the Minimum Stream Power (MSP) and the Maximum Sediment Transport (MST) hypothesis; this is, as far as I know, the first proof of the equivalence of hypotheses. Davies & Sutherland (1983) have conducted an exaustive examination of the different hypotheses, (the interested reader may find in it a complete discussion of the physical foundations of the hypotheses) showing the rough equivalence of most of them in the common situation of given water discharge.

International Conference on River Regime, edited by W.R.White.

The results become contradictory in the case of given water stage, where results derived from Maximum Friction Factor (MFF) hypothesis agree with the experience in contrast with those derived from the MSP and MEDR hypothesis. In (1983) I proposed the lemma, the essential part of the equivalence proof of the m-hypotheses and the narrowing-slope argument against these. They are represented in this occasion in a more complete form. Griffith note (1984) and the discussion about the notes [3] and [37] presented at the 2nd gravel-bed rivers workshop give evidence of the topical interest of the problem.

The present paper developes three arguments: -equivalence between the extremum hypotheses leading to the definition of the equivalence class that I call m-extremal, -arguments against the minimum dissipation rate principle and the m-extremum hypotheses as the narrowing-slope argument, -the proposal of a functional of the mixed type deriving from the transverse equilibrium theory.

## 2. The equivalence lemma

.. Let (n-m) independent relations between n variables be given

(1) $f_k( v_i ) = 0$ for i=1 to n and k=m+1 to n

and a partition of the n variables V in 2 sets, say V1 and V2, containing m and n-m variables. The condition 'm-1 variables in V1 being given, the remaining one is stationary (extremum)' defines a unique relation among the variables independent from the particular choice of the free variable in V1. The stationarity relation is equivalent to:

(2) $\det[ f_{k,l}] = 0$ for k and l = m+1 to n ..

This relation linked to the original ones (1) and to the given values defines normally all the n variables V.

With no loss of generality we may look at the partition between the 1st m ones (V1) and the remaining (V2) and, also, that the free variable in V1 is the 1st. Using the method of Lagrange multiplier, we define the functional L as:

$$L = \mu_1 \cdot v_1 + \sum_{j=2,m} \mu_j \cdot (v_j - c_j) + \sum_{k=m+1,n} \mu_k \cdot f_k(v_i)$$

The constrained stationarity of the 1st variable is equivalent to the stationarity of L all the variable beeing free and the 1st multiplier being not zero. As the derivatives of L with respect to the variables V are:

$$\mu_j + \sum_{k=m+1,n} \mu_k \cdot f_{k,j} = 0 \qquad \text{for } j=1 \text{ to } m$$

and

$$\sum_{k=m+1,n} \mu_k \cdot f_{k,l} = 0 \qquad \text{for } l=m+1 \text{ to } n$$

In order that the above written system of linear equations is satisfied by a set of admissible multiplier (the 1st must not be 0), it is necessary that the determinant of the matrix:

(3)
$$\left| \begin{array}{c:c} I & 0 \\ \hdashline f_{k,j} & f_{k,l} \end{array} \right|$$

is 0, or, as is trivially equivalent as a consequence of the structure of the matrix, that the determinant (2) is zero.

We would arrive to the same conclusion considering the system of relations:

(4)
$$\begin{array}{lll} dv_1 = 0 & & \text{: stationarity of the 1st variable} \\ dv_j = 0 & \text{for } j=2 \text{ to } m & \text{: constancy of data} \\ df_k = 0 & \text{for } k=m+1 \text{ to } n & \text{: satisfaction of relations (1)} \end{array}$$

and looking for a non trivial solution (not all $dv_i=0$) of this system.

As it is clear from the equivalent relation (2), the stationarity relation depends only on the original ones (1) and on the partition of the set of variables, being the conditions of stationarity (extremum) and constancy in fact equivalent.

## 3. An example

A channel section and a current being given, we may consider the relations defining the mean velocity and the specific head and the geometric relation between water stage and water area.

$$
(5)\qquad \begin{aligned} Q - A\cdot U &= 0 \\ H - (Y+U^2/2g) &= 0 \\ A - S(Y) &= 0 \end{aligned}
$$

The relevant variables involved are Q,H,Y,U,A . If we divide the set into {Q,H} and {Y,U,A} the above stated lemma support the equivalence of the two conditions defining the critical current, i.e. H is minimum for given Q or Q maximum for given H. This equivalence has nothing to do with the physics nor anybody believes that every current is critical, even if channel with almost critical slope are very frequent in mountain areas. If a physical meaning is requested for the condition {dQ,dH}=0 with {dA,dY,dU}≠0, I think every hydraulician would interpret it as the necessary condition for the exixtence of stationary waves in the channel. No law of motion is actually included in (5), but we refer to them when we look for a physical interpretation of the extremum condition, as for stationary flows {dQ,dH}=0 represents the mass and momentum balance, disregarding the small (over a wavelength) and counteracting effects of friction and channel slope.

## 4. The case of regime channels

In the case of a regime channel, disregarding any variation in sediments density and supposing the motion to be almost uniform, 3 relations hold between the seven variables Q,X,J,D,P,R,U , i.e.:

$$
(6)\qquad \begin{aligned} Q - P\cdot R\cdot U &= 0 &&: 5 \\ J - F_J(D,R,U) &= 0 &&: 6 \\ X - F_X(J,D,R,U) &= 0 &&: 7 \end{aligned}
$$

It is generally supposed that 3 of the 4 variables Q,X,J,D are given (one is not given depending on the time scale of the regime we are considering) so that it seems natural to divide the set af all variable into the 4 above mentioned characterizing the catchment area and the channel reach, and the 3 remaining P,R,U characterizing the local stream. No matter what its physical meaning could be, we are free to consider the stationarity condition of one, no matter which, of the 4 macrocharacteristics of the alluvial stream. The matrix of the derivatives of the relations (6) has the following zero ( ) non-zero (*) structure :

| | Q | X | J | D | : | P | R | U |
|---|---|---|---|---|---|---|---|---|
| 5 | * | | | | : | * | * | * |
| 6 | | | * | * | : | · | * | * |
| 7 | | * | * | * | : | · | * | * |

The stationarity (or extremum) condition, as a consequence, is equivalent to:

$$
(7)\qquad F_{J,R}\cdot F_{X,U} = F_{J,U}\cdot F_{X,R} \quad .
$$

This relation combined with relations (6) and the data determines normally all the 7 variables.

The 4 equivalent statements of the extremum hypothesis are:

1- given {X,J,D}, an alluvial channel adjusts {P,R,U} so that Q is minimum;
2- given {Q,J,D}, an alluvial channel adjusts {P,R,U} so that X is maximum;
3- given {Q,X,D}, an alluvial channel adjusts {P,R,U} so that J is minimum;
4- given {Q,X,J}, an alluvial channel adjusts {P,R,U} so that D is maximum.

Some of these statements are well known, as the 2-nd and the 3-rd which are respectively the MST principle (Singh 1961) and the principle of Minimum Slope (MS), tightly related to the MSP principle.

Some other equivalent extremum hypotheses can be derived from these tranforming the coordinates in the space spanned by the variables Q,X,J,D: the MSP principle (Chang 1980) which derives from the substitution of $Q \cdot J$ to Q or J, the MEDR principle (Brebner and Wilson 1967) which derives from the substitution of a combination of Q,X and J to one of these. Of some interest could be also the substitution of X or D with a relation between them characterizing perhaps more realistically the arrival and transport of sediments than X or D separately.

If the resistance and load relations make no reference to P, other formally equivalent extremum statement may be derived, because interchanging P with Q does not change the relevant part of the matrix. This cases seems not worth to be considered both because the situation of varying Q and fixed X doesn't seem realistic and because a secondary effect of the shape of the section (of the aspect ratio P/R) on the resistance and load relations should not be excluded.

Some other extremum hypoteses may be found in literature that are not equivalent to the set of the above mentioned. The best known is the principle of Minimum Unit Stream Power ($U \cdot J$) (MUSP) (Yang and Song 1979); another is the MFF (proportional to $R \cdot J/U^2$) principle (Davies and Sutherland 1983). They are characterized by a functional supposed to be extremum that combines a macro variable (in the mentioned cases J) with some local variables (in the case U and, in the second case, R). If this case is compared with the case of the extremum of the macro variable in it (J) the corresponding line of the matrix of derivatives is changed by the substitution of 1 with the appropriate derivative in the upper left diagonal block (which remains diagonal) and adding some terms in the upper right block altering the zero-non_zero structure of the matrix and so the relevant determinant. The two hypotheses are not even equivalent between them, as may be easily seen from the structure of the resultant matrix of derivatives.

As a consequence of this discussion about the equivalence of extremum hypotheses, I shall refer to the set of the extremum hypotheses of non mixed type referring to the partition between the macro variables (Q,X,J,D) and the local variables (P,R,U) as 'the' m-extremum hypothesis for alluvial regime channels.

Before leaving the formal discussion it is worth to evidence an exception to the normal existence of the solution for all the variables under the m-extremum hypothesis. In fact, if both the resistance and load relation are of the power type the determinant (7) results equal to

$$\frac{X \cdot J}{R \cdot U} \cdot \begin{vmatrix} \beta jr & \beta ju \\ \beta xr & \beta xu \end{vmatrix} ,$$

where β are the exponents of the variables R and U in the two relations giving J and X. The determinant is identically or never zero depending on the set of exponent assumed; in any case, either the solution doesn't exist or it is not unique.

As practical relations are not always of the power type but not far from that, it may be expected that the minimum problem is ill conditioned and one should mind about the numerical procedure in order to obtain the correct minimum. This could be the reason why the above stated equivalence is not always supported by numerical calculations.

## 5. Consequences of the m-extremum hypothesis

An obvious consequence of the m-extremum hypothesis is that however you may alter the section from the regime one, without altering the resistance and load relations, either you will decrease the sediment transport or the sediment dimension or you will increase the channel slope.

In practice we can consider the effect of narrowing the channel, without introducing artificial roughness (as groins would be). The consequence of the m-extremum hypothesis is in contrast with the conclusions of several authors (e.g. de Vries 1974), who stated that, water and sediment discharge remaining constant, narrowing of the channel from $B_0$ to $B_1$ would produce a variation of the slope at regime given by:

$$(8) \qquad J_1/J_0 = ( B_1/B_0)^{(\beta-3)/\beta} \qquad \text{if} \qquad X \cdot Q \sim B \cdot U^{\beta}$$

As, referring to almost any sediment transport formula, $\beta$ is greater than 3, narrowing the channel induces a decrement of the slope. The result has never been contradicted as far as I now.

We may consider also some real cases. One is the channeling started in 1844 of the Rhone river near Lyon within the Miribel Channel. The course was initially braiding and the talweg length of the reach covered by the works was 25 km; the river has been constricted in a channel 17 km long and 100 m wide. The fall over the reach was initially 15 m; as a consequence of the works an erosion of the uptream end occurred reaching 3 m and a deposit at the downstream end reaching 5 m; the fall was still reducing at a rate of 10 cm/year in 1952, 50 years after the end of works; after this date new works have substantially modified the evolution of the course. The slope of the new channel at grade would have been al least 25% less than the slope of the natural channel, as this was the reduction occurred in 1952 accounting for the reduced length of the channel.

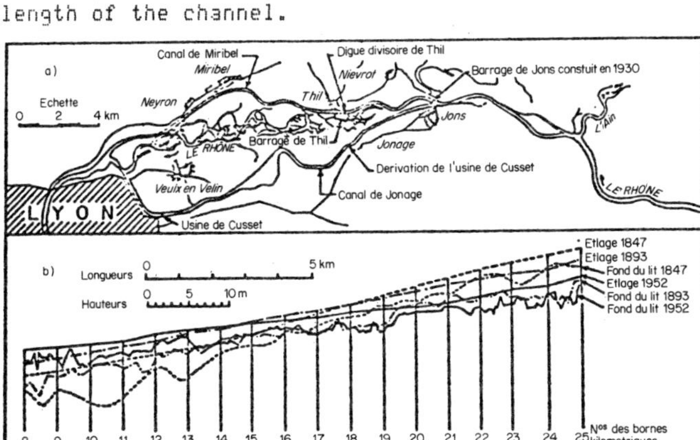


Fig. 1 The Miribel channel

A second case is the channel regulation of the river Rhein between Basel, Strasbourg and Sonderheim. The works begun in 1840 and produced a narrowing and shortening of the channel (see the monograph on the Rhein river [1]). The erosion of the upstream end was remarkable reaching 7 m in Chalampé. Other training works of the lower reach of the river using submergible groins seem to have produced no evident alteration of slope.

The training of the Po river from the confluence of the Adda to the confluence of the Mincio, obtained by a system of longitudinal defence and without altering the length of the reach, was followed by a strong erosion (3m) of the upstream end of the reach. The contemporary contruction of a barrage on the river at the upstream end (Isola Serafini) and the increase of the escavation from the bed of the river, does not allow to attribute all the erosion to the training works.

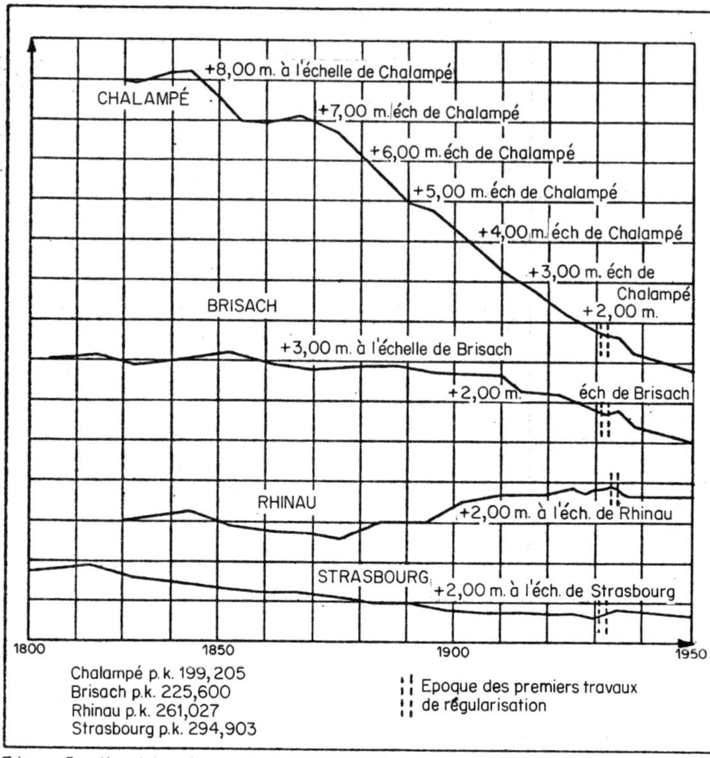


Fig. 2 Vertical evolution of the bed of the river Rhein

Real cases are generally the effect of more causes, and especially training works don't alter only the width of the river, but also the length or the sinuosity or the equivalent roughness, so that it is almost impossible to separate the effect of the narrowing; anyway no case is known of narrowing a graded river that has produced an increment in the grade slope.

## 6. Discussing the principle of minimum energy dissipation rate

As a contrast between well established opinions about the principles of analysis and about the actual behaviour of alluvial streams has been pointed out, the reliability of the MEDR principle could be profitably analyzed, as among the statements in the m-extremal class it is generally supported as the most fundamental.

According to Yang and Song 'the general theory of minimum rate of energy dissipation can be stated as: 'For a closed and dissipative system in stable dynamic equilibrium condition, its total rate of energy dissipation is at its minimum value which is compatible with the constraints applied to the system'.'

The statement is unfortunately not generally true, but happens to be true in some case, as was yet clearly shown by Howard (1972) and by Davies and Sutherland (1983). Howard, while examining the case of Couette flow, arguees that the principle is certainly false in cases where a choice between possible solutions exist, as, when the flow is laminar the dissipation rate is notoriously minimum among all velocities distributions satisfying the continuity equations but there is no choice, as this is the only solution satisfying equilibrium also; and at higher Reynolds numbers, when bifurcations occur, the actual dissipation rate is higher than in the case, always possible, of laminar flow: 'the observed occurrence of turbulence does rule out the hypothesis of minimum energy dissipation' is Howard conclusion. He also remarked that the actual dissipation rate behaves in the same way as the upper bound he demonstrated to exist, suggesting to propose a maximum dissipation hypothesis.

I want to show a couple more of simple counter-examples to the MEDR principle, taken from the field of every days Hydraulics.

For the first we shall consider a pipe-network. Let $h_r(Q)$ be the head loss function of a branch r in the network; we may define the functional L as:

$$(9) \qquad L = \sum_r L_r(Q_r) \qquad \text{with} \qquad L_r(Q) = \int_0^Q h_r(x)\cdot dx$$

The rate of energy dissipation is similarly defined as the sum over the branches of the network of

$$(10) \qquad \phi_r = h_r(Q_r)\cdot Q_r$$

The difference between L and $\phi$ is given by the substitution of the simple product in $\phi$ with an integral in L. It is easy to see that the $\{Q_r\}$ that minimizes L over all admissible discharges (satisfiing the node continuity equations) is the well known solution of the network problem, while thatone minimizing $\phi$ need not to be so.

In fact, the space of admissible discharges may be described as

$$(11) \qquad Q_r = Q_{0r} + \sum_m E_{rm}\cdot Q_m$$

where $Q_{0r}$: any set of discharges satisfiing the node continuity equations;

$E_{rm}$: 0,+1,-1 according to wether the branch r is not part of the loop m, or, wether it is part, if its direction agrees or not with the loop direction;

$Q_m$ is a set of (arbitrary when describing admissible discharges) loop discharges.

The minimum of L is reached when:

$$(12) \qquad \partial L/\partial Q_m = \sum_r E_{rm}\cdot h_r = 0 \quad .$$

We recognize in this equation the head balance over the loop, which is the physical condition to be imposed besides continuity. It is easy to see that if each head loss function is of power type with the same exponent, i.e. if

$$h_r(Q) = R_r\cdot Q\cdot|Q|^{n-1} \quad , \text{then} \qquad L = \phi/(n+1)$$

and the two minimization problems have in effect the same solution. But this is not the case if in some pipe the flow is laminar and in some other is turbulent, or if virtual pipes are included in the network joining couples of reservoirs. In these cases the two solutions are actually different and that one minimizing $\phi$ is not correct.

Another counter-example may be given by a ball or a trickle of water descending a surface as that described in fig 3. We can observe three possible trajectories, thatone in the middle having milder slope, and the two at the sides having greater slope. Slope is proportional to the dissipation rate. The middle trajectory is obviously stable in the upper part of the sloping surface, while lateral ones are stable in the lower part. At least in this case slope and stability are independent. It is worth to observe that the stability of a trajectory depends on the force components which, liing in the surface, are orthogonal to the equilibrium trajectory, and may in this way draw the body back to the trajectory or not, while the dissipation function depends on the component along the trajectory.

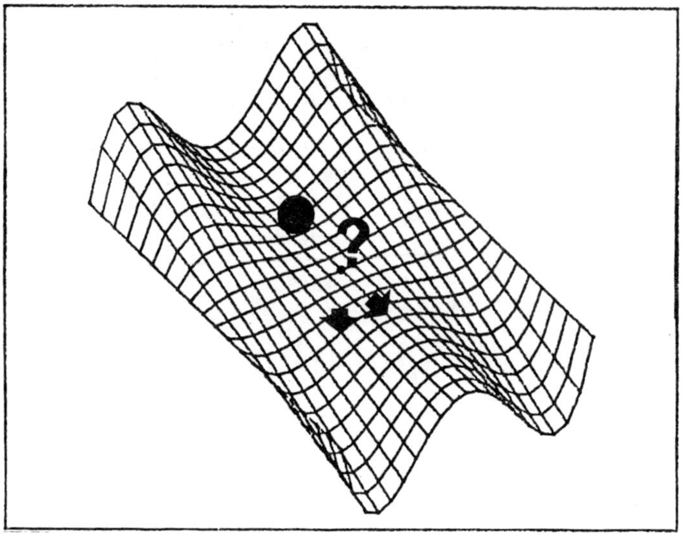

Fig. 3 Which is the trajectory the ball shall follow?

We may partially conclude that the MEDR principle is not a general principle of mechanics, as it fails to be true at least when 1-the stress-velocity relation is nonlinear or 2-the motion is multidimensional (the direction of forces is not the direction of motion).

## 7. Physical interpretation of the extremum hypotheses

We shall look before to any physical interpretation of the m-extremum hypothesis.

The principle is generally supposed to reflect a characteristic of nature (and/or of man); it is also known as the principle of minimum effort. It relies on the opinion that, in the case of two or more alternatives processes satisfying the same conditions and doing the same job, the one that spend the least effort in doing the job shall be able to use the spared effort and will ultimately prevail over the other. The essential features seems to be the competition between the tenders, the ability to use the spared effort and the finiteness of the resource divided between the tenders that will eventually cause the disappearence of all but one tender.

This principle may certainly be applied to the case of two or more parallel channels: if the load is equally divided between the two, the more efficient (the one able to carry the maximum load X with the same slope J) will be eroded and the less efficient will be silted, and eventually only the most efficient shall be there. On this basis we may correctly argue that a stable channel is at the root of a tree of choices performed according to efficiency and that it is more efficient than any of the silted ones, but not that the channel is the most efficient we may conceive. This argument seems also not to be applicable to the choice between different shapes of a single channel because of the absence of actual competition.

A possible interpretation of the m-extremum hypothesis could be given by the assumption of the existence of a stationary perturbation of the uniform flow; the stationarity of the perturbation could be related to the presence of an unerodible (or almost unerodible) obstacle. This interpretation, even if applicable to some real case, seems not pertinent to a pure alluvial channel.

The MUSP principle, although not equivalent to the m-extremum, is generally supported by the same qualitative considerations as the MSP or MEDR. The MFF principle is generally sustained by some kind of faith on the explosive on invading nature of disorder, meaning that once some level of disorder as turbulence or sedimentary forms is set up over a smooth flow it grows as far as possible; the friction factor, as ratio between friction and mean velocity or as a measure of the relative turbulent intensity, appears as a significative measure of the level of disorder. The MFF principle looks quite adherent to the physics of a turbulent fluid flow without any suspension load; it seems reasonable also when bedforms dimensions are increasing respect to flow intensity, as bedforms and turbulence will increase untill the particles mobility drops under the level necessary for the assumed transport. In the case of transition or upper regime the hypothesis seems no more adherent to reality as the increasing turbulence level cause particles suspension and a decrease of bedforms height and of disorder.

## 8. The transverse equilibrium

A substantially different approach to the regime theory of alluvial channel is followed by Glover & Florey (1951), Lane (1955), Simons (1957), Lundgren & Jonsson (1964), Hirano (1973), Parker (1978 and following papers) and Ikeda (1981 and following papers). The leading idea is that channels are stable as consequence of a static or dynamic (statistical) transverse equilibrium of the bed particles and therefore this equilibrium should be the object of the analysis.

Assuming an incipient motion all over the bed and banks the first authors define an equilibrium shape of the channel (at dominant water stage). The naive interpretation of the threshold theory of alluvial channels leads, however, to the stable channel paradox: tranport of bed material is incompatible with a stable width, as particles moving on the banks will fall toward the channel center and the channel must widen until a no_transport condition is reached. In any case the threshold theory, as commonly stated, do not define a unique channel shape as horizontal parts may be added everywhere in the usual transverse profile and parts may be also deleted, leaving the defining equation satisfied.

Parker's (1978) papers seem particularly devoted to come over the paradox, showing the effect of lateral dispersion of suspended motion (part 1) and the hiding effect of convexity of the bed and of the turbulent momentum tranfer towards the banks (part 2).

The problem is separated in two parts: 1-how stress is distributed along the bank and bed surface and 2-what should the cross section be in order that pick-up rate and sedimentation may be equal all over the bank and bed surface. The first problem, if the section Y(y) is taken as given, is reduced accordind to Lundgren & Jonsson to a 2-nd order self adjoint o.d.e. relating the stress depth $\delta = \tau/\rho g J$ to the actual depth and the lateral bed curvature (eq. 13 of Parker 1978 part 2); this kind of equation is notoriously equivalent to a variational problem (cmp. e.g. Mikhlin 1957).

It is not difficult to proof that also the 'partial' equilibrium problem, i.e. what should be the the pick-up distribution for a given tranverse displacement-dispersion distribution in order that sedimentation shall be equal to pick-up and hence the section remain stable, may be stated as an extremum problem. In fact pick-up and sedimentation must be equal to the eigenvector of the displacement-dispersion operator $\mathfrak{D}$ (the operator mapping pick-up to sedimentation as a consequence of the motion of particles) relative to the eigenvalue 1, i.e. $\mathfrak{D}\mathcal{E} = \mathcal{E}$ . This eigenvalue is the greatest and the associated eigenvector is the only everywhere positive and physically acceptable solution of the eigen-problem. As a consequence the partial equilibrium problem is equivalent to look for the maximum of the Rayleigh ratio $(\mathcal{E},\mathcal{D}\mathcal{E})/(\mathcal{E},\mathcal{E})$.

Unfortunately the full problem, i.e. find a section shape Y(y) and the corresponding stress distribution so that the pick-up distribution deriving from the stress distribution satisfy to 'partial' equilibrium (relating to the unknown section shape), is essentially nonlinear as both the displacement operator and the pick-up distribution depends on the depth-stress distribution. So it seems difficult to proof that the problem is exactly equivalent to some variational problem, nor this is strictly necessary as solutions may be found directly from the above mentioned relations, once the pick-up and the displacement-dispersion are related to local stress and tranverse bed slope. I will only show how the problem could be stated in a very symplified context, in order to make evident the relevant parameter and a possible structure of the functional.

Let us suppose that the transverse bed slope is small so that its square may be disregarded respect to 1 all over the channel width; as P don't differ substantially from y and so the depth from the normal depth, the depth-stress Lundgren & Jonsson's equation (Parker 1978 part 2 equation 13) becomes:

(13) $$\delta - d[Y^2 \cdot \Psi \cdot d\delta/dy]/dy \quad = \quad Y + 1/2 \cdot Y^2 \cdot d^2Y/dy^2$$

with the boundary condition $\delta=0$ at both channel sides. As $\Psi$ varies from 0.5 when Y/k=10 to 0.8 when Y/k=1000, the diffusion of momentum (2nd term in the left hand side) and the bed curvature have cooperating effects of the same order of magnitude; the functional, whose minimum defines the solution of the problem, is:

(14) $$\mathcal{J} = \int_B \{[\delta^2 + Y^2 \cdot \Psi \cdot (d\delta/dy)^2] + -2 \cdot [\delta \cdot Y + \delta \cdot Y \cdot (dY/dy)^2 + \tfrac{1}{2} \cdot Y^2 \cdot d\delta/dy \cdot dY/dy]\} \cdot dy$$

The actual stress distribution is somewhat flatter than Y(y) in the central part of the channel and steeper near the banks where Y is small. In any case the approximation $\delta=Y$ seems suited to a qualitative analysis for very wide channels as alluvial ones are; this is equivalent to neglect transverse derivatives appearing in the above written functional.

I shall follow in the main lines the reasoning proposed by Parker in the 1978 part 1 paper. Let us suppose the moving sediments divided in two phases, the suspended phase which is laterally dispersing and the bedload phase which is displaced towards major depths. We shall consider the mass balance for suspended load and for total load respectively:

(15) $$d\mathcal{F}_L/dy \;=\; \mathcal{E} - \mathcal{S} \qquad , \qquad \mathcal{F}_L + q_{BL} = 0$$

Let the dispersive flux of suspended sediments be proportional to the lateral gradient of the total-in-vertical suspended load, and let the lateral bedload transport be proportional the longitudinal tranport and the lateral bed slope:

(16) $$\mathcal{F}_L = -\epsilon_L \cdot d\zeta/dy \qquad , \qquad q_{BL} = \epsilon_B \cdot dY/dy$$

with $\zeta = \int_Y C(z)dz \simeq \epsilon_z \cdot C(0)/w$

The diffusion coefficients $\epsilon_L$ and $\epsilon_z$ are assumed proportional to the product of the friction velocity ($u_*^2$=g.Y.J) times a tipical length wich ranges from the surface roughness for pure bed load to depth for dominant suspended load. The coefficient $\epsilon_B$ is taken equal to the longitudinal bedload transport times an apparent friction coefficient about 0.58. Pick-up and sedimentation are taken respectively equal to:

(17) $$\mathcal{E} = w \cdot E(\vartheta) \qquad , \qquad \mathcal{S} = w \cdot C(0) = w^2 \cdot \zeta/\epsilon_z \; .$$

The last equality derives from the assumption of a constant vertical diffusivity and of the related exponential concentration distribution.

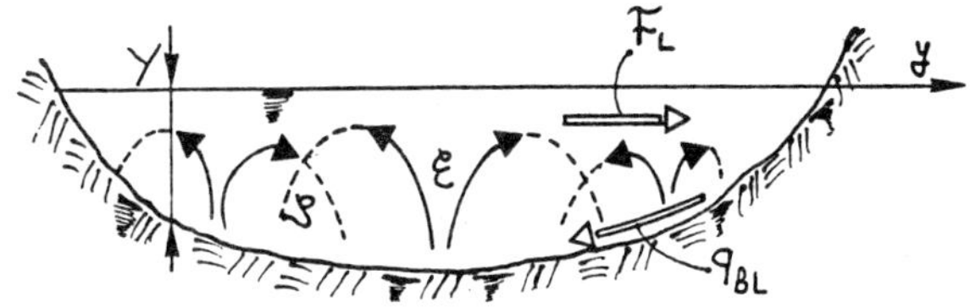

Fig. 4 A sketch of the section and tranverse sediment movement

As a consequence of the total load mass balance (15.2), of the similar relatioships (16) between the two tranverse fluxes appearing in it and the gradients of $\zeta$ and Y, as $\epsilon_B$ and $\zeta$ are both 0 untill bedload transport begins,

$$(18) \qquad \zeta = \int_0^Y \epsilon_B/\epsilon_L \cdot dY \quad \text{, or in nondimensional form} \quad \zeta/D = F(\vartheta)$$

$$\text{where} \quad F(\vartheta) = \int_0^{\vartheta} 4.5 \cdot \phi(\vartheta)/\vartheta^{3/2} \cdot (\Delta \cdot \vartheta/J)^{1-m} \cdot d\vartheta$$

and m ranges from 0 in the case of pure bed load to 1 in the case of dominant suspension. The combination of the suspended sediment balance and the relations (17) for pick-up and sedimentation give rise to the equation:

$$(19) \qquad d(D\cdot\phi(\vartheta)\cdot dY/dy)/dy = [-c_w \cdot E(\vartheta) + c_w^2 \cdot (D/Y)^m \cdot F(\vartheta)/\vartheta^{1/2}]/0.58$$

with the natural boundary conditons equivalent to $\mathcal{F}_L=0$ at the banks.

As Shields mobility index $\vartheta$ is proportional to Y, the equation is nonlinear, and no elementary variational problem is equivalent to it. But we may reasonably guess the shape of the Einstein tranport index $\phi$ distribution, so that it may be treated as a known function of y times the value at mid-channel $\phi_0$. A reasonable choice could be a parabolic one.

$$Y(y) = Y_0 \cdot \gamma(y)$$

$$(20) \qquad \phi(Y) = \phi_0 \cdot \varphi(y) \quad \text{, with } \phi_0 = \phi(Y_0)$$

The r.h.s. of (19), when the erosion and tranport formula are given, is a known function of the depth, say G(Y). Let us consider the functional:

$$(21) \qquad \mathcal{L} = \int_B \{[D\cdot\int_0^{Y_0} \phi(\xi)\cdot\xi\cdot d\xi]\cdot\varphi(y)\cdot(d\gamma/dy)^2] + [\int_0^{Y(y)} G(\xi)\cdot d\xi\,]\}\cdot dy$$

As the first variation of the functional, $\partial\mathcal{L} = \mathcal{L}(Y+u) - \mathcal{L}(Y)$, for small u is

$$(22) \qquad \partial\mathcal{L} = \int_B \{D\cdot\phi_0\cdot Y_0\cdot\varphi\cdot(d\gamma/dy)\cdot(du/dy) + G(Y)\cdot u\}\cdot dy \quad ,$$

the condition for a minimum of $\mathcal{L}$ ( $\partial\mathcal{L}=0$ for every small u ) is equivalent to the week form of eq. (19) with the approximation (20). If the funtion G(Y) is linearized as

$$G(Y) = a.\alpha(y) + b\cdot\beta(y)\cdot Y(y)$$

and the shape coefficients deriving from the integral of the shape functions over the width and of $\phi$ over the depth are disregarded, the functional looks like:

$$(23) \qquad \mathcal{L} \simeq [D\cdot\phi_0\cdot Y_0^2]/B + [J\cdot b\cdot Y_0^2/(\Delta\cdot D) + a\cdot Y_0]\cdot B$$

It is worth to observe that the functional is a combination of local variables, the width and the depth, and global variables, the channel slope, the sediment diameter and also the load through the mobility dependent parameters $\phi_0$, a and b ; as it is of the mixed type it should not be equivalent to the m-extremum as no evident cancellation occurs due to the load and resistance relations.

The above approach rely substantially on the mechanics of sediment motion, also if the result seems to be not very sensitive to the entity of motion, as scaling of the pick-up function and of the tranverse mean sediment transport and dispersion do not modify the solution and the section shape. On the contrary it is essential to the theory the absence of any curvature of the channel and the related uniformity of motion and the stationarity of the discharge and the associated water stage.

## 9. Conclusions

The set of equivalent extremum hypotheses containing MSP and MST, which I call the m-extremum, is shown to be in conflict with the commonly observed fact that a channel reacts to a restriction of the natural width reducing the regime slope.

The minimum unit stream power hypothesis, although of the mixed type and not exactly equivalent to the above mentioned ones, resembles so much to the MSP and the MEDR and rely on so similar considerations, that it shall probably leed to the same conflict.

The MFF hypothesis seems to be physically more founded in the case of growing bedforms, as the bedforms, which are a product of sediment movement, reduce the stress on the grains and the sediment transport while they grow, water discharge and channel slope remaining constant. The hypothesis, for the same reasons, seems not to apply to the upper (decreasing bedforms and plane bed) regime.

Between the functional of the mixed type the one I proposed as corresponding to lateral sedimentary equilibrium of the section is probably not amazing because it may look complex to use in qualitative analyses. In any case it is not better and only somewhat worse than the original differential equation describing the tranverse equilibrium.

In my opinion, there is no need of an extremum hypothesis, but there is need for more deep analyses of the tranverse equilibrium. As an example, in the Parker theory, which I followed except some detail, the discharge and the sediments are taken as unique, while in nature they are highly variable. It is reasonable to suppose that the variability of these alluvial flow characteristics may have some effect which we cannot account for by properly choosing a typical discharge and a typical sediment diameter. Similarly the flow is supposed uniform and secondary currents absent; straight reaches are rare and alternate bar are not well represented by a stochastic lateral motion of sediments; in any case even small channel curvature or low but horizontally extended bed forms may produce a proportionally high transverse sediment transport and modify substantially the section equilibrium. In this situation I fear that a supposed general principle, supported by estetical or simplicity more than physical motivations, may be obscuring and possibly misleading, as the actual underlying hypotheses are not made evident.


## ACKNOWLEDGMENTS

This study was supported by M.P.I. national research group 'Fenomeni di traspoto solido'.


## NOTATIONS

| | |
|---|---|
| $A$ | area of the water section |
| $B$ | surface width of the water section |
| $C$ | sediment concentration |
| $c_w$ | nondimensional fall verocity : $w/(g \cdot \Delta \cdot D)^{1/2}$ |
| $D$ | characteristic sediments diameter |
| $g$ | gravity acceleration |
| $H$ | specific head |

$J$ channel or friction slope
$P$ wetted perimeter
$Q$ liquid discharge
$q_B$ bedload trasport per unit width
$q_{BL}$ lateral bedload transport per unit length
$R$ hydraulic radius
$U$ mean water velocity
$X$ sediment load concentration
$Y$ water depth
$y$ transverse coordinate
$z$ vertical coordinate
$\mathcal{D}$ displacement-dispersion operator
$\mathcal{E}$ rate of bed erosion
$\mathcal{F}_L$ diffusive lateral flux of suspended sediments
$\mathcal{S}$ rate of sedimentation
$\delta$ stress depth ($\tau/\rho gJ$)
$\Delta$ relative sediment density
$\epsilon_L, \epsilon_z, \epsilon_B$ diffusion coefficients
$\xi$ total_in_vertical suspended load
$\vartheta$ Shields mobility factor
$\phi$ dissipation function, Einstein bedload factor

REFERENCES


[ 1] AA. VV.: La Houille Blanche, Fascicule no. 2, 1959
[ 2] AA. VV.: La Houille Blanche, Fascicule no. 2, 1964
[ 3] BETTES R. and W.R.WHITE: 'Extremal Hypotheses Applied to River Regime' in Sediment Transport in Gravel-bed Rivers edited by C.R.Thorn, J.C. Bathurst and R.D. Hey, John Wiley & Sons Ltd 1987
[ 4] CHANG H.H. and HILL J.C.: 'Minimum stream power for rivers and deltas' Journ. Hydraulics Division ASCE, Vol. 103 no. HY12, dec 1977
[ 5] CHANG H.H.: 'Geometry of rivers in regime' Journ. Hydraulics Division ASCE, Vol. 105 no. HY06, jun 1979
[ 6] CHANG H.H.: 'Stable alluvial canal design' Journ. Hydraulics Division ASCE, Vol. 106 no. HY05, may 1980
[ 7] CHANG H.H.: 'Geometry of gravel streams' Journ. Hydraulics Division ASCE, Vol. 106 no. HY09, sep 1980
[ 8] CHANG H.H.: 'Mathematical model for erodible channels' Journ. Hydraulics Division ASCE, Vol. 108 no. HY05, may 1982
[ 9] DAVIES T.R.H. and A.J. SUTHERLAND: 'Extremal Hypotheses for river behavior' Water Resources Res., vol. 19 no. 1, pp. 141-148, feb 1983
[10] DIPLAS P. and PARKER G.:'Boundary shear stress distribution and regime relations in straight channels' EUROMECH 215 Mechanics of sediment Transport in Fluvial and Marine Env., S. Margherita Ligure, sep 1987
[11] GHETTI A.: 'Forma delle sezioni degli alvei naturali e teoria del regime' in Proc. Seminar 'Moderne vedute sulla meccanica dei fenomeni fluviali', Bressanone, 1979
[12] GLOVER R.E. and FLOREY Q.L.: 'Stable channel profiles' U.S. Bur. Reclamation, Hydr. no. 325, 1951
[13] GRIFFITH G.A.: 'Extremal Hypotheses for River Regime: An Illusion of Progress' Water Resources Res., vol. 20 no. 1, pp. 113-118, jan 1984
[14] HIRANO M.: 'River bed variation with bank erosion" Proc. Japan Soc. Civil Engrs no 210 pp. 13-20, 1973
[15] HOWARD L.N.: 'Bounds on flow quantities' Ann. Rev. Fluid Mech. 4, pp. 473-494, 1972
[16] IKEDA S.: 'Self-formed straight channels in sandy beds' Journ. Hydraulics Division ASCE, Vol. 107 no. HY04, apr 1981

[17] IKEDA S.: 'Lateral Bed Load Transport on Sides Slopes' Journ. Hydraulics Division ASCE, Vol. 108 no. HY11, nov 1982 and Closure to the discussion, Vol. 110 no. HY02, feb 1984

[18] IKEDA S., Y. KIMURA and G. PARKER: 'Stable cross section of straight gravel rivers' Proceedings IAHR Congress Lausanne, pp. 339-344, 1987

[19] JAMMES G.: 'Travaux fluviaux' Ed. Eyrolles, Paris, 1974

[20] JANSEN P.Ph., VAN BENDEGOM L., VAN DEN BERG J., DE VRIES M. and ZANEN A.: 'Principles of river engineering' Ed. Pitman, London, 1979

[21] LAMBERTI A.: 'Sulla formulazione variazionale della teoria del regime' in Atti della Giornata di Studio Corsi d'Acqua con Letti di Materiale Grossolano ed Incoerente, Napoli 11 nov 1983

[22] LANE E.W.: 'Design of stable channels' Transactions ASCE vol. 120 no. 2776 pp. 1234-1279, 1955

[23] LANE E.W., LIN P.N. and LIU H.K.: 'The most efficient stable channel for compleately clean water in non cohesive material' Colorado State University Rep. CER 59 HKLS, 1959

[24] LI R.M., SIMONS D.B. and STEVENS M.A.: 'Morphology of cobble streams in small water-sheds' Journ. Hydraulics Division ASCE, Vol. 105 no. HY9, sep 1979

[25] LUNDGREN H. and JONSSON I.G.:'Shear and velocity distribution in shallow channels' Proc. ASCE J. Hydraulic Div. no 90 HY1 pp. 1-21, 1964

[26] PARKER G.: 'Self formed straight rivers with equilibrium banks and mobile bed. Part 1: the sand silt river. Part 2: the gravel river.' Journ. Fluid Mechanics, Vol. 89, nov 1978

[27] PARKER G.:'Hydraulic geometry of active gravel rivers' Journ. Hydraulics Division ASCE, Vol. 105 no. HY09, sep 1979

[28] PARKER G.: Discussion of 'Lateral bedload transport on side slopes' Journ. Hydraulics Division ASCE, Vol. 110 no. HY02, 1984

[29] RAMETTE M.: 'Une approche rationelle de la morphologie fluviale' La Houille Blanche no. 8, 1979

[30] QUESNEL B. 'Traite' d'hydraulique fluviale appliquée' Ed. Eyrolles, Paris, 2-me ed. 1970

[31] SIMONS D.B. 'Theory and design of stable channel in alluvial material' M.Sc. thesis Colorado State University, 1957

[32] SINGH B.: 'Bed load transport in channels', Irrigation and Power, Journal of the central Board of Irrigation and Power, Vol. 18 no. 5, pp.411-430, 1961

[33] SONG C.S. and YANG C.T.: 'Minimum stream power : theory' Journ. Hydraulics Division ASCE, Vol. 106 no. HY09, sep 1980

[34] SONG C.S. and YANG C.T. 'Application of variational principle to river flow' Conf. Applying Research to Hydraulic Practice, Jackson Mississippi, aug 1982

[35] WHITE W.T., BETTES R. and PARIS E. 'Analytical approach to river regime' Journ. Hydraulics Division ASCE, Vol. 108 no. HY10, oct 1982

[36] YANG C.T.: 'Minimum unit stream power and fluvial hydraulics' Journ. Hydraulics Division ASCE, Vol. 102 no. HY07, jul 1976

[37] YANG C.T.: 'Energy Dissipation Rate Approch in River Mechanics' in Sediment Transport in Gravel-bed Rivers edited by C.R.Thorn, J.C. Bathurst and R.D. Hey, John Wiley & Sons Ltd 1987

[38] YANG C.T. and SONG C.S.:'Theory of minimum rate of energy dissipation' Journ. Hydraulics Division ASCE, Vol. 105 no. HY07, jul 1979

[39] YANG C.T. and MOLINAS A.: 'Sediment transport and unit stream power function' Journ. Hydraulics Division ASCE, Vol. 108 no. HY06, jun 1982

International Conference on
**RIVER REGIME**
18-20 May, 1988

PAPER D1

# THE INFLUENCE OF BANK STABILITY ON REGIME GEOMETRY OF NATURAL CHANNELS

C.R. Thorne, Queen Mary College, Univ. London, London E1 4NS, UK
Akode M. Osman, University of Khartoum, Khartoum, Sudan

SUMMARY: Bank retreat occurs by a combination of flow erosion and mass failures. Processes and mechanisms of retreat depend on the bank material characteristics, but the rate of retreat is controlled by fluvial activity at the bank toe through the state of basal endpoint control. These considerations are used to explain how bank stability influences regime geometry.

## 1. INTRODUCTION

In a channel in regime the dependent variables defining the hydraulic geometry are adjusted to conform to the independent variables of the discharge of water and sediment, the bed and bank material characteristics and the valley slope. The hydraulic geometry is usually described by the width, depth, and channel slope with the mean velocity being given by continuity). However, to uniquely define the cross-sectional shape requires values for width (or wetted perimeter), depth (or hydraulic radius) and maximum depth. Also, as natural regime channels are rarely straight, it is necessary to define the planform as well as the cross-sectional shape. This requires two more parameters: the sinuosity (found from the ratio of valley to channel slopes) and the meander arc length (1).

The roles of the independent variables have received different amounts of attention in the course of research into river regime. The importance of discharge is universally recognized and considerable effort continues to be made into the best way to represent the range of flows experienced by a natural channel (2). The roles of sediment load and bed material size are closely related and have been subject to numerous theoretical, laboratory and field investigations, under the general headings of flow resistance and sediment transport studies. Valley slope is somewhat controversial, but its status as an independent variable is widely accepted at least during engineering timescales,and the comparison of valley slope to regime channel slope is useful in the prediction of sinuosity.

However, even though most engineers and scientists recognize that bank material characteristics belong in the list of independent variables, no concensus exists on the link between bank characteristics, bank processes and regime hydraulic geometry. Indeed, it has been argued that the success of regime analyses which ignore bank characteristics demonstrates that they must be at most of only secondary importance (3). Both theory and practical experience show otherwise. For example, Yu and Wolman (2) highlight the importance of bank material considerations on conceptual grounds, while Charlton (4), Bray (5), and Andrews (6), amongst many others, have reported field studies that clearly illustrate the importance of bank characteristics in establishing the stable hydraulic geometry. In this volume, Thorne et al. (7) present an analysis which relies on relations describing a bank characteristic (specifically, bank vegetation category) to produce satisfactory predictions of regime width.

Incorporation of bank characteristics into regime analyses requires

International Conference on River Regime, edited by W.R.White.

explanation of the processes and mechanisms responsible for bank retreat and the eventual establishment of the bank stability that is essential to the maintenance of a stable width. Specifically, it is necessary to show how bank processes tie into the sediment movement processes in the channel as a whole.

We first review bank erosion processes briefly. Widening of an alluvial channel occurs through bank retreat that is not matched by advance of the opposite bank through bar deposition. Retreat is the net result of lateral erosion by the flow and mass failure of the bank under gravity. The controls of these two components of bank retreat are different and so it is convenient to review them separately. Second, we show how the state of basal endpoint control links bank and channel processes regardless of the characteristics of the bank. Third, we demonstrate the influence of bank material characteristics, bank stability and the concept of basal endpoint control on the evolution of regime hydraulic geometry in a straight channel. Channel evolution with weak, non-cohesive banks is compared and contrasted to that with stiff, cohesive banks.

## 2. EROSION PROCESSES

Bank erosion consists of the detachment of grains or assemblages of grains from the bank surface, followed by their removal by entrainment into the flow. Flow forces of lift and drag may be entirely responsible for both detachment and entrainment, but more commonly grains are loosened and sometimes detached prior to entrainment by processes of sub-aqueous and sub-aerial weakening and weathering. On gentle banks, loosened grains remain <u>in situ</u>, to be entrained at the location of detachment, but on steep banks detached grains roll or slide to the bank toe before being removed by the flow. Entrainment occurs when the motivating forces due to the flow and to the downslope component of grain weight overcome the restoring forces tending to hold the grain in place. The nature of the grain or grain assemblage entrained depends on the engineering properties of the bank material and particularly whether the material is non-cohesive or cohesive (8).

### 2.1 Non-cohesive banks

Non-cohesive bank material is usually detached and entrained grain by grain. Stability depends on the balance of forces acting on individual particles. The motivating forces are the downslope component of submerged weight, and fluid forces of lift and drag. Restoring forces are the inter-particle forces due to friction and inter-locking acting under the slope normal component of submerged weight. The fluid forces can be represented by boundary shear stress when calculating the stability of non-cohesive grains, because this scales on the same parameters of flow intensity as lift and drag forces (9). Non-cohesive bank materials in alluvial channels are usually derived from in-channel deposits and as such show close packing and, in the case of non-spherical grains, imbrication. This can greatly enhance the resistance to erosion of these materials, and can allow banks in notionally non-cohesive materials to stand at steep bank angles. Reduction of the restoring forces on non-cohesive grains accompanies loss of close packing and imbrication, greatly increasing susceptibility to entrainment and decreasing the angle of repose by as much as 70% (10). The processes responsible are the removal of fine interstitial material from between the grains by wind or water, frost heave due to freezing of soil moisture, and weakening by positive pore water pressures leading in extreme cases to piping.

### 2.2 Cohesive banks

Cohesive bank material is usually eroded by the entrainment of aggregates or crumbs of soil rather than individual grains. No strong theory of erosion exists, but empirical studies show that resistance to entrainment depends more on the physio-chemical properties of the soil, pore and eroding fluids than on mechanical properties such as the soil's strength in compression or shear (11). Entrainment occurs when the motivating forces overcome the restoring forces of friction and cohesion. It is the cohesive forces, which depend on the dispersivity of the soil, the electrochemical bonds between particles and the chemistry of the pore and eroding fluids, that are most important in opposing detachment and entrainment. Field and flume experiments show that intact, undisturbed cohesive banks are much more resistant to fluvial entrainment than are non-cohesive banks (8).

The processes responsible for loosening and detaching grains and aggregates are mostly associated with moisture conditions in the soil close to the bank surface. Cohesive soils are often poorly drained and positive pore water pressures act to reduce effective cohesion and can lead to liquefaction, that is a complete loss of strength and a flow-type failure. The most favourable conditions for high positive pore pressures are during rapid drawdown in the channel following a high flow stage. Conversely, rapid immersion of a dry bank can lead to slaking, which is the detachment of aggregates by positive pressures in the pores due to the compression of trapped air. Cycles of wetting and drying are also important as they produce swelling and shrinkage, leading to cracking and ped formation. Cohesive forces between peds are much weaker than those within peds, to the point that a heavily desiccated soil may behave almost like a non-cohesive material. Surface processes may also be important, and grains may be detached by raindrop impact and surface run-off.

The physical state of soil moisture is also important. Freezing leads to needle ice formation and the detachment of surface grains and aggregates. Recent studies in the UK have demonstrated this to be highly effective in increasing the susceptibility of cohesive bank to flow erosion (12). At larger scales, ice lenses reduce cohesion by wedging peds apart and in riversthat freeze over in winter, serious damage can be done to the banks by cantilevers of ice attached to the banks and by ice rafts during the spring thaw (13).

### 2.3 Bank vegetation

Vegetation plays an important role in limiting the effectiveness of bank erosion by detachment and entrainment of individual grains or aggregates of bank material. Compared to the unvegetated state, banks vegetated with a good surficial covering experience a reduction in the effectiveness of erosion by the flow of between one and two orders of magnitude (10,14,15). Vegetation not only protects the soil surface directly, but also the roots and rhizomes of plants reinforce the soil and introduce extra cohesion (16). Also, vegetation reduces the near boundary velocity gradient, thereby reducing the shear stress on the soil. The presence of vegetation does not mean that a bank is immune from erosion, but the critical condition for erosion of a vegetated bank is the threshold of failure of the vegetation cover by scour of soil around the stems, by snapping, or by uprooting, rather than the detachment and entrainment of the bank material proper. Usually, the former is associated with a much higher level of flow intensity than the latter. In fact, in humid environments, active erosion by the flow is necessary to prevent the ingress of vegetation onto an unvegetated bank. In the lack of flow erosion, the other processes of erosion discussed here are insufficiently vigourous to maintain bank retreat at an appreciable level for any length of time (17). The absence of vegetation is then an indication that flow erosion is active on a bank, while its presence indicates at least temporary stability in that regard.

## 3. FAILURE MECHANICS

Erosion by the flow causes bank retreat directly by removing material from the bank face. It often also triggers mass instability of the bank, leading to retreat through the failure of a slab or block of bank material under gravity.

Failures come about when fluvial erosion leads to either bed scouring at the bank toe which increases the bank height, or under-cutting which increases the bank angle. The mechanics of the resulting failures depend on the size and geometry of the bank and the engineering properties of the bank material. Banks may be broadly classified as non-cohesive or cohesive in this respect.

### 3.1 Non-cohesive banks

Non-cohesive banks fail by the dislodgment of individual clasts or by shear failure along shallow, very slightly curved slip surfaces. Deep seated failures are rare because in non-cohesive banks shear strength usually increases more quickly than shear stress, with depth (18). In drained banks, failure occurs when the bank angle exceeds the friction angle. Hence, failures are associated with friction angle reduction due to loss of close packing and imbrication, and with bank angle increase due to basal scour and oversteepening. In poorly drained banks, a third possibility is failure due to an increase in pore pressure. This is possible following rapid drawdown in the channel or prolonged intense precipitation, although non-cohesive materials are usually highly permeable and are consequently well drained.

### 3.2 Cohesive banks

Cohesive banks often fail along a surface deep within the body of the soil. This is because in cohesive materials the shear strength increases less quickly than the shear stress with depth (18). The stability of the bank with regard to mass failure depends on the balance of motivating and resisting forces on the most critical potential failure surface. Failure may be brought about by an increase in motivating force or a decrease in resisting force.

The motivating force is the tangential component of the weight of the potential failure block. An increase in this force occurs when fluvial erosion leads to an increase in bank height or bank angle and catastrophic failure occurs when the critical value of height or angle is reached. Weight also increases when the moisture content of the soil increases. Of particular significance is the switch from submerged to saturated conditions that accompanies rapid drawdown. This can cause the bulk unit weight of the soil to double, and this often triggers failure even in the absence of significant excess pore water pressures.

A decrease in restoring force is associated with weakening of the soil due to reduction in either the cohesion or friction angle, or a reduction in the length of the intact potential failure surface. Cohesion is reduced by softening of the soil due to high moisture content or due to leaching of minerals from the soil by strong seepage. Friction is reduced by positive pore pressures during rapid drawdown. The length of the intact surface is reduced by tension cracking behind steep banks.

Most mass failures of cohesive banks occur following rather than during high flow in the channel. The saturated state during rapid drawdown is usually more critical than the submerged state when flow is actually attacking the bank, as the flow forces are at least an order of magnitude less than those of weight and friction. Consequently, failure is only indirectly related to discharge, and the conventional measures of flow intensity such as velocity or boundary shear stress.

The shape of the most critical failure surface depends mostly on the bank height. In low, steep banks this surface is almost planar, and a slab or block of soil slides downwards and outwards before toppling forwards into the channel. In high, gently sloping banks the failure surface is curved and the failure block tends to rotate backwards as its toe slides outwards into the channel. This occurs because the orientation of the principal stresses in gently sloping banks changes significantly with depth (18).

Rotational slips may be analyzed using well established procedures developed in geotechnical engineering (19,20), but most eroding river banks are

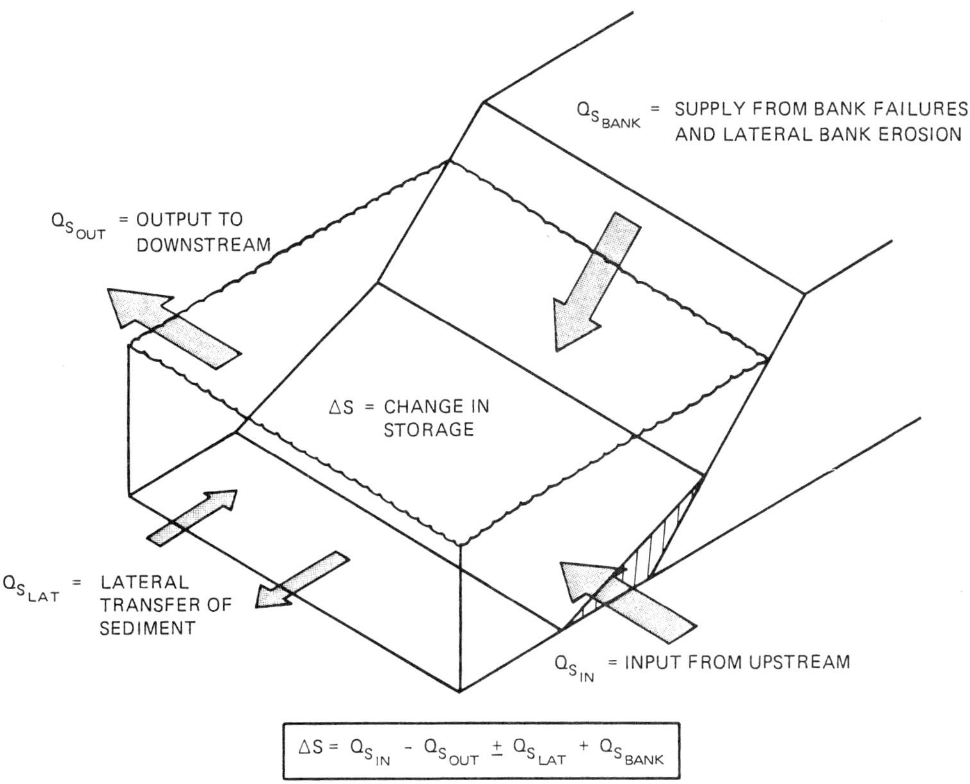


Fig. 1 Schematic diagram defining sediment fluxes in the basal area.

steep and the analyses for planar slip have received relatively little attention because of the more limited scope of application for planar analyses in geotechnical engineering. Recently, Osman and Thorne have published a new stability analysis specifically for steep, eroding river banks (21).

### 3.3 Bank Vegetation

Vegetation plays a dual role in affecting the stability of river banks, increasing stability under some circumstances and decreasing it in others. Much depends on the type of vegetation. Grasses and shrubs of low biomass enhance bank stability with regard to shallow slips because their roots and rhizomes reinforce the soil. Soil is strong in compression while the roots are strong in tension, and together soil and roots produce a type of reinforced earth which is much more stable than the soil alone. This reinforcement only extends down to the rooting depth of the plants, which for grasses and shrubs is limited to at most a few tens of centimetres. Consequently, vegetation of this type has little or no affect on stability with respect to deep seated failures.

Trees are more deeply rooted and can affect stability with regard to deep seated failures. Whether a given tree is a benefit or a liability to bank stability depends on a number of factors.

First, the type of tree is important. Willows, for example, have a dense and extensive system of roots which strongly bind the soil together. In contrast, sycamore roots are not good at reinforcing the soil and are much less beneficial to bank stability in this regard. Unlike grass and shrubs, the biomass of trees may be substantial and may produce a significant increase in the weight of the potential failure block, through surcharging. This is familiar to anyone who has seen a section of bank that has been dragged down by the weight of an overhanging tree. Generally, small trees such as dwarf willows are of more benefit than large woody species such as sycamore, with respect to bank stability.

Second, the geometry of the bank is important. Bank angle has a bearing because on steep, undercut banks the detrimental effect of the tree in increasing the downslope component of weight more than offsets any benefits through soil reinforcement so that the tree is a net liability. On gentle slopes the converse is the case and the tree is a benefit. Bank height is significant in relation to the rooting depth. If the rooting depth exceeds the bank height then the roots almost certainly cross the potential failure plane and reinforcement can be effective. However, once the height exceeds the rooting depth this benefit is lost even though the surcharge weight of the tree remains. Also, the flow is able to attack the bank below the rooting depth, where any enhancement of resistance to fluvial erosion offered by the roots is absent, promoting an increase in lateral erosion rate and failure due to under-cutting. Hence, the trees on banks of degrading streams switch from enhancing to reducing bank stability once the bed level falls below the rooting depth (22).

Third, the density of trees is significant. Single or small groups of trees are impediments to the flow, generating strong vorticity and sometimes separation in their wake. Hence, the flow is usually able to isolate and flank a hard point in the bank resulting from the effects of a single tree or group of trees. A continuous band of trees is far more effective in enhancing bank stability especially when it is several trees deep normal to the bank line. Trees in a band only one deep are usually lop-sided because they are grazed on the bankward side but not the riverward side and this decreases their stability with regard to toppling into the channel (23).

Fourth, the age and health of the trees is important. Not only do trees that have died tend to pull down the bank, but also, they leave relic roots in the bank that seriously weaken its structural integrity and provide ready pathways for seepage that promotes piping-type failure of the bank.

The role of vegetation in affecting bank stability is complicated and vegetation cannot be simply classed as a benefit or a liability without consideration of its type, age and density, and some knowledge of the geometry of the bank in question. This may explain the strong dichotomy of opinion amongst river engineers and fluvial geomorphologists regarding the effectiveness of vegetation in promoting bank stability .

At present there is no procedure to take the effects of bank vegetation into account explicitly when analyzing the stability of banks with respect to mass failure. Until such a procedure is developed, the best approach is to incorporate vegetation effects into the parameters used to represent the bank material's unit weight, effective friction angle and effective cohesion. In the remainder of this paper we adopt this convention, so that bank vegetation type and density is included under the broad term of "bank material characteristics".

## 4. BASAL ENDPOINT CONTROL

Basal endpoint control is of pivotal importance to understanding how bank processes are tied into sedimentary processes in the channel as a whole. Considerations of sediment continuity used to compute changes of bed elevation in aggradation/degradation models apply with equal justification to the bed and bank in the basal area: that is the portion of the bank below the water surface, extending into the channel a distance of one or two times the bank height (Fig.1). The flow supplies sediment to the basal area from upstream by longstream sediment transport and removes sediment downstream. There may also be a net input or output of sediment laterally due to cross-stream transport associated with transverse velocity components, especially in bends. Bank processes supply sediment to the basal area, as described in the sections 2 & 3.

Mechanical failures involve the movement of bank material directly from the intact bank to the basal area. Enroute, the material is disturbed, and it may come to rest in any state from a completely disaggregated slough deposit to an almost intact block, depending on the bank material characteristics, the degree of root binding and the type of failure. Similarly, detached grains that are not immediately entrained by the flow come to rest in the basal area. The removal of failed material from the basal area depends entirely on its entrainment by the flow, and the amount and duration of basal storage depends on the balance between the rates of supply from bank processes and removal by fluvial entrainment.

(i) Where there is insufficient transport capacity to remove the products of bank erosion processes and mass failures a wedge (or berm (24)) of failed material develops. This buttresses the bank, increasing its mass stability and protecting it from fluvial erosion. As a result the intensity of bank processes is reduced and the rate of bank retreat slows down. (ii) Where the rates of supply by bank processes and removal by the flow are just balanced, dynamic equilibrium holds, the bank is kept clear of buttressing sediment at its toe, and the shape of the bank profile is maintained. (iii) Where the flow has sufficient unfilled transport capacity to entrain and remove all of the debris and still continue to erode the intact bank, bank stability is further reduced, the frequency of failures is increased and bank retreat accelerates.

These three conditions of sediment balance at the base of the bank have been classified as three states of "Basal Endpoint Control" of the bank profile (25). Formally stated these are:

(i) *Impeded Removal.* Bank processes, plus inputs from upstream and lateral flow, supply material to the base at a higher rate than it is removed. Basal accumulation results, decreasing the bank angle and height. The rate of supply decreases, tending towards the second state.

(ii) *Unimpeded Removal.* Processes delivering material to the base and removing it from there are in balance. No change in basal elevation or slope angle occurs. The bank recedes by parallel retreat at a rate determined by the degree of fluvial activity at the base. If the sediment load at the base is zero, then the state of basal endpoint control is *static* and the bank retreat rate is zero.

(iii) *Excess Basal Capacity.* Basal scour has excess capacity over the supply from bank failures and fluvial inputs. Basal lowering and under-cutting occur, increasing the bank height and angle. The rate of supply increases, tending towards the second state.

The concept of basal endpoint control shows that the rate of bank retreat ultimately depends on the condition of the sediment balance in the basal area and that this is controlled by the dynamics of sediment exchange between the channel and the bank basal areas. Hence, the long-term retreat rate is fluvially controlled, regardless of the nature of the bank material, the erosion processes and the failure mechanisms responsible for bank retreat. That does not mean that these factors are irrelevant, however, because they are still responsible for determining limiting values for the bank height and bank angle.

The concept provides the link between bank stability and channel processes that allows the bank material characteristics to be taken into account when analysing the hydraulic geometry of natural channels.

## 5. INFLUENCE ON REGIME GEOMETRY

The establishment of regime geometry in a straight channel is recounted in a companion paper (7). In this paper, we consider the evolution of a channel toward stability, emphasizing the influence of the bank properties, bank stability, and state of basal endpoint control. The example used is of an initially under-sized channel with a trapezoidal cross-section and a mobile, non-cohesive bed (Fig.2, Stage 1). Two cases are considered: First, where the bank materials are similar to the bed material and; second, where the banks are formed in stiff, cohesive silty-clay that in its intact state is resistant to fluvial erosion.

### 5.1 Non-cohesive Banks

With non-cohesive banks enlargement begins by scouring of the bed and lateral erosion of the banks simultaneously, because the thresholds for incipient motion of bed and bank material are similar (26). Initially, the transport capacity of the flow is greater than the supply of sediment from upstream and so a state of excess basal capacity exists (Stage 2, Fig.2). Bed scouring adjacent to the banks results, generating shallow slip failures as soon as the bank angle is over-steepened to a value greater than the friction angle (see section 3.1). Hence, bank retreat by a combination of lateral erosion and mass failure is initiated immediately enlargement begins.

The increased supply of bank material due to the shallow slips changes the state of basal endpoint control to unimpeded removal, at the expense of rapid bank retreat and widening (Stage 3, Fig.2). Further lowering of the bed is limited, because as soon as scouring begins, the rate of bank failures and bank retreat is increased, satisfying the deficiency between sediment input and output and maintaining the state of unimpeded removal. In this way the channel cross-section is enlarged primarily by widening.

Widening continues so long as the flow in the basal areas is able to remove the sediment input from upstream plus that derived from lateral erosion and shallow slips. As the channel widens, this becomes increasingly difficult because the hydraulic efficiency of the channel decreases as its width to depth ratio and relative roughness both increase. The state of basal endpoint control switches to impeded removal when the reduced velocities and bank shear stresses in the widening channel can no longer remove all of the products of bank retreat (Stage 4, Fig.2). Material from the last generation of shallow slips accumulates on the bed and as a low angle wedge at the base of the banks and the rate of retreat (now due only to lateral erosion above the basal wedge), slows dramatically.

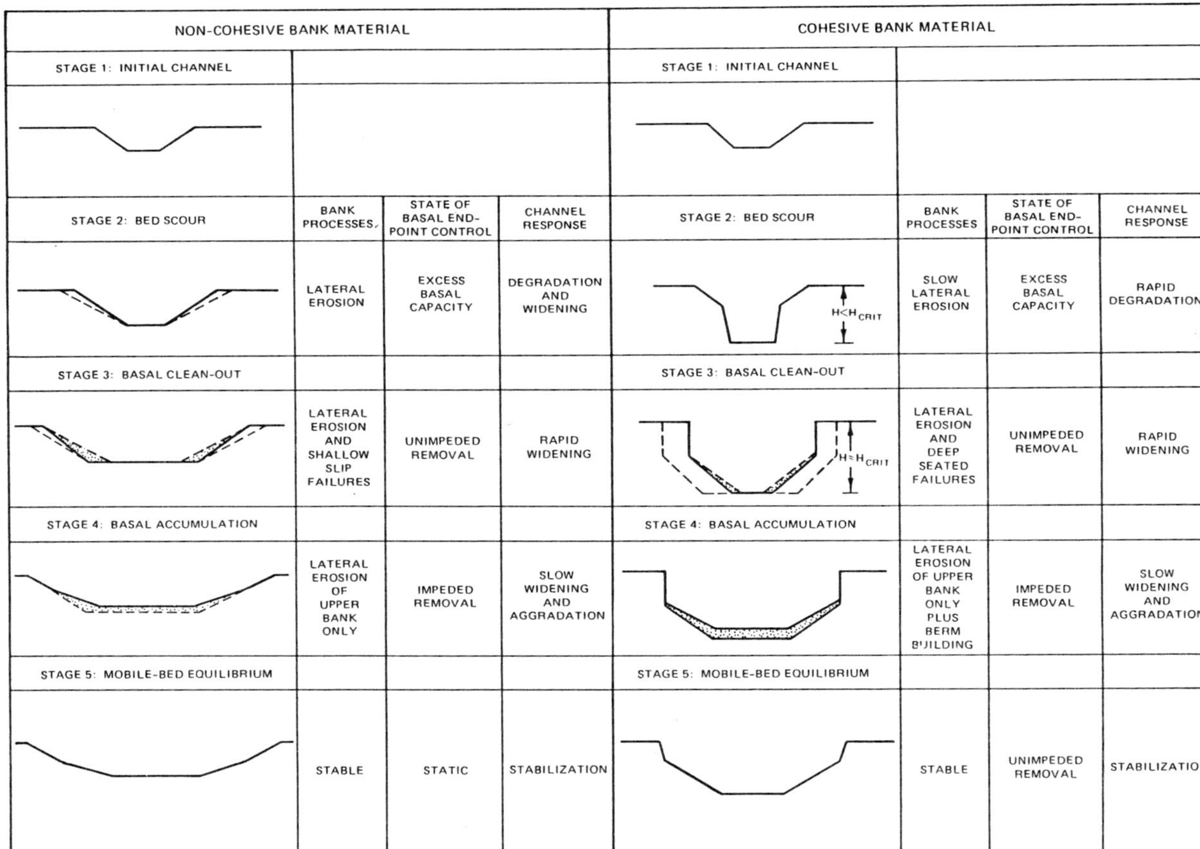


**Fig. 2** Influence of bank stability on channel evolution and regime geometry.

The regime width is established when the bank stresses in the slowly widening channel are reduced below the threshold for motion and the banks stabilize (Stage 5, Fig.2). The sediment flux on the banks must be zero, as the downslope creep of grains in motion must always produce widening in a straight channel with mobile banks (27). The basal endpoint is static.

If secondary currents produce significant lateral transport of sediment from the bank basal areas into the central portion of the channel, the state of unimpeded removal at the banks can be maintained, at the expense of overloading the transport capacity in the central part of the channel. This results in the building of a medial bar and continued widening, tending toward the development of a braided channel, as envisaged by Leopold and Wolman (28). If the secondary cells are asymmetrical, one bank may be kept in unimpeded removal while that opposite is in impeded removal. The result is development of an asymmetrical cross-section and skewing of the flow. If the sense of the cells reverses along the channel, alternate bars and a meandering thalweg form as precursors to meandering (29).

The regime channel, be it single thread or braided, is wide and shallow, with a mobile bed and immobile banks.

### 5.2 Cohesive Banks

With cohesive banks enlargement begins by scouring of the bed, bank erosion being minor because of the high resistance to erosion of the intact cohesive sediments, as described in section 2.2. The basal areas are in a state of excess basal capacity and scouring continues to degrade the bed, increasing the bank height, but the banks remain stable so long as the critical height for mass failure is not exceeded (see section 3.2). A state of excess basal capacity is maintained for some time, and the channel incises with little widening to produce a narrow, deep channel (Stage 2, Fig.2). The degraded channel has a low width to depth ratio and low relative roughness, giving it a high hydraulic efficiency and a large bankfull capacity, quite unlike that in the channel with non-cohesive banks at this stage.

If the bed is formed in gravel, degradation may be limited by bed armouring at this stage, so that the banks do not reach the critical height for mass failure. In this case, if the bank material is highly erosion resistant, then stage 2 may terminate channel evolution. The resulting regime channel is narrow and deep, with both the bed and bank materials immobile at the formative discharge. There will still be a sediment load of "throughput" material derived from channel and catchment processes upstream, moving over an immobile bed of static armour.

However, if degradation continues, an important threshold is reached when the bank height reaches the critical value for mass failure. Slab or rotational failures then occur, suddenly initiating rapid widening and supplying large volumes of disaggregated bank material to the basal areas (Stage 3, Fig.2). The state of endpoint control immediately switches to unimpeded removal and degradation ceases. Channel incision is in this case limited by the critical bank height, which depends on the bank material properties.

When basal clean-out is completed the flow again erodes the bed and banks leading to further failures under gravity and initiating another cycle of bank retreat. Bank retreat by parallel retreat then occurs, with the basal zone in unimpeded removal when considered over the whole cycle of over-steepening, failure and clean-out. At this stage the channel may well be over-sized because of rapid widening by mass failure of the banks and depending on the ease with which the bank sediment derived from mass failures is transported away. Disturbed bank debris often disintegrates quickly into fine wash load, and river flows usually posses abundant transport capacity for material of this calibre. With the transport capacity unfilled, erosion can be maintained and an over-large channel results. This shows that the influence of the bank material is not limited to affecting only the bank height or angle, and hence the width of the bank zones, but that it can significantly alter the whole nature and extent of channel evolution.

Eventually, as the channel widens, the flow at the banks is no longer able to remove all of the slump debris (Stage 4, Fig.2). Also, as the time taken for the flow at the banks to remove the slump debris lengthens, vegetation is able to become established on it. This tends to increase its resistance to erosion and also traps fine material from the sediment load, leading to berm building (24). Bank erosion above the berm is slowed and in the lack of basal clean-out the bank is stable with respect to mass failure. Surface erosion and local shallow failures may continue for sometime, resulting in minor increases of top bank width until vegetation becomes fully established. At this stage a major flood may destroy the berm and re-activate bank retreat, but except for this special case the banks will stabilize and the regime width will be established (Stage 5, Fig.2). Berm building can result in the stable width being somewhat less than its stage 4 maximum.

Braiding is unlikely in a channel with resistant banks and a narrow, deep cross-section, but as in the case of the non-cohesive banks, significant secondary currents can lead to opposite banks being maintained in different states of basal endpoint control, leading to the development of a meandering planform. In fact meandering is more likely in the second case, because the cohesive banks are better able to withstand the high outer bank shear stresses in developing bends where non-cohesive banks would simply retreat more rapidly, increasing channel width and tending towards a braided condition.

The regime channel, be it straight or meandering, is narrower and deeper than that with non-cohesive banks. The banks are much steeper, so that the water surface width changes less with stage in this channel than in that with non-cohesive banks. The bed is mobile and it is possible for bed material in the basal areas to be mobile if the shear stresses there are above the level for bed material movement, but below that for bank material entrainment. Hence the active width of the bed may actually be close to that of the channel with non-cohesive banks, even though the total width is considerably less.

## 5.3 Discussion

In this account of the evolution of a straight channel to a regime condition, we have concentrated on the influence of bank material characteristics, bank stability, bank processes and the state of basal endpoint control. Our purpose is not to over-emphasize the importance of bank factors, but to highlight their roles in the development of the stable geometry, which have been relatively neglected in the past.

Bank stability alone does not guarantee channel stability and time invariant hydraulic geometry. It is not even solely responsible for establishing the stable width, which depends also on the discharges of water and sediment, and the bed material characteristics (1). There must be sufficient width to match the conveyance to that required for the channel to contain the discharge from upstream, and the width of the active bed must be able to transport the bed load input from upstream, given that the input from bank retreat is zero when the banks are stable. The relative mobility of the bed versus the bank materials determines whether the channel responds to disequilibrium primarily depth or width adjustment. Still, for width to be steady in a straight channel, the banks must be stable. Bank stability is a *necessary condition* for the establishment of a straight regime channel, but it is not in itself the *sufficient condition* for channel stability. In the case of a meandering regime channel, width is steady because bank retreat at the outer bank in bend is just matched over time by advance of the inner bank by point bar growth. We have shown in a recent paper (30) that the stability of the outer bank is also important in determining regime geometry under these circumstances.

Channel evolution has been presented as a series of channel adjustments that lead directly to establishment of a regime condition. Of course, this is a simplification. In nature, disequilibrium channels display complex response in evolving towards a stable state, in which the stability of the banks plays an integral role. It may be hypothesized that the critical height for mass failures of cohesive banks is an intrinsic geomorphic threshold of equal importance to the threshold bed slope for incision in generating complex response. This hypothesis is explored quantitatively in a recent paper by Thorne and Osman (30).

Periods of channel adjustment under formative flows are separated by much longer periods when the form of the channel largely determines the dynamics of flow and sediment movement. During these periods, bank processes operate differently above and below the water surface, resulting in profile changes that would not occur under a steady flow. For example, during low flows sub-aerial erosion delivers bank material to the basal area, where it collects due to temporarily impeded removal during low flows. In the long-term, however, basal accumulation and the form of the banks adjusts to the state of basal endpoint control associated with the dominant discharge.

Verification of the model presented here requires detailed accounts of channel evolution to a regime condition. Such accounts are rare, but the results of an aggradation/degradation model incorporating bank stability criteria developed by Osman (30), quantitatively replicate the main features of the conceptual model proposed here. Studies on gravel-bed rivers lend support to the case of non-cohesive banks (27) while a recently reported field study provides strong support for the case of cohesive banks (24). Watson and Harvey (24) and, independently, Simon and Hupp (31), have produced multistage models of channel evolution in the Bluff-line streams of North Mississippi and West Tennessee, based on geomorphic field observations and channel morphometry in the sand-bed, cohesive bank channels. Their models exhibit the same characteristics as that developed here on the basis of hydraulic and geotechnical considerations and strongly the concept of a critical bank height which must be exceeded for rapid widening to occur.

Certainly, further testing and development of our model is required, but we believe it to be a useful first step in illustrating way in which bank processes should be integrated into the channel processes responsible for the development and maintenance of regime hydraulic geometry in natural channels.

## 6. ACKNOWLEDGMENTS

This paper was written while the first author was a visiting scientist at the US Army Waterways Experiment Station, Vicksburg, Mississippi, USA, working on a project sponsored by the Lower Mississippi, Vicksburg District, US Army Corps of Engineers. Support and advice from Phil Combs and David Biedenharn at the District, and from Bobby Brown and John Ingram at WES are gratefully acknowledged.

## 7. REFERENCES

1. Hey, R.D. and Thorne, C.R. "Stable channels with mobile gravel beds" Jour. Hydrau. Eng., ASCE, 112(8), 1986, pp 671-689.
2. Bofu Yu and Wolman, M.G. "Some dynamic aspects of river geometry" Water Resources Research, 23 (3), 1987, pp 501-509.
3. Bettess, R. and White, W.R. "Extremal hypotheses applied to river regime" In, Sediment Transport in Gravel-Bed Rivers, Thorne,C.R. Bathurst,J.C. and Hey,R.D. (Eds), J. Wiley and Sons, Chichester, UK, 1987, pp 767-789.
4. Charlton, F.G. "River stabilization and training in gravel-bed rivers" In, Gravel-Bed Rivers, Hey,R.D. Bathurst,J.C. and Thorne,C.R. (Eds), J.Wiley and Sons, Chichester, UK, 1982, pp 635-657.
5. Bray, D.I. "Study of channel changes in a reach of the North Nashwaaksis stream, New Brunswick" Proc. Sixth Canadian Hydrotech. Conf., Ottawa, Ontario, Canadian Soc. Civ. Eng., pp 107-127.
6. Andrews, E.D., "Bank stability and channel width adjustment, East Fork River, Wyoming" Water Resources Research, 18, 1982, pp 1184-1192.

7. Thorne, C.R., Chang, H.H. and Hey, R.D. "Prediction of hydraulic geometry of gravel-bed streams using the minimum stream power concept" In, Proc. Int. Conf. on River Regime, Hydrau. Res., Wallingford, May 18-20, 1987.
8. Thorne, C.R. "Processes and mechanisms of river bank erosion" In, Gravel-Bed Rivers, Hey,R.D., Bathurst,J.C. and Thorne,C.R. (Eds), J. Wiley and Sons, Chichester, UK, 1982, pp 227-271.
9. Wiberg, P.L. and Smith, J.D."Calculations of the critical shear stress for motion of uniform and heterogeneous sediments" Water Resources Research, 23(8), 1987, pp 1471-1480.
10. Carson, M.A. and Kirkby, M.J. Hillslope Form and Process, Cambridge Univ. Press, Cambridge, UK, 1972, 475p.
11. Arulanandan, K., Gillogley, E. and Tully, R. "Development of a quantitative method to predict critical shear stress and rate of erosion of natural undisturbed cohesive soils" Report GL-80-5, US Army WES, Vicksburg, Mississippi 39180, USA, 1980.
12. Lawler, D.M. "River bank erosion and the influence of frost: a statistical examination" Trans. Inst. British Geographers, NS 11, 1986, pp 227-242.
13. Church, M. and Miles, M.J. "Riverbank stability in arctic and subarctic rivers" In Gravel-Bed Rivers, Hey,R.D., Bathurst,J.C. and Thorne,C.R. (Eds), J. Wiley and Sons, Chichester, UK, 1982, pp 259-268.
14. Kirkby, M.J. and Morgan, R.P.C. Soil Erosion, Wiley-Interscience, London and New York, 1980, 312p.
15. Smith, D.G. "Effects of vegetation on lateral migration of anastomosed channels of a glacial meltwater" Geol. Soc. Am. Bull., 87,1976,pp 857-869.
16. Weaver, J.E. "Effects of roots of vegetation in erosion control" In, Symposium, Aspects of Veget. Erosion Control, Am. Soc. Agronomy, Dec.1937
17. Thorne, C.R. and Lewin, J. "Bank processes, bed material movement and planform development in a meandering river" In, Adjustments of the Fluvial System, Rhodes,D.D. and Williams, G.P. (Eds), Kendall/Hunt, Dubuque, Iowa, USA, 1979, pp 117-137.
18. Terzaghi, K. and Peck, R.B. Soil Mechanics and Engineering Practice, J. Wiley and Sons, New York, 1948, 510p.
19. Bishop, A.W. "The use of the slip circle in the stability analysis of slopes" Geotchnique, 5, 1955, pp 7-17.
20. Fredlund, D.G. "Slope Stability analysis using PC-Slope" Geo-Slope Programming Ltd., 7927 Silver Springs Rd, Calgary, Alberta, Canada, 1987.
21. Osman, A.M. and Thorne, C.R. "River bank stability analysis" Jour. Hydrau. Eng., ASCE, 114, 2, 1988.
22. Duijsings, J.J.H.M. "A sediment budget for a forested catchment in Luxembourg and its implications for channel development" Earth Surface Processes and Landforms, 12, 1987, pp 173-184.
23. Newson, M.D. and Leeks, G.J. "Transport processes at the catchment scale" In, Sediment Transport in Gravel-Bed Rivers, J Wiley and Sons, Chichester, UK, 1987, pp 187-223.
24. Watson, C.C. and Harvey, M.D. and Garbrecht, J. "Geomorphic-hydraulic simulation of channel evolution" In, Proc. Fourth Interagency Sedimentation Conf., Las Vegas, Nevada, March 1986, vol. 2, pp 5.21-5.30.
25. Thorne, C.R. "Processes of bank erosion in river channels" unpubl. Ph.D thesis, Sch. Env. Sciences, Univ. East Anglia, Norwich, UK, 1978, 447p.
26. Raudkivi, A.J. Loose Boundary Hydraulics, Pergammon Press, New York, 1976.
27. Parker, G. "Self-formed rivers with equilibrium banks and mobile bed: Part II, the gravel river"Jour. Fluid Mechanics, 89(1), 1978, pp 127-148.
28. Leopold, L.B. and Wolman, M.G. "River channel patterns; braided, meandering and straight" US Geol. Survey, Prof. Paper 262B, 1957.
29. Hey,R.D. "Geometry of river meanders" Nature 262, 1976, pp 482-484.
30. Thorne, C.R. and Osman, A.M. "River bank stability analysis: II Applications" Jour. Hydrau. Eng., ASCE, 114, 2, 1988.
31. Simon, A. and Hupp, C.R. "Channel evolution in modified Tennessee channels" In, Proc. Fourth Interagency Sedimentation Conf., Las Vegas, Nevada, March 1986, vol. 2, pp 5.71-5.82.

International Conference on
RIVER REGIME
18-20 May, 1988

PAPER D2

# ON THE WIDTH OF REGIME CHANNELS

R Bettess, Hydraulics Research, Wallingford
W R White, Hydraulics Research, Wallingford
C E Reeve, Hydraulics Research, Wallingford

SUMMARY: There have been numerous attempts both to explain and to predict the width of regime channels. In this paper a number of approaches are discussed including limiting tractive force, bank slope stability and extremal hypotheses such as minimum stream power or maximum sediment concentration. The methods are compared with field data. An unsuccessful attempt is made to relate an extremal hypothesis to the distribution of shear stress on the bed and banks of a channel.

## 1 Introduction

Regime theory developed out of the need to be able to predict the dimensions and slope of a channel flowing through an alluvium carrying a certain discharge and sediment load. A particular variable of interest was the stable channel width. The first approach to this problem was purely empirical. Data was collected from canals which appeared to be stable or 'in regime' and empirical equations were derived which related the variables specifying the channel, such as the width, to the discharge. More recently attempts have been made to derive regime relationships by using descriptions of the fundamental processes involved. This has been termed 'rational regime theory'. There has been almost universal agreement that the processes of sediment transport and alluvial friction are significant and should be included in such an approach. These, however, do not by themselves enable any determination of channel width to be made. This can only be supplied by a further relationship. At the moment there is no concensus about what relationship should be used to determine channel width, in fact, opinion differs markedly. In this paper a number of possible approaches to channel width are discussed.

## 2 Rational regime theory

To set the context for the following discussion we will briefly describe the approach of rational regime theory.

It is assumed that for the purposes of regime theory the channel system can be described by the following variables

| | |
|---|---|
| $Q$ | discharge |
| $X$ | sediment concentration |
| $B$ | channel width |
| $V$ | mean velocity |
| $d$ | depth |
| $S$ | slope |
| $D$ | sediment size |
| $\rho_s$ | sediment density |

International Conference on River Regime, edited by W.R.White.

For a particular channel the sediment details are known and fixed, that is D and $\rho_s$ are specified and known. The remaining six variables Q, X, B, V, d, and S can be related by a number of equations

(1) From the definitions of Q, B, V and d we have that

$$Q = B\ V\ D \tag{1}$$

(2) A sediment transport equation implies a relationship of the form

$$X = X\ (V, d, S, D, \rho_s) \tag{2}$$

(3) A relationship for alluvial friction implies an equation of the form

$$V = V\ (d, S, D, \rho_s) \tag{3}$$

For natural and man-made channels two of these variables are normally imposed, for example Q and X. These equations however, are not by themselves sufficient to determine the width of the channel. It is the specification of some suitable equation involving explicitly or implicity width which is now considered.

## 3 Relationships for channel width

### 3.1 Overview of various width relationships

#### 3.1.1 Empirical regime theory

Empirical regime theory has little firm theoretical basis but, by the nature of its derivation and from its history over more than eighty years, we know that it presents an approximate representation of the dominant aspects of channel form and shape.

The Lacey equations (Lacey, 1929, 1933) are approximately of the form

$$S = 0.00055\ f^{5/8}\ Q^{-1/6} \tag{4}$$

$$R = \left(\frac{Q}{3.07f}\right)^{1/3} \tag{5}$$

where f is the Lacey silt factor. R is the hydraulic radius and Q the discharge all expressed in imperial units.

Also

$$P = 2.67\ \sqrt{Q} \tag{6}$$

where P is the wetted perimeter

Whatever other methods are suggested for determining regime conditions we know that they must be broadly comparable with equations of this form.

### 3.1.2 Bank stability

It has been postulated that the channel shape is controlled by the stability of the banks. In this approach the soil mechanics of the bank material is used to determine a critical bank height which cannot be exceeded. If the bank is of a greater height then bank failure will take place. There are a number of possible failure mechanisms depending upon the nature of the situation. For the purposes of illustration we give here an equation quoted by Thorne (1982) for the critical bank height for plane-slip failure, which is of the form

$$H_c = \frac{4c}{\gamma} \frac{\sin\theta\cos\phi}{[1-\cos(\theta-\phi)]} \tag{7}$$

Where $\theta$ is the bank angle and $\phi$ is the friction angle. We are aware that in practice it would be rare for a river bank to fail under plane-slip but this equation is of a particularly simple form which illustrates the method. A particular property of these and similar equations is that they are independent of the water flow. Assuming that the range of variation of the bank angle is limited then this equation would suggest that for a given bank material, that is, for a given friction angle and c value, there is a maximum permissible bank height or channel depth. This implies that for large discharges the channel depth should be independent of the discharge. The Lacey regime relations suggest the opposite, that depth increases with discharge. In the comparison with observations detailed below the critical bank height was calculated using Taylor's curves, (Taylor 1937).

### 3.1.3 Critical bank shear

Singh (1983) suggested that there is a limiting maximum shear which the banks can support and that this controls the width of the channel. For a given sediment, therefore, the maximum shear stress on the banks should be constant. Singh approximated the maximum shear stress on the banks $\tau_w$ by the equation

$$\tau_w = 0.75\,\gamma_w\,d\,S \tag{8}$$

It follows, therefore, that for regime channels in a given sediment (dS) should be constant.

Combining the Lacey regime relations for R and S given above, however, we have

$$g\,RS = \frac{g\;0.00055\;f^{7/24}\;Q^{1/6}}{(3.07)^{1/3}} \tag{9}$$

where g is the acceleration due to gravity. Thus we see that empirical regime relations suggest that (RS) increases as the discharge increases.

### 3.1.4 Extremal hypotheses

The extremal hypothesis is based on the assumption that the regime width is that width which maximises or minimises some variable. A number of extremal hypotheses have been proposed (Singh, 1961; Ramette, 1979; Yang et al, 1981; White et al 1982), such as maximum sediment concentration or minimum stream power. The extremal hypothesis is similar to the empirical regime theories and differs from those other methods so far considered in that it just presents a method of calculating channel width while not suggesting a mechanism by which this is achieved.

Comparisons with observations of channel widths described by White et al (1982) and Wang Shiqiang et al (1986) indicate that extremal hypotheses provide a relatively reliable method of assessing channel width.

### 3.1.5 Parker (1978)

It may seem obvious to remark that if the discharge down an alluvial channel increases then there is a tendency for bank erosion to take place and the channel width to increase. This is incorporated either explicitly or implicitly in all the theories for channel width considered so far. It should not, however, be forgotten that if the discharge is reduced, for example downstream of a reservoir, then there is a tendency for deposition to take place and for the channel width to decrease. It can thus be seen that the regime width is achieved only as a balance between opposing mechanisms tending to cause erosion and hence increase channel width and tending to cause deposition and so decrease channel width. It is the fact that there are two opposing mechanisms that must be considered that is missing in some of the explanations of channel width discussed above.

Parker (1978) has introduced a theory for self-formed sand-silt rivers in equilibrium that explicitly includes mechanisms for bank erosion and for deposition. Bank erosion is ascribed to lateral bed load from the banks to the bed of the river generated by gravity and related to the longitudinal bed load. Deposition is provided by the lateral diffusion of suspended sediment generated by the non-uniform distribution of suspended sediment across the width.

While in essence the theory developed by Parker gives a complete regime theory, in practise the only parameter that can be determined easily is the channel depth. In considering this approach, therefore, we depart from a strict consideration of channel width and consider the depth prediction, aware of the close connection between width and depth. Parker quotes an equation for the centreline depth of a regime channel, which, in our notation is

$$\frac{d}{D} = 85.1 \left[\frac{\omega}{((\rho_s/\rho - 1)\, g\, D)^{\frac{1}{2}}}\right]^{\frac{1}{2}} S^{-\frac{1}{2}} \qquad (10)$$

where $\rho$ and $\rho_s$ are the densities of water and sediment respectively and $\omega$ is the fall velocity of the sediment. Thus for a given sediment

$$d \propto S^{-\frac{1}{2}} \qquad (11)$$

If we use the empirical regime expression for S,

$$S = 0.00055\, f^{5/8}\, Q^{-1/6} \qquad (12)$$

then we have that

$$d \propto Q^{1/12} \qquad (13)$$

This contrasts with empirical regime theory in which the exponent on the discharge in the expression for channel depth is in the range 0.3 to 0.4. Comparison of Parker's theory predictions with those of other regime theories does indeed show that Parker's theory provides larger depths for smaller channels and smaller depths for larger channels. As the prediction of depth and width are linked it is probable that a similar sort of discrepancy would arise in the prediction of channel width. Though this discrepancy is disappointing it need not suggest that the basic approach to the problem is at fault. The derivation of the theory relies on expressions for the suspended load and the bed load in the area adjacent to the bank. It is in just such areas where the lateral bed slope and the spatial variation of flow are significant that there is, at present, most uncertainty as to the sediment mechanics involved. To add, perhaps, a little understanding to the approach of Parker it may be worth recording, that our calculations indicate that for a fixed discharge and slope, the total lateral bed load and the total lateral flux of suspended sediment are both monotonicaly decreasing functions of width, in the range of interest. As one would expect, for widths less than the 'regime' value the lateral bed load exceeds the lateral flux of suspended load, that is, there is a tendency for the channel to widen while for widths in excess of the regime value the lateral flux of suspended sediment dominates, that is, there is a tendency for the channel width to decrease.

## 3.2 Discussion

### 3.2.1 Distribution of boundary shear stress

Implicit in any method of determining the channel width is an assumption about the distribution of shear stress on the bed and the banks as the width to depth ratio of the channel varies. Unfortunately there is little data available on this and not all of it is consistent. Nor is it is clear whether one should consider the mean shear stress or maximum shear stress developed.

Davidian and Cahal (1963) present data on the distribution of shear in a smooth rectangular channel for two different aspect ratios and for a range of Froude numbers varying from 0.1 to over 1. Ghosh and Roy (1970) performed experiments in smooth and artificially roughened rectangular and trapezoidal channels. Trapezoidal channels with three different side-slopes were used. This is probably the most comprehensive data set available for trapezoidal channels but all the results are for relatively small width to depth ratios. Experiments were carried out by Kartha and Leutheusser (1970) in a sequence of rectangular channels with width to depth ratios varying from 1 to 12.5. Interestingly these show that the ratio of the maximum shear stress on the side walls to the total shear stress is almost independent of the width to depth ratio for aspect ratios greater than 10. Komura (1973) has also reported results on shear stress distribution in a sequence of trapezoidal channels of varying side slope. A number of analytic studies of shear stress distribution have been made, notably by Lane (1955) and Lundgren and Jonsson (1964).

Knight et al (1982) performed experiments in smooth flumes and showed that the wall shear stress $\tau_w$ and the bed shear stress $\tau_b$ are related to BS and dS respectively. The equations suggested by Knight are of the form

$$\overline{\tau_w} = C_1 \, BS \qquad \overline{\tau_b} = C_2 \, dS, \tag{14}$$

where $C_1$ and $C_2$ are constants with depend upon the width to depth ratio.

It still remains, however, that the information as yet available is insufficient to enable the development of a comprehensive regime theory based on the distribution of boundary shear stress.

### 3.2.2 Relationship between shear-stress distribution and extremal hypotheses

We have investigated whether the principle of maximum concentration is equivalent to a simple assumption about the distribution of shear. We hoped, for example, that the maximum sediment concentration corresponded to a minimum in the stress on the walls or some similar expression. If one solves the Ackers and White sediment relationships with a given discharge, sediment concentration and sediment diameter for a sequence of widths and then plots BS and dS against width then one obtains Figure 1. In this case the extremal hypothesis of maximum sediment concentration or minimum slope gives a width of 154m. BS has a minimum considerably less than the regime width and, over the range of interest, dS decreases monotonicaly. These functions, however, do not take account of the varying aspect ratio. When the Knight expressions for $C_1$ and $C_1$ are included one obtains the results plotted in Figure 2. In the range of interest $\tau_b$ and $\tau_w$ are both monotonicaly decreasing. Comparison of Figures 1 and 2 suggest that the existence and location of a minimum in $\tau_w$ is very dependent upon the expression used to take account of the dependence of $C_1$ upon the aspect ratio.

## 4 An initial approach to channel width

To investigate if there was any relationship between the width of regime channels carrying small sediment loads and conditions for the initiation of motion the following analysis was carried out. Two assumptions were made, firstly that if the channel is carrying a small sediment load then conditions at the banks of the channel are close to those for initiation of motion and secondly that the ratio of the shear stress on the banks to the total shear stress is a function of the width to depth ratio, that is,

$$\tau_w = \tau_t \, F(B/d) \tag{15}$$

for some function F, where B is the channel width, d is the channel depth, $\tau_w$ is the shear stress on the banks and $\tau_t$ is the total shear stress.

We will express conditions for initiation of motion using Shields equation. This takes two different forms for the extreme cases of fine and coarse sediments and we consider these separately.

### 4.1 Fine sediments

For fine sediments the Shield's relationship is of the form

$$\frac{\tau_{cr}}{\gamma_s D} = \frac{\text{constant}}{\dfrac{(\tau_{cr})^{\frac{1}{2}} D}{(\rho^{\frac{1}{2}})\, \nu}} \tag{16}$$

If we assume that critical conditions occur on the banks of the channel we have

$$\frac{\tau_w}{\gamma_s D} = \frac{\text{constant}}{\dfrac{\tau_w^{\frac{1}{2}} D}{\nu \rho^{\frac{1}{2}}}} \tag{17}$$

Expressing $\tau_w$ in terms of $\tau_t$ and F(B/d) we have

$$\frac{(\tau_t) F}{\gamma_s D} = \frac{\text{constant } \nu \rho^{\frac{1}{2}}}{(\tau_t^{\frac{1}{2}}) F^{\frac{1}{2}} D} \tag{18}$$

This implies that

$$\tau_t = \frac{C}{F^{3/2} (B/d)} \tag{19}$$

where C is a constant. To test this one can plot values of $\tau_t$ against B/D. It is difficult to obtain a comprehensive set of appropriate field data so as a substitute values were used derived from the White et al regime theory. This theory has been shown to produce values which agree with observations over a wide range of conditions. Assuming $\tau_t$ is proportional to (dS), values of (dS) are plotted against B/d in Figure 3 for a sediment concentration of 10ppm. It can be seen that the data for sediment sizes of 0.5mm, 0.3mm and 0.2mm plot together. Values of a sediment size of 1mm plot slightly separately as the transition area is entered where the assumed form of Shield's curve is no longer valid.

## 4.2 Coarse sediments

For coarse sediments the Shield's relationship is of the form

$$\frac{\tau}{\gamma_s D} = \text{constant} \tag{20}$$

If we assume that critical conditions occur on the banks of the channel we have

$$\frac{\tau_w}{\gamma_s D} = \text{constant} \tag{21}$$

Expressing $\tau_w$ in terms of $\tau_t$ and f(B/d) we have

$$\frac{\tau_t F}{\gamma_s D} = \text{constant,} \tag{22}$$

that is

$$\frac{\tau_t}{D} = \frac{E}{F} \tag{23}$$

for some constant E. This was again tested using data from the White et al regime theory and the results are plotted in Figure 4.

The above analysis is based on the assumption that the channel has a small sediment load and has been tested using data for low concentrations. One may consider the effect of larger sediment loads. Figure 5 shows a plot for concentrations of 10 and 100ppm. It can be seen that the two curves are related, the higher concentration one being transposed upwards and tilted. The change in slope results from the consideration of curves for constant sediment concentrations. Though the convenient way to express sediment concentration is in terms of ppm the more natural measure is to use a dimensionless sediment transport rate, for example, Ackers and White $G_{gr}$. If curves of equal $G_{gr}$ are considered then they are effectively parallel.

We make no pretence that this even approaches a full analysis of the problem but it provides an indication of what is required from a more complete solution. It seems clear that any mechanism for determining channel width should implicity or explicitly take account of the different form of behaviour of fine and coarse sediments. Further, and more controversially, the rational regime theory based on extremal hypotheses suggests that the width relationship also depends upon the sediment load in the channel.

## 5 Comparison with observations

The predictions of the various approaches described above were compared with data from canals in India (ICID, 1966). Where possible the comparison is on the width of the channel. For some theories, for example Parker's, the prediction of width is difficult in which case other parameters were compared, in this case depth. In other cases, principally the theories relating to bank stability and critical bank shear, the theories utilised parameters which could not be determined from the observed data. Best estimates for these parameters were used to allow the comparison to be made. It is acknowledged that this introduces a degree of uncertainty into the predictions. It may be fair to comment however that parameters such as the soil shear strength would be difficult to determine for the field conditions.

The comparisons of predictions and observations are made in Table 1. It is clear that the best predictions are provided by the two methods not linked to a particular physical mechanism for determining channel width, that is, empirical regime theory and rational regime theory using an extremal hypotheses.

## 6 Conclusions

There is at present no satisfactory description of the mechanisms which determine the regime width of channels. The two most reliable predictive methods that are available, empirical regime theory and rational regime theory based on an extremal hypothesis, both eschew from any description of the physical processes involved. Too many approaches only consider mechanisms for channel enlargement without the awareness that the regime width is the result of a balance between different processes involving both erosion and deposition. A full description of the development of channel width may require a knowledge of the distribution of shear stress over the boundary of a channel. It can only be a source of regret that so little information is available on what is such a fundamental subject.

## 7 References


Ackers P and White W R, 1973, Sediment transport: new approach and analysis, ASCE, JHD, 99, HY11, pp2041-2060

Davidian J and Cahal D I, 1963, Distribution of shear in rectangular channels, US Geol Survey Paper 475-C

Ghosh S N and Roy N, 1970, Boundary shear distribution in open channel flow, ASCE, JHD, 96, HY4

Kartha V C and Leutheusser H J, 1970, Distribution of tractive force in open channels, ASCE, JHD, 96, HY7

Knight D W , et al, 1982, Boundary shear stress distributions in open channel and conduit flows, Euromech 156; Mechanics of sediment transport. Editors B Mutlu Sumer and A Muller

Komura J, 1973, Hydraulic resistance to flow in channels, Proc 15 IAHR Congress, Istanbul

Lacey G, 1929, Stable channels in alluvium Proc ICE, 229, pp259 et seq

Lacey G, 1933, Uniform flow in alluvial rivers and canals, Proc ICE, 237, pp421 et seq

Lane E W, 1955, Design of stable channels, ASCE, 120, pp1234-1260

Lundgren H, and Jonsson I G, 1964, Shear and velocity distribution in shallow channels, ASCE, JHD, 90, HY1

Parker G, 1978, Self-formed straight rivers with equilibirum banks and mobile bed. Part 1 The sand-silt river, JFM

Ramette M, 1979, Une approche rationelle de la morphologie fluviale, La Houille Blanche, No 8, pp491-498

Singh B, 1961, Bed load transport in channels, Irrigation and Power, J of Central Board of Irrigation and Power, 18 no 5, pp411-430

Singh B, 1983, Self adjustment of alluvial streams, Proc 2nd International Symposium on River Sedimentation (continuation), Nanjing, China, pp295-303

Taylor D W, 1937, Stability of earth slopes, J Boston Soc. Civ Engrs 24, pp197-246

Thorne C R, 1982, Processes and mechanisms of river bank erosion, in Gravel-bed Rivers edited by R D Hey, J C Bathurst and C R Thorne

Wang Shiqiang et al, 1986, A rational approach to river regime Proc. Third International Symposium on River Sedimentation, Jackson, Mississippi, pp167-176

White W R, Bettess R and Paris E, 1982, Analytical approach to river regime ASCE, JHD, 108, HY10, pp1179-1193

Yang C T et al, 1987, Hydraulic geometry and minimum rate of energy dissipation, Water Resources Res.17 no 4, pp1014-1018

**Table 1: Comparison of observed and predicted regime parameters**

| OBSERVATIONS | | | | | | PREDICTIONS | | | | |
|---|---|---|---|---|---|---|---|---|---|---|
| Discharge $m^3/s$ | Width m | Depth m | Slope $x10^4$ | Sediment diameter (mm) | Sediment diameter (ppm) | Empirical regime theory P (m) | Bank stability d (m) | Critical bank shear B | Extremal hypothesis B | Parker d (m) |
| 262 | 56.7 | 2.03 | 1.667 | 0.528 | 126 | 78.2 | 3.98 | 479 | 63.7 | 2.37 |
| 8.3 | 11.0 | 1.28 | 1.08 | 0.196 | 100 | 13.9 | 3.98 | 12 | 9.38 | 1.03 |
| 0.42 | 1.83 | 0.36 | 4.129 | 0.230 | 172 | 3.15 | 3.98 | 1.05 | 2.4 | 0.65 |

P is wetted perimeter

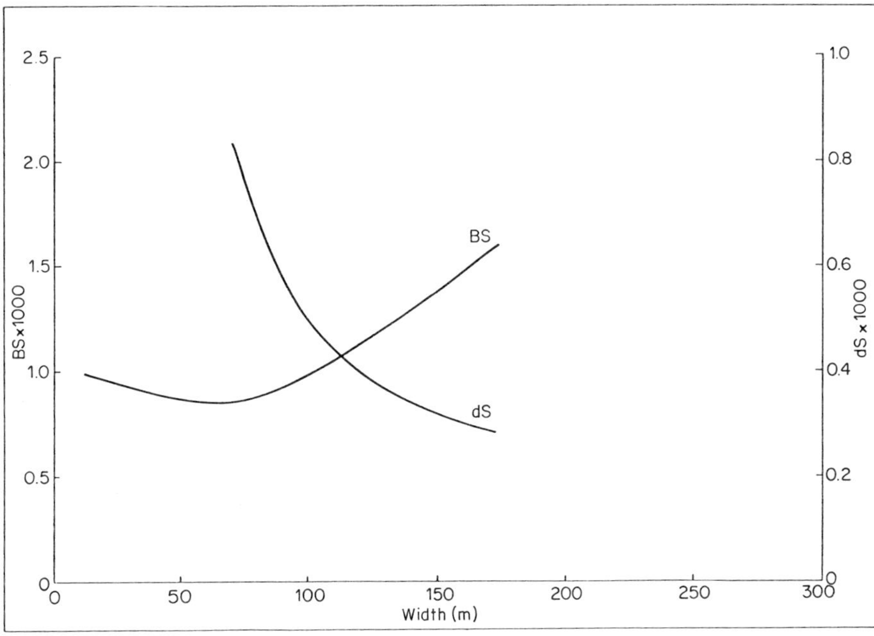


Fig 1 BS and dS against width

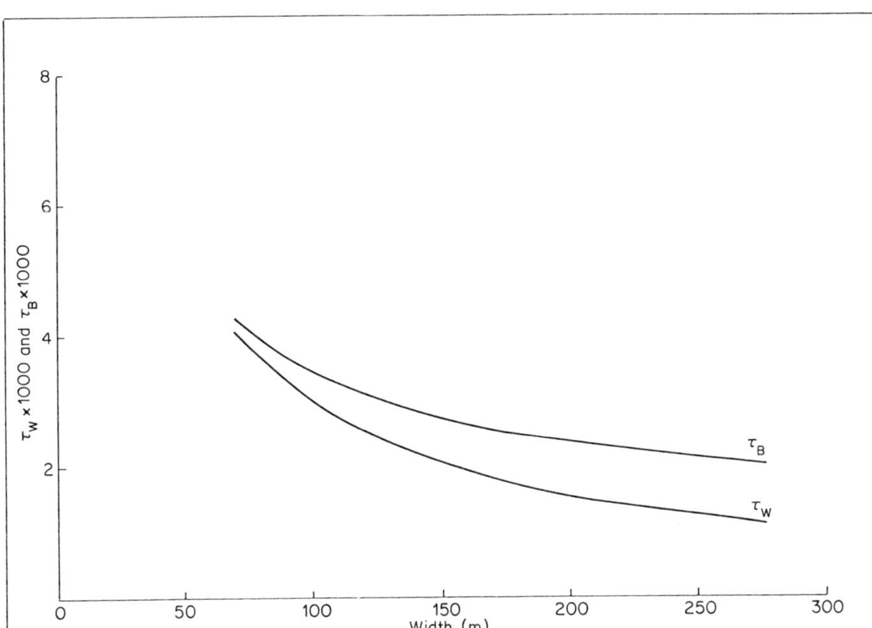


Fig 2 $\tau_W$ and $\tau_B$ against width

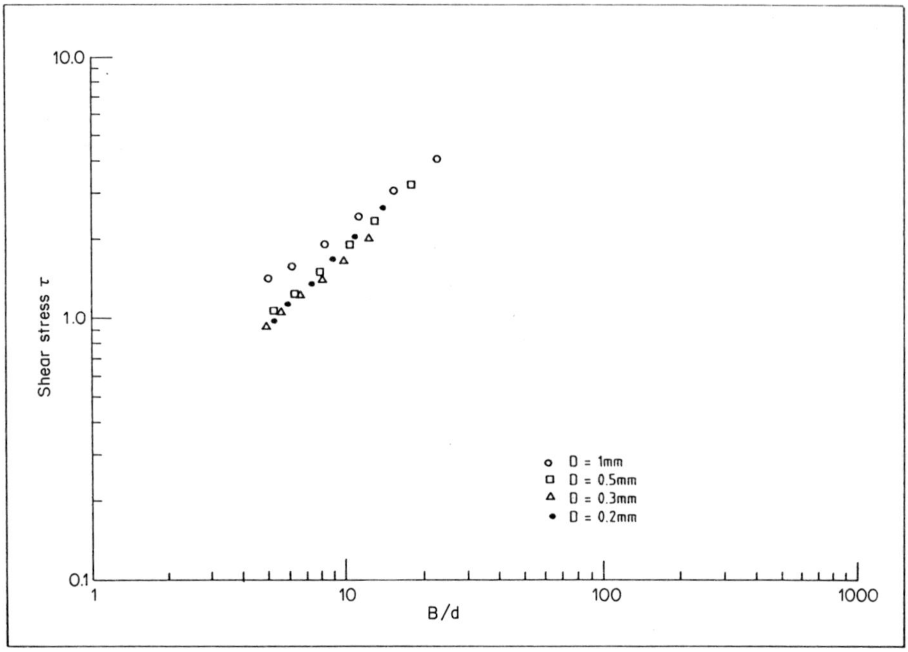


Fig 3 Shear stress against B/d

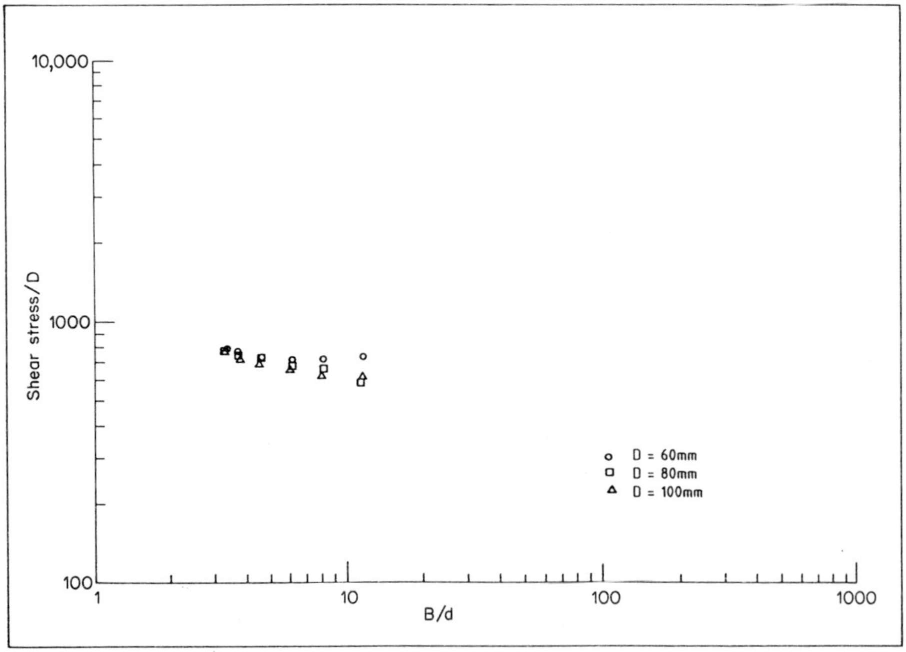


Fig 4 Shear stress/D against B/d

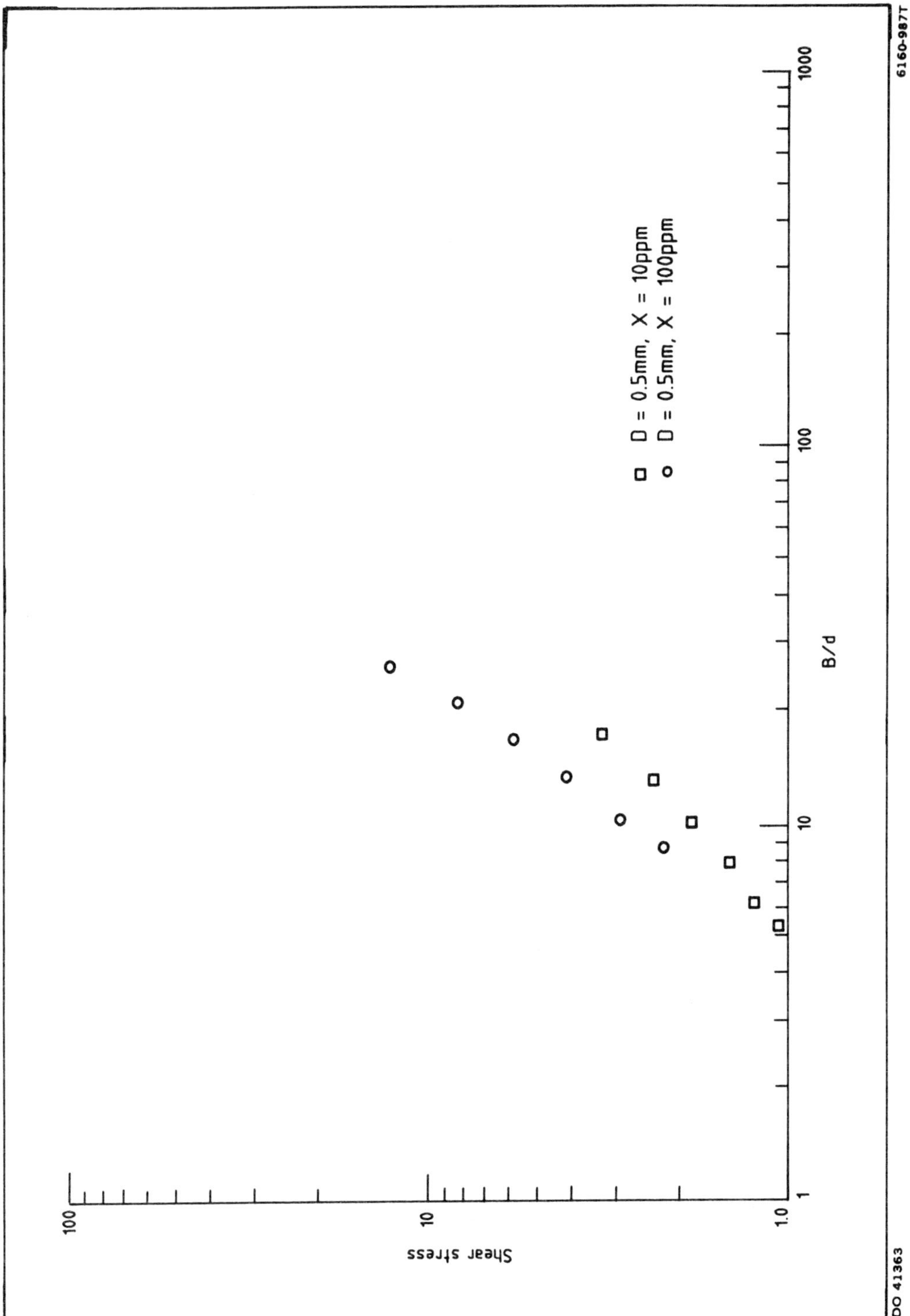


Fig 5 Shear stress against B/d

International Conference on
RIVER REGIME
18-20 May, 1988

PAPER D3

# EFFECTS OF GRADATION OF BED MATERIALS AND TRANSPIRATION THROUGH A BED ON BED LOAD TRANSPORT

Hiroji NAKAGAWA*, Tetsuro TSUJIMOTO** and Shogo MURAKAMI*
*Department of Civil Engrg., Kyoto University, Japan
**LHYDREP, E.P.F.L., Lausanne, Suisse

SUMMARY: In general, a gravel bed is composed of widely graded size materials and it has high permeability. Hence, on prediction of bed load transport rate, the effects of such characteristics cannot be neglected. In this study, the effects of gradation of bed materials and transpiration through a bottom due to the high permeability on bed load transport process are investigated, based on the mechanism of bed load motion. Although these effects are often interrelated, they are independently treated in the present study. Here, a bed load transport model constituted by pick-up rate and step length plays a powerful role, because it is still applicable for non-equilibrium condition and it quite well represents the essential mechanism of bed load motion. The results obtained in this study may improve the accuracy of prediction of sediment transport rate discharge in gravel bed rivers.

## 1. INTRODUCTION

In most of gravel streams, the size distribution of bed materials is very wide, and the permeability of the layer of bed materials is so high as to accommodate a subsurface flow. These facts bring about various effects on bed load transport. For example, in the case of sediment mixtures, bed load motion for each grain size has particular characteristics and consequently a non-equilibrium situation and sediment sorting become apparent. Here, the term "non-equilibrium" is used for the situation where sediment transport is in a "relaxation process" to respond to the change of its driving force (for example the hydraulic condition). Meanwhile, the existence of seepage flow in the layer of bed materials means active mass and momentum exchange between surface flow and subsurface flow through a bottom, and it degenerates the surface flow structure, and the bed load transport process driven by it is eventually degenerated. In fact, however, these effects are complicated, and moreover they are interrelated. Thus, in the present study, these effects are separately or independently investigated, but some perspective view points are intended to be kept for the future coupling of the respective aspects.

The characteristics of sediment transport of mixture is that the respective size fractions show different behaviour, and usually, transport rate formula for each grain size is used. The difference of transport behaviour due to size brings about a sediment sorting process and then the transport process becomes inevitably non-equilibrium. Thus, here we develop a model composed of pick-up rate and step length, which was originally established by Einstein (Ref 1) but specially adjusted as one applicable even to non-equilibrium condition by us (Ref 2), for sediment mixtures, in which sediment pick-up rate and step length are defined for each grain size. When we have such a transport model for each grain size of sediment mixture applicable even to non-equilibrium process, we can describe the sorting process accompanying the transport process.

International Conference on River Regime, edited by W.R.White.

The influence of the presence of subsurface flow under the bottom is not so easy to be investigated, particularly when mass and momentum exchange between surface flow (over a porous medium) and subsurface flow (in the porous medium) are active. Certainly, if there is net transpiration (suction or injection) stationarily through a bottom, it is relatively easy to guess that it changes the surface-flow structure and affects bed load motion directly and indirectly (through the change of surface-flow structure). Since such effects may be non-linear, we can expect some residual effect even for the situation without net transpiration. In the present study, such kind of residual effect will not be discussed, but the effect of steady transpiration on surface-flow structure and that on bed load transport process (more concretely saying, on sediment pick-up rate and step length) are investigated.

The effects of gradation of bed materials and transpiration velocity through a permeable bottom on flow structure and bed load transport process are also here investigated respectively with laboratory experiments. They provide us a confirmation that the present respective analytical treatments are available and they may suggest the possibility to analyze both effects simultaneously.

## 2.BED LOAD TRANSPORT OF SEDIMENT MIXTURES

### 2.1 Flow over a bed composed of sand mixtures

It is available for study on bed load transport of sediment mixtures to consider quantities for each grain size. But, flow velocity profile over a bed composed of sediment mixture is common for each grain size, and it is important to determine the representative size for the "roughness".In previous studies, $d_m$ (mean diameter), $d_{65}$ and other characteristic diameters were used as $k_s$("equivalent sand roughness") for convenience sake without any physical background.

When the flow properties as expressed by velocity profile over a mixed sand bed is "equivalent" to that over a uniform sand bed of diameter $d_e$, the tractive forces acting on both beds per unit area must be identified, and the following equation should be valid.

$$\varepsilon_o C_{D0}(\rho u_{bo}{}^2/2)A_2 d_e{}^2/(A_2' d_e{}^2) = \Sigma\,[\varepsilon_i C_{Di}(\rho u_{bi}{}^2/2)A_2 d_i{}^2 p_i/(A_2' d_i{}^2)] \tag{1}$$

in which d=sand diameter; ε=sheltering coefficient; ρ=mass density of fluid; $A_2$, $A_2'$,=areal geometrical coefficients of sand ($A_2 \neq A_2'$); $C_D$=drag coefficient; $u_b$=local flow velocity near the bed material particles; $p_i$=volume fraction of sands with the diameter $d_i$; and the subscript 0 and i indicate the quantities for the uniform sand and those for the i-th fraction of sediment mixure with the diameter $d_i$. If it is assumed that $C_{Di}=C_{D0}$ and $\varepsilon_i=\varepsilon_0$ for simplicity,and the logarithmic law for rough turbulent flow velocity distribution is applied, Eq.(1) can be written by the following equation.

$$[\ln(30.1a/\beta)]^2 = \int_{\eta_c}^{\infty} \{\ln[30.1a\eta/(\beta\eta_e)]\}^2 p(\eta)\,d\eta \tag{2}$$

in which z=ad=height of the center of the particle from the theoretical wall of velocity distribution ($a \cong 0.5$); $\beta=k_s/d$ ($\beta=1\sim3$); $\eta=d/d_{50}$; $\eta_e=d_e/d_{50}$; $d_{50}$=median diameter of sand mixture; $\eta_c=\beta\eta_e/30.1a$; and $p(\eta)$=density distribution of sediment size. In Eq.(2), the drag forces acting on the particles existing below the theoretical wall are assumed to be zero. In the other words, the sands finer than ($k_s/30.1a$) are neglected in the force balance. From Eq.(2), the following equation with respect to $\eta_e$ is obtained:

$$A(\eta_e)\ln^2\eta_e - 2B(\eta_e)\ln\eta_e + C(\eta_e) = 0 \tag{3}$$

in which

$$A(\eta_e)=1-P(\beta\eta_e/30.1a) \quad (4)$$

$$B(\eta_e)=[1-P(\beta\eta_e/30.1a)]\ln 30.1a+E[\ln\eta] \quad (5)$$

$$C(\eta_e)=2(\ln 30.1a/\beta)E[\ln\eta]+E[\ln^2\eta]-P(\beta\eta_e/30.1a)\ln^2(30.1a/\beta) \quad (6)$$

and $E[x]=\int_{\eta_c}^{\infty} xp(\eta)d\eta$ ; $P(\xi)=\int_{0}^{\xi} p(\eta)d\eta$.

In Fig.1, the calculated value of $\eta_e$ for sand mixture with log-normal type size distribution is shown against the geometric standard deviation of grain size distribution defined as $\sigma_g=\sqrt{d_{84}/d_{16}}$. According to this figure, $d_e$ does not vary so much compared with $d_{65}$ and $d_m$ against $\sigma_g$.

2.2 Characteristics of Individual sediment Motion for Each Grain Size

Since sediment transport of graded materials becomes non-equilibrium easily, ordinary transport formula established under assumption of equilibrium cannot be applied, but the model constituted by step length and pick-up rate is still applicable for such a non-equilibrium situation and it is also applicable to description of sorting process as mentioned later. The bed load transport model composed of sediment pick-up rate and step length was first proposed as a kind of stochastic model by Einstein (Ref 1), but it was much improved to more applicable version by Nakagawa-Tsujimoto (Ref 2). In such a model, sediment pick-up rate is defined as the probability density per unit time for a bed material particle to be dislodged from a bed by fluid action, while the step length is defined as the distance for a dislodged particle to travel as a bed load to the next definite stop on the bed.

As for the incipient motion of bed materials, the most different aspect between sediment mixtures and uniform sands is the critical tractive force. According to Egiazaroff (Ref 3), the dimensionless critical tractive force for each grain size of sediment mixture is given as follows:

$$\tau_{*ci}/\tau_{*c0}=[\ln(30.1a)/\ln(30.1a\zeta_i)]^2 \quad (7)$$

in which $\tau_{*c}$=dimensionless critical tractive force; and $\zeta_i=d_i/k_s$. The mean diameter $d_m$ is often used instead of $k_s$, but $k_s$ should be used because it is related to the velocity profile represented by log-law. Furthermore, for the finer sands, the above equation should be modified as follows according to Ashida-Michiue (Ref 4):

$$\tau_{*ci}/\tau_{*c0} = 0.8/\zeta_i \quad (8)$$

For uniform bed materials, Nakagawa-Tsujimoto (Ref 2) proposed a following formula to evaluate the pick-up rate:

$$p_{s*}=p_s\sqrt{d/(\sigma/\rho-1)g} = F_0\tau_*(1-k_2\tau_{*c}/\tau_*)^m \quad (9)$$

in which $\sigma$=mass density of sand; g=gravitational acceleration; $F_0$=0.03; $k_2$=0.7; and m=3 (Ref 2). The original form of the above equation was derived quite theoretically, but the above expression was modified for simplicity.

If we use $\tau_{*i}$ and $\tau_{*ci}$ instead of $\tau_*$ and $\tau_{*c}$, respectively, we can easily evaluate the pick-up rate for each grain size of sand mixture as follows:

$$p_{s*i}=p_{si}\sqrt{d_i/(\sigma/\rho-1)g} = F_0\tau_{*i}(1-k_2\tau_{*ci}/\tau_{*i})^m \tag{10}$$

In Fig.2, the experimental data as for the pick-up rate for each grain size are shown compared with Eq.(10).

On the other hand, the mean step length of bed load transport of uniform sand is about 80~250 times sand diameter and is less affected by the flow parameter than the pick-up rate according to the experimental data (Ref 2). The step length is in general composed by several successive irregular saltations and thus it is not so easy to evaluate analytically, based on the mechanics of particle motion, even for uniform sand. According to the data collected by Nakagawa et al. (Ref 5), the mean step length of each grain size of sand mixture can be expressed as follows:

$$\Lambda_i=\lambda_i d_i \tag{11}$$

in which $\lambda_i$ is almost constant and about 20~40, which is slightly smaller than uniform sand. For non-equilibrium bed load transport process, the statistical distribution of step length plays an important role (Ref 2), and according to the experimental data (see Fig.3) the step length for each grain size follows an exponential distribution as that for uniform sand bed does.

## 2.3 Non-Equilibrium Transport of Sediment Mixtures and Armouring Propagation

Non-equilibrium transport of graded materials in general accompanies sorting process. Here, the case of the so-called "parallel degradation" defined by Gessler (Ref 6) is treated (see Fig.4), in which formation and propagation processes of armour coat are observed, by applying the bed load transport model constituted by pick-up rate and step length defined for each grain size.

For convenience' sake, the alluvial bed downstream of the rigid bed (without sediment supply) is divided into intervals of finite length $\Delta x$ (see Fig.5). This interval length should be sufficiently longer than the maximum sand diameter and should be shorter than the mean step length of the minimum diameter of significant sand fractions. In the following, the quantities have two subscripts, i (1~N) and k (1~K), and they represent the size fraction of sands and the interval of the stream (see Fig.5), respectively. For example, $n_{ik}$ means the number of particles of the i-th fraction of size exposed at the surface of the k-th interval. Then, the number basis size distribution $q_{ik}$ and the volume basis size distribution $p_{ik}$ are expressed as follows:

$$q_{ik}(t)=n_{ik}(t)/\sum_{j=1}^{N} n_{jk}(t) \tag{12}$$

$$p_{ik}(t)=q_{ik}(t)d_i/\sum_{j=1}^{N} [q_{jk}(t)d_j] \tag{13}$$

Using the concepts of pick-up rate, the number of particles of the i-th fraction dislodged from the k-th interval during the time interval $\Delta t$, $\Delta M_{ik}(t)$, is written as follows:

$$\Delta M_{ik}(t)=n_{ik}(t)p_{sik}(t)\Delta t \tag{14}$$

Meanwhile, the number of particles of the i-th fraction depositing on the k-th interval during $\Delta t$, $\Delta Q_{ik}$, is expressed by making use of the concept of the distribution of step length as follows.:

$$\Delta Q_{ik}(t) = \sum_{s=1}^{k-1} [\Delta M_{is}(t)\mu_{i,k-s}] \tag{15}$$

in which $\mu_{i,k-s}$ is the probability that the step length of the i-th fraction is longer than $(k-s)\Delta x$ and shorter than $(k-s+1)\Delta x$.

As the result of incessant repetition of dislodgement and deposition, the lower layer is newly exposed at the surface in case of degradation, and it is now assumed that the lower layer material has the initial size distribution, $p_{io}$. Considering the aforementioned events, the number of particles of the i-th fraction exposed at the bed surface of the k-th interval at $(t+\Delta t)$ can be written as

$$n_{ik}(t+\Delta t) = n_{ik}(t) - \Delta M_{ik}(t) + p_{io} \sum_{j=1}^{N} \{[\Delta M_{jk}(t) - \Delta Q_{jk}(t)](d_j/d_i)^2\} \tag{16}$$

This can be rewritten by the following differential equation after substituting Eqs.(14) and (15).

$$\dot{n}_{ik}(t) = - \sum_{j=1}^{N} \{p_{io}(d_j/d_i)^2 - \delta_{ij}\} \sum_{s=1}^{k} [\tilde{\mu}_{j,k-s} p_{sjs}(t) n_{js}(t)]\} \tag{17}$$

in which $\delta_{ij}$=Kronecker's delta; $\tilde{\mu}_{i,k-s}=\mu_{i,k-s}$ for (s<k), -1 for (s=k); and 0 for (s>k). Based on the above-deduced equation, the temporal and spatial variation of gradation of surface layer of the bed can be calculated in succession.

The transport rate is also treated for each grain size, as already suggested by Einstein (Ref 7), Ashida-Michiue (Ref 4) and others. However, the transport process accompanying sorting is always non-equilibrium, and thus the previous treatment without consideration on non-equilibrium process cannot be applied to such a situation. Based on the model proposed here, the non-equilibrium transport rate for each grain size can be reasonably expressed as a function of time and space, as follows.

$$q_{Bi}(t, K\Delta x) = (A_3 d_i/A_2) \sum_{k=1}^{K} [p_{ik}(t) p_{sik}(t) \Delta x \sum_{s=K-k}^{\infty} \mu_{i,s}] \tag{18}$$

in which $q_B$=bed load transport rate; and $A_3$=volumetric geometrical coefficient of sand.

In Fig.6, the experimental data for armour coat propagation process obtained by Ashida-Michiue (Ref 4) are shown with calculated prediction based on the present model. The agreement between them is reasonable.

## 3. EFFECT OF STEADY TRANSPIRATION THROUGH A BOTTOM ON BED LOAD TRANSPORT

### 3.1 Free Surface flow Structure with Transpiration

In the region where the so-called wall law is applied, the following equation is valid for the flow with transpiration.

$$\tau_0 = -\rho\overline{uv} - \rho v_0 U \tag{19}$$

in which $\tau_0=\rho g h I_e$=bed shear stress; $I_e$=energy slope; u, v=longitudinal and vertical components of velocity fluctuation of free surface flow; U(y)=free surface flow velocity profile; $v_0$=transpiration velocity through a permeable bottom (injection to the surface flow is taken positive). The first and the second terms of the right-hand side are the Reynolds stress and the contribution of the transpiration velocity to the momentum transfer, respectively. Since the bed composed of granular materials and thus with high permeability is generally hydraulically rough, the viscous term has been neglected. When the so-called mixing length theory is applied to the term of the Reynolds stress and the mixing-length is assumed to be $l=\kappa_0 y$ ($\kappa_0$=Karman constant), Eq.(19) is rearranged in the following version.

$$(\kappa_0 y^+/v_0^+)[d(1+v_0^+U^+)/dy^+] = \sqrt{1+v_0^+U^+} \tag{20}$$

in which $U^+=U/u_*$; $v_0^+=v_0/u_*$; $y^+=yu_*/\nu$; $u_*$=shear velocity; $\nu$=kinematic viscosity of fluid; and $(1+v_0^+U^+)>0$ is imposed here. Solving Eq.(20) with respect to $U^+$ gives the following velocity distribution formula.

$$U^+=[v_0^+/(4\kappa_0^2)]\ln^2(y/k_s)+[(1+v_0^+D_r/2)/\kappa_0][\ln(y/k_s)]+(1+v_0^+D_r/4)D_r \tag{21}$$

in which $D_r=(1/\kappa_0)\ln(k_s/y_0)$; and $y_0$=height at which the free surface flow velocity becomes theoretically zero in case of flow without transpiration. This is called "bi-logarithmic law" for flow over a hydraulically rough bed. Dorrance-Dore (Ref 8) first derived it for a smooth bed. If $v_0=0$, Eq.(21) gives the ordinary logarithmic law.

As an index of the effect of transpiration on the flow resistance, $\Omega=U_{mn}/U_m$ is here introduced. The subscript n indicates the value for flow without transpiration but with the same surface flow discharge and energy gradient; while the subscript m indicates the depth-averaged values. $\Omega$ also means $(h/h_n)$ and $(f/f_n)^{1/3}$, in which h=flow depth; and f=friction factor. Based on the average flow velocity formula obtained by integrating Eq.(21) along the flow depth, we obtain the following relation between $\Omega$ and $v_0^+$ with $f_n$ as a parameter.

$$(v_0^+/4)[(2/f_n)+(\ln^2\Omega+1)/\kappa_0]+(\ln\Omega)/\kappa_0=\sqrt{2/f_n}\{\Omega^{-3/2}-[v_0^+/(2\kappa_0)]\ln\Omega-1\} \tag{22}$$

The flow measurements were carried out in the laboratory flume 33cm wide and 8m long with adjustable slope (see Fig.7), under the following hydraulic conditions: h=3.78~7.79cm; $i_b$(=bed slope)=1/500~1/300; Fr(=Froude number)=0.41~0.57; and $v_0$=-0.74~0.55cm/s. Two kinds of uniform sands were used as bed materials, of which diameter are 0.164 and 0.136cm. The measured velocity profiles are shown in Fig.8, and the present model can well explain the effect of transpiration. Furthermore, the relation between $\Omega$ and $v_0^+$ obtained by the experiments is shown in Fig.9 with the theoretically predicted curve, and a good agreement is recognized here.

### 3.2 Effect of Transpiration on Pick-up Rate

The forces acting on a sand particle on a permeable bed with the longitudinal slope $\alpha$ are the drag force D $(=\varepsilon_0 C_D A_2 d^2 \rho U_d^2/2)$ and the lift force L $(=k_L D)$ due to the free surface flow; the subsidiary drag force due to the transpiration velocity itself $D_s$ $(=\varepsilon_0' C_D' A_2 d^2 |v_0| v_0/2$; the weight of sand particle W $(=\sigma g A_3 d^3)$; the buoyant force B $(=\rho g A_3 d^3)$; and the frictional force $f_r$ (see Fig.10); in which $\varepsilon_0'$ and $C_D'$=sheltering coefficient and drag coefficient related to transpiration. The longitudinal magnitude of the resultant force, F, is expressed by a dimensionless version as follows:

$$F/W=[1+(k_L+D_s/D)\mu_f](D/W)-[\mu_f(\cos\alpha-B/W)-\sin\alpha] \tag{23}$$

in which $\mu_f$=static frictional coefficient of sand. F=0 corresponds to the so-called critical condition for incipient motion of bed material particles.

Assuming that a sand particle is to be dislodged in form of sliding motion in case of $F>0$, the equation of motion is expressed in terms of $M\dot{v}_g=F$; in which $M=\rho(\sigma/\rho+C_M)A_3d^3$; $C_M$=added mass coefficient of sand: $\dot{v}_g$=acceleration at incipient motion. Associating the solution of the equation of motion with the effect of fluctuating bed shear stress, sediment pick-up rate, which is defined as the probability density per unit time for a bed material particle to be dislodged from a bed, can be evaluated. Using a similar technique by which Nakagawa-Tsujimoto (Ref 2) derived an approximating formula of pick-up rate for flat bed without transpiration, the pick-up rate of a sand particle on a permeable bed with transpiration can be expressed as follows:

$$p_s{*}=(F_0/\Psi)\tau{*}[1-(k_2\Psi\tau{*}_{c0}/\tau{*})]^m \quad (24)$$

$$\Psi=(1+k_L\mu_f)/\{\phi_0{}^2[1+(k_L+\phi_1)\mu_f]\}\ ; \quad \tau{*}_c=\Psi\tau{*}_{c0} \quad (25)$$

$$\phi_0=(U_d{}^+/U_{dn}{}^+)/\sqrt{\Omega}\ ; \quad \phi_1=[\varepsilon_0'C_D'/(\varepsilon_0C_D)]v_0{}^+|v_0{}^+|/U_d{}^{+2} \quad (26)$$

$$\tau{*}_{c0}=\{(\sigma/\rho)/[(\sigma/\rho-1)(1+k_L\mu_f)\Xi_0]\}[(\cos\alpha-\rho/\sigma)\mu_f-\sin\alpha] \quad (27)$$

$$\Xi_0=\varepsilon_0[C_DA_2/(2A_3)](U_{dn}/u{*}_n)^2\ ; \quad \tau{*}=u{*}^2/[(\sigma/\rho-1)gd] \quad (28)$$

in which $U^+=u_d/u{*}$; and $\Omega=(u{*}/u{*}_n)^2$. The drag coefficients are given by the following expressions for the whole range of Reynolds number.

$$C_D=0.4+24/(U_d{}^+Re{*})\ ; \quad C_D'=0.4+24/(|v_0{}^+|Re{*}) \quad (29)$$

in which $Re{*}=u{*}d/\nu$. According to Eq.(24), the effect of the transpiration is represented by a parameter $\Psi$, and it is determined by $\phi_0$ and $\phi_1$. The former represents the change of the free surface flow structure, while the latter expresses the drag effect directly caused by the transpiration velocity.

Fig.11 shows the relation between $p_s/p_{sn}$ and $v_0{}^+$ with $\eta$ as a parameter, in which $\eta=\tau{*}/\tau{*}_{c0}$. In general, the injection promotes the dislodgement of bed material particles. In order to verify the present theoretical derivation, the laboratory experiments were carried out in the flume as shown in Fig.7, in which the sediment pick-up rate was measured by video-film analysis. The experimental data are plotted in Fig.11 with the theoretical results. Moreover, the calculated $p_s$ and measured $p_s$ are compared in Fig.12. These figures suggest that the derived formula can reasonably estimate the pick-up rate for flow with transpiration.

3.3 Effect of Transpiration on Step Length of Bed Load

One step length is the total length of a succession of several irregular saltations. In other words, some collisions with a bed are included in a step length. When the mean particle's speed of 'small steps' involved in a "step length" is represented by $u_p$ and the number of succession of small steps $N_s$, the step length is proportional to the product of them. The change of the surface-flow structure due to transpiration affects the particle speed; while the residual part of the step length must be subjected rather to the geometrical irregularity of bed surface (of the order of sand diameter) and thus it may be independent of transpiration itself. In other words, the dimensionless mean step length can be written as a product of $\Psi_\Lambda$ and $\phi_p$ as written by the following equation, and $\Psi_\Lambda$ is independent of $v_0{}^+$.

$$\lambda=\Lambda/d=\Psi_\Lambda(\tau{*})\,\phi_p(\tau{*}) \quad (30)$$

in which $\phi_p=u_p/u{*}$; $\Psi_\Lambda=\lambda/\phi_p$; and $\Lambda$=mean step length.

Without transpiration, the mean step length can be expressed as a function of bed shear stress as follows: Under equilibrium condition, the bed load transport model gives the following relation:

$$\lambda=(A_2/A_3)q_{Be}*/p_s* \ ; \quad q_B*=q_B/\sqrt{(\sigma/\rho-1)gd^3} \tag{31}$$

in which the subscript e indicates the equilibrium value. In general, the equilibrium bed load transport without transpiration is written by the following version.

$$q_{Bne}*=A_0\tau*_n{}^n(\tau*-\tau*_{c0})^s \tag{32}$$

in which $A_0$, n and s are the constants for each bed load transport formula. Then, $\lambda_n$ is written by the following form.

$$\lambda_n=(A_0A_2/A_3)\tau*_n{}^{n+s-1}(1-\tau*_{c0}/\tau*_n)^s/(1-k_2\tau*_{c0}/\tau*_n)^m \tag{33}$$

If it is assumed that the mean particle's speed within a small step can be represented by the equilibrium speed of sliding motion, $\phi_p$ can be related to the flow velocity profile, as follows:

$$\phi_p=U_d/u*-\sqrt{\mu_k[2A_3/(C_DA_2)]}[\sqrt{(\sigma/\rho-1)gd}/u*] \tag{34}$$

in which $\mu_k$=kinematic frictional coefficient of a sand on a bed.

By considering the case without transpiration, $\Psi_\Lambda$ is determined as $\Psi_\Lambda=\lambda_n/\phi_{pn}$. Consequently, the change of the mean step length due to the transpiration velocity can be expressed by

$$\gamma=\lambda/\lambda_n=[\Psi_\Lambda(\tau*)/\Psi_\Lambda(\tau*_n)]\ [\phi_p(\tau*)/\phi_p(\tau*_n)] \tag{35}$$

in which $(\tau*/\tau*_n)$ is expressed by $\Omega$, and $\Omega$ has been given by Eq.(22). $\Psi_\Lambda$ and $\phi_p$ have been here evaluated. Fig.13 shows the relation between $\gamma$ and $v_0{}^+$ with $\eta$ as a parameter, where Bagnold formula (Ref 9) is applied ($A_0$=4.25; n=0.5; and s=1.0), and it is assumed that $\mu_k$=0.4 and $C_D$=0.4. When $\tau*$ is not so large, the injection suppresses the step length.

In order to verify the present derivation of step length, the experimental data were collected in the laboratory. The alluvial part where uniform transpiration was present was 1.5m long, and made of uniform sand (d=0.199cm). The data of the step length were obtained by direct measurements by eyes, where more than hundred and fifty samples were measured for each experimental run and their statistical properties were investigated. The values of $\gamma$ based on the measured step length are plotted in Fig.13 with the theoretical curves, and they show relatively good agreements. Meanwhile, the measured distribution of step length is shown in Fig.14 in the version of the normalized probability density, and it suggests that the step length on the flat bed with transpiration still approximately follows an exponential distribution.

## CONCLUSIONS

In this study, the effects of gradation of bed materials and transpiration through a bed on bed load transport have been investigated separately both theoretically and experimentally.

At first, the equivalent sand roughness of bed composed of sediment mixture has been reasonably evaluated against the standard deviation of gradation distribution of sediment size. Then, we can apply the logarithmic law reasonably for flow over a bed composed of sediment mixture, and sediment pick-up rate for each grain size can be theoretically evaluated using the knowledge for uniform sand. Considering on non-equilibrium transport of sediment mixture, bed load transport model constituted by pick-up rate and step length has been developed to apply to description of sorting process inevitably brought about by non-equilibrium transport of sediment mixtures. These results have been verified by the experimental data.

Next, the effect of transpiration velocity through a bottom, such as injection and suction, on the surface-flow structure has been investigated, and "bi-logarithmic law" for turbulent flow over a rough permeable bed has been deduced. Using the bi-logarithmic law instead of the logarithmic law, the bed load transport model can be applied to the situation with transpiration. Considering the applicability to non-equilibrium situation and furthermore to sediment-mixture transport including sorting process, the changes of pick-up rate and mean step length for the situation with transpiration from those without transpiration have been reasonably evaluated, and the theoretical derivations have been verified by the laboratory experiments.

In gravel bed rivers, these effects may play important roles respectively and furthermore both effects act simultaneously. Through the theoretical development in this study, it is obviously possible to couple these two effects. For example, sediment pick-up rate can be evaluated for each grain size of sediment mixture using the bi-logarithmic law as a velocity profile for the situation with transpiration instead of the ordinary logarithmic law. However, the experimental verifications still remain.

## REFERENCES


Einstein, H.A. : "Der Gescheibetrieb als Wahrscheinlichkeitsproblem". Mitteilung der Versuchsanstalt für Wasserbau an der Eidgenössische Technische Hochscule, Zürich, Verlag Rascher, 1937.

Nakagawa, H. and T.Tsujimoto : "Sand bed instability due to bed load motion". Journal of the Hydraulics Division, Proc. ASCE, Vol.106, HY12, 1980, pp.2029-2051.

Egiazaroff, I.V. : "Calculation of nonuniform sediment concentration". Journal of the Hydraulics Division, Proc. ASCE, Vol.91, HY4, 1965, pp.225-247.

Ashida, K. and M. Michiue : "Study on hydraulic resistance and bed-load transport rate in alluvial streams". Proc. JSCE, No.206, 1972, pp.59-69 (in Japanese).

Nakagawa, H., T. Tsujimoto and S. Nakano : "Characteristics of sediment motion for respective grain sizes of sand mixtures". Bulletin, Disaster Prevention Research Institute, Kyoto University, Vol.32, Part 1, 1982, pp.1-32.

Gessler, J. : "Self-stabilizing tendencies of alluvial channels". Journal of the Waterways, Harbors and Coastal Engineering Divisions, Proc. ASCE, Vol.96, WW2, 1970, pp.235-248.

Einstein, H.A. : "The bed load function for sediment transportation in open channel flows". Technical Bulletin, U.S. Department of Agriculture, Soil Conservation Service, No.1026, 1950.

Dorrance, W.H. and F.J. Dore : "The effect of mass transfer on the compressible turbulent boundary layer skin friction and heat transfer". Journal of Aeronautical Science, Vol.21, 1954, pp.404-410.

Bagnold, R.A. : "The flow of cohesionless grains in fluids". Philosophical Transactions, Royal Society of London, Series A, Vol.249, 1957, pp.235-297.

International Conference on
RIVER REGIME
18-20 May, 1988

PAPER E1

# CHANNEL CHARACTERISTICS OF THE BRAIDING JAMUNA RIVER, BANGLADESH

by

G.J. Klaassen and K. Vermeer

DELFT HYDRAULICS, P.O. Box 152
8300 AD Emmeloord, The Netherlands

SUMMARY

Channel characteristics of the braiding Jamuna River in Bangladesh were studied within the framework of a feasibility study for a bridge across this large braiding sand-bed river. Planform characteristics were studied on the basis of LANDSAT-images. The variation of some topological and geometrical braiding indices over the years was studied for in total six reaches. Also the planform of channel bends was studied. Cross-sectional characteristics were derived from bed level soundings made by the Bangladesh Water Development Board since 1966/67. From the analysis of the available data it is concluded that over the last decades any lateral shifting of the Jamuna River is not noticeable. This is in contrast with the behaviour of the river since the avulsion at the end of the 18th century, which was followed by a Western movement of up to 10 km in some reaches.

## 1. Introduction

The Jamuna River in Bangladesh is the lowest reach of the Brahmaputra River upstream of its confluence with the Ganges River (see Figure 1). The Jamuna River is a large, braiding sand-bed river. The average discharge during floods is about 50,000 $m^3/s$, the maximum width during floods about 17 km, the bed material is quite uniform and varies from 0.25 mm near the Indian border to 0.16 mm at the confluence with Ganges River. The valley slope in Bangladesh decreases gradually from 0.10 to 0.06 m/km. More details on the river characteristics can be found in e.g. Coleman (1969) and, more recently, in RPT/NEDECO/BCL (1987), notably appendix C.

About two centuries ago the Brahmaputra River shifted from its original course (presently named the Old Brahmaputra River, see Figure 1) to its present course, now known as the Jamuna River. The reason for this avulsion is not discussed here, as it is outside the scope of this contribution. Readers interested in this phenomenon are referred to Morgan and McIntyre (1967) and Coleman (1969). Since its avulsion the Jamuna River has increased its total width and has gradually shifted in Western direction, in some reaches up to 10 km (see Coleman (1969) or IECO (1969)). Within the framework of a feasibility study for a bridge across the Jamuna river, to be located slighty downstream of Sirajganj (see Figure 1), an important issue is whether or not the river has reached an equilibrium condition in the centuries that have passed since

International Conference on River Regime, edited by W.R.White.

the avulsion. This was especially important because the risk of outflanking of the bridge had to be evaluated as part of the feasibility study.

During this evaluation a detailed analysis was made of all the available data on the Jamuna River. In this process a number of interesting data on the characteristics of this large braiding sand-bed river became available. In this paper a summary of these results is given. Broadly speaking, a distinction can be made in planform data and cross-sectional data. They are discussed here seperately in Chapters 2 and 3, respectively. Some preliminary conclusions on the river's present state are given in Chapter 4.

## 2. Planform characteristics

### 2.1 General

Planform characteristics of braided streams can be divided in:
(i) characteristics of the network of channels and
(ii) characteristics of the individual channels.
Howard et al (1970) have introduced a number of topological and geometrical indices for networks of braided channels. Their approach and definitions were used in the analysis of the Jamuna planform characteristics. Details are provided in Section 2.3. Planform characteristics of individual channels relate to radii of curvature of bends in relation to channel widths, the total arc length of bends and channel directions. These parameters and their correlation are discussed in Section 2.4. The data used in the analysis are described in Section 2.2.

### 2.2 Available data

For the analysis discussed here the main source of information are satellite images, notably LANDSAT-scenes. These images are available from 1972 onward. The smallest details that can be distinguished on these images have approximate dimensions of 80 x 80 $m^2$. For this large river with channel widths in the order of kilometers this is quite sufficient. A selection of relevant LANDSAT-images was made. For the present study in particular the scenes 148-42 and 148-43 (for Landsat 4 and 5 138-42 and 138-43, respectively) are of relevance. Images from different years, covering the full period between 1973 and 1987, were used in the present analysis.

As the satellite images cover a rather short period only, attempts were made to use older data too. In particular old maps were consulted. Looking at the drawn channel configurations it had to be concluded, however, that most probably the maps were not based on quantitative information but should be characterized as a rough impression of what the river looked like from the banks. These maps were subsequently discarded. No older aerial photographs were obtained either, leaving the LANDSAT-images as the only source of reliable information. Figure 2 provides a print of a satellite image of the lower reaches of the Jamuna River in February 1973. Also the other selected satellite images are related to low flow conditions, allowing for a mutual comparison on realistic terms. A disadvantage of using low-flow images is that the width that can be observed on the scenes does not correspond to bankfull stages.

### 2.3 Braiding indices

Braiding indices are used to characterize channel networks of braided streams. Howard et al (1970) have proposed a number of parameters, partly from a topological and partly from a geometrical nature. In doing this, they made a distinction between channel segments, nodes where segments branch or join, and islands enclosed by segments. In the present paper the following parameters are used:

E average number of segments bisected by the crosslines at the ends and interior of a river reach (-)

$E_i$ excess segments index, defined via $E_i = E - 1$ (for a purely meandering river $E_i = 0$) (-)

N total number of segments entirely within the river reach and entering the reach from a lower numbered reach (-)

C average width of the stream between the outermost segments within the section (km).

Howard et al (1970) have proposed additional parameters (including average segment length, average wave length, etc.) and have also used derivative scale free parameters. Here these parameters are not used, because it is felt that in particular observed channel widths are subject to a considerable bias as they are related to low flow conditions.

For the present analysis the total length of the Brahmaputra (Jamuna) River in Bangladesh from the Indian border to the confluence with the Ganges River (see Figure 1) was divided in six reaches, each with a length of about 30 km. For each of these river reaches braiding indices were determined on the basis of LANDSAT-images. The derived indices for the parameters defined before are listed in Table 1 for four years (1973, 1978, 1981 and 1987). No data are available for the upper reaches in the years 1973 and 1978, as no satellite image was available.

The following observations can be made regarding the data provided in Table 1:

- The degree of braiding decreases in downstream direction. This may be in line with the observed decrease in valley slope (conform Leopold et al, 1967 and Struiksma and Klaassen, 1988).
- After 1981 the degree of braiding has increased, while the total width of the braiding pattern has not changed substantially. In particular reach 4, just upstream of Sirajganj (see Figure 1), has experienced substantial changes. The 1986 data (not included in Table 1) support the change in braiding characteristics. At present there is no explanation for the change. Attempts are being made to correlate the braiding pattern to the flood discharges in previous periods.

### 2.4 Bend characteristics

The LANDSAT-images were also used to study the characteristics of bends in individual channels, being part of the braided network. Some results are provided in the Figures 3 and 4. Figure 3 related to the relation between radii of curvature, channel widths (B) and arc lengths ($\psi$). A plot of radii of curvature versus channel widths for the year 1977 (see Figure 3 (a)) shows considerable scatter. In the figure also a relation derived from Leopold et al (1964) for meandering channels was plotted. This relation reads:

$$R = 2.35\ B^{0.99} \tag{2.1}$$

Leopold etal (1967) have not suggested this particular relation but it can be derived from their empirical relations between meander length, channel width and radius of curvature. The large deviation between Equation (2.1) and the present observations may be explained at least partly from the use of low flow data. As is shown in Section 3.5, the width of the channels varies with the discharge to the power of about 0.5. Average low flow discharges in the Jamuna River are a factor 10 lower than bankfull discharges (RPT/NEDECO/BCL , 1987), thus a consistent deviation of about 3 can be explained by the use of low flow data. In Figure 3(a) a corrected relation is plotted taking into account this factor 3. It may be concluded that for the present braided river the radii of curvature are on the average smaller than for meandering channels. This analysis does not take into account the possibility that the radii of curvature may change during the transition from flood to low flow stages.

A much better relationship is obtained when the arc length of the bends is plotted against the radii of curvature. For the low flow period of 1983/1984 this relationship is plotted in Figure 3 (b). From this figure it may be concluded that rather extended bends are only possible for small radii of curvature, and thus for smaller channels only (according to Equation (2.1), which does approximately hold for bankfull discharges). Thus the major channels are less curved than the smaller ones.

Channel directions for the year 1983/84 are presented in Figure 4. Here the observed channel directions (according to the definition shown in the figure) are plotted against the radii of curvature, as it was anticipated that the channel directions (expressed as the deviation from the overall valley slope) would decrease when the conveyance of the channels would increase. Taking into account the relationship between R and B (and thus the conveyance, see also Section 3.5), this is substantiated by Figure 4. This figure may have a significant bearing on the design of the river training works for the bridge. When near the bridge the different channels are combined to one major channel, the channel direction of this channel may differ only slightly from the direction of the valley slope.

## 3. Cross-sectional characteristics

### 3.1 General

Apart from planform data, also cross-sectional characteristics of the Jamuna River were studied. This was done on the basis of soundings which have been carried out by the Bangladesh Water Development Board (BWDB) since 1966/67. The analysis carried out was aiming mainly at the identification of any consistent changes in channel location, but as side product also the variation in wet area and in conveyance in longitudinal direction and over the years was obtained. Furthermore regime equations were derived for the channels in the Jamuna River and a comparison was made with a regime equation derived for (smaller) stable channels with low sediment transport rates.

The structure of this Chapter is as follows. The available basic data (i.e. BWDB soundings) are described in Section 3.2. Data on channel conveyance and wet area and their variation in place and time are provided and discussed in Section 3.3. In Section 3.4 the horizontal and vertical stability are studied. Section 3.5 is considering the dimensions of individual channels (notably widths and average depths), and an attempt is made to correlate them to the bankfull discharge of the channels.

The present analysis expands on recent work by Ullah (1987), who compared soundings of 1965/66 with more recent ones from 1983/83. By including soundings from other years, here a better insight is obtained into the fluctuations over the years in relation to any trends in bed levels and channel locations.

## 3.2 Available data

Cross-sectional soundings of the Jamuna River have been made since 1965/66. Initially these measurements were done every year, but they were interrupted in 1970 due to the Independence War. In 1975/76 the soundings were taken up again, although not every year the Jamuna River was included in the country-wide morphological measurements. In the period between 1965/66 and 1986/87 for most of the cross-sections in total some 10 soundings are available for analysis.

The soundings were carried out during the low flow season. Both the submerged areas and the dry islands (locally called "chars") in between the braids were sounded. The bed levels are related to the local Public Works Datum (PWD). The horizontal distance was measured from monuments. Sometimes the monuments got lost either by bank erosion or by other causes. Consequently for a number of cross-sections it is not possible to make a quantitative comparison of channel location in the horizontal direction.

Initially the distance between the cross-sections was 4 miles, but gradually more cross-sections have been introduced over the years. Figure 5 provides an overview of the sounded cross-sections.

In the present study twelve cross-sections were selected for detailed analysis considering:
1) conveyance
2) width (of channels and chars)
3) depth (average and maximum)
4) average cross-section level.

Cross-sections J5, J5-1, J6, J6-1, J8 were selected in view of their location close to the proposed bridge site (J6-1), the others to investigate possible changes in channel characteristics upstream and downstream of the bridge site. Figure 6 provides the results of soundings of two cross-sections (notably J13-1 and J6-1) for a number of years. Other cross-sections may have a total width of more than 15 km.

In a braided river the channel directions may deviate considerably from the direction of the valley slope (see Section 2.4), while furthermore this direction will change over the years. Consequently the channel width of individual channels may be too large when derived from the sounded cross-sections.

## 3.3 Processing of cross-sectional data

The cross-sectional data contain important information on channel characteristics and horizontal and vertical stability. To isolate this information, the cross-sectional data have to be processed further. A computer programs was developed for this purpose. This program uses the sounded data as input, and subsequently the following processing is being done per cross-section:

- for three discharge levels (44,000 $m^3/s$ approximately corresponding to bankful stage, 12,000 $m^3/s$ and 4,000 $m^3/s$ representing average low flow conditions in the Jamuna River) stages are plotted in the sounded cross-

sections, and channels and sand bars (chars) are differentiated;
- the total wet area for each of the three discharge levels is computed;
- going from the left (Eastern) bank to the right (Western) bank monument the cumulative conveyance is computed as a function of the distance to the left bank monument; in a second step this relationship is made dimensionless by dividing it by the total conveyance for that particular cross-section and stage (for an example of the obtained results see Figure 8);
- the width and the average levels of the individual channels in each cross-section is determined, together with the average char level, the average channel level averaged over all channels and finally the average bed level of the whole sounded cross-section, including both the parts above the water and the submerged parts.

It is of interest to study the results of the computation of the wet area and the conveyance of the river for bankfull conditions. In Figure 7 the computed values of these two parameters are plotted for some cross-section. From this figure it may be concluded that both the wet area and the conveyance do vary considerably over the years. There is no trend in the computed values, so apparently the variation is inherent to the river's behaviour. Including the direction of the different channels and thus reducing the computed values does not result in a reduction of the scatter.
It should be stressed here that the underlying assuption for the figures in Figure 7 is that the low flow cross-sections are representative for bankfull stages also. The scarce information available on changes of the cross-section through the year seems to support this assumption.

### 3.4 Horizontal and vertical stability

A major item is the stability of the Jamuna River in its present course. A distinction can be made betweeen the horizontal and the vertical stability.

Horizontal stability is considered on the basis of the relative conveyances computed according to the method outlined in the previous section. The relative conveyances for a sequence of years were plotted for all cross-sections considered. Some examples are presented in Figure 9, notably the cross-sections J6 and J5-1. Any significant horizontal shift in river location would show up in the figure as a consistent shift in the relative conveyance in either direction. Such a trend is not detectable. The relative conveyance do show a substantial scatter, and this is due to the phenomenon that gradually minor channels become more important and vice versa. That this occurs gradually (see also Figure 6) is different from what is stated by Coleman (1969). Coleman's data relate to thalweg locations and these may vary considerably in one year, while the actual changes in depth of different channels in one cross-section may be rather small. Other analyzed cross-sections show a similar picture as Figure 9. It seems therefore that over the last two decades there is no significant shifting of the Jamuna River in either direction.

Vertical stability was studied by considering the average bed levels of the channels, the chars and of the total cross-section over the years. In Figure 10 the average bed level over the whole width is plotted for all years considered and for the different cross-sections analyzed. Substantial changes may occur over subsequent years. At least partly this may be due to the absence of a criterion as to which part of the chars (= flood plain) level should be included when computing the average bed level. However, considering Figure 10 it may be concluded that on the average there is no consistent trend in the level of the Jamuna River either. So it seems that the Jamuna River is stable in the vertical direction also. The changes in planform in recent years (see Section 2.3) can therefore not be explained by the advance of an alluvial fan. Summarizing, it may be stated that it seems that over the last decades the Jamuna River is stable both in the horizontal and in the lateral direction. This contradicts the findings of Ullah (1987), who based on a comparison of 1965/66 and 1983/84 data the conclusion that substantial changes have occurred. Apparently concluded is not substantiated when more years are included in the analysis.

## 3.5 Regime equations for Jamuna channels

Regime equations have originally been developed for stable canals on the Indian subcontinent. Lacey (1930,1947) derived the following equations for the wetted perimeter p and the hydraulic radius R of a stable channel:

$$p = 2.67\ Q^{1/2} \tag{3.1}$$

$$R = 0.47\ (\frac{Q}{f})^{1/3} \tag{3.2}$$

where Q = bankfull discharge and f = silt factor, defined via:

$$f = 1.59\ D_{50}^{1/2} \tag{3.3}$$

where $D_{50}$ = diameter of the bed sediment in mm. All the other parameters are expressed in imperial units. Simons and Albertson (1963) extended the regime equations to include the effect of the soil properties of the banks. For canals with sand banks and beds they found:

$$p = 3.3\ Q^{0.512} \tag{3.4}$$

pointing at slightly larger widths for channels with sandy banks. Some others, e.g. Stevens (1986), have shown that the above equations hold for small sediment charges only. For larger charges the width may be larger.

It is of interest to attempt to compare the characteristics of the individual channels of the braiding Jamuna River to these regime equations. This was done for bankfull discharge (44,000 $m^3/s$). When it is assumed that the roughness and the energy slope of each channel in a cross-section is about the same, the total discharge in a cross-section can be distributed over the channels according to their conveyance. Next, the width B and the average depth $\bar{h}$ of the individual channels can be plotted against their discharge. The result is presented in Figure 11. The plotted data can be described by the following relations:

$$\bar{h} = 0.23\ Q^{0.32} \tag{3.5}$$

$$B = 16.1\ Q^{0.53} \tag{3.6}$$

Both relations are in S.I - units. The Lacey equations, expressed in these units and assumming p = B and R = $\bar{h}$ for these very wide channels, read:

$$\bar{h} = 0.47\ (\frac{Q}{f})^{1/3} \tag{3.7}$$

$$B = 4.81\ Q^{1/2} \tag{3.8}$$

As $D_{50} \cong 0.2$ mm, f = 1.13, so Equation (3.8) can be written as:

$$\bar{h} = 0.45\ Q^{1/3} \quad (3.9)$$

Comparison of the Equations (3.7) and (3.9) to the Jamuna Equations (3.5) and (3.6) yields the following conclusions:

- the exponent of the discharge is approximately the same in both comparable equations;
- the Jamuna channels are substantially wider and more shallow than the stable channels of Lacey; also the relatively small increase of the width indicated by Simons and Albertson (1963) is irrelevant for the Jamuna channels.

Apparently large sediment loads (in the Jamuna River up to 10,000 ppm for the total load, wash load included) result in very wide, shallow channels. The aspect ratio of the Jamuna River is about 750, which is extremely high.

Furthermore the following remarks are made:

i The substantial scatter in Figure 11 is at least partly due to the channel direction not being perpendicular to the cross-seciton. However, also other factors will play a rôle.

ii Lacey proposed a third equation for the slope of a stable channel. Introducing the relevant figures in this equation results in a slope which is 20% smaller than the actual slope of the Jamuna River.

Average depths and widths can also be derived for lower stages. For this, and also for the previous analysis, it is assumed that the cross-sections sounded during low flow conditions do not substantially differ from cross-sections for higher stages (see Figure 11). Under this assumption the average depth and width are described by:

$$\bar{h} = 0.56\ Q^{0.23} \quad (3.10)$$

$$B = 18.9\ Q^{0.51} \quad (3.11)$$

It is remarkable how similar these equations are to the equations which relate to bankfull discharge only (Equation (3.5) and (3.6)). Apparently the Jamuna channels can to some extent be considered as regime channels for lower stages also, possible indicating that morphological processes in the Jamuna River are relatively fast.

## 4. Conclusions

Both planform characteristics and cross-sectional characteristics of the braiding Jamuna River in Bangladesh were studied. It seems that this river, which was subject to a major avulsion about 2 centuries ago, has stabilized in its present position. Recently, the braiding of the river seems to have increased, for which no explanation is available at present.

Radii of curvature and bend characteristics are approximately similar to meandering rivers, which may indicate that individual channels of a braiding river may to some extent be comparable to the single channel of meandering rivers.

Regime equations for individual channels of the Jamuna River deviate considerably from regime equations for sand-bed canals and rivers. A possible explanation for this may be the high sediment load of the Jamuna River.

## 5. Acknowledgements

The present investigations were carried out within the frame-work of the Jamuna Bridge Appraisal Study. The study is being funded by the United Nations Development Programme (UNDP) through the World Bank. The Jamuna Multi-purpose Bridge Authority (JMBA) is furthering the studies for and the construction of the bridge. The present study is carried out by Rendel, Palmer and Tritton, NEDECO and Bangladesh Consultants Ltd. Mr. T. Swaneborg carried out the analysis of the planform data. Mr. Kamal collected the BWDB soundings and analyzed monument data.

## REFERENCES

COLEMAN, J.M. (1969), Brahmaputra River: Channel processes and sedimentation, Sedimentary Geology, Vol. 3, pp. 129-239

HOWARD, A.D., KEETCH, M.E. and VINCENT, C.L. (1970), Topological and geometrical properties of braided streams, Water Resources Research, Vol. 6, no. 6, pp. 1974-1688

IECO (International Engineering Compancy) (1969), Bank-line movement of the Brahmaputra - Jamuna River, Prepared for East Pakistan Water and Power Development Board

KLAASSEN, G.J. (1987), On the scaling of braided rivers, Proc. NATO Workshop on scaling of movable bed models, Delft Hydraulics, De Voorst office, The Netherlands

LACEY, G. (1930), Stable channels in alluvium, Minutes of Proc. Instn. of Civ. Eng., Vol. 229, pp. 259-292.

LACEY, G. (1947), A general theory of flow in alluvium, Journal supplement, Instn. of Civ. Eng., no. 8, paper 5518, pp. 425-451.

LEOPOLD, L.B., WOLMAN, M.G. and MILLER, J.P. (1964), Fluvial processes in geomorphology, San Francisco, W.H. Freeman and Cy.

MORGAN, J.P. and McINTIRE, W.G. (1959), Quaternary geology of the Bengal basin - East Pakistan, India, Bull Geol. Soc. Am., Vol. 70, pp. 319-342.

RPT/NEDECO/BCL (1987), Jamuna Bridge Appraisal study, Phase I, Prepared for World Bank/UNDP/Jamuna Multi-purpose Bridge Authority.

STEVENS, M.A. (1986), Width of straight alluvial channels, Journ. Hydr. Engng., ASCE, Vol. 112, no. HY 9, pp.

SIMONS, D.B. and ALBERTSON, M.L., (1960), Uniform water conveyance channels in alluvial material, Journ. Hydr. Div. ASCE, Vol. 86, no HY 5, pp. 33-71.

STRUIKSMA, N. and KLAASSEN, G.J. (1988), On the treshold between meandering and braiding, Proc. Intern. Conf. on River Regime, Wallingford, England.

ULLAH, M.H. (1987), Computer modelling of river channel changes in alluvial conditon. First Interim Report: Analytical study of channel changes of the Brahmaputra, Dhaka, Bangladesh University of Engng. and Technology, Institute of Flood Control and Drainage Research.

| Date | | 21 Feb., 1973 | | | | 22 Feb., 1978 | | | | 27 Dec., 1981 | | | | 27 Feb., 1987 | | | |
|---|---|---|---|---|---|---|---|---|---|---|---|---|---|---|---|---|---|
| Parameter | | E (-) | $E_i$ (-) | N (-) | C (km) | E (-) | $E_i$ (-) | N (-) | C (km) | E (-) | $E_i$ (-) | N (-) | C (km) | E (-) | $E_i$ (-) | N (-) | C (km) |
| Reach | 1 | - | - | - | - | - | - | - | - | 3.2 | 2.2 | 16 | 7.5 | 3.7 | 2.7 | 21 | 8.0 |
| | 2 | 3.2 | 2.2 | 19 | 8.5 | 3.0 | 2.0 | 20 | 8.0 | 3.3 | 2.3 | 19 | 12.0 | 3.8 | 2.8 | 21 | 8.0 |
| | 3 | 3.3 | 2.3 | 31 | 8.5 | 3.3 | 2.3 | 19 | 8.0 | 3.5 | 2.5 | 26 | 10.0 | 3.8 | 2.8 | 27 | 10.0 |
| | 4 | 2.8 | 1.8 | 12 | 10.5 | 2.8 | 1.8 | 16 | 8.0 | 2.9 | 1.9 | 12 | 10.0 | 4.3 | 3.3 | 25 | 10.0 |
| | 5 | 1.9 | 0.9 | 18 | 4.5 | 1.6 | 0.6 | 9 | 5.0 | 1.7 | 0.7 | 8 | 6.5 | 2.7 | 1.7 | 22 | 6.0 |
| | 6 | 1.9 | 0.9 | 8 | 6.0 | 2.2 | 1.2 | 14 | 6.5 | 1.8 | 0.8 | 4 | 6.0 | 2.6 | 1.6 | 9 | 8.0 |

Table 1 Braiding indices of the Jamuna River, Bangladesh over the period 1973-1987

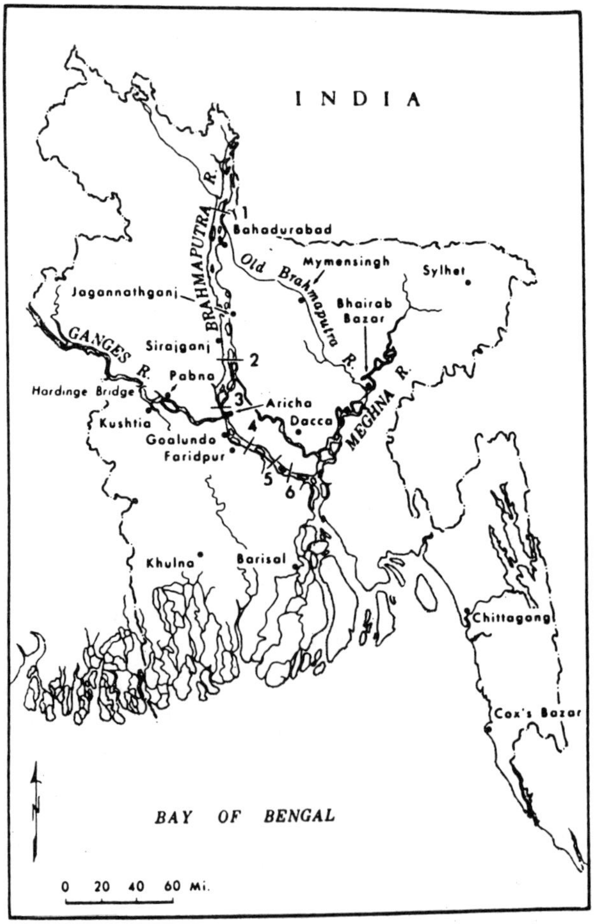


Figure 1 Index map of Bangladesh with major rivers (Coleman, 1969)

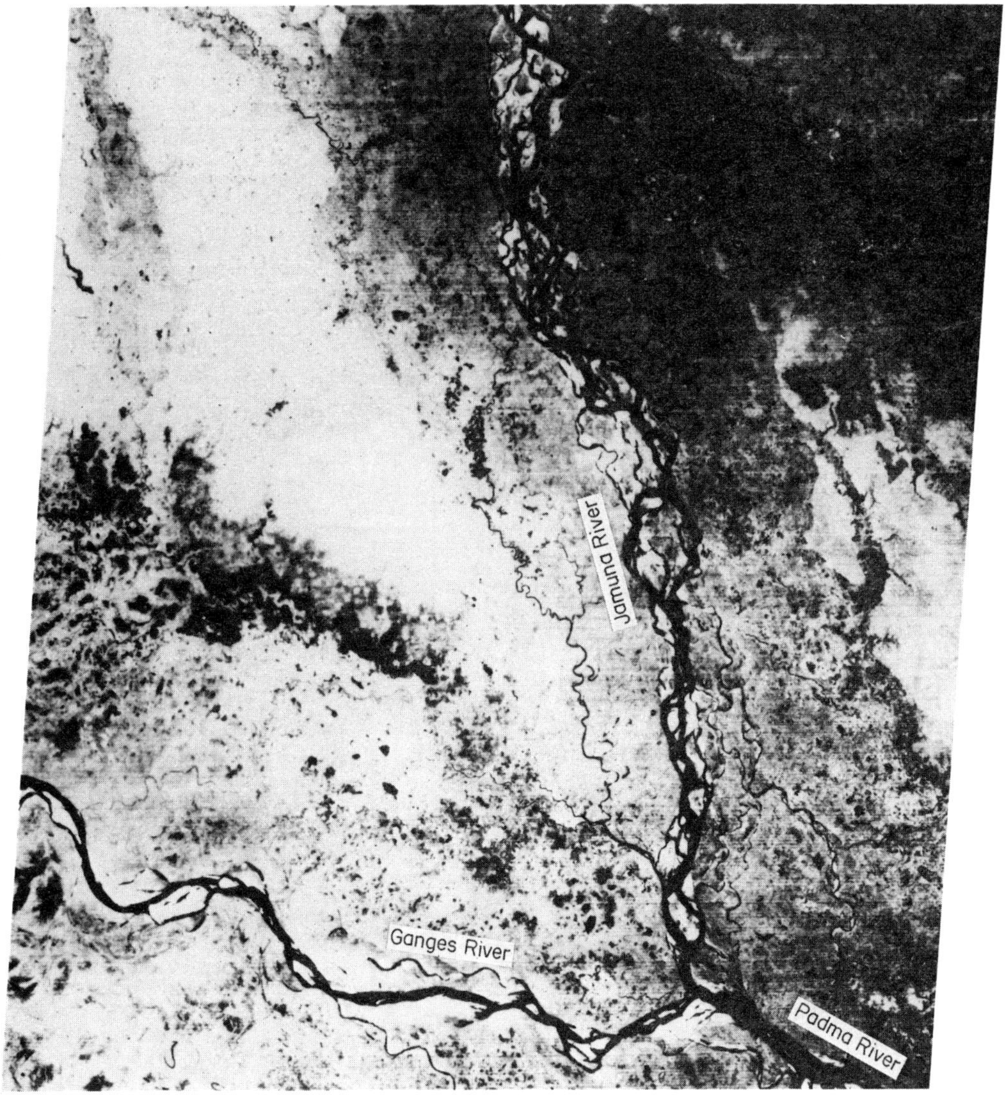


Figure 2 LANDSAT-image lower reach of the Jamuna River, date 27-2-1973
(scale 1:1,000,000)

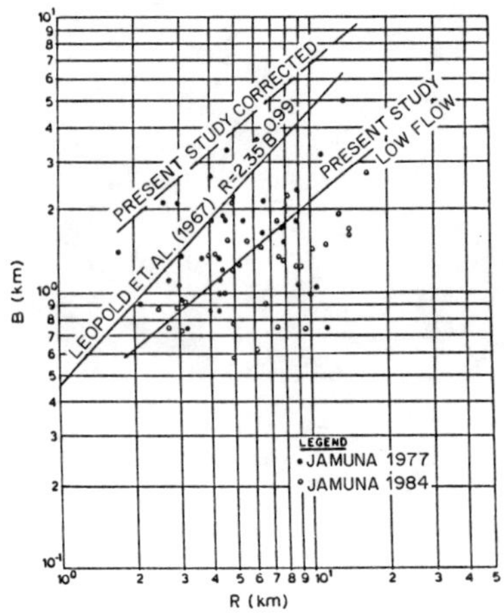


(a) Width versus radius of curvature for low flow data 1977

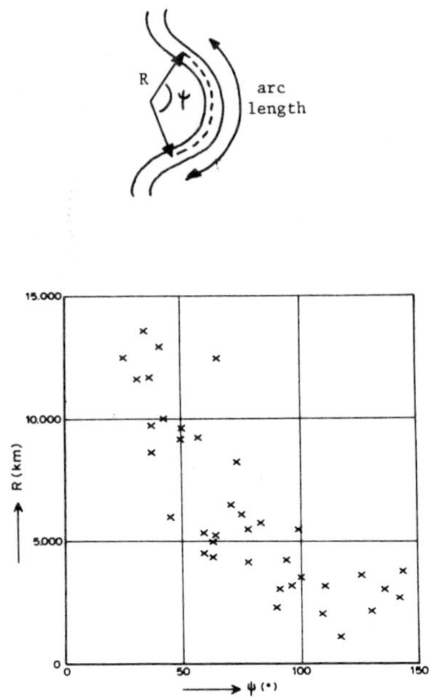


(b) Arc length versus radius of curvature for low flow data 1983/1984

Figure 3 Bend characteristics of the Jamuna River

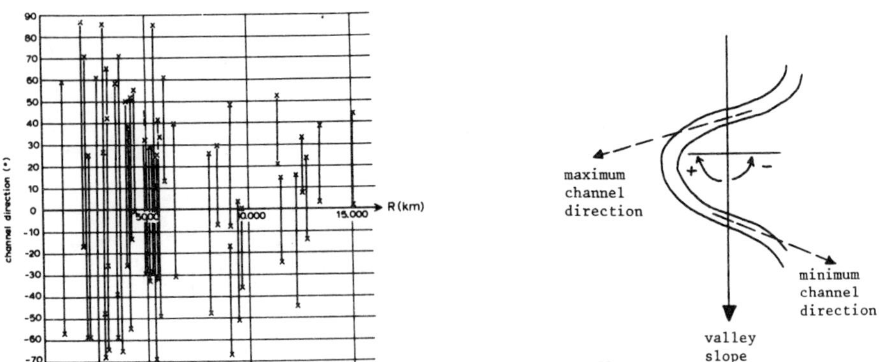


Figure 4 Channel direction versus width for the Jamuna River for the low flow season 1983/1984

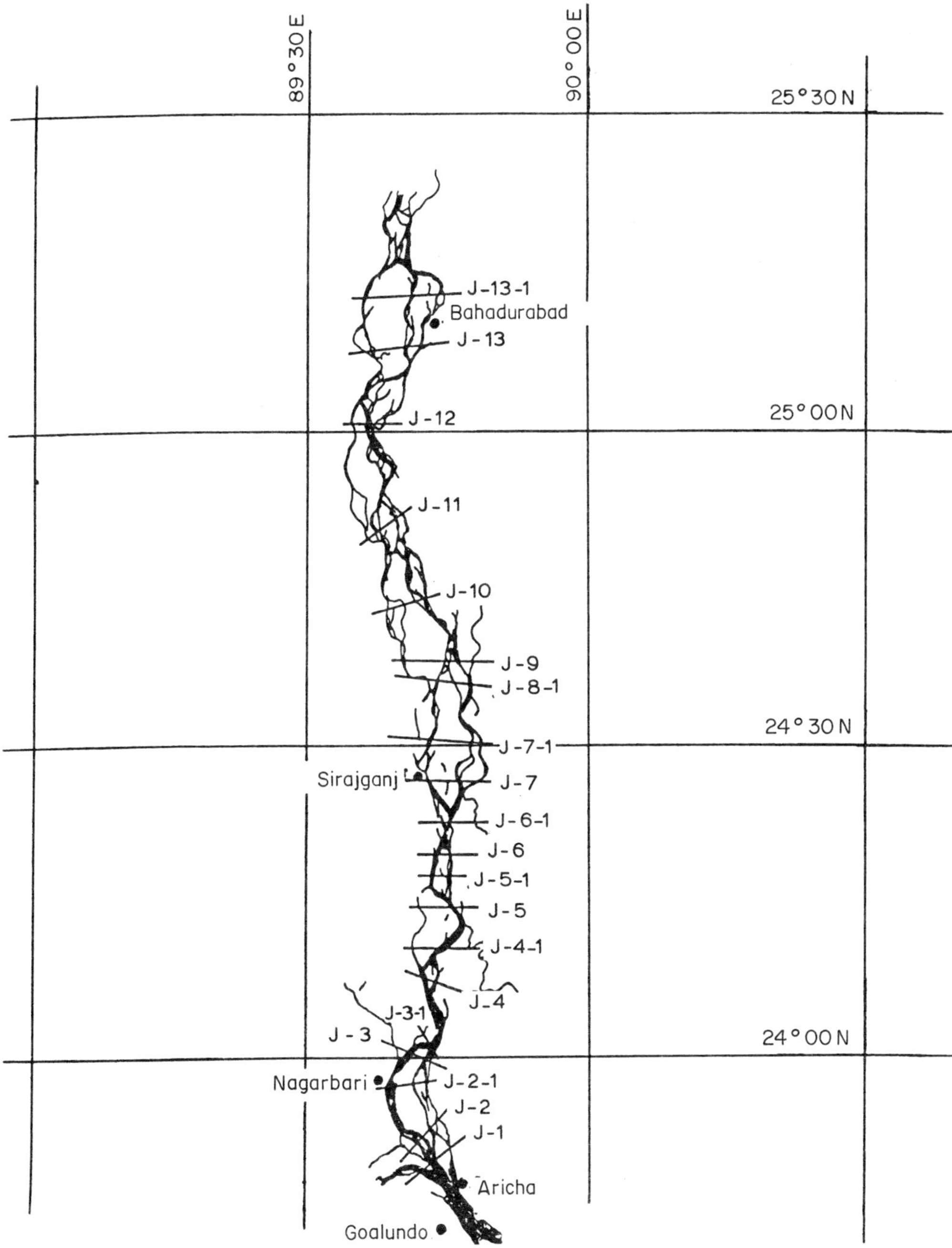


Figure 5 Cross sections measured by BWDB in the period 1966-1987

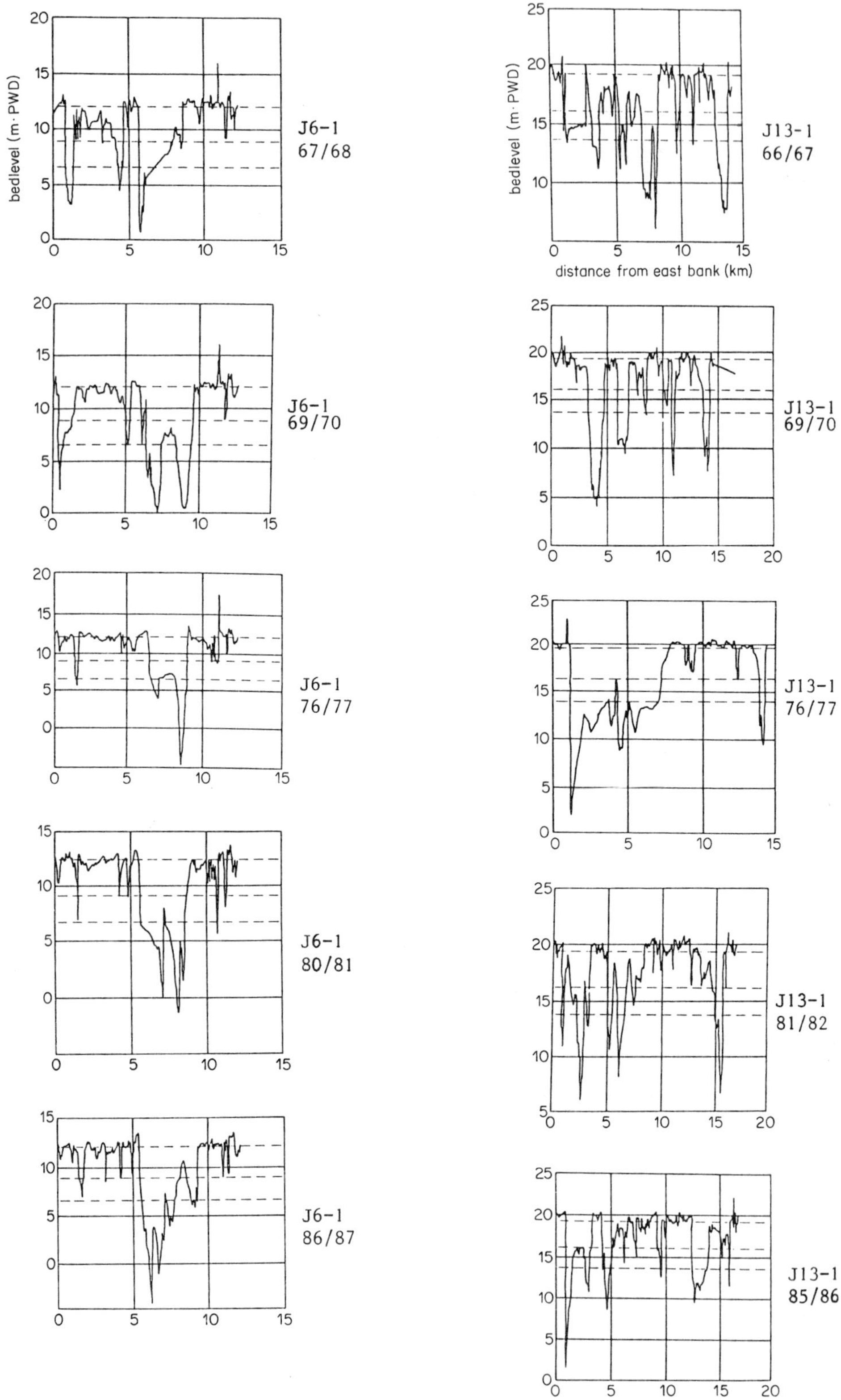


Figure 6 Two cross sections of the Jamuna River and their variation in time

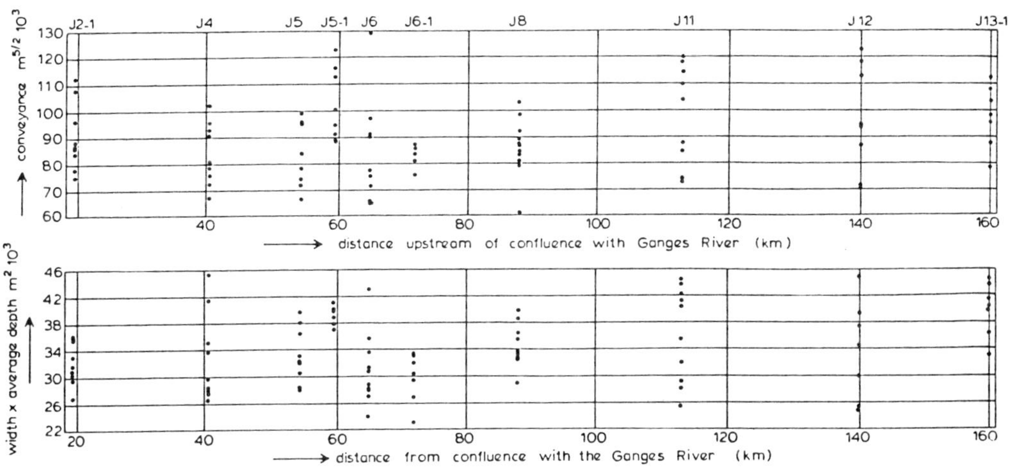


Figure 7 Variation of wet cross sectional area and conveyance along the Jamuna River.

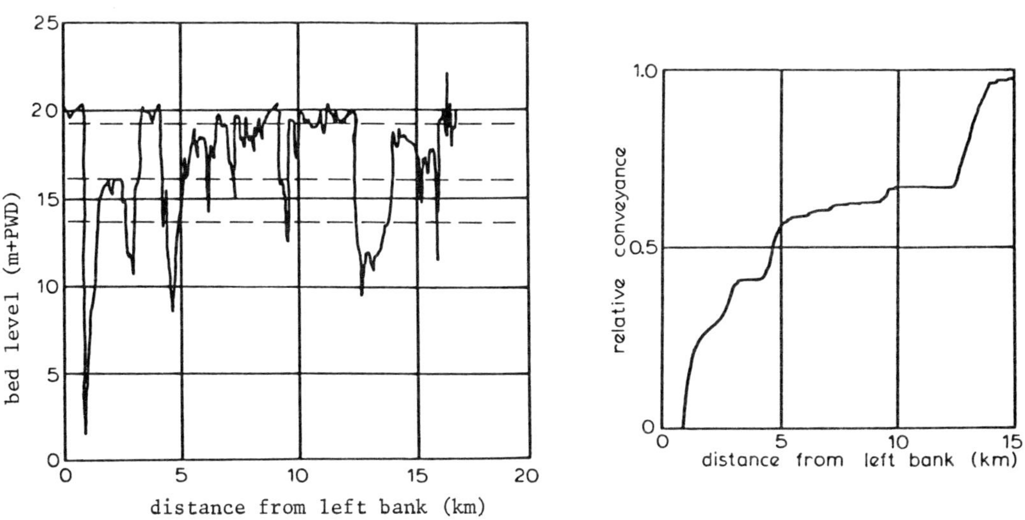


Figure 8 Cross section J13-1 in the year 1985/1986 and the relative conveyance as function of distance from the left bank.

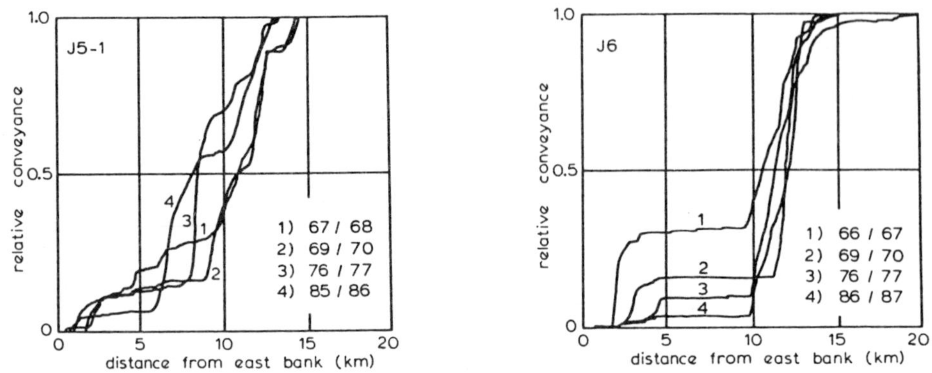


Figure 9 Two examples of the variation of the relative conveyance in time

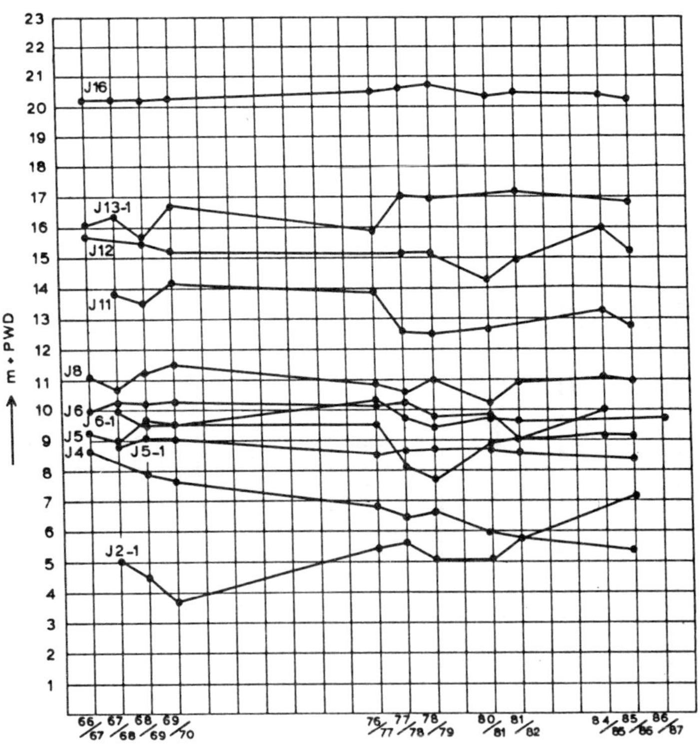


Figure 10 Average bed level of the Jamuna River cross-sections as function of time

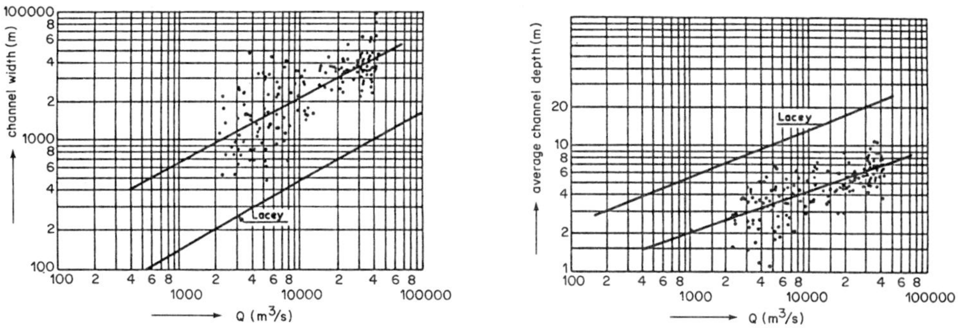


(a) width versus discharge (b) average depth versus discharge

Figure 11 Width and depth of Jamuna River channels at bankful discharge ($44000\ m^3/s$)

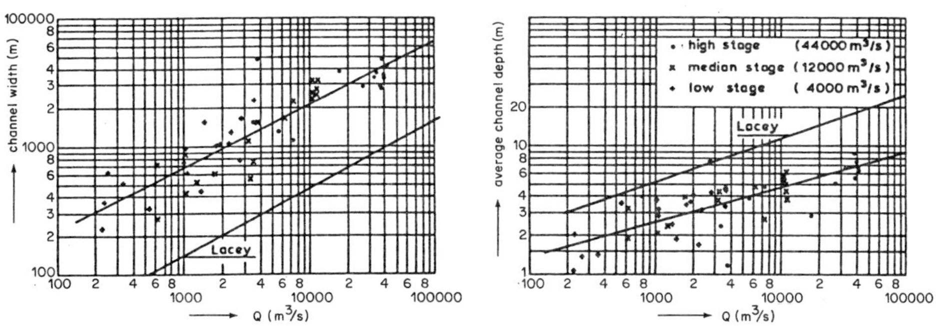


(a) width versus discharge (b) average depth versus discharge

Figure 12 Channel width and depth at cross section J6-1 at three defined stages

International Conference on
RIVER REGIME
18-20 May, 1988

PAPER E2

# CHARACTERISTICS OF FLOW AT THE INTERSECTION OF RIVER MEANDERS AND FLOOD RELIEF (CUT-OFF) CHANNELS

J.G. Herbertson and Y.R. Fares
Department of Civil Engineering, University of Glasgow,
Glasgow, Scotland.

SUMMARY: A 2D mathematical model is presented for the prediction of the effects of a flood relief (cut-off) channel on the flow characteristics in a wide gentle channel bend. Results are presented for various flood relief channel geometries showing that a flood channel is more effective in reducing water level in the bend than in reducing discharge.

## 1. Introduction.

Fisk (Ref. 1) identified two types of cut-off, namely chute cut-off and neck cut-off, as shown in Fig. 1. Although neck cut-off occurs naturally, full or partial neck cut-off can be used as a means of flood regulation as indicated by Thorn (Ref. 2). Winkley (Ref. 3) states that opinion is divided on the benefits of flood relief (cut-off) channels. The difference centres on whether the local effects, often assumed detrimental, at the intersection of the channels outweigh the benefits of the more stable regime produced in the main part of the meander loop. The present paper reports on the early stages of an investigation into the characteristics of the flow at the upstream end of the junction of the channels from which it is hoped that the magnitude of the local effects may be assessed.

## 2. The Mathematical Model.

Before proceeding to the modelling of the flow at the intersection of a channel bend and a flood relief channel it was necessary to develop and test a model for the flow in the bend alone. This model not only provides a base with which changes produced by the inclusion of the flood relief channel can be compared but is also necessary to establish the flow conditions just upstream of the flood channel.

### 2.1 Model for flow in a channel bend.

The model is restricted to gentle wide channel bends which as previously shown by Chow (Ref. 4), Rozovskii (Ref. 5) and Soliman and Tinney (Ref. 6) allows the vertical distribution of the pressure field to be considered hydrostatic. For the purposes of the model the channel is taken to be of uniform rectangular section throughout the bend.

When dealing with flow following a curved path it is necessary to consider both the radial and tangential velocity components. In each case the distribution of the velocity component may be obtained by applying the appropriate momentum equation.

In the case of the radial velocity component the momentum equation is -

$$-\frac{u_\theta^2}{r} + gS_r = \frac{\partial}{\partial z}\left(\upsilon\frac{\partial u_r}{\partial z}\right) \qquad (1)$$

International Conference on River Regime, edited by W.R.White.

in which $u_\theta$ is the tangential velocity at depth z

$r$ is the local bend radius

$g$ is the gravitational acceleration

$S_r$ is the radial water surface slope, $\beta \frac{u^2_m}{gr}$

$\beta$ is the momentum flux correction factor

$u_m$ is the depth averaged tangential velocity obtained from the Chezy equation

$\upsilon$ is the kinematic eddy viscosity

and $u_r$ is the radial velocity component.

As shown by Rozovskii (Ref. 5), Kalkwijk and De Vriend (Ref. 7) and Hussein and Smith (Ref. 8) the assumption of a wide gentle bend allows the non-linear inertia terms in the above equation to be neglected thus simplifying its solution. After applying the boundary conditions –

(i) that there is no net flow in the radial direction

and (ii) that there is no shear stress at the water surface

– a solution to equation 1 may be obtained. Expressed in functional form this is –

$$u_r(\frac{z}{h}) = f_n[\frac{h}{r}, u_m, C, g, k, \varphi_1 (\frac{z}{h}), \varphi_2 (\frac{z}{h})] \qquad (2)$$

in which $h$ is the total water depth at the section

$C$ is the Chezy coefficient

$k$ is Von-Karman constant = 0.4

$\varphi_1$ and $\varphi_2$ are integration functions the distribution of which are shown in Fig. 2.

In the case of the tangential velocity the momentum equation is

$$u_r.\frac{\partial u_\theta}{\partial r} + \frac{u_\theta}{r}.\frac{\partial u_\theta}{\partial \theta} + u_z \frac{\partial u_\theta}{\partial z} + \frac{u_r u_\theta}{r} = g\, s_\theta + \frac{\partial}{\partial z}(\upsilon.\frac{\partial u_\theta}{\partial z}) \qquad (3)$$

in which $u_z$ is the vertical component of velocity

$S_\theta$ is the longitudinal water surface slope.

Taking equation 3 along with the continuity equation

$$\frac{\partial u_z}{\partial z} = -(\frac{\partial u_r}{\partial r} + \frac{u_r}{r} + \frac{1}{r}\frac{\partial u_\theta}{\partial \theta}) \qquad (4)$$

and by substituting equation (4) into equation (3) and integrating the resulting equation vertically from 0 to h, the general equation of the longitudinal variation of the depth averaged velocity can be obtained, this is,

$$\frac{1}{r}.\Delta_5.\frac{\partial}{\partial r}(r.u_m^2.h^2) + \beta \frac{\partial}{\partial \theta}(h.u_m^2) = g s_\theta.r.h - \frac{u_m^2}{C^2}.g.r \qquad (5)$$

where

$$\Delta_5 = \Delta_1^2[\Delta_4 - \beta \frac{C^2}{g} - 0.591\Delta_3 + \frac{0.744}{k^2}] \qquad (6)$$

and $$\Delta_1 = \frac{\sqrt{g}}{kC}\ ; \quad \Delta_2 = 1 + \Delta 1; \quad \Delta_3 = \frac{2\Delta_2 C}{k\sqrt{g}}; \quad \Delta_4 = \frac{\Delta_2{}^2 C^2}{g} \tag{7}$$

The set of equations, 5, 6 and 7 represent the general model equations for the horizontal distribution of the depth averaged tangential flow velocity. The equations are solved using the finite difference technique. Once the depth averaged tangential velocity has been found the longitudinal and transverse water surface slopes can be found.

2.2 Verification of the bend model.

The bend model allows prediction of three properties of the flow.

(i) the vertical distribution of the secondary (radial) currents,

(ii) the horizontal distribution of the depth averaged tangential velocity at sections along the bend,

and (iii) the longitudinal and transverse water surface slopes.

Attempts were made to check each of these using data from the literature before the next step of introducing the flood relief (cut–off) channel was taken.

Fig. 3 shows the comparison of the theoretically predicted vertical distributions of the secondary currents with those of experimental results obtained by Rozovskii (Ref. 5) for one bend geometry, i.e. with radius to width ratio of 2.5 and width to depth ratio of 8.0.

Fig. 4 shows the comparison of transverse water surface slopes using data given by Yen and Yen (Ref. 9).

No suitable data has yet been found to allow comparison of predicted and measured horizontal distributions of depth averaged tangential velocities nor of longitudinal water surface slopes. Existing data is mainly for sharp bends. However since satisfactory comparisons were obtained with the radial velocities and surface profiles, in the latter case at different sections along the bend, it seemed reasonable to assume that satisfactory agreement would also apply in the case of the tangential velocities and longitudinal surface profiles. On this basis it was considered justified to proceed to the modelling of the flood relief channel.

2.3 Modelling of the flood relief channel.

The flow entering the flood relief channel is modelled as a flow over a side weir with a long base or broad crest. Previous work by Balmforth Sarginson (Refs. 10, 11) has shown that in a straight channel the performance and various coefficients of a sharp edged weir are sensibly independent of the orientation of the weir i.e. whether the weir is placed normal to or along the direction of the flow. If this holds for sharp edged weirs there seems little reason why it should not also be true for broad crested weirs. This assumption has been made and further extended to include the case of a curved channel. In the latter case, allowance for the bend effect on the horizontal distribution of water depth and depth averaged velocity (and hence discharge) has been included in the governing flow equations. To accomplish this, the bend cross section has been divided into a series of concentric strips in each of which the longitudinal variation of the flow characteristics such as discharge, water depth, depth average velocity and Froude Number have been calculated.

Based on the appropriate British Standard BS 3680 (Ref. 12) and the above assumptions the discharge, $dQ_{fR}$, over the weir per step length $dC_w$ is –

$$dQ_{fR} = \left(\frac{2}{3}\right)^{3/2} C_D . C_v . dC_w \sqrt{g} . (h - C_h)^{3/2} . f \tag{8}$$

in which $C_v$ is a non-dimensional coefficient which allows for the effect of the velocity of approach.

$C_w$ is the weir width.

$C_h$ is the weir crest height.

f is the drowned flow reduction factor.

Values of $C_D$, $C_v$ and f are given in the British Standard.

By using the momentum equation in the longitudinal direction at the intersection, the equation for the spatially varied flow pertaining along the width of the weir is

$$\frac{dh}{rd\theta} = \frac{S_o - S_f + \frac{1}{gA} . \frac{dQ_{fR}}{dC_w} [u_m + \frac{Q_{fR}}{A}](2\beta - 1)}{1 - \beta \frac{u_m B}{gA^2} [Q + Q_{fR}]} \tag{9}$$

in which $S_o$ is the channel bed slope.

$S_f$ is the slope of the total energy line in each strip.

A is the area of each channel strip.

$\beta$ is the momentum flux correction factor.

B is the width of each concentric channel strip.

Q is the strip discharge.

Equations 8 and 9 represent the general model equations of the flow variations at the intersection. The simultaneous solution of these equations is obtained by using a standard numerical technique.

## 3. Effect of the flood relief channel.

The mathematical model described in outline in the previous sections has been used to determine the effect of a flood relief channel on the flow round a channel bend and also to determine the effect of varying the dimensions of the flood channel. Many test runs have been carried out but two particular cases have been chosen to illustrate the effect of varying the width and bed level of the flood channel, modelled respectively as side overflow weir crest width ($C_w$) and crest height ($C_h$) respectively. These are (1) $C_w$ = 0.61 x main channel width (B) and (2) $C_w$ = 1.22 x B. In each case runs were made with several values of flood channel to main channel water depth ratios ($H_w/H$). In all cases the length of the flood channel (i.e. the weir crest length in the direction of flow in the flood channel, $C_L$) was taken to be the same as the channel (weir) width, $C_w$, and flow over the weir was taken as modular.

The start of the flood relief channel was positioned at a bend angle of 25°. This position allows significant development of secondary currents and superelevation.

The results obtained from the computer model are given in Tables 1 and 2 and sample results are shown on Figs. 5 and 6. Results are presented as percentage differences i.e. the percentages difference between the bend alone and the bend with a flood relief channel. In Tables 1 and 2, a positive difference means the flood channel gave rise to an increase and a negative difference implies that the flood channel resulted in a decrease in the parameter concerned.

The results may be summarised as follows –

(1) For all bar the inner bank section at high flood channel bed levels, there is a significant decrease in the longitudinal water surface profile and in discharge. As might be expected the greatest reduction in depth occurs around the outer bank due to the reduction in the superelevation effect.

(2) The reduction in discharge is generally less than the reduction in water depth due to the local increase in the tangential velocity component.

(3) The redistribution of depth averaged tangential velocity results in a considerable flattening of the transverse water surface profile, the effect becoming more marked as the bed level of the flood channel is lowered.

4. Conclusions.

1. A 2–D mathematical model has been developed for flow around a wide gentle channel bend of uniform section which as can be seen from Figs. 3 and 4 is capable of successfully predicting the vertical distribution of secondary currents and the transverse water surface slopes at sections around the bend.

2. Combining the bend model with the equation for a broad crested side weir it has been possible to simulate the effect of introducing a flood relief (cut–off) channel into the bend.

3. It has been shown that for all ratios of flood channel bed level to water depth in the main channel, the flood channel is more effective in reducing water depth in the bend than in reducing discharge. This is brought about by a local increase in the tangential velocity.

4. As might be expected the reduction in the water depth is much more significant around the outer bank due to the redistribution of the depth averaged tangential velocity and the virtual elimination of the superelevation effect.

5. Although so far restricted to study of local changes at intersection of the bend and flood relief channel it is considered that the model is capable of extention to the study of conditions downstream of the intersection.

5. Nomenclature.

| | | |
|---|---|---|
| A | = | strip area in the channel bend. |
| B | = | strip width in the channel bend. |
| C | = | Chezy coefficient. |
| $C_D$ | = | Coefficient of the flood-relief channel (weir) discharge. |
| $C_h$ | = | flood-relief channel (weir) crest height. |
| $C_w$, $C_L$ | = | width and length of the flood-relief channel (weir) respectively. |
| $C_v$ | = | velocity coefficient of the flow entering the flood-relief channel. |
| f | = | drowned flow reduction factor (= 1.0). |

| | | |
|---|---|---|
| $g$ | = | gravitational acceleration. |
| $h$ | = | local water depth in the channel bend. |
| $H$ | = | mean depth in the channel bend. |
| $H_w$ | = | water head above the weir crest (= $H - C_h$). |
| $k$ | = | Von-Karman constant (= 0.4). |
| $Q$ | = | strip discharge in the channel bend. |
| $Q_{fR}$ | = | water discharge taken by the flood-relief channel. |
| $r$ | = | local bend radius. |
| $s_o$ | = | bend channel bed slope. |
| $s_f$ | = | strip slope of the total energy line in the channel bend. |
| $S_r$, $s_\theta$ | = | radial and longitudinal surface slopes in the bend respectively. |
| $u_m$ | = | local depth-averaged velocity in the channel bend. |
| $u_r$, $u_\theta$, $u_z$ | = | radial, tangential and vertical velocity components in the bend respectively. |
| $z$ | = | local vertical coordinate of water depth. |
| $\beta$ | = | momentum flux correction factor. |
| $\Delta_1$, $\Delta_2$, ... $\Delta_5$ | = | coefficients defined in equations 6 and 7. |
| $\upsilon$ | = | turbulent kinematic eddy viscosity of the flow. |
| $\frac{\partial}{\partial r}$, $\frac{\partial}{\partial \theta}$, $\frac{\partial}{\partial z}$ | = | gradient operators in the radial, tangential and vertical directions in the bend respectively. |
| $\varphi_1$, $\varphi_2$ | = | mathematical integration functions defined in equation 2. |

## 6. References.


1. Fisk, H.N.: "Geological investigation of the alluvial valley of the lower Mississippi river". Mississippi River Commission, Vicksburg, Miss., 1944, pp 1–78.

2. Thorn, R.B.: "River engineering and water conservation works". Butterworths & Co. (Publishers) Ltd., London, 1966, pp 239–271.

3. Winkley, B.R.: "Practical aspects of river regulation and control". From Shen, H.W.: "River mechanics", Chapter 19, Published by the editor, Colorado, U.S.A., 1971, pp 1–78.

4. Chow, V.T.: "Open channel hydraulics". McGraw Hill Book Company, Inc. New York, N.Y., 1959, pp. 439–460.

5. Rozovskii, I.L.: "Flow of water in bends of open channels". Israel program for scientific translation, 1961, pp 11–68 (available from the Office of Technical Services, U.S. Department of Commerce, PST Cat. No. 363)

6. Soliman, M.M. and Tinney, E.R.: "Flow around bends in open rectangular channels". ASCE, J. Hydr. Div., 94, 1968, pp 893–908.

7. Kalkwijk, J.P.TH. and De Vriend, H.J.: "Computation of the flow in shallow river bends". J. Hydr. R., 18, 1980, pp 327–342.

8. Hussein, A.H. and Smith, K.V.H.: "Flow and bed deviation angle in curved open channels". J. Hydr. R., 24, 1986, pp 93–108.

9. Yen, C.L. and Yen, B.C.: "Water surface configuration in channel bends". ASCE, J. Hydr. Div., 97, 1971, pp. 303–321.

10. Balmforth, D.J. and Sarginson, E.J.: "Discussion of experimental investigation of flow over side weirs" by Ek–Khashab and Smith. ASCE, J. Hydr. Div., HY8, 1977, pp 941–943.

11. Balmforth, D.J. and Sarginson, E.J.: "A Comparison of methods of analysis of side weir flow". J.Inst. of Munic. Eng., 105, 1978, pp 273–279.

12. British Standard Institution: Part 4B, "Triangular profile weirs", 1986, and Part 4E, "Free over–fall weirs of finite width (rectangular broad–crested weirs)", 1980. Methods of measurements of liquid flow in open channels, BS 3680, BSI.

| BEND ANGLE | RADIAL SECTION | $H_W/H = 0.20$ (1) % DIFF. OF H | (2) % DIFF. OF Q | $H_W/H = 0.40$ (3) % DIFF. OF H | (4) % DIFF. OF Q | $H_W/H = 0.60$ (5) % DIFF. OF H | (6) % DIFF. OF Q |
|---|---|---|---|---|---|---|---|
| 25° Beginning of F.R.C. | INNER BANK | - | - | - | - | - | - |
| | MID BEND RADIUS | - | - | - | - | - | - |
| | OUTER BANK | - | - | - | - | - | - |
| 30° | INNER BANK | +1.84 | +1.74 | -4.86 | -2.99 | -15.30 | -9.49 |
| | MID BEND RADIUS | -4.56 | -3.01 | -11.74 | -8.49 | -22.42 | -15.40 |
| | OUTER BANK | -9.02 | -5.84 | -16.49 | -12.13 | -27.34 | -19.60 |
| 35° End of F.R.C. | INNER BANK | +2.61 | +3.00 | -9.39 | -4.91 | -28.77 | -14.28 |
| | MID BEND RADIUS | -8.04 | -5.23 | -20.45 | -13.79 | -39.91 | -22.96 |
| | OUTER BANK | -15.35 | -10.35 | -27.91 | -19.71 | -47.39 | -29.04 |

TABLE (1) - Percent variation of the water head (H) and discharge (Q) in the channel bend due to the flood-relief channel (F.R.C.) for $C_W = 0.61B$

| BEND ANGLE | RADIAL SECTION | $H_w/H = 0.20$ (1) % DIFF. OF H | (2) % DIFF. OF Q | $H_w/H = 0.40$ (3) % DIFF. OF H | (4) % DIFF. OF Q | $H_w/H = 0.50$ (5) % DIFF. OF H | (6) % DIFF OF Q |
|---|---|---|---|---|---|---|---|
| 25° Begin-ing of F.R.C. | INNER BANK | - | - | - | - | - | - |
| | MID BEND RADIUS | - | - | - | - | - | - |
| | OUTER BANK | - | - | - | - | - | - |
| 30° | INNER BANK | +1.83 | +1.82 | -5.15 | -2.66 | -10.32 | -5.54 |
| | MID BEND RADIUS | -4.60 | -2.91 | -12.18 | -8.02 | -17.61 | -11.11 |
| | OUTER BANK | -9.00 | -5.68 | -17.04 | -11.52 | -22.69 | -14.86 |
| 35° | INNER BANK | +2.59 | +3.09 | -9.88 | -4.31 | -19.17 | -8.58 |
| | MID BEND RADIUS | -8.10 | -5.06 | -21.13 | -12.97 | -30.60 | -17.15 |
| | OUTER BANK | -15.44 | -10.09 | -28.77 | -18.68 | -38.36 | -22.89 |
| 40° | INNER BANK | +3.70 | +4.65 | -13.00 | -4.63 | -25.22 | -9.35 |
| | MID BEND RADIUS | -10.75 | -6.68 | -27.43 | -16.01 | -39.32 | -20.23 |
| | OUTER BANK | -20.27 | -13.61 | -36.73 | -23.26 | -48.27 | -27.17 |
| 45° End of F.R.C. | INNER BANK | +5.15 | +6.51 | -14.77 | -4.01 | -28.81 | -8.75 |
| | MID BEND RADIUS | -12.76 | -7.97 | -31.66 | -17.91 | -44.37 | -21.75 |
| | OUTER BANK | -23.97 | -16.56 | -41.87 | -26.41 | -53.36 | -29.74 |

TABLE (2) - Percent variation of the water head (H) and discharge (Q) in the channel bend due to the flood-relief channel (F.R.C.), for $C_w = 1.22B$.

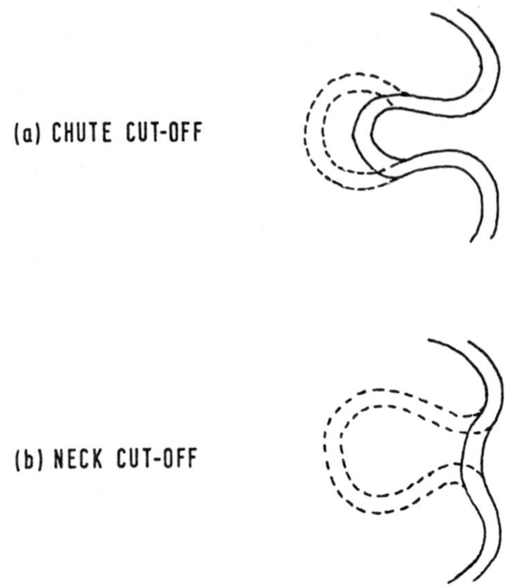


FIG.1 - TYPES OF CUT-OFF AFTER FISK (Ref. 1)

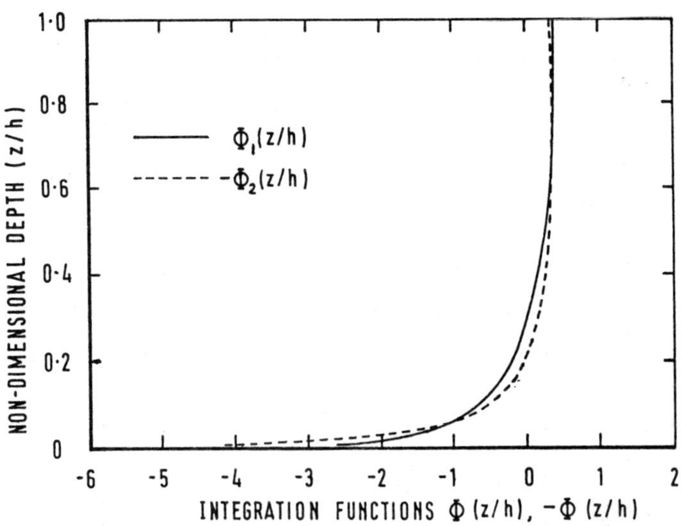


FIG.2 - VERTICAL DISTRIBUTION OF INTEGRATION FUNCTIONS.

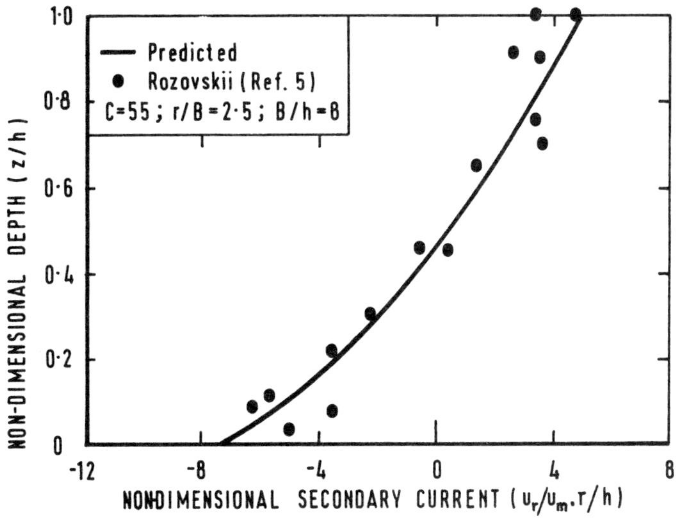


FIG.3 - VERTICAL DISTRIBUTION OF SECONDARY CURRENTS.

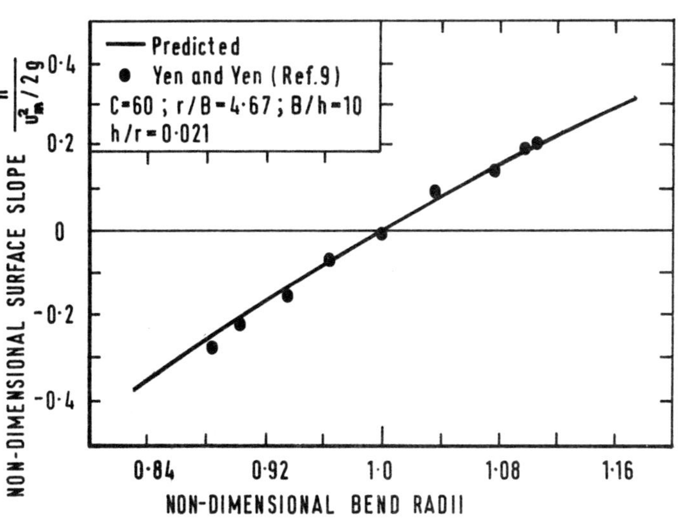


FIG.4 - TRANSVERSE FLOW SURFACE PROFILES (BEND ANGLE 45°).

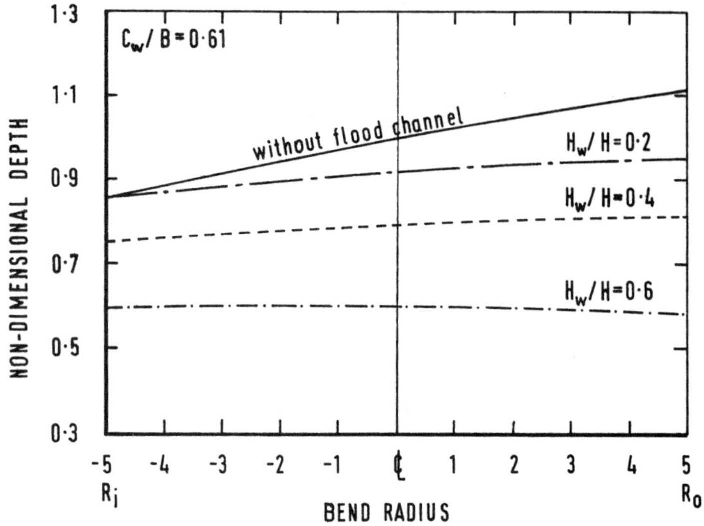


FIG.5 - EFFECT OF FLOOD RELIEF CHANNEL ON RADIAL WATER SURFACE PROFILES AT BEND ANGLE OF 35°

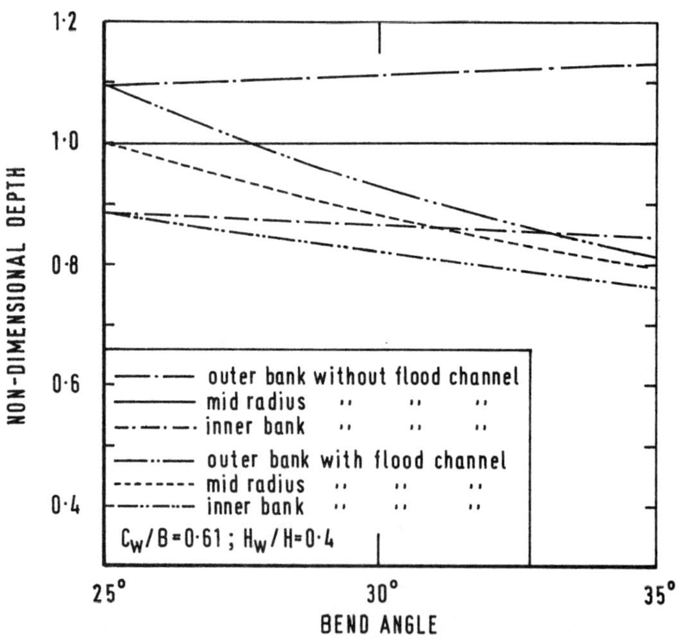


FIG.6 - EFFECT OF FLOOD RELIEF CHANNEL ON LONGITUDINAL WATER SURFACE PROFILES.

International Conference on
RIVER REGIME
18-20 May, 1988

PAPER E3

# REGIME BED SLOPES OF CHANNELS WITH A BUOYANT BOUNDARY

S. P. Chee, Ph.D., C.Eng., P.Eng.
Professor of Civil Engineering
University of Windsor, Canada

M. R. Haggag, Ph.D., P.Eng.
Formerly Research Associate
University of Windsor, Canada

SUMMARY: This paper describes the results of laboratory observations of the bed slopes of channels with an erodible sand bed in the presence of a buoyant cover made up of polyethylene plastic pellets which simulate an ice cover. An analytical model has also been included to predict regime slopes.

## 1. INTRODUCTION

### 1/1.0 Ice-Covered Channels

Rivers close to the arctic circle in Canada's north flow for about half the year with an ice cover and the remaining warmer months with a free surface. Such rivers develop two sets of regime conditions cyclically every year. While regime flow in open channels have been widely studied, the parallel case in which a floating boundary such as an ice cover is present is less known and remains to be investigated more carefully.

The authors (Ref. 1) have developed relationships concerning floating boundary channel flow which will be used in this paper. The relevant equations will subsequently be mentioned as required. In order to facilitate the understanding of the theoretical development here, some of the essential concepts and assumptions are discussed here.

A channel with a floating cover can be divided into two subsections as defined in Fig. 1. The flow in subsection (1) is influenced by the bed and sides while subsection (2) is controlled by the cover. The two subsections are divided by a separation surface which represents the locus of no shear and filaments of maximum velocity with reference to a vertical. The equations of energy, continuity, and momentum are applicable to each subsection individually as well as to the entire channel cross-section.

## 2. THEORETICAL ANALYSIS

### 2/1.0 Preamble

The problem which is to be analysed is based on the following model. The ice cover of a river is considered as a floating boundary. The underside of the cover has ridges or undulations which are known. The channel has an erodible sand bed which could self-adjust to the flow regime and the under ice cover configuration. The objective is, therefore, to attempt to predict the bed profile and mean slope.

International Conference on River Regime, edited by W.R.White.

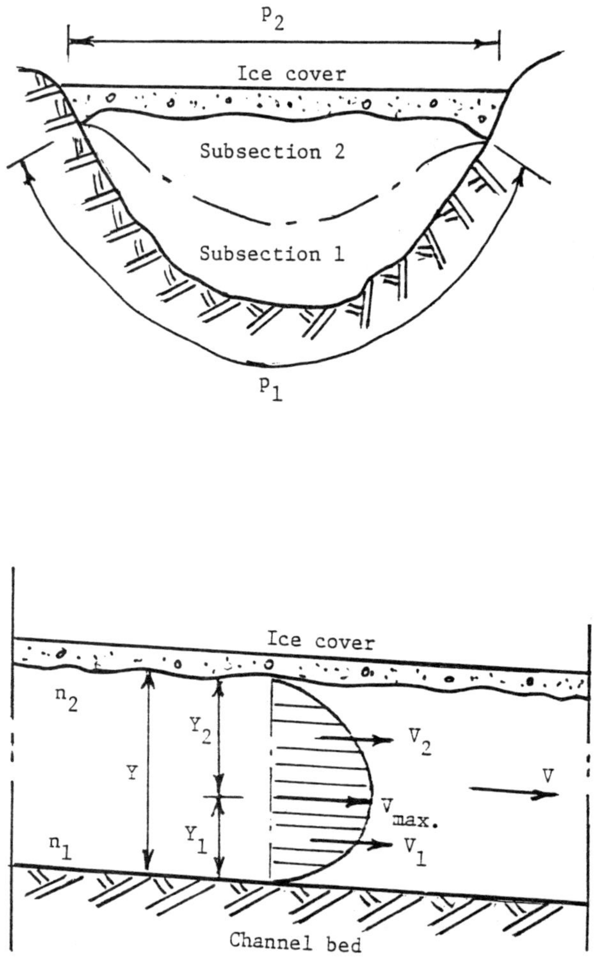


Fig. 1. Definition Sketch

2/1.1 The Bed Profile

In Fig. 2 a typical longitudinal profile of a channel with a buoyant cover is shown. Various variables to be used in the analysis are also defined.

The continuity equation for a unit width for the entire channel can be written as

$$q = V\,Y \tag{1}$$

where q = discharge per unit width; V = mean velocity; Y = flow depth. Differentiation of Eq. (1) with respect to x will yield

$$dq/dx = V\,dY/dx + Y\,dV/dx \tag{2}$$

Application of the principle of the conservation of energy at the section provides

$$E = \eta_b + V^2/2g + Y + S_g\,\eta_s \tag{3}$$

in which E = total energy at this section; $\eta_b$ = vertical distance of the erodible bed from a horizontal datum; $S_g$ = specific gravity of the cover material; $\eta_s$ = cover thickness. The equation for the local slope, $S_f$, of the total energy line can be obtained by differentiating Eq. (3) to give

$$S_f = -\frac{dE}{dx} = -\left(\frac{d\eta_b}{dx} + \frac{V}{gY}\frac{dq}{dx} - F_n^2\frac{dY}{dx} + \frac{dY}{dx} + S_g\frac{d\eta_s}{dx}\right) \tag{4}$$

where $F_n = V/\sqrt{gY}$ is the local Froude number.

Based on experimental and field observations, it seemed reasonable to assume that the mean slope of the erodible bed is sensibly equal to that of the cover top surface slope. With reference to the horizontal datum, the height of the top surface of the cover, H, can be expressed as

$$H = \eta_b + Y + \eta_s \tag{5}$$

Equation (5) can be differentiated to give the cover or bed slope, $S_o$, in the form of

$$S_o = -\frac{dH}{dx} = -\left(\frac{d\eta_b}{dx} + \frac{dY}{dx} + \frac{d\eta_s}{dx}\right) \tag{6}$$

Using Eq. (2) through Eq. (6), the bed profile could be written as

$$\frac{d\eta_b}{dx} = \left[-S_f + (1+F_n^2)\,S_o - \frac{V}{gY}\frac{dq}{dx} - \frac{d\eta_s}{dx}(F_n^2 - S_{gl})\right] / F_n^2 \tag{7}$$

where $F_{gl} = (1 - S_g)$. The flow depth of the entire channel can also be written as

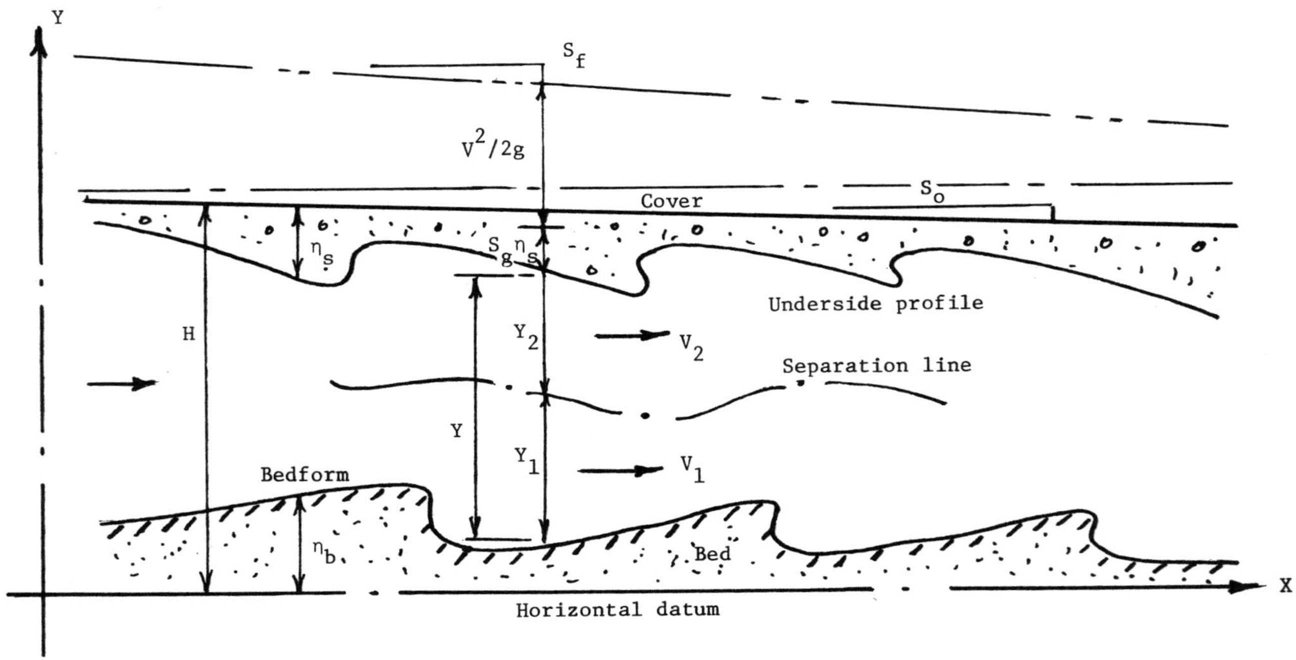


Fig. 2. Bed and Cover Profile

$$\frac{dY}{dx} = \left( -S_f + S_g S_o - S_{g1} \frac{d\eta_b}{dx} + \frac{V}{gY} \frac{dq}{dx} \right) / (S_{g1} - F_n^2) \tag{8}$$

Equations (6) and (7) relate the bedform to the cover underside profile for different flow conditions. In order to solve these equations, expressions for the local friction slope, $S_f$, has to be derived.

### 2/1.2 The Energy Slope

The presence of bedforms and the undulating cover underside will give rise to local variations of energy dissipation along the longitudinal stream channel from station to station. Hence, the local value of the energy slope at any given station has to be used in the analysis. Since for a channel with a floating cover where two subsection flows have been postulated, the total energy loss is, therefore, the sum of the losses in each subsection. Using this argument, then, the local energy slope at a given section of the channel would be the sum of the two individual subsection energy slopes, that is,

$$S_f = S_{f1} + S_{f2} \tag{9}$$

in which $S_f$ = local total friction slope; $S_{f1}$ and $S_{f2}$ represent energy slopes of subsections 1 and 2, respectively.

The friction slope is related to the boundary shear, $\tau$, by the equation

$$S_{fi} = \tau_i/\gamma Y_i, \qquad i = 1, 2 \tag{10}$$

where i = 1, 2 refers to subsections (1) and (2); $\gamma$ = unit weight of water. In order to evaluate Eq. (10), expressions for the boundary shear will next be developed.

### 2/1.3 Boundary Friction

The boundary shear will be determined employing the momentum principle for each subsection of the channel cross-section.

#### The Cover Subsection

Figure 3 describes the forces acting on a sectional element of the cover subsection. The application of the momentum equation to this element in the horizontal direction will result in

$$P_{2x} - P_{2(x+dx)} - \tau_{2x}\, dx = \rho q_2 \left(V_{2(x+dx)} - V_{2x}\right) \tag{11}$$

in which $P_{2x}$, $P_{2(x+dx)}$ = total pressure per unit width at the upstream and downstream sections of the element, respectively; $\tau_{2x}$ = boundary cover shear; dx = length of the element; $\rho$ = density of water; $q_2$ = unit discharge at subsection 2; $V_{2(x+dx)}$ and $V_{2x}$ = mean velocities at the downstream and upstream sections of the element, respectively.

The total pressure per unit width, which is taken equal to the static pressure, is

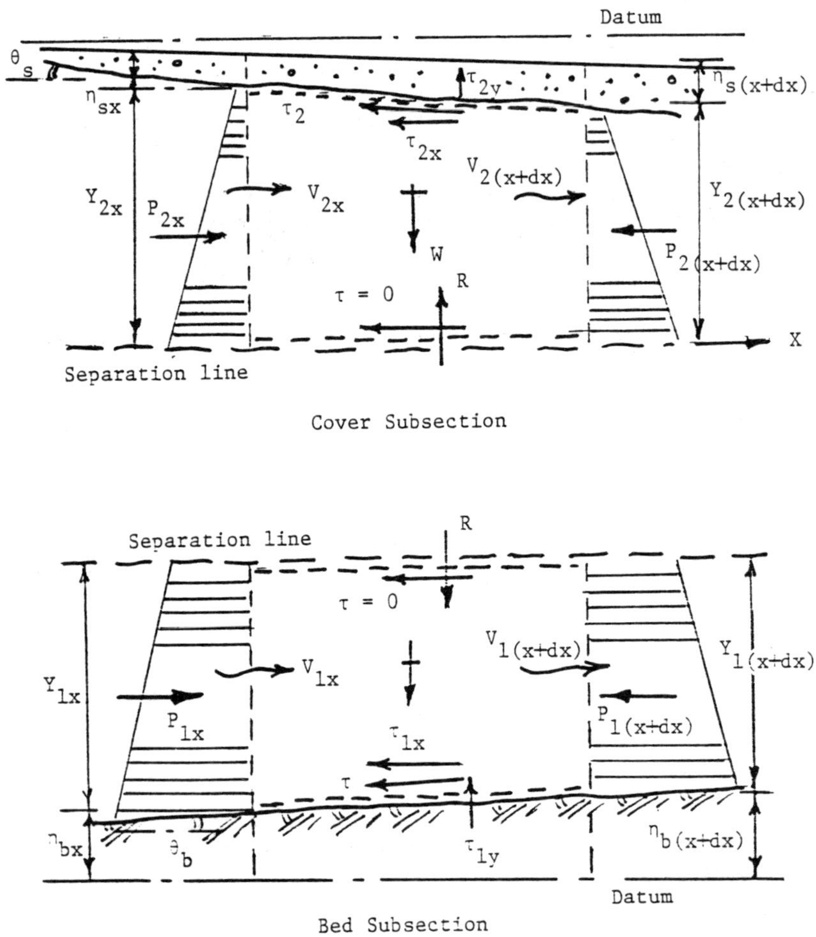


Fig. 3. Forces on Subsections

$$P_{2x} = C_p \gamma Y_2 (S_g \eta_s + \frac{1}{2} Y_2) \tag{12}$$

where $C_p$ = pressure coefficient.

The continuity equation gives

$$q_2 = Y_2 V_2 \tag{13}$$

Hence, the horizontal shear force acting on the cover can be expressed as

$$\tau_{2x} = - \left[ \gamma C_p S_g Y_2 \frac{d\eta_s}{dx} + \frac{\gamma}{g} V_2 \frac{dq_2}{dx} + \gamma \frac{dY_2}{dx} \left( C_p S_g \eta_s + C_p Y_2 - \frac{V_2^2}{g} \right) \right] \tag{14}$$

Applying the momentum equation in the vertical direction will provide

$$\tau_{2y} = \gamma S_g \eta_s \tag{15}$$

where $\tau_{2y}$ is the vertical component of the cover shear force. The total cover shear, $\tau_2$, can be expressed as

$$\tau_2 = \tau_{2x} \cos \theta_s - \tau_{2y} \sin \theta_s \tag{16}$$

in which $\theta_s$ is the slope of the cover underside. The following assumptions can be made for practical purposes:

$\cos \theta_s = 1$ and $\sin \theta_2 = d\eta_s/dx$.

The authors (Ref. 1) have derived an expression relating the subsection depth, $Y_2$, to the entire flow depth, Y, as

$$Y_2 = Y \lambda / (1+\lambda) \tag{17}$$

where $\lambda$ = hydraulic subsection radii ratio = $R_2/R_1$.

The cover subsection friction slope is obtained by combining these relationships with Eq. (10) to give

$$-S_{f2} = S_g \left( C_p + \frac{\eta_s}{Y} \frac{1+\lambda}{\lambda} \right) \frac{d\eta_s}{dx} + \frac{V_2}{gY} \frac{1+\lambda}{\lambda} \frac{dq_2}{dx}$$

$$+ \frac{dY}{dx} \left( C_p S_g \frac{\eta_s}{Y} + C_p \frac{\lambda}{1+\lambda} - \frac{V_2^2}{gY} \right) \tag{18}$$

The Bed Subsection

A similar type of analysis for the friction slope of the bed subsection for the forces shown in the lower portion of Fig. 3 will give

$$-S_{f1} = C_p S_g \frac{d\eta_s}{dx} + (1+\lambda)\left[1 + \frac{S_g \eta_s}{Y}\right]\frac{d\eta_b}{dx} + (1+\lambda)\frac{V_1}{gY}\frac{dq_1}{dx}$$

$$+ \frac{dY}{dx}\left[C_p S_g \frac{\eta_s}{Y} + C_p + \frac{\lambda}{1+\lambda} C_p - \frac{V_1^2}{gY}\right] \tag{19}$$

The total friction slope, $S_f$, as given by Eq. (9) is then the sum of subsection friction slopes, $S_{f2}$ and $S_{f1}$, as given by Eq. (18) and Eq. (19), respectively.

The friction slope can be determined once the values of the hydraulic radii ratio, $\lambda$, and the mean subsection velocities, $V_1$ and $V_2$, are known. These quantities can be obtained from the authors' earlier paper (Ref. 1) and are reproduced below.

Using the Manning equation,

$$V_i = \frac{1}{n_i} R_i^{2/3} S_o^{1/2}, \qquad i = 1, 2 \tag{20}$$

the hydraulic radii ratio, $\lambda = R_2/R_1$, is expressed by the relation

$$\frac{R^{1/6}}{n_1 g^{1/2}} = \frac{0.44}{\kappa} \frac{(\lambda^{1/2}-1)}{(1-n_1\lambda^{3/2}/n_2)} \left[\alpha + (1-\alpha)\lambda\right]^{1/6} \tag{21}$$

where $R$ = hydraulic radius of the entire channel cross-section; $n_1$, $n_2$ = Manning's roughness coefficient for subsections (1) and (2), respectively; $\kappa$ = Von-Karman's constant and equals 0.4; $\alpha = p_1/p$ = ratio of the subsection (1) wetted perimeter ($p_1$) to that of the total covered channel wetted perimeter ($p$).

The subsection velocities $V_1$ and $V_2$ can be found from the equation

$$V_1 - V_2 = \frac{-2}{3\kappa}(V_{*1} - V_{*2}) \tag{22}$$

where $V_{*i} = (g\, R_i\, S_o)^{1/2}, \qquad i = 1, 2$ (23)

3. EXPERIMENTAL INVESTIGATION

3/1.1 Floating Cover

The underside profile of ice-covers in nature frequently show undulations resembling ripples and dunes as reported by Ashton and Kennedy (Ref. 2), Carey (Ref. 3 and 4) and Larsen (Ref. 5 and 6). In order to produce the undercover ripples in the laboratory, polyethylene plastic pellets with a specific gravity equal to 0.92, which is close to that of natural ice, were used to simulate the ice-cover. The pellets have a disc shape measuring 4 mm in diameter, 2 mm in core thickness, and 1 mm at the sides. The buoyant loose cover has the ability to self-form ripples on its underside.

### 3/1.2 Mobile Bed

The sand used for the mobile bed consisted of the Ottawa Silica 90 and the Ottawa Flint Shot with mean grain sizes varying from 0.20 mm to 0.60 mm. The fluvial bed was able to form its own slope and bedform. Since the flow depth and bed slope were able to self-adjust, the channel is considered to have two degrees of freedom by Blench (Ref. 7).

### 3/1.3 Laboratory Flumes

Laboratory flumes with plexiglass sides with widths measuring 1.2 m, 0.46 m, and 0.15 m were used in the experimental observations. Besides the above flumes, channels fabricated in sand-cement mortar were also employed. In all cases, centrifugal pump-motor units were utilized to circulate the flow.

## 4. DISCUSSION OF RESULTS

### 4/1.1 Solution of Equations

The bed profile can be obtained by solving Eq. (7) using Eqs. (18) and (19) at each section of the stream channel for any given underside cover configuration. In general, direct integration of these equations is not feasible and a numerical solution is required.

In solving the differential equation, the direct step method was adopted. In using this technique, the value of $d\eta_s/dx$ is assumed to be known along the channel. At the starting point, $X_o$, the flow depth, $Y_o$, is assumed known and the value of $(dY/dx)_o$ can be computed and the expected Y at $(X_o+dx)$ can be determined. The solution progresses along the channel to obtain the bed profile.

### 4/1.2 Experimental Observations

Comparison of analytical results with test observations were made in two types of channels. In the first type, clear water was allowed to circulate in a covered channel with a plane sand bed until equilibrium conditions were attained. In the second case, a sand bed was built up from an initial rigid covered channel by depositing sand from a hopper at the upstream end until an aggraded equilibrium bed was established.

The experimentally observed mean bed slopes were in good agreement with those computed from the theoretical formulations. The bedforms found in the tests were in fair agreement generally with those computed from the equations.

## 5. CONCLUSIONS

Differential equations were formulated to predict regime bed profiles in two degrees of freedom channels with sand beds in the presence of a floating cover. The theory was able to explain the presence of bedforms in the mobile bed for any known undercover configuration. The investigation should be extended not only to larger stream channels in nature but also to the case of three degrees of freedom in which self-adjustment of the sides of the cross-section takes place.

## 6. NOMENCLATURE

The following symbols have been used in the paper. Each term has also been defined as it first appeared.

A — Total channel cross-section area
$A_1$, $A_2$ — Covered channel subsection cross-section areas (1) and (2), respectively
$C_p$ — Pressure coefficient
d — Differential operator
E — Total energy
$F_n$ — Froude number $= V/\sqrt{gY}$
g — Acceleration due to gravity
H — Height of the top surface of the cover from the datum
i = 1, 2 — Symbol i = 1, 2 refers to the covered channel subareas (1) and (2), respectively
n — Manning's roughness coefficient
$n_1$, $n_2$ — Bed and cover Manning's coefficients, respectively
P — Total pressure
p — Total channel wetted perimeter
$p_1$ — Wetted perimeter of bed subsection (1)
q — Discharge per unit width
$q_1$, $q_2$ — Flowrate per unit width of covered channel subsections (1) and (2), respectively
R — Total hydraulic radius = A/p
$R_1$, $R_2$ — Hydraulic radii of subsections (1) and (2), respectively
$S_f$ — Total local energy slope
$S_{f1}$, $S_{f2}$ — Local energy slopes of subsections (1) and (2), respectively
V — Mean velocity of entire cross-section
$V_1$, $V_2$ — Average velocities of subsections (1) and (2), respectively
$V_{*1}$, $V_{*2}$ — Shear velocity $(g\, R_i S_o)^{1/2}$ of subsections (1) and (2), respectively
x — Distance along axis X
Y — Local flow depth; distance along axis Y
$Y_1$, $Y_2$ — Local depths in bed and cover subsections (1) and (2), respectively
$\alpha$ — Wetted perimeter ratio $= p_1/p$
$\gamma$ — Unit weight of water
$\kappa$ — Von Karman's constant
$\eta_b$, $\eta_s$ — Thickness of bed and cover layers, respectively
$\lambda$ — Hydraulic radii ratio $= R_2/R_1$
$\rho$ — Density of water
$\tau$ — Shear stress

## 7. ACKNOWLEDGEMENTS

The research grant provided by the Natural Sciences and Engineering Research Council Canada is gratefully acknowledged.

## 8. REFERENCES

1. Chee, S P and Haggag, M R I: "Flow resistance of ice-covered streams". Canadian Journal of Civil Engineering, 11, 4, December 1984, pp. 815-823.

2. Ashton, G and Kennedy, J F: "Ripples on underside of river ice covers". Journal of the Hydraulics Division, ASCE, HY9, September 1972, pp. 1603-1624.

3. Carey, K L: "Observed configurations and computed roughness of the underside of river ice, St. Croix river, Wisconsin". U.S. Geological Survey Professional Paper 550B, Geological Research, 1966, pp. B197-B198.

4. Carey, K L: "The underside of river ice, St. Croix river, Wisconsin". U.S. Geological Survey Research, 1967, pp. C195-C199.

5. Larsen, P A: "Head losses caused by an ice cover on open channels". Boston Society of Civil Engineers, 1966, pp. 45-67.

6. Larsen, P A: "Hydraulic roughness of ice covers". Journal of the Hydraulics Division, ASCE, HY1, January 1973, pp. 111-119.

7. Blench, T: "Mobile-bed fluviology". University of Alberta, Edmonton, Canada, 1966, pp. 54-56.

International Conference on
RIVER REGIME
18-20 May, 1988

PAPER F1

# SEDIMENT SIZE AND CHANNEL CHANGES IN BRAIDED AND MEANDERING GRAVEL-BED STREAMS, UPPER DEESIDE, SCOTLAND

Judith Maizels
Department of Geography
University of Aberdeen AB9 2UF

SUMMARY: Contrasts in channel sedimentology between two upland gravel-bed streams appear to be reflected in different patterns and scales of channel change. A coarse-grained, braided reach, with extensive surface armouring, imbrication and clast clusters, has exhibited significantly less change than a finer-grained, meandering reach.

## 1. INTRODUCTION

Channel change represents the adjustment of the channel to a wide range of independent variables, each operating over different timescales and within different threshold conditions. Longterm, largescale changes in channel characteristics reflect the culmination of successive episodic, short-term, local changes responding to varying combinations of the controlling variables. The variables that dominate the processes of channel change are the hydrology of the catchment, the hydraulics, morphology and sedimentology of the stream network and local channel reaches. Maximum change in the channel system occurs when the greatest number of process thresholds are exceeded during a given event, while less permanent or significant channel changes occur during exceedance of some, but not all, process thresholds, the latter acting as constraints or feedback controls on further change.

Many recent studies have demonstrated a wide range of channel changes in response to rainfall and stream flow events of differing magnitude and frequency, as well as to a variety of sediment transport events, patterns of bar development, bed and bank scour, and small-scale patterns of sedimentation and bedform development (eg Refs 1, 2, 3, 4). However, few models have yet been developed that can provide a convincing link between the processes that are operating throughout the hierarchy of spatial scales. In particular, no models yet exist that can explain larger scale channel behaviour in relation not only to the larger scale magnitude-frequency relations of flow and sediment transport events, but also to the local and smaller scale controls of sediment entrainment, erosion and deposition. A major obstacle in developing such a model for a natural fluvial environment is the lack of field instrumentation and a detailed sedimentological data base over a sufficiently long time period to allow large scale channel changes to be identified.

This paper explores the initial stages of a field project which aims ultimately to develop and test such a model. The objective of the model will be to test the extent to which hydrologic, hydraulic, morphologic and sedimentologic controls on channel change vary in significance in contrasting gravel-bed streams. The immediate objective of this paper is to test whether the contrasts in channel sedimentology between two upland gravel-bed streams are reflected in contrasting patterns and scales of channel change in response to a given range of stream flow events, or whether these contrasts represent an equilibrium adjustment by each stream, such that the responses

International Conference on River Regime, edited by W.R.White.

reflected in channel change are similar in each stream environment.

## 2. CONTROLS ON CHANNEL CHANGE

Channel change is defined as any change in the planform, morphology, pattern or sedimentology of the channel, and occuring over any timescale, ranging from seconds through to millenia. Such change may also occur on any spatial scale, ranging from the movement and re-deposition of individual bed particles, through to largescale channel planform or pattern changes (eg Refs 3, 5). A model of channel change must identify the mechanisms by which processes operating at all these spatial and temporal scales may act to control every level of channel change (Ref 5). Hey (Refs 6, 7, 8) identifies 9 channel variables that respond to changes in the independent controlling variables. These variables will be explored in the Dee-Quoich model as hydrologic and hydraulic controls, and as the effects of bed and bank conditions, on channel change. Channel change is identified here as changes in channel planform shape or position, in development or removal of bars and channel, in changes of cross-sectional morphology, and in raising, lowering or removal of banks. The timescale for channel change in this study extends from the period of the first Ordnance Survey maps (1860's) through to the present day. The hydrologic variables include the magnitude-frequency relations of flow events and the catchment controls on these relations; stream power, velocities and bed shear stress; local bed and bank conditions as represented by bed particle size and fabric variation, pebble cluster and bedform development, bed roughness and local bed topography, bar form, riffle and pool development, bank cohesiveness and bend resistance to flow. This paper concentrates on the links between observed channel changes and local variations in bed sediment characteristics.

## 3. BACKGROUND TO STUDY AREA

### 3.1 Morphologic Conditions

Two contrasting streams have been selected from the upper River Dee catchment in the Grampians of Scotland (Fig. 1). The 2.5km reach of the Upper Dee forms a low gradient (averaging 0.0033), meandering channel (thalweg sinuosity of 1.31), characterized by a downstream sequence of fine gravel point bars. The channel banks are largely fine sandy alluvium, intercalated with humic horizons and gravel beds. The Upper Dee catchment lies largely within an area of upland terrain rising to Ben Macdui at 1309m, and includes the source of the Dee, and covers an area of c 290km$^2$. The downstream end of the Dee study reach is marked by its confluence with the River Quoich, which forms a more steeply graded (averaging 0.0147), low sinuosity (1.10), braided channel system characterized by non vegetated, coarse gravel and cobble bars (Table 1). The catchment area of 60km$^2$ lies within an upland moorland glen, with heather slopes, and valley floor infilled with coarse glaciofluvial sediments, and with Moine gneiss bedrock exposed along much of the middle Quoich channel course.

### 3.2 Hydrologic Conditions

Discharge records for the upper Dee have only been collected since late 1982 (by N.E.R.P.B), providing continuous daily discharge data for only a 5-year period (Fig 2A). As yet, no continuous records are available for the Quoich, but individual gaugings of the Quoich have allowed a Dee-Quoich discharge relation to be established with an $r^2$ value of 0.88 (Fig 2B). These data indicate that the mean annual discharge of the upper Dee is c. 12.5 $m^3s^{-1}$, compared with c. 3 $m^3s^{-1}$ on the Quoich (Table 1).

However, these data mask the significance of instantaneous high peak discharges which, during rapidly rising 'flash floods' (Ref 9), may reach over 4 times the magnitude of the mean daily discharge, and almost 7 times that of the minimum daily discharge, reflecting rapid response rates in both catchments. The absence of a perfect correlation between the two discharge records clearly reflects, however, contrasting catchment responses to varying rainfall durations and intensities. These contrasts cannot be explored until continuous discharge records are available for the Quoich (in late 1987).

Hence, current data suggest that the two channel environments are not only subject to similar kinds of flow regime, but also experience almost identical mean stream power conditions. Stream power estimates based on the mean annual flood ($Q_{2.33}$) average 3040 and 3097 W $m^{-1}$ on the Dee and Quoich, respectively, a similarity also reflected in specific power values in the active channel zones averaging 40.8 and 42.5 W $m^{-2}$ respectively (Table 1). The contrasting stream reaches therefore appear to have developed under very similar mean stream power conditions, both lying very close to Richards (Ref 10) meandering-braided channel pattern threshold of 50 W $m^{-2}$. Examination of the gradient-discharge relations for the two reaches also indicates that both channels exceed Ackers and Charlton's (Ref 11) thresholds for channel pattern change (0.0008 and 0.0098, respectively); the Quoich falls below the threshold predicted by Ferguson (Ref 12, after Parker Ref 13) (0.0336), while the Dee lies above this threshold (0.0185); but both reaches conform to the patterns predicted by Leopold and Wolman's (Ref 14) slope-discharge relationship (0.0042 and 0.0080, respectively). These contradictory results suggest that both channel patterns may be transitional forms, perhaps reflecting long-term adjustments to inherited channel systems and sedimentology (Ref 12), or possibly also reflecting the inadequacy of the short discharge record in computing recurrence intervals.

## 4. METHODOLOGY

### 4.1 Measurements of channel change

Channel planform changes over historic (post-1860's) and recent (1966-1987) time periods have been determined by comparison of Ordnance Survey maps published in 1985/86, 1900 (1:10560) and 1969 (1:10000), and of aerial photographs taken in 1955, 1964, 1966 (1:12000) and 1987 (1:4900). Indicators of channel change from these sources include changes in channel course, initiation of new channels, re-activation or abandonment of channel reaches, new areas of sedimentation, areas of floodplain stabilisation, and mean annual rates of bank retreat and of bar accretion. However, since these longer term changes between the often widely speculated surveys do not allow direct correlations either with the hydrologic record or with sedimentologic data, more recent, short-term channel changes have been monitored from repeated ground surveys of channel cross-sections and bank surveys following a range of flow events between November 1986 and October 1987. These observations are intended to represent the initial reference conditions at the start of the longer term field programme.

### 4.2 Measurements of bed sediment characteristics

Five major sedimentological parameters were examined as potential controls on rates and patterns of largerscale channel changes.

(1) Bar morphology, since bars represent the major sediment storage elements between high flows, and hence act as the main source both of sediment for transport, and channel change, and also of resistance to flow, during these events. Bar location, planform shape, microtopography and bar unit components were all examined.

(2) Maximum clast sizes across bar top surfaces were measured in order to determine the largest sizes of material which would have to be transported in order to allow local bed changes to occur.

(3) The size distribution of material on bar surfaces was determined, since this plays a significant role in affecting the degree of armour development, and the degree to which individual clasts may be exposed to entrainment (Ref 15).

(4) The degree of imbricate clast cluster and pebble bedform development, particularly transverse ribs (Refs 16, 17, 18, 19), was determined, largely as a means of determining local resistance of the bed to disturbance by high flows. Extensive development of imbricate clast clusters significantly raise the threshold conditions for entrainment (Refs 16, 17) and increase the stability of the sediment surface. By contrast, an increased proportion of clasts protruding above the armour layer may even enhance the chances of entrainment (Ref 15), with all size grades becoming mobile under the same

flows (Refs 20, 21), and thereby decreasing the stability of the sediment surface. The microtopography formed by groups of clasts on a bar surface was determined from a Clast Topography Index. This index represents the ratio of clast surface length to the horizontal bed distance occupied by the group of clasts, as determined from ground-level photographs (eg. see Fig 3A).

(5) Finally, the degree of imbrication and inter-clast contacts were determined, as means of determining the resistance of the bed to entrainment, the extent of development of the armour layer, and the relative exposure of the sub-armour layer to further disturbance. A Clast Contact Index was derived, based on measurements of the proportion of the clast boundary length in the a-b plane which is in direct contact with adjacent clasts, for the 10 largest clasts at each site (eg see Fig 3B).

## 5. SEDIMENTOLOGICAL CONTROLS ON CHANNEL CHANGE IN THE DEE-QUOICH REACH

Bar forms and channel sediments within the two stream reaches differ significantly in a number of respects. First, the Dee is characterized by a sequence of point bars comprising fine pebble, gravel and sand mixtures (eg Fig.4A). These bars are normally attached to the bank, or have been partially dissected by a slough chute channel, filled with standing water during low flows and blocked at the downstream outlet by sands. The point bars commonly form simple bar units, exhibiting downstream and lateral variation in clast dimension (Fig. 5A), with little microtopographic relief across the bar surface, which otherwise exhibits a high total relief of <3m from bar top to channel bed. Where chute channels have developed, transverse flows across the bar have created minor chutes, and areas of dune and rippled sand accumulation. The bar deposits in the Quoich, by contrast, exhibit highly complex morphologic and sedimentologic variations, but superimposed on a system exhibiting a low total relief of <1.5 m (Fig 5B). This local variability reflects the composite origin of individual bars, associated with successive periods of channel shifting, sediment reworking, and channel infilling during and following high flow events. Thus, while overall downstream decreases in clast size are apparent on the Quoich bars concentrations of large clasts occur on former bar head locations, while fines are located particularly in lee sides of such concentrations, at the downstream ends of transverse chute channels, and newly accreted bar units in sloughs and adjacent to low water pools (cf Refs 18, 22).

Second, material on the Quoich bars is significantly coarser than that on the Dee bars (Fig 5D). Samples from c 900 bar top sites on the Dee and c 750 sites on the Quoich indicate mean maximum clast sizes (b-axis) of 7.78 $\pm$ 1 standard deviation of 3.47 cm and 13.39 $\pm$ 3.62cm, respectively. However, the distribution of clast sizes on the bars varies not only across the bar surface as a whole, but exhibits significant local variations both horizontally and vertically. Thirdly, since the Quoich sediments include such a wide range of clast sizes (Fig. 5D) there is a higher probability of obstacle clasts occurring and remaining relatively stable on Quoich bar sediments than on the finer grained Dee bar surfaces (cf Ref 23). Indeed, Quoich bar surface clasts also exhibit significantly greater degrees of imbricated fabric, not only forming relatively stable pebble clusters, but also creating a more distinct clast topography (Fig 3A). Comparison of clast topographies for Quoich and Dee bar top sediments indicates that the Quoich sediments exhibit significantly larger values of clast topography, with indices averaging 1.42 $\pm$ 0.32 on the Quoich bars, compared with only 1.27 $\pm$ 0.18 on the Dee bars.

Fourthly, the wide size distribution of clasts present on the Quoich bar surfaces has also led to the extensive development of a surface armour layer, forming a closely packed, relatively stable and resistant channel boundary. Clast contact indices are significantly higher on the Quoich bars, averaging 0.69 $\pm$ 0.15, whilst those on the Dee bars averaged only 0.37 $\pm$ 0.19. These data indicate that the armour layer is significantly more continuous on Quoich bar surfaces, while on those of the Dee the armour layer is less complete, with areas of clasts separated by numerous patches and pockets of fine gravels and sands (Fig. 3B). Hence, sediment transport is likely to be more continuous during events in which bar surfaces of the Dee are submerged, assuming a readily available sediment supply (Ref 24).

Finally, the Quoich bar deposits also exhibited significant layer areas of transverse ribs and transverse clast dams (Ref 18), which occur in the majority of the chute channels cut into the Quoich bar surfaces. The beds of these chute channels were universally composed of cobbles and coarse pebbles embedded in a coarse sandy matrix. By contrast, the Dee chute channels are largely infilled with medium sand and fine gravel, and exhibit low dune and ripple bedforms.

## 6. CHANNEL CHANGES

### 6.1 Historic channel changes

Examination of map and air photo evidence indicates that the Upper Dee - Quoich flood plain zone has been one of the most dynamic and unstable on the River Dee. The flood plain extends for c1.5km between steep valley-side slopes, and has formed an area of poorly drained marshland and ponds, which since the mid-19th century have been subject to extensive land drainage. Much of the floodplain is traversed by artificial drainage channels, levees have been constructed along the Dee, and the banks have been locally re-inforced or protected. However, the map evidence, together with records of flood history, on the Dee indicate that regular large floods have breached levees, flooded across the floodplain, created major areas of bar development and stabilization, and initiation of new channels (Fig 4A). These changes have partly been related to downstream meander migration through bank retreat, and partly to channel avulsion during floods. Maximum channel change has occurred through truncation of flood plain 'promontories' extending into the channel, through enlargement of meander embayments, through occupation of adjacent older, abandoned channels, and through initiation of a new thalweg channel aross a major bar, leading to abandonment of the former divided main channel. Maximum rates of bank retreat and channel movement reached 0.29 to 1.23m $a^{-1}$ between 1866 and 1969 (Ref 25), while accretion reached 0.03-1.23m $a^{-1}$ over the same period (Table 1).

Rapid longterm channel changes on the Quoich, by contrast, were very marked only until the mid-1950's, since when the active channel system has become increasingly confined to a narrow zone and progressively less sinuous (Ref 26). Small channels extending around floodplain 'islands' have been infilled, vegetated and abandoned, while the thalweg and bar locations have remained relatively constant. Following initial drainage of the Quoich marshland and Lochan (in the 1860's), the Quoich extended for a further 150m west of its present channel as a series of braided stream courses forming a large embayment cut into the fine alluvial sediments (Fig 4B) of the Dee floodplain and marshes. The present channel course appears to have been established between 1900 and 1955, and may represent a response to the construction of the bridge, and consequent channelling of the river beneath it, where the Quoich emerges from the glen on to the floodplain. Levee construction from local gravels may also have encouraged to the stream to maintain its present course for the latter period.

The location of the confluence zone has remained constant throughout the period of historic records, despite migration of the main Quoich channel itself. However, bar deposits accumulating in the Quoich at the junction with the Dee have been extending progressively southeastwards into the main Dee channel, at a rate averaging 1.9 to 5.7m $a^{-1}$.

### 6.2 Recent channel changes 1966-1987

Major and dramatic changes in channel course have occurred on the River Dee during the 1966-1987 period. The two most marked changes have been the northward migration of the most upstream meander by 194m, representing an average rate of bank retreat of c 9.2m $a^{-1}$. This migratory trend appears to have been triggered by the transverse breaching of the central gravel bar ('A' in Fig 4A) between 1955 and 1964. This northward expansion of the river has been matched by similar rates of gravel accretion on to the expanding and increasingly stable southern bar, which has long sustained an undisturbed 'core island' of the original floodplain surface ('B' in Fig 4A).

The second major change in channel course has been accomplished at the outer bank of the meander lying one wavelength downstream (c 430 m). This part of the channel has again cut northwards, creating a 112-m wide breach in a long finger of the floodplain ('C' in Fig 4A) which formerly separated the main Dee from an artificial channel which ran parallel with the Dee for c 400m. This breaching has resulted in major transformation of the downstream reach. The main thalweg now follows a significantly widened (by c 3 times, from 12m to 34m) former drainage course ('D'), while the former thalweg has been abandoned and forms a swathe of partly vegetated gravels c 100m broad ('E') lying between the floodplain itself in the south and the remnants of the narrow floodplain segment which bounded the artificial channel in the north. The former bars and scour pools of the abandoned Dee river bed are recognised as coarse-grained topographic highs and water filled pools, respectively, over a distance of c 400m. The upstream segment of the abandoned channel has since been reworked, and gravel has been accreted by the newly developed north-flowing river, such that new point bar units have become attached to the floodplain outliers to north and south.

By contrast, relatively little change in channel course and pattern have emerged in the Quoich over the same 21-year period. Any significant channel changes have been limited to the central 250m length of the reach, where the active channel zone has re-expanded westwards into the formerly active zone (see above) by 52m (from c 60m to c 112m), representing an overall rate of bank recession of c 2.5m $a^{-1}$ (Table 1). This westerly embayment ('A' in Fig 4B), forming a broad zone of shallow, expanding flow, has been infilled with a large, attached lateral bar. The former large attached bar on the east has been dissected both by transverse and by bank-foot chute channels, producing a more complex braided system of composite bar units. Hence, this central channel zone has therefore expanded to form the major sedimentation zone for the lower Quoich.

The channel zones lying both upstream ('B') and downstream ('C') of the central active zone have remained confined and relatively stable 'transport reaches' (Refs 27, 28). In these reaches, bar locations have not changed significantly, while many small side channels in the flood plain have become further vegetated and abandoned, thereby increasing yet further the resistance of the bank to channel widening.

Changes in the confluence zone have repeated the earlier pattern of downstream expansion of the eastern lateral bar of the Quoich along the northern bank of the main stream of the Dee to form a second, smaller, sedimentation zone for the lower Quoich ('$D_1$'). This expansion by 20-30m between 1966 and 1987, averaging c 1-1.4m $a^{-1}$, has acted to deflect the Dee thalweg southwards away from the flow separation zone near Quoich inlet, resulting in additional scour of the southern bar, which formerly extended some 30m further downstream ('E'). The location and morphology of large point bars lying upstream along the Dee have remained unchanged throughout this period, apart from small-scale headward expansion of the bank-foot chute channels.

6.3 Short-term channel changes

Examination of changes in cross-sectional profiles across the Quoich (Fig 6) following flood events of different magnitude, frequency, and seasonal occurrence, indicated that significant changes, through localized bed scour of <0.35m and vertical bar accretion of <0.2m, only occurred in divided reaches during the November 1986 floods which exceeded 11 $m^3s^{-1}$. At other cross-sections no significant changes were recorded over the 9-month period March-October 1987, despite 3 floods exceeding 9 $m^3s^{-1}$. The most upstream cross-section exhibited no changes in bedlevel at all within the 10-month observation period (Fig 6A) despite peak flows exceeding 14 $m^3s^{-1}$. The bed material at this cross-section exhibits a smooth, densely packed, armoured bed.

Although sediment tranport measurements have not yet been undertaken, measurements of changes in sediment size distribution of bar top materials following these recent flood events showed that there had been no change of individual particles on the Quoich 'main bar' deposit while, on Dee 'bar 1', for example, extensive reworking of sediments had been accomplished (Fig 5C).

## 7. CONCLUSIONS

This paper has summarised some very preliminary results from the initial stages of a longer term programme aimed at developing a model of channel change that will link macro- and micro-scale catchment, hydrologic, hydraulic and sedimentary variables. The results of the present study suggest that while large flood events are responsible for maximum changes in the both fine and coarse gravel channel environments, the thresholds for change are highly variable through time and space. In particular, the thresholds for local sediment transport, and hence the controls on longerterm and largerscale channel change, depend largely on the relationship between flow competence, sediment supply and bed conditions. Hence, stability of the system will depend not only on variations in sediment input from upstream, but also on the sequence of high flows both during a single storm event and over a period of years, allowing variable periods for bed armour restabilisation (Ref 29). Thus while major changes in channel course have occurred on the Dee, the changes apparent in the Quoich have remained relatively insignificant, and confined only to local expansions of the active channel zone, and downstream extension of the confluence sedimentation zone. The contrasts in scale of channel change, although possibly reflecting varying hydrologic, hydraulic bank conditions and sediment input controls, together with local floodplain morphology, seem particularly to reflect the increased resistance to entrainment of the armoured, clustered, coarse-grained Quoich sediments.

Secondly, the types of change in channel morphology and planform have also been highly variable. The changes on the Dee include overbank avulsion, bar entrenchment, and rapid bank retreat to produce downstream meander migration; occupation and enlargement of smaller channels; and abandonment of the former major thalweg zone, while changes on the Quoich include progressive confinement of the active channel zone over the 1866-1966 period, followed by its small scale and local re-expansion; dissection of former attached bars; and downstream extension of the confluence bar zone. The patterns of change in each channel environment are hence also distinctive, and are also likely to reflect the mechanisms and rates of bed sediment removal and downstream transport. This paper has suggested some tentative relations between a macro-scale (channel change) and a selected micro-scale (bar surface sediment conditions) parameter. Future stages of this research programme will concentrate on obtaining hydrologic, hydraulic, bank and bed sediment data at a sufficiently detailed level to allow rigorous testing of the suggested relationships, and to form the basis of a channel change model which incorporates a wider range of inter-related variables operating within the catchment-channel system.

Acknowledgements

The author would like to thank Mr H Ross and Mrs K Leiper for assistance with field work, Mr W Ekin for the recent aerial photography, and Mr J Warren and Mr D Fraser of the North East River Purification Board for access to, and discussion on, the river discharge records.

References


1. Harvey, A.M.: "Event frequency in sediment production and channel change". In: River channel changes. Ed by K.J. Gregory. Wiley. 1977, pp 301-315.
2. Harvey, A.M., Hitchcock, D.H. and Hughes, D.J.: "Event frequency and morphological adjustment of fluvial systems in upland Britain". In: Adjustments of the Fluvial System. Ed by D.D. Rhodes and G.P. Williams. Allen & Unwin. 1979, pp 139-167.
3. Thorne, C.R. and Lewin, J.: "Bank processes, bed material movement, and planform development in a meandering river". In: Adjustments of the Fluvial System. Ed by D.D. Rhodes and G.P. Williams. Allen & Unwin. 1979, pp 117-137.
4. Carling, P.A.: "Bed stability in gravel streams, with reference to stream regulation and ecology". In: River channels. Environment and Process. Ed by K. Richards. Blackwell. 1987, pp 321-347.
5. Ruh-Ming Li and Simons, D.B.: "Dynamic modelling of channel responses". In: Gravel-Bed Rivers. Ed by R.D. Hey, J.C. Bathurst and C.R. Thorne. Wiley & Sons. 1982, pp 469-485.
6. Hey, R.D.: "Gravel-bed rivers: form and processes". In: Gravel-Bed Rivers. Ed by R.D. Hey, J.C. Bathurst and C.R. Thorne. Wiley & Sons. 1982, pp 5-13.
7. Hey, R.D.: "Design equations for mobile gravel-bed rivers". In: Gravel-Bed Rivers. Ed by R.D. Hey, J.C. Bathurst and C.R. Thorne. Wiley & Sons. 1982, pp 553-580.
8. Hey, R.D. and Thorne, C.R.: "Flow processes and river channel morphology". In: Catchment Experiments in Fluvial Geomorphology. Ed by T.P. Burt and D.E. Walling. Geo Books Norwich. 1984, pp 489-514.
9. Werritty, A.: "Stream response to flash floods in upland Scotland". In: Catchment Experiments in Fluvial Geomorphology. Ed by T.P. Burt and D.E. Walling. Geo Books Norwich. 1984, pp 537-560.
10. Richards, K.S.: "Rivers: Environment, process and form". In: River channels. Environment and Process. Ed by K. Richards. Blackwell. 1987, pp -13.
11. Ackers, P. and Charlton, F.G. 1970: "The geometry of small meandering streams". Proc. Inst. Civil Engrs., 12, 1970, 289-317.
12. Ferguson, R.: "Hydraulic and sedimentary controls of channel pattern". In: River channels. Environment and Process. Ed by K. Richards. Blackwell. 1987, pp 129-158.
13. Parker, G.: "Hydraulic geometry of active gravel rivers". Jl. Hydr. Div. Proc. Amer. Soc. Civil Engr., 105, 1185-1201.
14. Leopold, L.B. and Wolman, M.G.: "River channel patterns - braided, meandering and straight". U.S. Geol. Surv. Prof. Paper 282B, 1957.
15. Andrews, E.D.: "Entrainment of gravel from naturally sorted riverbed material". Bull. Geol. Soc. Amer., v. 94. 1983, pp 1225-1231.
16. Koster, E.H.: "Transverse ribs: their characteristics, origin and paleohydraulic significance". In: Fluvial Sedimentology. Ed A.D. Miall. Canad. Soc. Petrol. Geol. Memoir 5. 1978, pp 161-186.
17. Brayshaw, A.C.: "Characteristics and origin of cluster bedforms in coarse-grained alluvial channels". In: Sedimentology of Gravels and Conglomerates. Ed by E.H. Koster and R.J. Steel. Canad. Soc. Petrol. Geol., Memoir 10. 1984, pp 77-85.
18. Bluck, B.J.: "Bed forms and clast size changes in gravel-bed rivers". In: River Channels. Environment and Process. Ed by K. Richards. Blackwell. 1987, pp 159-178.
19. Naden, P.S. and Brayshaw, A.C.: "Small- and medium-scale bedforms in gravel-bed rivers". In: River Channels. Environment and Process. Ed by K. Richards. Blackwell. 1987, pp 249-271.
20. Parker, G., Klingeman, P.C. and McLean, D.G.: "Bedload and size distribution in paved gravel-bed streams". Jl. Hyd. Div. Proc. Amer. Soc. Civil Engr., 108, HY4, 1982, pp 544-571.

21. Bathurst, J.C.: "Measuring and modelling bedload transport in channels with coarse bed materials". In: River Channels. Environment and Process. Ed by K. Richards. Blackwell. 1987, pp 272-294.
22. Bluck, B.J.: "Texture of gravel bars in braided streams". In: Gravel-Bed Rivers. Ed by R.D. Hey, J.C. Bathurst and C.R. Thorne. Wiley & Sons. 1982, pp 339-355.
23. Laronne, J.C. and Carson, M.A.: "Interrelationships between bed morphology and bed-material transport for a small, gravel-bed channel". Sedimentology, 23, 1976, pp 67-85.
24. Gomez, B.: "Temporal variations in bedload transport rates: the effect of progressive bed armouring". Earth Surf. Processes and Landforms, 8, 1983, 41-54.
25. Maizels, J.K. : "The physical background of the River Dee". In: The Biology and Management of the River Dee. Ed. by D. Jenkins. Inst. Terrestr. Ecol., NERC, 1985, pp 7-22.
26. McEwan, L.J.: River channel planform changes in upland Scotland, with specific reference to climatic fluctuations and landuse changes over the last 250 years. Unpubld. Ph.D. thesis, Univ. of St. Andrews. 1985.
27. Church, M.: "Pattern of instability in a wandering gravel bed channel". In: Modern and ancient fluvial systems. Spec. Publn. No. 6, Int. Assoc. Sedimentol., Blackwell. 1983. pp 169-180.
28. Church, M. and Jones, D.: "Channel bars in gravel-bed rivers". In: Gravel-Bed Rivers. Ed by R.D. Hey, J.C. Bathurst and C.R. Thorne. Wiley & Sons. 1982, pp 291-338.
29. Klingeman, P.C. and Emmett, W.W.: "Gravel bedload transport processes". In: Gravel-Bed Rivers. Ed by R.D. Hey, J.C. Bathurst and C.R. Thorne. Wiley & Sons. 1982, pp 141-179.

Table 1 Summary of channel and sediment parameters, Dee-Quoich study reach

| Parameter | Dee reach | Quoich reach | Confluence zone |
|---|---|---|---|
| Length of reach (m) | 2450 | 710 | 200 |
| Catchment area ($km^2$) | 290 | 60 | |
| Mean gradient | 0.0033 | 0.0147 | |
| Channel sinuosity | 1.31 | 1.10 | |
| Bankfull channel width (m) | $76.5 \pm 29.3$ | $52.0 \pm 30.1$ | $66.8 \pm 19.9$ |
| Mean bankfull width in active channel zone (m) | 74.6 | 72.9 | |
| Mean annual discharge ($m^3\ s^{-1}$) | 12.5 | 3.0 | (15.5) |
| Most probable annual flood ($Q_{1.58}$)($m^3\ s^{-1}$) | 86.0 | 19.5 | (105.5) |
| Mean annual flood ($Q_{2.33}$)($m^3\ s^{-1}$) | 94.0 | 21.5 | (115.5) |
| Stream power ($W\ m^{-1}$) | 3040 | 3097 | |
| Specific stream power ($W\ m^{-2}$) | 39.7 | 59.6 | |
| Specific stream power in active channel zone ($W\ m^{-2}$) | 40.8 | 42.5 | |
| Mean maximum clast diameter (mm) | $77.8 \pm 34.7$ | $133.9 \pm 36.2$ | |
| Clast topography index | $1.27 \pm 0.18$ | $1.42 \pm 0.32$ | |
| Clast contact index | $0.37 \pm 0.19$ | $0.69 \pm 0.15$ | |
| Maximum rates of bar/channel movement ($m\ a^{-1}$) | 0.29-9.2 | <2.5 | 1-5.7 |

List of Figures

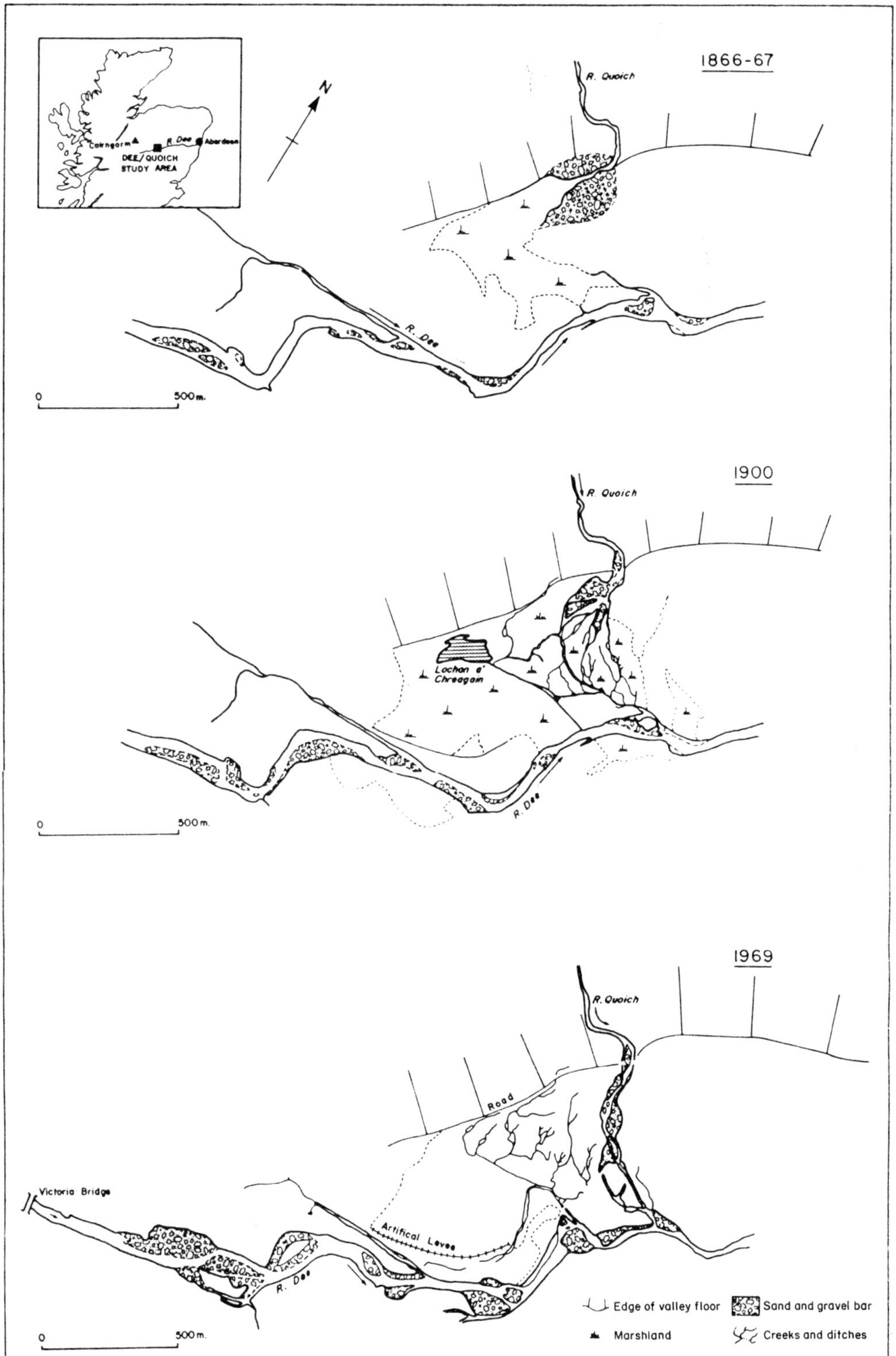
1866-67
R. Quoich
N
Cairngorm
R. Dee
Aberdeen
DEE/QUOICH STUDY AREA
R. Dee
0
500m.
1900
R. Quoich
Lochan a' Chreagain
R. Dee
0
500m.
1969
R. Quoich
Road
Victoria Bridge
Artifical Levee
R. Dee
Edge of valley floor
Sand and gravel bar
Marshland
Creeks and ditches
0
500m.

**Figure 1**

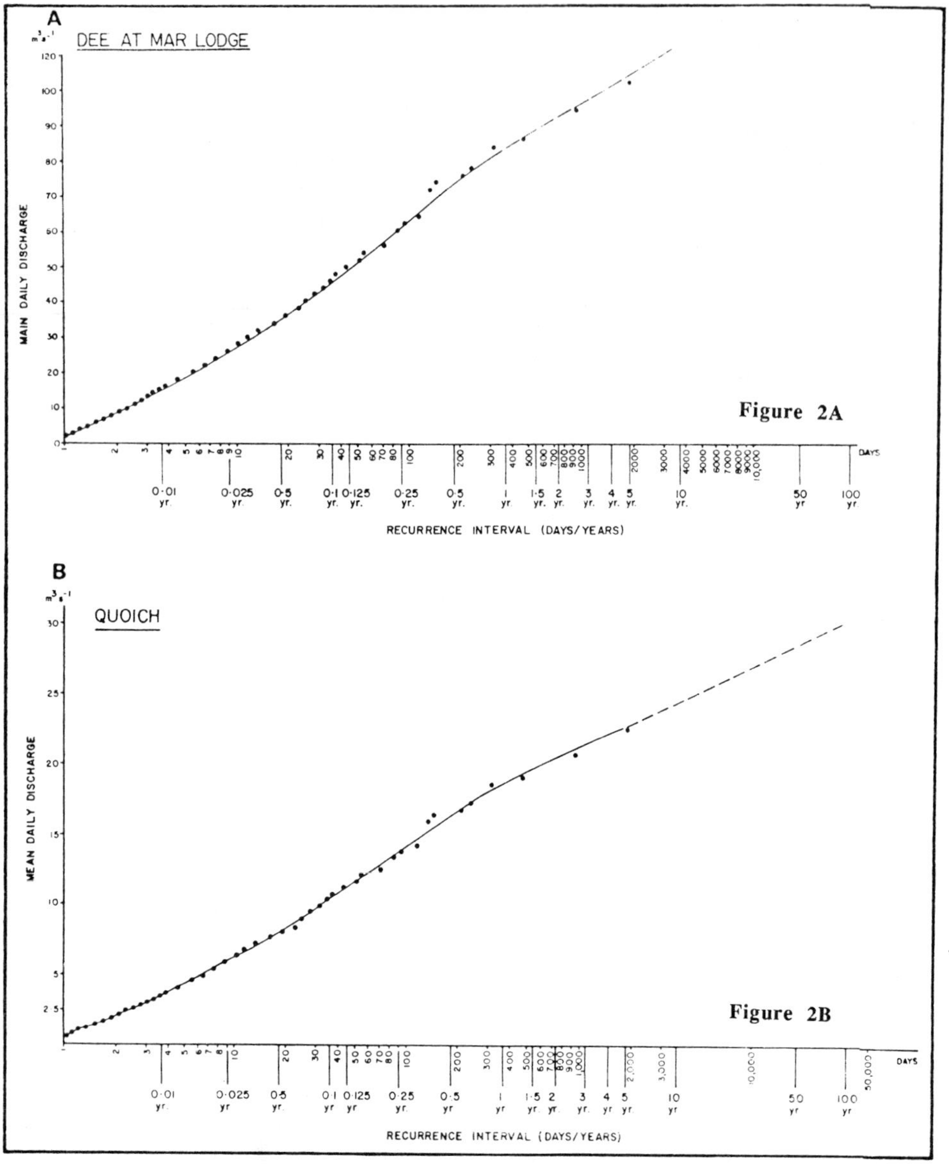


Figure 2A

Figure 2B

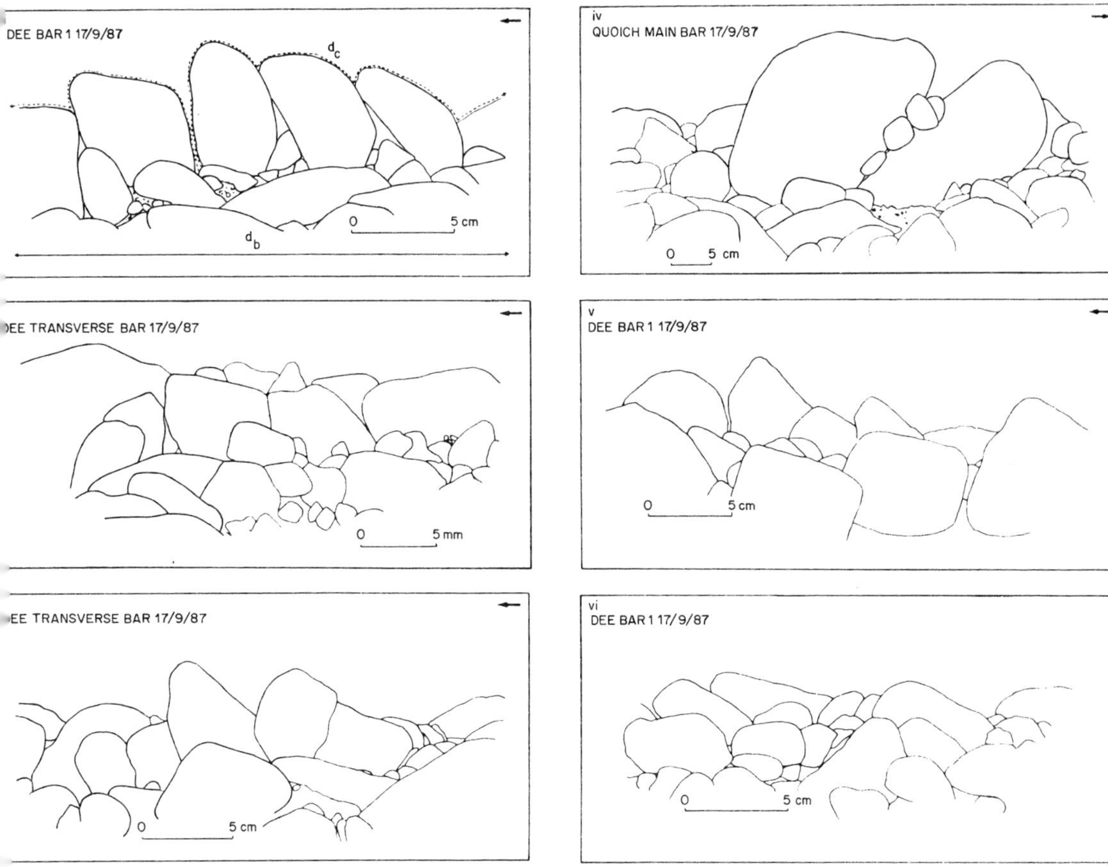


Figure 3A

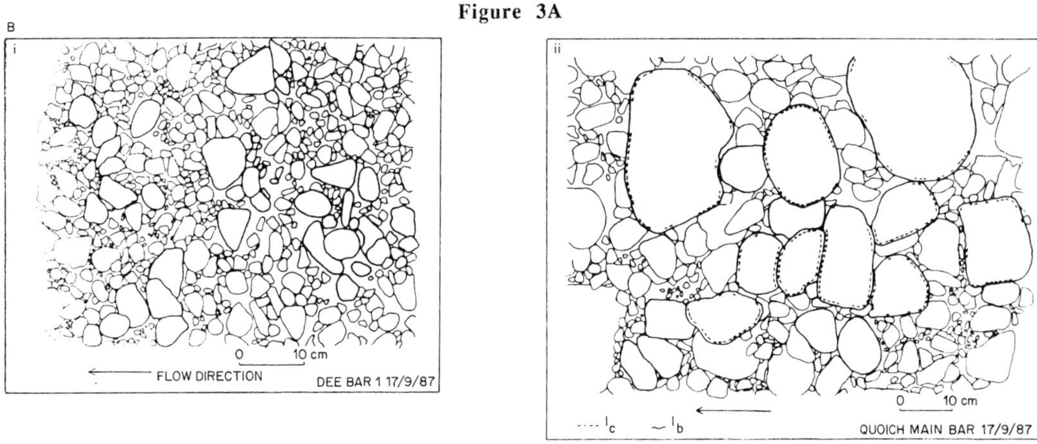


Figure 3B

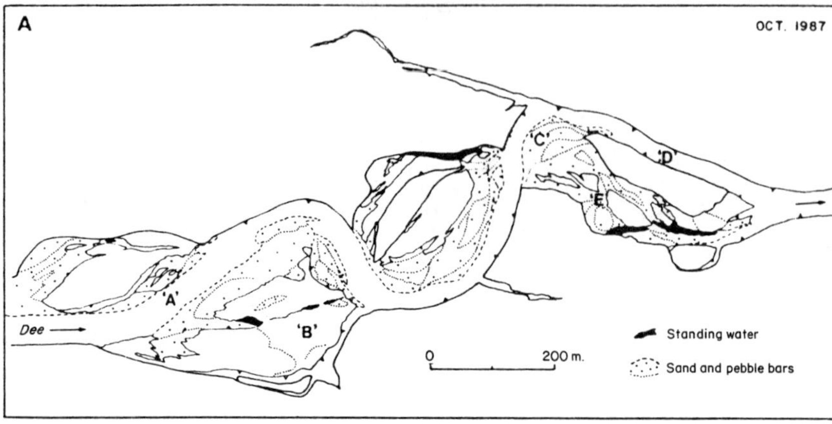


Figure 4A

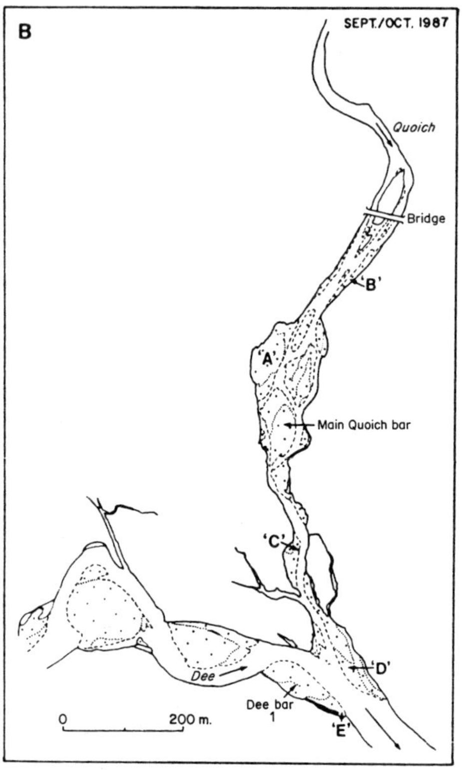


Figure 4B

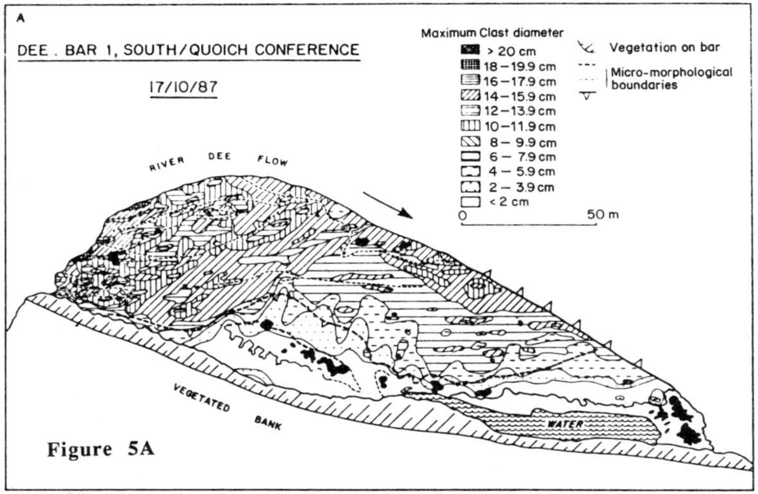


Figure 5A

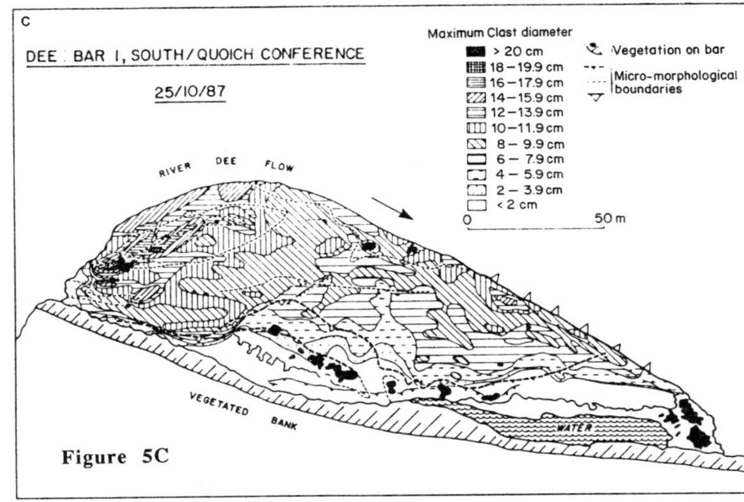


Figure 5C

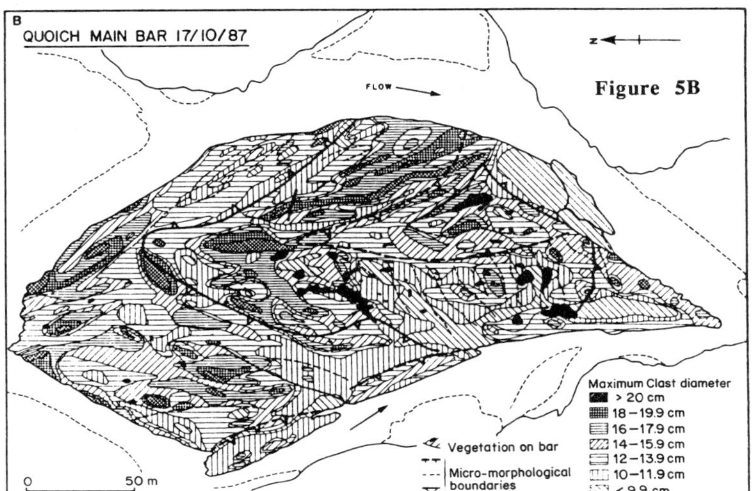


Figure 5B

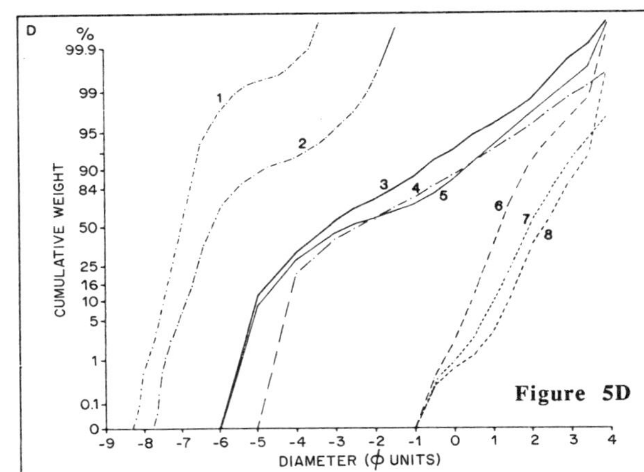


Figure 5D

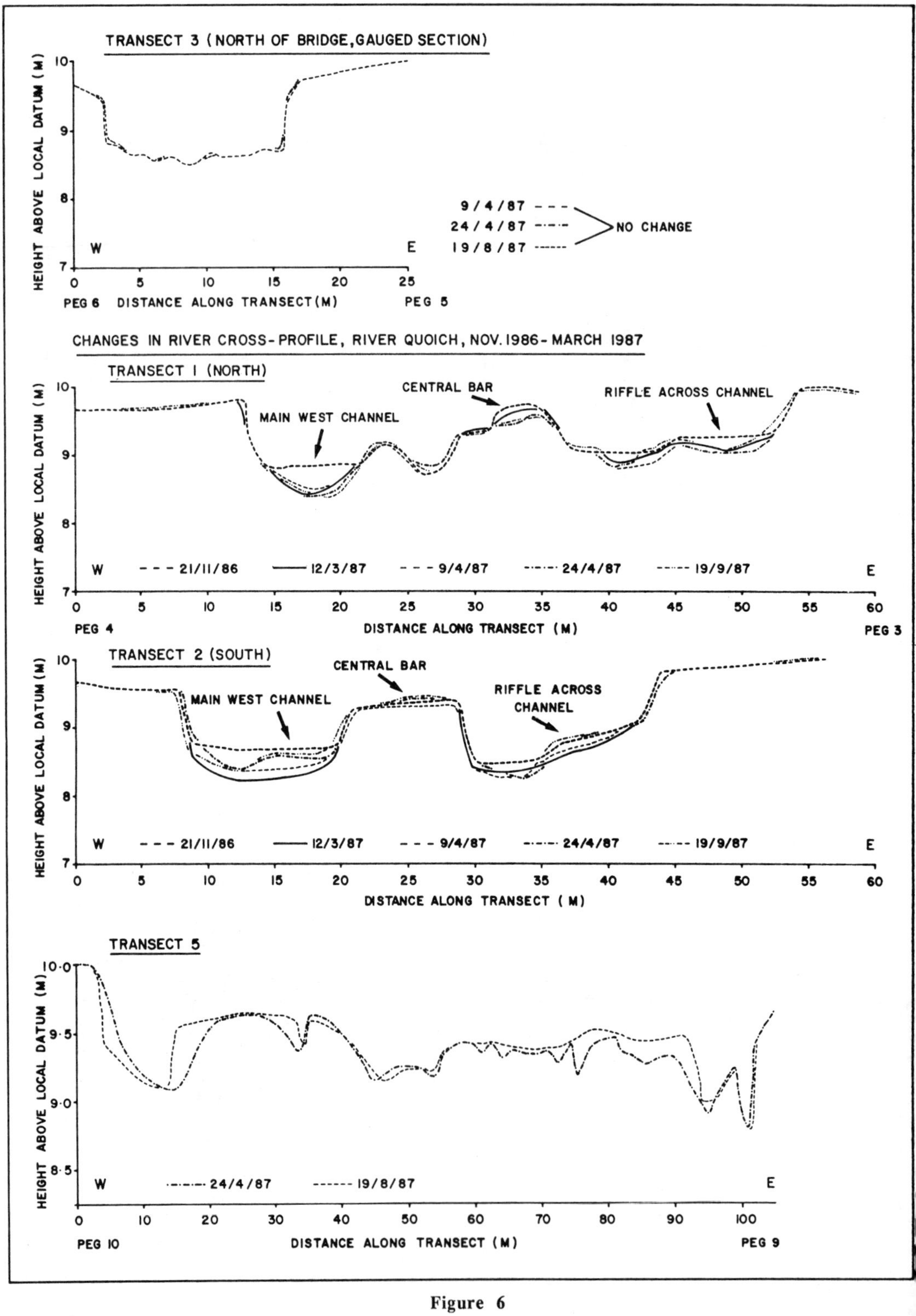
TRANSECT 3 (NORTH OF BRIDGE, GAUGED SECTION)
HEIGHT ABOVE LOCAL DATUM (M)
W
E
PEG 6
DISTANCE ALONG TRANSECT (M)
PEG 5
9/4/87
24/4/87
19/8/87
NO CHANGE
CHANGES IN RIVER CROSS-PROFILE, RIVER QUOICH, NOV. 1986 - MARCH 1987
TRANSECT 1 (NORTH)
MAIN WEST CHANNEL
CENTRAL BAR
RIFFLE ACROSS CHANNEL
21/11/86
12/3/87
9/4/87
24/4/87
19/9/87
PEG 4
DISTANCE ALONG TRANSECT (M)
PEG 3
TRANSECT 2 (SOUTH)
MAIN WEST CHANNEL
CENTRAL BAR
RIFFLE ACROSS CHANNEL
TRANSECT 5
24/4/87
19/8/87
PEG 10
DISTANCE ALONG TRANSECT (M)
PEG 9

Figure 6

International Conference on
**RIVER REGIME**
18-20 May, 1988

PAPER F2

# CHANGE OF RIVER REGIME AND GRADED-SEDIMENT DISCHARGE DURING FLOOD

Koichi SUZUKI, Dr. Eng. and Masanori MICHIUE, Dr. Eng.

Department of Civil Engineering, Tottori University
Koyama-Minami 4-101, Tottori, 680 JAPAN

SUMMARY: Sediment discharge characteristics of nonuniform sand bed during the increase and decrease periods of flood discharge are discussed experimentally in connection with the formation and destruction of an armor coat.

## 1 Introduction

Sediment discharge in alluvial channels with coarse and nonuniform bed materials depends strongly on the bed surface regime. Sediment discharge is very small when the bed surface is covered with an armor coat before and after flood, while it increases sharply when the armor coat is destroyed during flood. Therefore, in order to estimate sediment discharge of sand mixture it is important to know the grain size distribution of bed surface.

In this paper, sediment discharge characteristics of nonuniform sand bed during the increase and decrease periods of flood discharge are discussed experimentally in connection with the formation and destruction of an armor coat. And also a method for the estimation of grain size distribution of river bed surface is proposed under the assumption of an armor coat formation.

## 2 Summary of experiments

### 2.1 Experimental conditions

Experiments about sediment discharge of sand mixture were carried out, using nonuniform sand with a grain size distribution curve as shown in Fig.1 with mean diameter $d_m$ of 0.415 cm and standard deviation of 2.74. An experimental flume 8 m long and 0.4 m wide was used. Three series of experiments for different bed slopes I=1/80, 1/100 and 1/115 were carried out with various flow discharges as shown in Table 1, where Q is the discharge, h is the flow depth, $u_*$ is the shear velocity, $q_B$ is the total sediment discharge per unit width, $d_m$ is the mean diameter of bed surface sand which is sampled and mixed at equally spaced seven places along the flume and $d_m'$ is the mean diameter of transported sediment which is gathered at the downstream end of the flume.

There was no sediment supply from the upstream end of the flume. Each experimental run was finished in about ten minutes so that the influence of no sediment supply could not reach the downstream region. The experimental conditions as shown in Table 1 cover the flow conditions both with and without armor coat.

International Conference on River Regime, edited by W.R.White.

## 2.2 Sediment discharge of sand mixture

Many formulae about sediment discharge for uniform sand have been proposed. Here, one of these formulae with a concept of the critical shear stress of sand such as Meyer-Peter & Müller's formula(Ref.1) or Ashida & Michiue's formula(Ref.2) is modified for nonuniform sand by substituting the critical shear stress of each grain size of sand mixture $\tau_{ci}$ for the critical shear stress $\tau_c$ of uniform sand. This procedure has been proposed by Ashida & Michiue(Ref.3), Hirano(Ref.4) and Parker(Ref.5). But there still remains room for the modification of $\tau_{ci}$ which is calculated from the modified Egiazaroff's equation(Ref.6) by Ashida & Michiue(Ref.3). Ashida et al.(Ref.7) pointed out that there is difference between the critical shear stress which is defined as the minimum shear stress for the initiation of sand particles at rest and the critical shear stress which is defined as the minimum shear stress to stop the moving sand particles. And the definition of the critical shear stress is rather subjective and not clear. Here, the critical shear stress based on sediment discharge is defined by the shear stress when the sediment discharge is almost zero, i.e. $q_{*i}=10^{-3}$, where $q_{*i}=i_B q_B / i_b u_* d_i$, $i_B$ and $i_b$ are the frequencies of transported sand $d_i$ and bed surface sand $d_i$, respectively. Fig.2 shows the relationship between $\tau_{ci}/\tau_{cm}$ and $d_i/d_m$, where $\tau_{cm}$ is the critical shear stress of uniform sand with mean diameter $d_m$ of sand mixture. The following equation can be proposed experimentally for the critical shear stress $\tau_{ci}$ of sand $d_i$ based on the sediment discharge;

$$\tau_{ci}/\tau_{cm} = (d_i/d_m)^{0.07} \tag{1}$$

Fig.3 shows the relationship between $q_{*i}$ and $\tau_{*i}$, where $\tau_{*i}=u_*^2/sgd_i$, g is the acceleration due to gravity, $s=\sigma/\rho-1$, $\sigma$ and $\rho$ are the densities of sand and water, respectively. Solid lines are modified Ashida & Michiue's formula expressed by

$$q_{*i} = 17\tau_{*i}(1-\tau_{*ci}/\tau_{*i})(1-\sqrt{\tau_{*ci}/\tau_{*i}}) \tag{2}$$

which is almost equal to Meyer-Peter & Müller's equation(Ref.1) or Einstein's equation(Ref.8) in the range of $\tau_{*i}<\sim 1$, where $\tau_{*ci}=\tau_{ci}/sgd_i$, $\tau_{*cm}=\tau_{cm}/sgd_m$. $\tau_{cm}$ is given by well known Shield's curve for the initiation of sand particles and $\tau_{ci}$ is given by eq.(1). Eq.(2) seems to fit the experimental data fairly well.

## 2.3 Change of sediment discharge during flood

Fig.4 shows the relationship between flow discharge Q and sediment discharge per unit width $q_B$. There is a range of Q where different sediment discharges appear for the same Q, which means that the grain size compositions of bed surface are different among different sediment discharges even for the same Q. When the bed surface is covered with an armor coat, the sediment discharge is very small until the armor coat is destroyed. The sediment transport rate of each grain size can be estimated by eq.(2), only if the grain size distribution of bed surface $i_b$ is known. Therefore it is necessary to estimate $i_b$ during flood for the calculation of sediment transport rate of sand mixture.

# 3 Estimation of grain size frequency of bed surface

## 3.1 Process of armor coat formation

The armor coat formation has been considered as a result of selective transport of smaller particles from the bed surface of sand mixture under nonequilibrium state with a bed shear stress smaller than the critical shear stress of the maximum grain size. Then the armor coat formation should propagate from upstream towards downstream. But, according to the experiments, the armor coat seems to be formed almost at the same time throughout the whole length of the flume. The essential mechanism of the armor coat formation may be somewhat different from what has been considered. Here a hypothesis is proposed that the armor coat is formed by trapping smaller particles beneath the larger particles when the shear stress is smaller than the critical shear stress of the larger particles. If this hypothesis is admitted, then the grain size frequency of the bed surface $i_b$ may be estimated as follows; When the tractive force is equal to the critical shear stress of grain size $d_c$, some particles smaller than $d_c$ are assumed to submerge into the space under the particles larger than $d_c$, in proportion to the composition of each grain size particles smaller than $d_c$. If the space formed by a particle $d_i$ at rest is assumed to be $\alpha(1-\pi/6)d_i^3$, then the rate of the space formed by particles $d_i$ can be expressed by dividing $\alpha(1-\pi/6)d_i^3 i_{bo}$ by the volume $(\pi/6)d_i^3$ of sand $d_i$ $(=\alpha(6/\pi-1)i_{bo})$, where $\alpha$ is the coefficient and $i_{bo}$ is the frequency of sand $d_i$ in the original sand mixture. Therefore the total space rate formed by particles larger than $d_c$ is written by

$$\sum_{d_c}^{d_L} \alpha(6/\pi-1)i_{bo} \tag{3}$$

where $d_L$ is the diameter of the maximum grain size. The rate of submerged sand of $d_i$ smaller than $d_c$ in the space expressed by eq.(3) becomes

$$(i_{bo}/\sum_{d_s}^{d_c} i_{bo})\sum_{d_c}^{d_L}\{\alpha(6/\pi-1)i_{bo}\} \qquad (d_i \leq d_c) \tag{4}$$

where $d_s$ is the diameter of the minimum grain size. Then, the rate of sand $d_i$ which is smaller than $d_c$ and is still moving on the bed surface $i_b'$ is written by

$$\left.\begin{aligned} i_b' &= i_{bo}\{1-\Lambda_{ci}\} && (i_b'>0) \\ i_b' &= 0 && (i_b'<0) \end{aligned}\right\} \tag{5}$$

where

$$\Lambda_{ci} = \alpha(6/\pi-1)(\sum_{d_c}^{d_L} i_{bo}/\sum_{d_s}^{d_c} i_{bo})$$

The frequency $i_b$ of sand $d_i$ after the armoring is finally calculated by

$$i_b = i_b'/\sum i_b' \tag{6}$$

The above discussion is for the decreasing period of flow discharge.

After the armor coat is formed by small discharge, it can not be easily destroyed with the increase of discharge. Therefore, it can be assumed that the grain size composition does not change for the increasing period of discharge until the armor coat is destroyed.

### 3.2 Critical condition of armor coat destruction

Fig.5 shows the relationship between $d_m/d_{mo}$ and $\tau/\tau_{cm}$, where $d_m$ and $d_{mo}$ are the mean diameters of armored bed surface and sand mixture without armoring, respectively, $\tau$ is the bed shear stress and $\tau_{cm}$ is the critical shear stress of mean grain size of armored bed surface. When the flow discharge decreases, $d_m/d_{mo}$ increases continuously, while it does not change until $\tau/\tau_{cm}$ is nearly equal to 1.1 and decreases suddenly with small increase of $\tau/\tau_{cm}$ at $\tau/\tau_{cm}$=1.1 when the flow discharge increases. Thus, the critical condition of the armor coat destruction can be written by the following equation(7).

$$\tau/\tau_{cm} = 1.1 \qquad (7)$$

### 3.3 Numerical results of grain size frequency

Fig.6 shows the observed and calculated grain size distribution curves of bed surface sand. The calculated curves seem to fit the observed date both for the increase and decrease periods of flow discharge.

## 4 Graded-sediment discharge during flood

As the grain size frequency $i_b$ can be estimated by eqs.(5) and (6) and the destruction condition of an armor coat is given by eq.(7), the graded-sediment discharge with increase and decrease of flow discharge can be calculated by eq.(2) with the critical shear stress given by eq.(1). Fig.7 shows the change of sediment discharge $q_B$ with the change of shear velocity $u_*$, under the influence of the armor coat formation and destruction. Calculated curves (solid lines) seem to fit the observed change of sediment discharge with the change of shear stress fairly well, although there is some difference between observed and calculated curves mainly due to the uncertainty of the critical condition of the armor coat destruction as determined by eq.(7).

## 5 Discussion and conclusions

Sediment discharge for each grain size of sand mixture can be calculated, modifying already proposed sediment transport formula of uniform sand by substituting $\tau_c$ with $\tau_{ci}$ given by eq.(1), if the grain size frequency $i_b$ of the bed surface is known. The frequency $i_b$ varies with the bed shear stress and can be estimated by eqs.(5) and (6). The critical conditions of the armor coat destruction is given experimentally by $\tau/\tau_{cm}$=1.1, where $\tau$ is the bed shear stress and $\tau_{cm}$ is the critical shear stress of the mean particles of the armor coat. There is difference of sediment discharge between increase and decrease periods of flow discharge due to the gap of the shear stress between formation and destruction of the armor coat. A method to estimate the sediment discharge under the influence of armor coat formation and destruction was proposed by synthesizing the above results and proved to be reasonable.

References


1) Meyer-Peter,E. and Müller,R.:" Formulas for bed-load transport", Proc. IAHR, 2nd Congress, 1947, pp.61-62.
2) Ashida,K. and Michiue,M.:" Study on flow resistance of movable bed and bed-load discharge", Proc. JSCE, No.206, 1972, pp.59-69 (in Japanese).
3) Ashida K.and Michiue,M.:"An investigation of river bed degradation downstream of a dam", Proc. IAHR, 14th Congress, Vol.3, 1971, pp.247-255.
4) Hirano, M.:"On bed degradation with armoring", Proc. JSCE, No.195, 1971, pp.55-65(in Japanese).
5) Parker,G., Klingeman,P.C. and Mclean,D.G.: "Bed load and size distribution in paved gravel-bed streams", Proc. ASCE, Vol.108, HY4, 1982, pp.544-571.
6) Egiazaroff,I.V.:"Calculation of nonuniform sediment concentrations", Proc. ASCE, Vol.91, HY4, 1965, pp.225-247.
7) Ashida,K., Takahashi,T. and Mizuyama,T.:"Study on bed-load discharge in rivers of mountainous area", New Sabo, No.107, 1978, pp.9-17(in Japanese).
8) Einstein,H.A.:"The bed-load function for sediment transport in open channel flows", Tech.Bull., No.1026, USDA, 1950, pp.1-71.

Table 1 Experimental conditions and results.

Run 1 I=1/80

| Q (ℓ/s) | h (cm) | $u_*$ (cm/sec) | $q_B$ (cm³/cm-sec) | dm (cm) | d'm (cm) |
|---|---|---|---|---|---|
| 15.3 | 4.99 | 7.82 | 0.4710 | 0.450 | 0.385 |
| 14.3 | 4.80 | 7.67 | 0.4030 | 0.453 | 0.374 |
| 13.3 | 4.67 | 7.56 | 0.2770 | 0.464 | 0.370 |
| 12.3 | 4.32 | 7.27 | 0.2710 | 0.469 | 0.405 |
| 11.3 | 4.01 | 7.01 | 0.2110 | 0.484 | 0.336 |
| 10.3 | 3.87 | 6.89 | 0.1350 | 0.465 | 0.381 |
| 9.3 | 3.85 | 6.87 | 0.1780 | 0.506 | 0.371 |
| 8.3 | 3.56 | 6.60 | 0.0602 | 0.491 | 0.400 |
| 7.3 | 3.49 | 6.54 | 0.0560 | 0.488 | 0.355 |
| 6.3 | 3.30 | 6.36 | 0.0073 | 0.556 | 0.281 |
| 7.3 | 3.66 | 6.70 | 0.0060 | 0.580 | 0.270 |
| 8.3 | 3.85 | 6.87 | 0.0120 | 0.580 | 0.356 |
| 9.3 | 3.97 | 6.97 | 0.0241 | 0.588 | 0.433 |
| 10.3 | 4.09 | 7.08 | 0.0602 | 0.570 | 0.461 |
| 11.3 | 4.27 | 7.23 | 0.0950 | 0.526 | 0.399 |
| 12.3 | 4.34 | 7.29 | 0.1720 | 0.534 | 0.459 |
| 13.3 | 4.54 | 7.46 | 0.3030 | 0.488 | 0.423 |

Run 2 I=1/100

| Q (ℓ/s) | h (cm) | $u_*$ (cm/sec) | $q_B$ (cm³/cm-sec) | dm (cm) | d'm (cm) |
|---|---|---|---|---|---|
| 14.3 | 4.75 | 6.82 | 0.0020 | 0.627 | 0.537 |
| 15.3 | 5.02 | 7.01 | 0.0722 | 0.590 | 0.446 |
| 16.3 | 5.61 | 7.42 | 0.1788 | 0.625 | 0.466 |
| 17.3 | 5.65 | 7.44 | 0.2961 | 0.498 | 0.368 |
| 18.3 | 5.97 | 7.65 | 0.3900 | 0.467 | 0.409 |
| 17.3 | 5.60 | 7.41 | 0.3470 | 0.468 | 0.380 |
| 16.3 | 5.34 | 7.23 | 0.2952 | 0.444 | 0.363 |
| 15.3 | 5.15 | 7.10 | 0.2889 | 0.448 | 0.334 |
| 14.3 | 4.90 | 6.93 | 0.2197 | 0.459 | 0.351 |
| 13.3 | 4.83 | 6.88 | 0.2227 | 0.498 | 0.372 |
| 12.3 | 4.82 | 6.87 | 0.1869 | 0.472 | 0.352 |
| 11.3 | 4.35 | 6.53 | 0.0664 | 0.510 | 0.370 |
| 10.3 | 4.14 | 6.37 | 0.0433 | 0.539 | 0.340 |
| 9.3 | 4.11 | 6.35 | 0.0110 | 0.557 | 0.263 |
| 8.3 | 3.82 | 6.12 | 0.0014 | 0.585 | 0.187 |

Run 3 I=1/115

| Q (ℓ/s) | h (cm) | $u_*$ (cm/sec) | $q_B$ (cm³/cm-sec) | dm (cm) | d'm (cm) |
|---|---|---|---|---|---|
| 18.3 | 6.25 | 7.30 | 0.2252 | 0.463 | 0.395 |
| 17.3 | 5.60 | 6.91 | 0.1490 | 0.459 | 0.393 |
| 16.3 | 5.51 | 6.85 | 0.1469 | 0.465 | 0.396 |
| 15.3 | 5.40 | 6.78 | 0.1204 | 0.462 | 0.388 |
| 14.3 | 5.31 | 6.73 | 0.0677 | 0.509 | 0.373 |
| 13.3 | 5.20 | 6.66 | 0.0325 | 0.514 | 0.336 |
| 12.3 | 5.05 | 6.56 | 0.0036 | 0.517 | 0.264 |
| 13.3 | 5.34 | 6.75 | 0.0 | 0.590 | – |
| 14.3 | 5.36 | 6.76 | 0.0039 | 0.590 | 0.338 |
| 15.3 | 5.54 | 6.87 | 0.0039 | 0.573 | 0.348 |
| 16.3 | 5.99 | 7.14 | 0.0090 | 0.583 | 0.392 |
| 17.3 | 6.09 | 7.20 | 0.0090 | 0.576 | 0.438 |
| 18.3 | 6.09 | 7.20 | 0.0969 | 0.581 | 0.451 |
| 19.3 | 6.64 | 7.52 | 0.0867 | 0.496 | 0.437 |
| 21.3 | 6.71 | 7.56 | 0.2889 | 0.510 | 0.422 |
| 23.3 | 7.29 | 7.88 | 0.4439 | 0.522 | 0.437 |
| 25.3 | 7.91 | 8.21 | 0.6113 | 0.455 | 0.408 |
| 23.3 | 7.51 | 8.00 | 0.4008 | 0.428 | 0.392 |
| 21.3 | 6.96 | 7.70 | 0.3774 | 0.446 | 0.393 |
| 19.3 | 6.38 | 7.37 | 0.2590 | 0.458 | 0.414 |

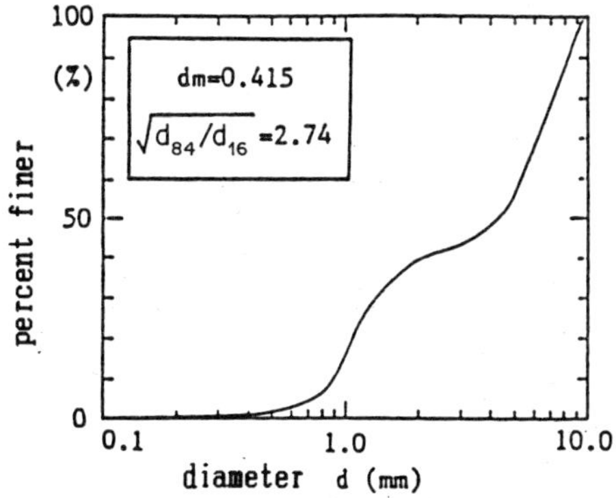


Fig.1 Grain size distribution curve.

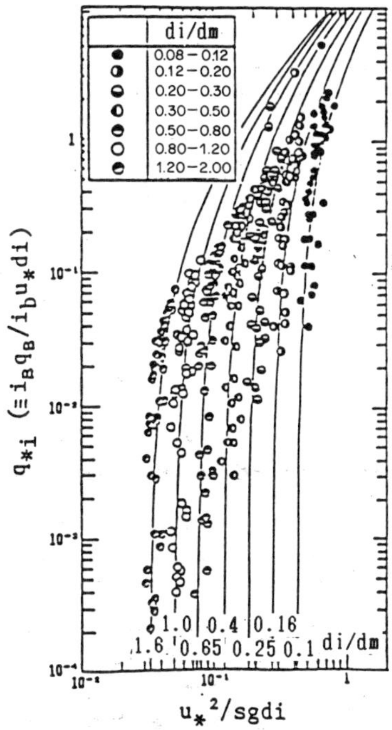


Fig.3 Sediment transport rate of each fraction of sand mixture.

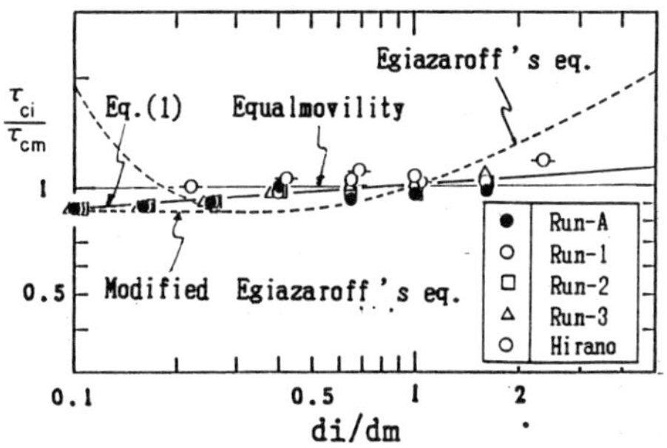


Fig.2 Critical shear stress of each grain size of sand mixture.

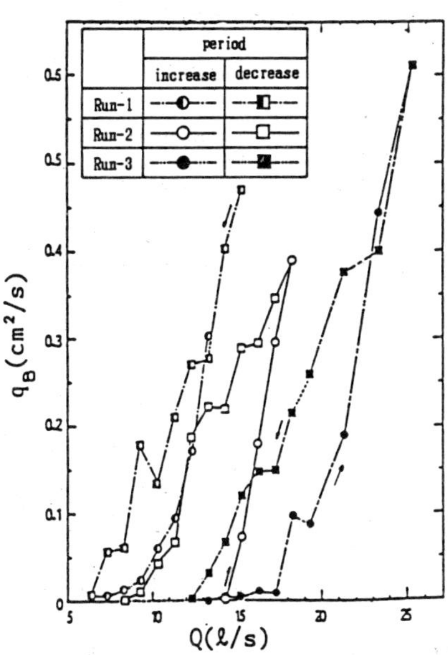


Fig.4 Sediment discharge of increase and decrease periods of flow discharge.

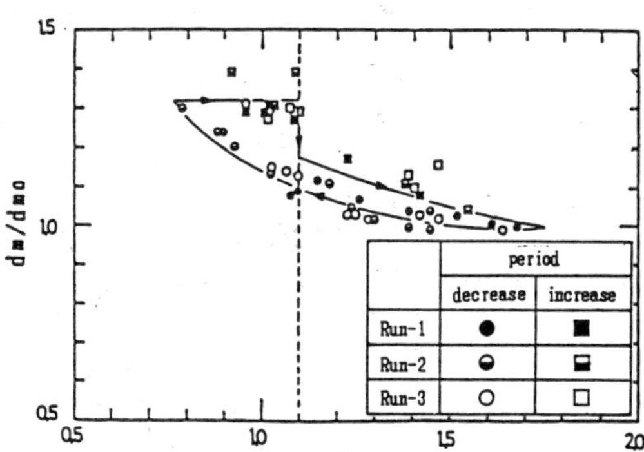


Fig.5 Mean diameter of bed surface sand during increase and decrease of flow discharge.

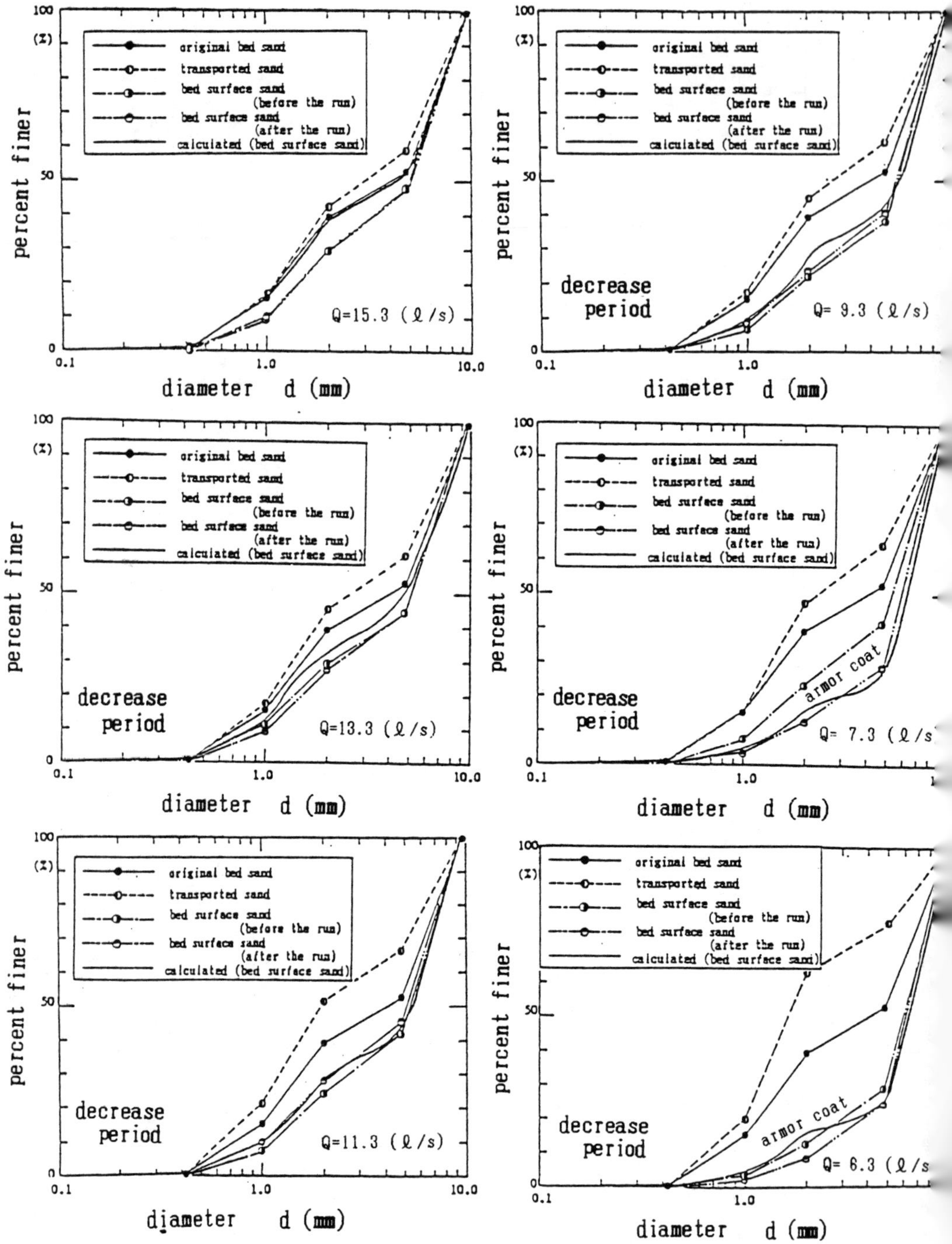


Fig.6(1) Observed and calculated grain size distribution curves of bed surface sand (decrease period of flow discharge).

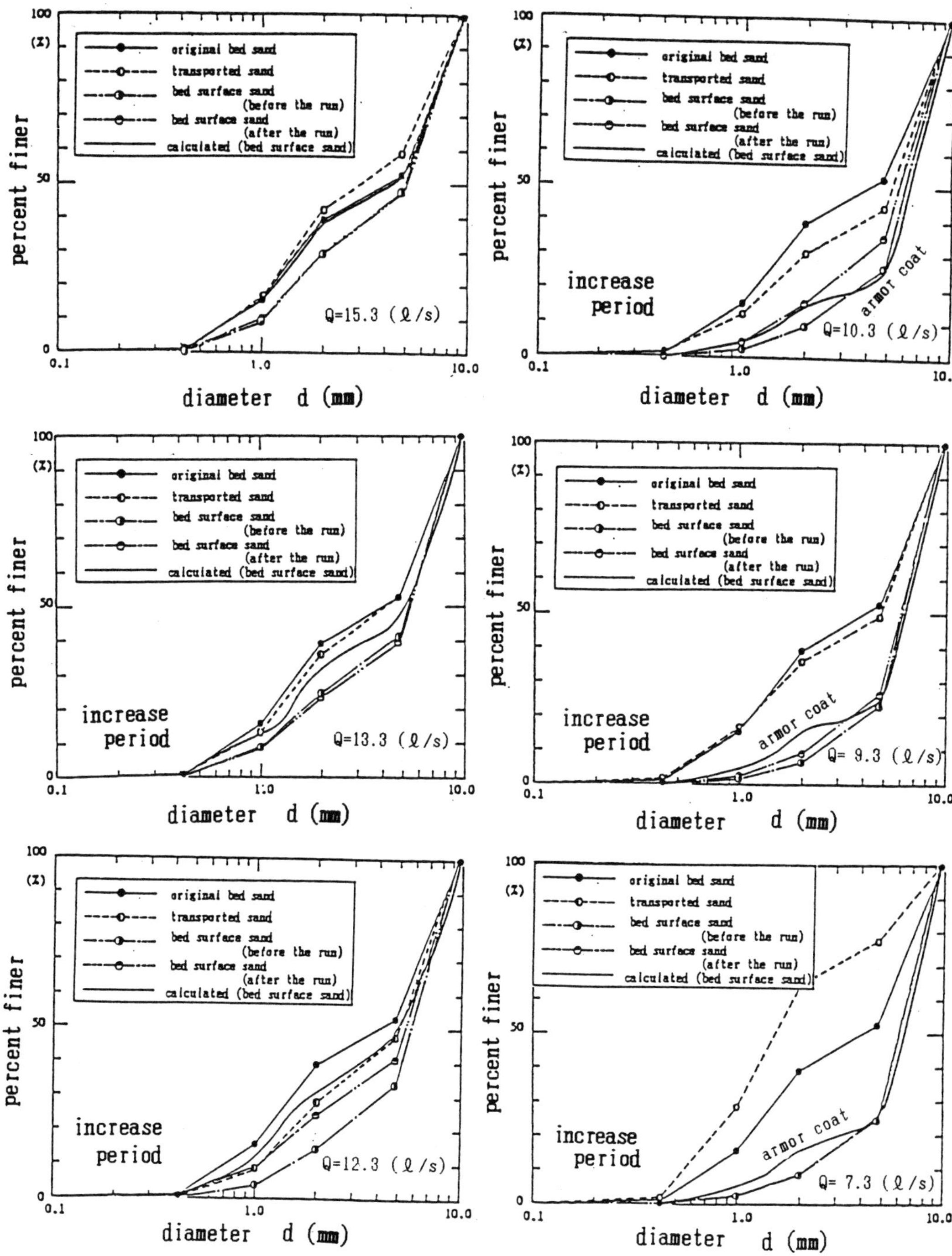


Fig.6(2) Observed and calculated grain size distribution curves of bed surface sand (increase period of flow discharge).

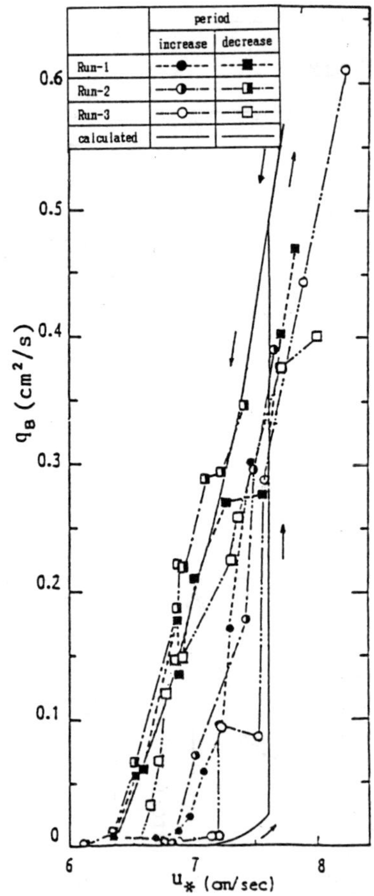


Fig.7 Observed and simulated variation of sediment discharge during increase and decrease periods of flow discharge.

International Conference on
RIVER REGIME
18-20 May, 1988

PAPER F3

# A STUDY ON THE STAIRCASE PROFILE AND ARMOURING EFFECTS OF RIVER BED FOR TORRENTIAL STREAMS IN TAIWAN

Chih-Wu Ho

Graduate Institute of Soil & Water Conservation
National Chung Hsing University, Taichung 40227, Taiwan, R.O.C.

SUMMARY: This paper proceeds theoretically with the analysis on the relation between sediment decrease and the river response caused by the staircase profile and armouring effects. In addition, the experimental results and field data are applied to investigate the effects. At the same time, it suggests a concept to ground sill design, and considers the possibility for use under actual condition.

## INTRODUCTION

Taiwan is a narrow-shaped island; besides, a continuous high mountain runs through the central part. So the rivers in this island are generally short in length, steep in gradients and coarse in bed materials. Under the special condition, the staircase profile and armour coat may be established on the river bed naturally. This paper analyzes the effects of the staircase profile to sediment decrease, on the basis of formation condition of the phenomenon and equilibrium theoretics of river system, besides, coordinates with indoor flume experiments and surveyed field data to verify the accurance. On the armouring effects, this paper utilizes numerical method to solve the sediment continuous equation and the variation equation of grain size distribution which describes the river bed variation. Simultaneously, indoor flume experiments and field data are adopted to investigate the results gained theoretically.

## 1. THEORETICAL BASIS OF FORMATION OF THE STAIRCASE PROFILE

Ashida (Ref 1) suggested the formation conditon of the staircase profile can be expressed as follows:

(i) $h/dm < (6.0 + 5.75 \log h/\alpha dm)^2 (\sigma/\rho - 1)\tau_{*m}$ ............ (1)

(ii) $\tau_{*m} > \beta\tau_{*cm}$ ............ (2)

(iii) $\tau_{*m} < \tau_{*cmax}$ ............ (3)

where, h = water depth; dm = mean grain diameter; $\tau_{*m}$ = dimensionless shear stress of mean grain; $\tau_{*cm}$ = critical dimensionless shear stress of mean grain; $\tau_{*cmax}$ = critical dimensionless shear stress of maximum grain.

After the staircase profile appears, the sediment will decrease and the decrease extents can be solved by the view of stream power. The relation between stream power and sediment suggested by Yang, C.T. (Ref 2) can be expressed as

$$Qs = a(\tau U)^b \quad \text{............} \quad (4)$$

where, Qs = sediment discharge per unit width; $\tau U$ = stream power; a and b are both coefficients. From Fig. 1, we obtain a = 0.5268 and b = 1.6766.

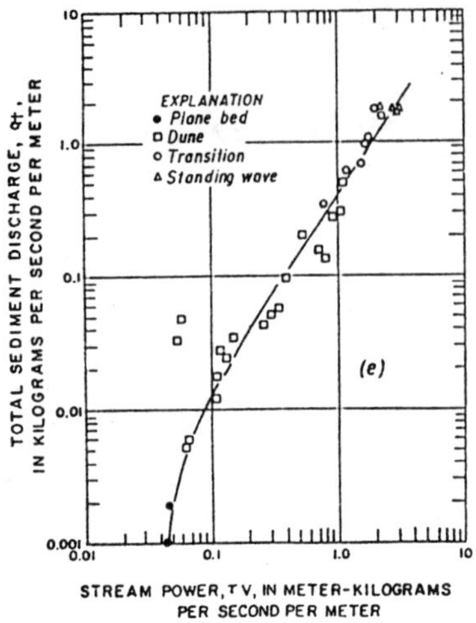


Fig. 1 The Relations of Sediment and Stream Power

Then

$$Qs = 0.5268\ (\tau U)^{1.6766} \quad \cdots\cdots \quad (5)$$

The Manning's formula is

$$Q = \frac{1}{n} B \cdot h^{5/3} \cdot S^{1/2} \quad \cdots\cdots \quad (6)$$

Where Q = flow discharge; n = Manning's roughness; h = water depth; B = river width; S = slope of the energy gradient. When Q, n, and B keep constant, Eq. (6) can be written as

$$h = K\ S^{-3/10} \quad \cdots\cdots \quad (7)$$

From Eqs. (5) & (7), the tractive force and mean velocity formulas are

$$Qs = K'\ S^{1.6766} \quad \cdots\cdots \quad (8)$$

where $K' = 0.5268\ (\frac{\rho Q}{B})^{1.6766}$. According to the deposit slope data of sabo dams in Taiwan, the ratio between deposit slope and original slope of river bed is about 0.2 - 0.8. But the range will become smaller for the serial sabo dams, where the ratio is about 0.2 - 0.5. We think the staircase profile is similar as smaller serial sabo dams, and substitute the slope ratios of serial sabo dams as above to Eq. (8). Then we have

$$Qs' / Qs = 0.067 \text{ -- } 0.313 \quad \cdots\cdots \quad (9)$$

Eq. (9) means that the sediment decrease ratio is within 69% - 93%.

## 2. VERIFICATION OF THE STAIRCASE PROFILE

### 2/1. Experimental Results

For the purpose of investigating the effect to the sediment transport from staircase profile , we proceed the flume experiments in Graduate Institute of Soil and Water Conservation, National Chung Hsing University. The experiments is classified in three groups on different slope, which is 3%, 5%, and 7%, respectively. And the experimental parameters are referred to the field of the Wo Thuo creek, Nantou, Taiwan. In this experiment, added sediment yield is estimated by Schoklitsch bed load model, and we recorded these data from experiments with the staircase profile established and not established on the river bed, respectively (Ref 3). Meantimes, the rating curve may be expressed as rating equation in the form of $Qs = K Q^n$ . The K and n Values of three groups experiments are listed in Table 1. In accordance with the results as shown in Table 1, the decrease ratio due to the effects of staircase profile is 85-95%, Fig.2 shows the results.

Table 1. The K and n Value of $Qs = KQ^n$ with Two Different Bed Profiles.

| Group | S1 (3%) | | S2 (5%) | | S3 (7%) | |
|---|---|---|---|---|---|---|
| | Normal | Staircase | Normal | Staircase | Normal | Staircase |
| K | 0.20602 | 0.10408 | 0.14366 | 0.00433 | 0.56503 | 0.14689 |
| n | 1.33275 | 1.57625 | 1.36672 | 1.13673 | 1.78478 | 2.1101 |
| R | 0.9985 | 0.9979 | 0.9937 | 0.9968 | 0.9931 | 0.9434 |

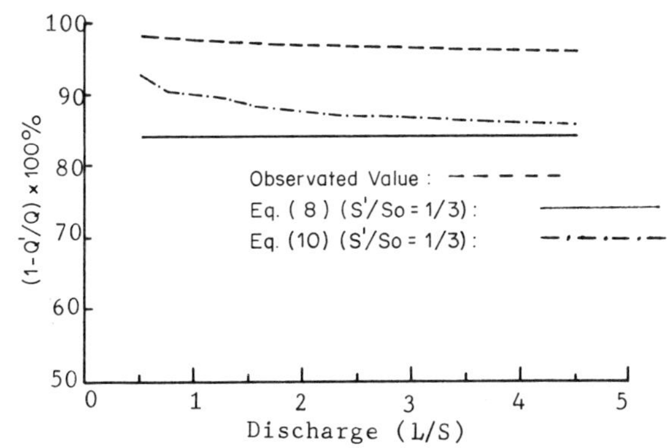


Fig. 2 (a) The Decrease Ratios by Experimental Results and Computed Results. (Group S1)

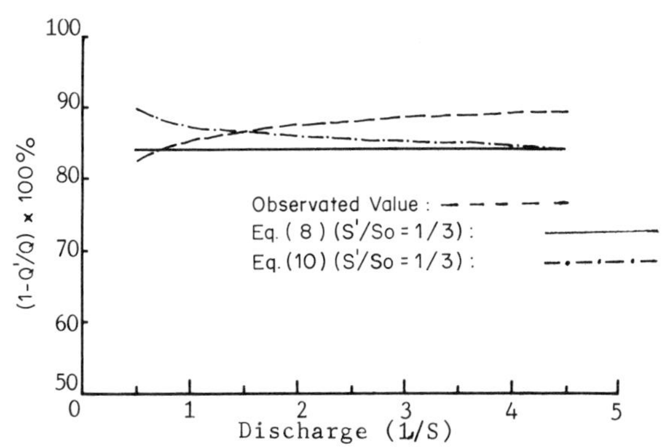


Fig. 2 (b) The Decrease Ratios by Experimental Results and Computed Results. (Group S2)

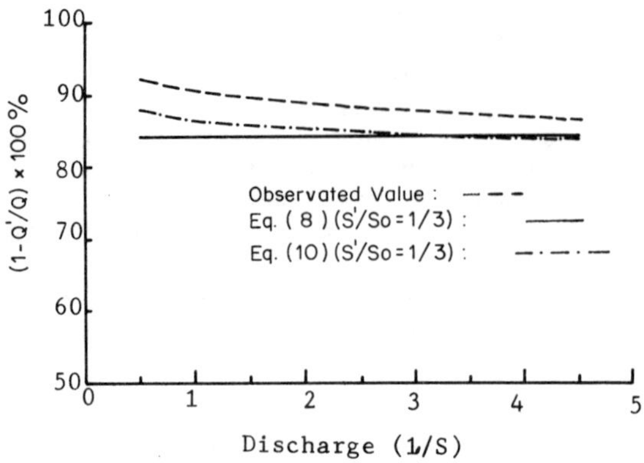


Fig. 2 (c) The Decrease Ratios by Experimental Results and Computed Results. (Group S3)

Also the experimental results are close to those analyzed theoretically. Nevertheless, the sediment through river bed with staircase profile is not only dependent on slope of river bed, but also flow discharge, the height and interval distance of the staircase profile. Hence, this paper takes the regression on the experiments data and obtains two types of regression equations as follows:

$$q_s'/q_s = 208.9296\ (qS_0)^{0.106}\ (S_0 - S')^{1.22} \qquad (r = 0.9062) \quad \cdots\cdots (10)$$

$$q_s'/q_s = 13.490\ (qS_0)^{0.114}\ S'^{1.36} \qquad (r = 0.9440) \quad \cdots\cdots (11)$$

where $q_s$ = sediment per unit width on original bed; $q_s'$ = sediment per unit width with staircase profile; $q$ = flow discharge per unit width; $S_0$ = slope of original river bed; Hs, Ls = the height and interval distance of staircase profile; $S'$ = the slope with the staircase profile.

2/2. Results of Field Observation

Two surveyed field data are considered in order to investigate the actual effects to sediment transport when the staircase profile has been existed. One field site is the Wo-Thuo creek (situated in central Taiwan), and another field site is the Tong-Hsi Keng creek (situate in southern Taiwan). Both river reaches are considered as that the phenomenon of staircase profile is present on the spot-investigation. In other hand, the analysis of field data is fitted to the condition of formation about the phenomenon (as shown in Table 2).

Table 2. The Formation Condition of Staircase Profile.

| River Name | S | B (m) | Dm (mm) | Dmax (mm) | n | Qp (cms) | h (m) | Fr | $\tau_0/\tau_{cm}$ | $\tau_0/\tau_{cmax}$ |
|---|---|---|---|---|---|---|---|---|---|---|
| Wo-Thuo Creek | 0.0405 | 20.4 | 121.3 | 450 | 0.0305 | 68.6 | 0.614 | 2.233* | 3.283* | 0.884* |
| Tong-Hsi-Keng Creek | 0.0512 | 12.3 | 195.3 | 820 | 0.0323 | 53.3 | 0.705 | 2.426* | 2.958* | 0.704* |

Remarks: * indicates matched value.

After analyzing five-year field data from the Wo-Thuo creek, we can find that the more adaptable bed load models are Schoklitsch model with representative grain size of D85, A1 model (derived by the author) with representative grain size of D50, and Meyer-Peter and Müller model with representative grain size of D50. When the staircase profile has been existed in the field, the actual sediment decreased (Ref 4). The ratios between actual amount and sediment estimated by models are in the range of 82% - 94%. The results are extremely corresponded to theoretical and experimental analyses. In addition, the model A1 and A2 (derived by the author) with representative grain size of D35 is the most adaptable in the Tong-Hsi-Keng creek. Comparing the two different sediment transports with and without the staircase profile, the sediment transport decreases on the river reach is about 79% - 91% (Ref 5). In other words, it means that the theoretical and experimental results are close, too.

## 3. ANALYSIS OF ARMOURING EFFECTS

### 3/1. Armouring Effects in Torrential Stream

According to the equilibrium theoretics of river system, an equilibrium river bed will lose the equilibrium state when it is lack of sediment yield or a sabo dam has been built in upstream. Gradually, it will tend to a new equilibrium. During the process to tend to a new equilibrium state, the bed materials on the surface of river become coaser in grain size because of the functions of hydraulic sorting. And it is just named the armouring of river bed. In general, the armouring effects will affect sediemnt transport of river bed on decreasing. As to the armouring process of river bed, this paper coordinates with indoor flume experiments and field data to proceed with analysis and investigation for more accurate results.

### 3/2. Theoretical Aspects

#### 3/2.1 Mixed Layer Theory (Ref 6)

The water depth to stabilize the river bed of uniform grain size can be derived by Manning's formula, Strickler's roughness formula, and Einstein flow intensity formula. The equilibirum water depth under bed materials with grain diameter d is:

$$De = h\,(\,q/5.64\,d^{1/3}\,)^{6/7} \qquad (12)$$

If the bed materials is composed of non-uniform grain size and is below the equilibrium water depth with grain size of d , there exists a scouring depth. Gessler gave the expression as follows:

$$Dsc = (2/3) \times fs \times d\,/\,P \qquad (13)$$

where, Dsc = scouring depth; $f_s$ = safety factor; P = the fraction of bed particles which is coarser than grain d.

From Eqs.(12) and (13),the equilibrium water depth with mixed grain size of bed materials can be obtained:

$$Deq = De + Dsc \qquad (14)$$

$$T = Deq - h \qquad (15)$$

where T = the thickness of mixed layer

3/2.2 Sediment Continous Equation

The continuous properties of sediment transport can be written as

$$\partial Z/\partial t + 1/(1-\lambda)(\partial Q_B/\partial X) = 0 \quad \cdots\cdots (16)$$

where $Q_B$ = total sediemnt; Z = the elevation of river bed; $\lambda$ = porosity of bed materials; t = time; Z = distance.

3/2.3 The Theory about the Continuity of Grain Variation of Mixed Layer

In the descending process of river bed, the particles within some thickness of mixed layer are scoured and become bed load. The mixed layer will be recomposed and have new grain distributions of particles in mixed layer. The variation situation can be represented by the following equation:

$$I_b' = I_b + \Delta T/T\,(I_{b_0} - I_b) \quad \cdots\cdots (17)$$

where $I_b'$ = new fraction of grain size d i of mixed layer in $\Delta$ t time; $I_b$ = original fraction of mixed layer; $I_{b_0}$ = original fraction of grain size $d_i$ of bed materials; $\Delta$ T = the descendant thickness of river bed; T = the thickness of mixed layer. From Eqs.(16) and (17) and $Q_B = \sum_i Q_{Bi}$, we can obtian

$$\partial I_b/\partial t + \{\partial Q_{Bi}/\partial X - I_{b_0}\,\partial Q_B/\partial x\}/T(1-\lambda) = 0 \quad \cdots\cdots (18)$$

By utilizing difference methods to solve Eqs (16) and (18) step by step, the solution to the elevation of river bed and the grain distribution variation of bed materials are calculated for the case of river bed descending.

Eqs.(16) and(18) can be written as the form of difference equation (Ref 7)

$$\text{Fr} > 1: \quad Z_j^{n+1} = Z_j^n - \frac{2\Delta t}{(1-\lambda)(B_j + B_{j+1})\Delta X}(Q_{Bj+1}^n - Q_{Bj}^n) \quad \cdots\cdots (19)$$

$$\text{Fr} < 1: \quad Z_j^{n+1} = Z_j^n - \frac{2\Delta t}{(1-\lambda)(B_{j-1}+B_j)\Delta X}(Q_{Bj}^n - Q_{Bj-1}^n) \quad \cdots\cdots (20)$$

$$I_{bj}^{n+1} = I_{bj}^n - \frac{2\Delta t}{T(1-\lambda)(B_{j-1}+B_j)\Delta X}[(Q_{Bi\,j}^n - Q_{Bi\,j-1}^n) - I_{b_0}(Q_{Bj}^n - Q_{Bj-1}^n)] \quad \cdots\cdots (21)$$

the $\Delta$ t and $\Delta$ X of Eqs. (19), (20) and (21) must satisfy the following constraint

$$0 < \frac{\Delta t}{\Delta X} \le \min\left[\left|\left(\frac{dx}{dt}\right)_z^{-1}\right|, \left(\frac{dx}{dt}\right)_{Ib}^{-1}\right] \quad \cdots\cdots (22)$$

4. EXPERIMENTAL AND FIELD STUDY ON ARMOURING EFFECTS

4/1. Experimental Study

In order to analyze the variation of river bed and bed materials in the case, a simulative variation is built by flume experiment under the condition without upstream sediment yield. The experiment is taken place in the experimental flume of Graduate Institute of Soil and Water Conservation, National Chung Hsing University. After the results calculated by numerical difference method on the basis of experiments data, the following three items can be indicated:

(a) According to experimental results, the scouring depth in each interval of time reveals the apparent tendency for river bed to descend; and the descendant rate of river bed is decreased as time runs through.

(b) In the end of each experiments, there is armour coat on the surface of bed apparantly. If we apply the mathematic method to estimate the variation of grain size, it is yet corresponded to the grain distribution curve surveyed from experiments.

(c) In accordance with sediment in each interval and accumulated amount, it shows that actual sediment surveyed from experiments is more than the sediment estimated by each models. That is, the actual equilibrium time is shorter than that estimating by each model. Coordinating with the analysis results obtained by estimation and experimental measurement from all these experiments, the Sato-Kikawa-Ashida and A2 models have better effects.

4/2. Field Study

This paper takes these data from Yuan-San creek and Hsung-Lo creek as the basis of investigation. The calculation interval of distance ($\Delta X$) of these two creeks is pretended to be 50m and 206m, respectively. The maximum time interval ($\Delta t$ max) constrainted by Eq.(22) are 7153 sec and 10413 sec., respectively. So we take $\Delta t$ for 3600 sec. Such that, it can be more conventive for the value to be adapted to discharge's calculation. After we have solved the equation to estimate the variation of river bed due to armouring effects, we take it to compare with the results which is surveyed from the two fields. As a result, the comparisons of them are shown in Fig. 3, and Table 3.

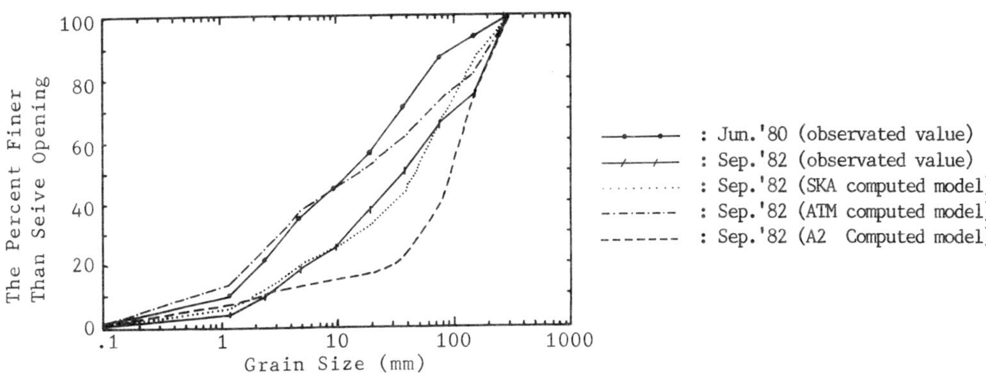


Fig. 3 (a) The Comparison of Grain Size Distribution Between Field Data and Computed Results in the Yuan San Creek.

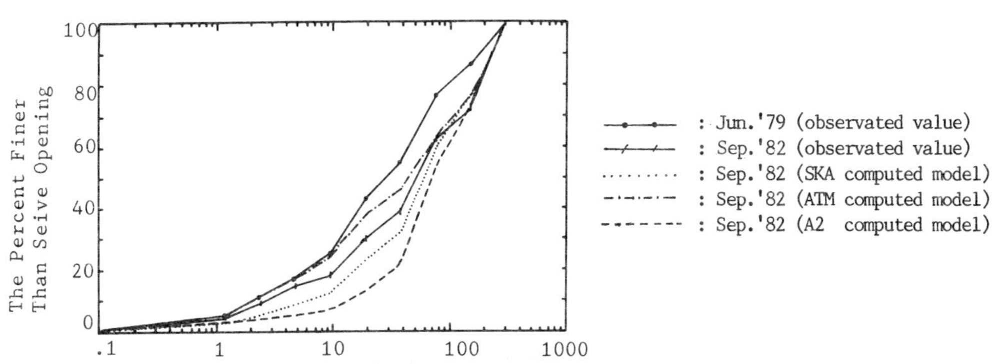


Fig. 3 (b) The Comparison of Grain Size Distribution Between Field Data and Computed Results in the Hsung Lo Creek.

Table 3. The Comparison of Synthetic Results between Field Data and Computed Data in the Yuan-San Creek and Hsung-Lo Creek.

| MODEL | | | SKA | | ATM | | A2 | |
|---|---|---|---|---|---|---|---|---|
| RIVER NAME | PERIODS | FIELD DATA | COMPUTED | ERROR | COMPUTED | ERROR | COMPUTED | ERROR |
| | | | AVERAGE SCOURED THICKNESS (M) | | | | | |
| YUAN SAN CREEK | I | -0.108 | -0.079 | 26.1 | -0.145 | -35.0 | -0.111 | -3.3 |
| | II | -0.353 | -0.288 | 18.5 | -0.744 | -110.5 | -0.458 | -29.8 |
| | III | -0.148 | -0.093 | 37.5 | -0.358 | -141.8 | -0.216 | -45.9 |
| | IV | -0.112 | -0.070 | 37.4 | -0.114 | -1.3 | -0.105 | 6.8 |
| | V | -0.772 | -0.530 | 26.5 | -1.362 | -88.7 | -0.891 | -23.4 |
| HSUNG LO CREEK | VI | -0.306 | -0.235 | 23.1 | -0.297 | 19.5 | -0.271 | 11.3 |
| | VII | -0.025 | -0.016 | 35.9 | -0.017 | 29.9 | -0.016 | 36.8 |
| | VIII | -0.330 | -0.251 | 24.1 | -0.294 | 20.1 | -0.287 | 13.2 |
| | | | TOTAL SCOURED VOLUME ($M^3$) | | | | | |
| YUAN SAN CREEK | I | -2964 | -2188 | 26.1 | -4002 | -35.0 | -3063 | -3.3 |
| | II | -9729 | -7929 | 18.5 | -20483 | -110.5 | -12624 | -29.8 |
| | III | -4082 | -2549 | 37.5 | -9870 | -141.8 | -5955 | -45.9 |
| | IV | -3097 | -1939 | 37.4 | -3138 | -1.3 | -2886 | 6.8 |
| | V | -19871 | -14606 | 26.5 | -37492 | -88.7 | -24527 | -23.4 |
| HSUNG LO CREEK | VI | -41982 | -32275 | 23.1 | -37996 | 19.5 | -37240 | 11.3 |
| | VII | -3411 | -2154 | 35.9 | -2390 | 29.9 | -2154 | 36.8 |
| | VIII | -45393 | -34463 | 24.1 | -40387 | 20.1 | -39395 | 13.2 |
| | | | AVERAGE SLOPE | | | | | |
| YUAN SAN CREEK | 1979.7 | 5.385 | 5.385 | -- | 5.385 | -- | 5.385 | -- |
| | 1979.9 | 5.312 | 5.330 | 0.34 | 5.282 | -0.56 | 5.308 | -0.08 |
| | 1980.6 | 5.069 | 5.131 | 1.22 | 4.773 | -5.84 | 4.994 | -1.48 |
| | 1981.4 | 4.968 | 5.068 | 2.01 | 4.534 | -8.78 | 4.849 | -2.40 |
| | 1981.8 | 4.890 | 5.019 | 2.64 | 4.457 | -8.85 | 4.778 | -2.29 |
| HSUNG LO CREEK | 1980.6 | 2.343 | 2.343 | -- | 2.343 | -- | 2.343 | -- |
| | 1982.7 | 2.298 | 2.308 | 0.43 | 2.302 | 0.17 | 2.298 | 0.22 |
| | 1982.9 | 2.294 | 2.306 | 0.52 | 2.299 | 0.22 | 2.294 | 0.26 |
| | | | MEAN GRAIN SIZE (mm) | | | | | |
| YUAN SAN CREEK | 1979.7 | 37.24 | -- | -- | -- | -- | -- | -- |
| | 1981.8 | 78.98 | 76.09 | -3.66 | 48.87 | -38.12 | 120.76 | 52.90 |
| HSUNG LO CREEK | 1980.6 | 57.86 | -- | -- | -- | -- | -- | -- |
| | 1982.9 | 89.73 | 94.18 | 4.96 | 84.75 | -5.55 | 109.65 | 22.20 |

Remarks: I : 1979.7 -- 1979.9 ; II : 1979.9 -- 1980.6 ; III : 1980.6 -- 1981.4
IV : 1981.4 -- 1981.8 ; V : 1979.7 -- 1981.8 ; VI : 1980.6 -- 1982.7
VII : 1982.7 -- 1982.9 ; VIII : 1980.6 -- 1982.9 .
"-" indicates scouring.

After the Sabo-dams of the Hsung Lo creek and the Yuan San Creek have been built, it affects the variation situation of these two river beds. And the following two discussions are suggested

(a) Yuan-San Creek

In this paper, we consider these data of bed profile surveyed from, June, 1979 to August, 1981. When the Sabo-dam has been full of sediment yield, the elevation of river bed descended about 72.2cm between the interval from No.1 Sabo-dam to Jun-Hsin Bridge, a distance for 303m. Besides, the total scoured volume is 19871.2$m^3$. After analyzing from surveyed data and calculated results, the mean relative error is smaller with A2 model, and it is around -23.4%. With regard to the bed materials' distribution variation, it is apparant that the surface of river bed is armoured by coarse grain particles. When we estimate the results in algorithm model, we find that the Sato-Kikawa Ashida bed load model is better than others. The error between them is about 3.7%.

(b) Hsung-Lo Creek

After the Sabo-dam of Hsung-Lo creek was built in June, 1980. The dam was full in September, 1982. The descendant depth is measured as 33cm; total scoured sediment volume is 45392.8$m^3$, and the slope of river bed becomes 0.02294 but 0.02343 originally. Besides, the mean grain size of bed materials is 57.86mm before the sabo-dam was built, and 89.73mm after sabo-dam was built. Among estimation of these algorithm models, the A2 model is more corresponded to actual suveyed value. As to the calculated part, the descendant depth is 28.7cm, total scoured sediemnt volume is 39394.8$m^3$, and the mean relative error is 13.2%. As the result of grain size distribution of river bed, the more adaptable model is the Sato-Kikawa Ashida model. The mean grain size of bed materials is 94.18mm under the sabo-dam for 206m distance.

## 5. POSSIBLE APPLICATIONS

By the preliminary research and analysis, the phenomenon is clear that the staircase profile has two functions. They are stabilizing river bed and decreasing the hydraulic energy due to hydraulic jump through the special profile. According to the results, we may try to apply the idea of the staircase profile with the design of ground-sill engineering in order to improve them. So, if we constructed the ground-sill similar to the staircase profile in properities. As a result of this, the stability of the sediment must be decreased and become armoured on the river bed meantimes. Such that, the armouring effect will be also constributed to the design. Of course, it is helpful for our purposes.

## CONCLUSIONS

(1) Due to the properties such as narrow in shape, steep in gradients, and the Froude Number of flow being large in torrential stream of Taiwan, the staircase profile is extremely easy to appear in this island. After analysis by some theories, the results are quantified. the discount ratio to decrease sediemnt is about 69% - 93%. In addition, we analyze two regression equation about these factors such as deposit, flow discharge, and the height and interval distance of the staircase profile in advantage of application. (See Eqs. (9) and (10) ).

(2) By analyzing and investigating the part of two field data and experimental results, the sediment decrease ratio is 85% - 95% in experimental part, and it is 82% - 94% and 79% - 91%, when verified in the Wo-Thuo creek and the Tong Hsi Keng creek, respectively. That is, the results with the two parts of verification is very corresponded to the sediment decrease derived theoretically.

(3) About the armouring effects on river bed, this paper also design flume experiments in order that we can analyze and observe the variation process of mixed layer under non-sediment yield condition. When we compare with actual observated field data nad estimated value by algorithm models, it is shown that all they are unanimous briefly.

(4) If we apply the idea about the effects due to the staircase profile and armouring function to ground-sill design. The advantage is worthy of investigating. Since the equilibrium slope referred to the staircase profile has been considered, it should be helpful to stablize the constructions and protect from disaster due to flood and sediment.

ACKNOWLEDGEMENTS

The author would like to express his thanks to Prof. C.H. Tuan, for his helpful discussions, Assistants Miss C.W. Chen, Mr. C.C. Wang, former graduate student Mr. H.L. Moh, graduate students Mr. C.S. Chen and Mr. P.C. Li, for their helpful assistance to the work of field observation and data analysis of this study. Finally this research supported by the National Science Council of the R.O.C. is also gratefully acknowledged.

REFERENCES

1. Ashida K., S. Egashira & N. Ando: "Origin and geometry of step-pool bed form". In: Proc. 28th Japanese Conference on Hydraulics, 1984, pp.743 -749. (in Japanese).

2. Yang, C.T. : "Sediement transport and unit stream power". ASCE, Vol. 18, No. 10, 1972, PP.1805 - 1926.

3. Moh, H.L. : "A study on the effect of armouring process and step-pool phenomenon to bed-load discharge evaluation". Master thesis, Graduate Institute of Soil and Water Conservation, National Chung Hsing University, Taichung, Taiwan, R.O.C., 1987, pp. 39 -76. (in Chinese).

4. Ho, C.W. & Wang, C.C. : "A study on the method of river sediment observation and the estimation of upstream of Cho-Shui stream (IV)". Report of project supported by National Science Council, 1987, pp.29 - 33. (in Chinese).

5. Ho, C.W. : "A study on the Investigation of sediment source of the watershed in Tseng-Wen Reservoir (I)". Report of project supported by the Bureau of Administration of Tseng-Wen Reservoir, 1987, pp.39 - 43. (in Chinese)

6. M. Fazle Karim & Forrest M. Holly: "Armouring and sorting simulation in alluvial rivers". ASCE, Vol. 112, No. 8, Jun. 1986, pp.705 - 716.

7. Dale A. Anderson , John C. Tannehill & Richard H. Pletcher : "Computational fluid mechanics and heat transfer". 1985, pp76.

International Conference on
RIVER REGIME
18-20 May, 1988

PAPER G1

# ESTIMATION OF FRICTION FACTOR IN NATURAL STREAMS

N.N.J. Higginson, Department of Agriculture for Northern Ireland
H.T. Johnston, The Queen's University of Belfast

SUMMARY:

The estimation of friction factor is of great importance in assessing the flow capacity of a channel. A study relating flow to channel geometry and bed material has been undertaken on Northern Ireland rivers. The Einstein-Barbarossa, and Engelund and White methods of predicting channel capacity have been tested and are discussed. A new prediction method based on the Strickler formula and analysis of residuals is presented.

## 1/1 Introduction

The estimation of the capacity of a natural channel is of great importance to the drainage engineer in assessing the risk of flooding. Over the last two centuries several semi-empirical formulae have been proposed for the velocity or discharge in a channel. The accuracy of these formulae is highly dependent on the channel roughness or friction factor. The most commonly used formulae are the Chezy, Manning and Darcy-Weisbach equations which relate velocity to hydraulic mean depth, channel slope and a measure of channel roughness. The most widely used of these equations is the Manning formula for which published data on the roughness coefficient is readily available.

Engineers have to be cautious in applying published data for use in a river or channel having different hydraulic characteristics and the data should only be used as a guide. An alternative method of estimating 'n' was proposed by Strickler (Ref. 1) which depends only on a measure of bed material size. This method was extended by Einstein and Barbarossa (Ref. 2) who proposed dividing the hydraulic radius into two components, one associated with grain resistance and the other with bed form resistance. Using flow data from several rivers in North America, they derived empirically a curve relating the ratio of total mean velocity to velocity associated with bed form resistance and a grain shear stress parameter '$\psi$'. The purpose of this curve was to estimate sediment transport and to determine channel resistance for use in the design of irrigation canals and river training works.

The work by Engelund (Ref. 3) looked at river bed configurations under different hydraulic conditions and proposed that the hydraulic resistance of alluvial rivers followed a simple similarity principle. This enabled him to predict the stage discharge relationship for a stream with known cross-section, slope and bed material. This work used data obtained from experiments conducted in a laboratory flume.

International Conference on River Regime, edited by W.R.White.

White et al (Ref. 4) reviewed available methods of estimating channel roughness and tested them on a range of laboratory and field data. The data used in this comparison of methods was mainly for channels having beds of sands or fine gravels. They derived a new method based upon flume experiments with sand for estimating the roughness coefficient for use in the Darcy-Weisbach equation which can be related to the Manning coefficient. This method was based upon the sediment transport analysis derived by Ackers and White (Ref. 5) and they found it to be an improvement on existing methods.

A better estimate of Mannings friction factor 'n' may be obtained by field measurements. It is known that the value of 'n' is not only dependent on the absolute roughness of the bed but also on the depth of flow and channel geometry. Therefore numerous field measurements covering a range of flaws are required at any river section to give a good estimate of 'n'. Field investigations of this type are time consuming and therefore a better method of estimating 'n' is required.

A comprehensive field study has been undertaken by the Department of Agriculture for Northern Ireland at some 80 sites on both major rivers and minor streams in Northern Ireland. This ongoing study programme has involved establishing permanent flow gauging stations at these sites and the development of rating equations based upon numerous velocity-area gaugings. This data has been supplemented by sampling and classifying the bed material at these sites. This data base (see Appendix I) has been used to test the Einstein-Barbarossa and Engelund methods and to investigate the relationship between field measurements of 'n' and channel and flow properties.

## 2/1 Field Data

The total area of Northern Ireland is about 13,400 $km^2$ and it is situated on the north-eastern corner of Ireland. Virtually the whole surface of the lowlands of Northern Ireland is covered by thick glacial drift which often stretches far up the mountain slopes. The main drainage systems are into Lough Foyle and through Lough Erne and Lough Neagh with the latter system contributing to over 40% of the total land area. Most of the Lough Neagh basin lies below the 100 m contour, and more than 90 per cent of the total area of Northern Ireland lies below the 300 m contour. The topography of the area shows typical glacial features with the summits of hills and mountains well rounded, wide U-shaped river valleys and swarms of drumlins over much of the eastern parts of the country.

The gauging sites chosen by the Department of Agriculture cover the majority of the river catchments of Northern Ireland with the exception of a few minor coastal streams. The sites range from several rivers draining areas in excess of 1000 $km^2$ to small streams having catchments less than 20 $km^2$. Twelve of these sites were excluded from this study because the rating equations had not been well established. The locations of the remaining stations are shown in Fig. 1. together with their sub-catchments.

The river sites chosen have a wide variety of bed materials ranging from sands and silts to coarse gravels. A histogram of mean grain diameters, $d_{50}$ (defined as equal to the sieve size passing 50% by weight) for these rivers is shown in Fig. 2(a). Fig. 2(b) shows the range of bank-full mean velocities which were up to 3.8 m/s. The bed slopes, shown in Fig. 2(c), range from gentle slopes less than 1 in 1000 to steep slopes approaching 1 in 100. The resulting calculated 'n' values for bank-full conditions are highly variable covering a range between 0.0018 and 0.08 as shown in Fig. 2.(d)

### 3/1 Einstein-Barbarossa Diagram

For each channel site in the present investigation three flows were chosen representing low, medium and bank-full discharges. These flows were used in the computation of the dimensionsless parameters of $\frac{V}{V''_f}$ and $\psi'$ proposed by Einstein and Barbarossa (Ref. 2). The approach adopted was to subdivide the hydraulic radius R into R' (related to grain resistance) and R" (related to form resistance).

With $V'_f = (gR'I)^{0.5}$ and $V_f'' = (gR''I)^{0.5}$ (1)

the Einstein-Barbarossa eqn is

$$\frac{V}{(gR'I)^{0.5}} = 5.75 \, \text{Log}_{10} \frac{[12.2 \, R']}{K_s} \qquad (2)$$

in which $K_s = 2 \times d_{65}$

This can be solved for R' with known values of V (velocity), I (channel slope) and $K_s$ (grain size).

R" is solved from

$$R = R' + R'' \qquad (3)$$

$\psi'$ is defined as

$$\psi' = \frac{1.68}{R'I} d_{35} \qquad (4)$$

These parameters are shown plotted in Fig. 3 which also shows the curve suggested by Einstein and Barbarossa and an envelope indicating the spread of their original data used to derive this curve. It is evident that the Northern Ireland rivers produce a considerably greater spread of values and no real correlation could be observed with the Einstein-Barbarossa curve. It is therefore not considered practical to apply this method to Northern Ireland rivers.

4/1 <u>Engelund Method</u>

Engelund (Ref. 3) proposed that the stage discharge relationship for a stream could be derived from the cross-section, slope and bed material. He presented a relationship between submerged relative density of bed material (S-1), the channel slope (I), the grain diameter corresponding to 35% being finer by weight ($d_{35}$) and a dimensionless bed shear parameter θ'. This equation is

$$\frac{D}{d} \times \frac{1}{\theta'} = \beta\frac{(S-1)}{I} \tag{5}$$

in which β was found to be 1.6.

The dimensionless shear parameter was related to the hydraulic mean depth corresponding to grain resistance proposed by Einstein-Barbarossa in the equation

$$\theta' = \frac{R' \, I}{(S-1)d_{35}} \tag{6}$$

This mean velocity of flow was then related to R" by the Einstein-Barbarossa eqn 2 in which $K_s = 2 \times d_{65}$.

Equations 1, 2 and 3 were used to compute mean velocity for a range of depths up to bank-full condition for each of the sites in Northern Ireland. Some of the predicted velocities were found to be negative and in the majority of cases the predicted velocities were considerably lower than those measured in the field. The percentage difference between actual and predicted values at bank-full conditions is shown in Fig. 4. This shows that for beds giving $d_{50}$ less than 35 mm there was no correlation between actual and predicted values. For bed materials having $d_{50}$ greater than this value the velocities were under-predicted by approximately 85%.

This Engelund method may therefore be used to estimate the flow capacity of a channel provided the bed material size $d_{50}$ is greater than 35 mm. The flow capacity is estimated by calculating the bank-full mean velocity using this method and dividing by 0.15. However this very large amplification factor is only approximate and may therefore lead to large errors.

5/1 <u>White Method</u>

White et al (Ref. 4) used the dimensionless grain size number, $D_{gr}$, and the sediment mobility, $F_{gr}$ developed by Ackers and White to evaluate a mobility number for fine grain sediments, $F_{fg}$'. A ratio of the general form to the fine grain form of the mobility number was related to the sediment size number by a best fit curve using sand flume data. This best fit curve can then be used to evaluate the mean velocity of the flow enabling the roughness coefficient to be calculated.

An examination of the data on Northern Ireland rivers showed that the majority of bed sizes may be classified as medium and coarse gravels (see Fig. 2a). A better method of estimation of the flow capacity was therefore sought which would not have the restriction on bed material size or require major scaling of values.

6/1 Analysis of Data

A study of the variation of 'n' with discharge in the rivers in this investigation showed that at low flows the value of 'n' is very high. This decreased with increasing discharge becoming sensibly constant at high and medium flows. The variation of Mannings 'n' with discharge for the river Kells is shown on Fig. 5. This behaviour is typical of that observed for all of the rivers in the study. The rivers in Northern Ireland generally exhibit a pool-riffle formation. The variation in 'n' over a pool-riffle sequence which included the gauging station on the river Kells was examined at different discharges assuming a hydraulic gradient equal to the average for the reach, see Fig. 6. This showed that there was a considerable variation in 'n' between the pool and riffle at low discharges. The differences decreased with increasing discharges and the effect of the pool and riffle became insignificant at higher flows. At these high flows the effect of the pool or riffle on the hydraulic mean depth became small.

At each station as the bank-full condition was approached the effect of the bed formation became small and the flow condition became sensibility uniform. The Manning formula assumes uniform flow which was not satisfied at low flow conditions and therefore data on low flows was discarded. The value of 'n' corresponding to bank-full conditions was used in the study to minimise local variations due to bed formation. This selection was justified by the fact that the drainage engineer is primarily interested in channel capacity.

A method was sought whereby the friction factor at any site could be related to the channel properties and the bed material size. The Stickler formula

$$n = 0.01355\ K_s^{1/6} \qquad (7)$$

where $K_s$ is the grain roughness in mm, has been widely used to evaluate 'n'. This formula takes account of bed material but ignores the effect of the channel properties. Engelund proposed that $K_s$ should be taken as twice $d_{65}$ and this was used as an initial estimate of 'n'.

The difference between the value of 'n' obtained using eqn (7) and that derived experimentally was calculated for each site. These residuals were related to the channel properties $d_{35}$ and mean velocity. A residual plot of 'n' against velocity is shown in Fig. 7. Regression analysis was used to obtain a best fit equation for this residual. This produced the relationship

$$n - n_s = \frac{I^{0.02922} R^{0.03978}}{d_{35}^{0.0063} V^{0.1025}} \qquad (8)$$

in which $d_{35}$ is in mm, I is in cm per 100 m of channel and $n_s$ is the Strickler estimate of 'n'. This equation was able to explain 86.8% of the variability of the residuals.

This equation has been derived as an empirical best fit equation for use with Manning formula to evaluate bank-full mean velocity. The equation is dependent upon velocity and hence may admit more than one solution or no unique solution for any given site. The proposed method of estimating the channel friction coefficient is to plot the variation of 'n' with velocity for the given channel cross-section and slope on semi-log graph paper. The equation 8 will plot as a straight line of slope equal to −0.1025. For exact solution this line should be tangential to the curve obtained using the Manning formula. The relationship with velocity has been derived using data from all sites and hence the slope of the line is assumed the best fit. However, uncertainties in the estimate of $n_s$, the hydraulic gradient or $d_{35}$ may cause a constant shift of this line. This will result in the line and curve intersecting in two positions or not at all. The predicted value of 'n' is obtained by transferring the line to become tangential to the curve, the point of intersection giving the estimated value.

### 7/1 Sample Calculation

River No. 41 Channel data: $R = 1.49$ m $\quad I = 38$ cm per 100 m

$d_{35} = 10$ mm $\quad d_{65} = 36$ mm

For channel section $n = \frac{R^{0.67} I^{0.5}}{V}$

This plots as a curve shown in Fig. 8.

With $K_s = 2d_{65}$ eqn 7 gives $n_s = 0.0276$

Eqn 8 is then $n = 0.0276 + \log_{10} \frac{(1.1137)}{(V^{0.1025})}$

This plots as a straight line with slope of 0.1025 as shown in Fig. 8, (as $n_s$, I, R and $d_{35}$ are constant for any given site). The line is transferred parallel to become tangential to the curve, the point of intersection having an n value of 0.0445. The true value for this site was found to be 0.0476.

## 8/1 Comparison of Methods

The Einstein and Barbarossa method was not included in the comparison of methods as no best fit curve could be obtained for the data. The Engelund and White methods were tested on the data for Northern Ireland rivers and were compared with the statistically derived method. In each case the value of Mannings coefficient was calculated for every river and related to observed values (see Fig. 9).

The basis of comparison which has been used is the % of calculated values which lie within a factor of 2 of the observed values. The Engelund method gave 21% of calculated values lying in this region. This shows poor prediction obtained by this method. The White method was considerably better and gave 79% lying in this region, with some values being up to a factor of 8 times the observed. The new method achieved 89% of values within a factor of 2 and in all cases was less than a factor of 4 off the observed values.

Further comparison of the White and new method was made in which values lying between 4/5 and 5/4 times the observed value were examined. In each case 33% of values lay between these values.

## 9/1 Conclusions

The dimensionless Einstein-Barbarossa bed form resistance parameters were evaluated for the rivers in Northern Ireland but departed very considerably from the Einstein-Barbarossa curve.

The Engelund method of estimating a channel stage discharge relationship was applied to local rivers but give poor predictions. With bed materials having $D_{50}$ sizes less than 35 mm no prediction could be made. Above this threshold estimates were found to be approximately 15% of true values at bank-full conditions.

A prediction method for bank-full discharge has been derived based on the Strickler formula and bed and channel properties. This method explains 86% of the residuals between the actual values of n and estimated using the Stickler formula.

The new method has been shown to give a better estimation of roughness coefficient than the White method when applied to Northern Ireland rivers.

## References


1. Strickler, K. "Beitrage zur Frage der Geschwindigkeitsformel und der Rauhigkeitszahlen fur Strom Kanale und geschlossene Leitungen", Mitt No. 16 des Eidgenossische Amtes fur Wasser wirtschaft, Bern, Switzerland 1923.
2. Einstein, H.A. and Barbarossa, N.L., "River channel roughness". A.S.C.E. Trans. Paper No. 1528. 1951, pp 1121-1137.
3. Engelund, F. "Hydraulic resistance of alluvial streams" A.S.C.E. Proc. March 1966, Hy2 paper 4739, pp 315-326.
4. White, W.R., Paris, E. and Bettess, R., "A new method for predicting the frictional characteristics of alluvial streams". Hydraulics Research Station, Report No. IT 187, July 1979.
5. Ackers, P. and White, W.R., "Sediment transport: new approach and analysis". Proc. ASCE JHD, 1973, 99, HY11, pp 2041-2060.

## Nomenclature

| | |
|---|---|
| $d_{35}$, $d_{50}$, $d_{65}$ | Sieve size passing 35%, 50% amd 65% by weight (mm) |
| $D_{gr}$ | dimensionless sediment size number |
| $F_{fg}$ | fine grain sediment mobility number |
| $F_{gr}$ | general sediment mobility number |
| $g$ | gravity (9.81 $m/s^2$) |
| $I$ | channel slope |
| $K_s$ | grain roughness (mm) |
| $n$ | Manning friction coefficient |
| $n_s$ | Strickler estimate of n |
| $R$ | hydraulic radius (m) |
| $R'$ | hydraulic radius relating to grain resistance (m) |
| $R''$ | hydraulic radius relating to bed form resistance (m) |
| $S$ | submerged relative density of bed material |
| $V$ | mean velocity (m/s) |
| $V'_f$ | shear velocity related to grain resistance |
| $V''_f$ | shear velocity related to best form resistance |
| $\beta$ | constant (value = 1.6) |
| $\theta'$ | dimensionless shear parameter (Engelund) |
| $\psi'$ | Einstein grain shear parameter |

## Acknowledgements

The authors wish to thank the Department of Agriculture (NI) for access to the data used in this study.

Appendix I

DATA FROM NORTHERN IRELAND RIVERS
AT BANK-FULL CONDITIONS

| No | $\times 10^4$ Slope | $D_{35}$ m | $D_{50}$ m | Surface Width m | R m | 'n' observed | Vel m/s |
|---|---|---|---|---|---|---|---|
| 1 | 26.3 | 0.160 | 0.18 | 15.3 | 1.07 | 0.0301 | 1.78 |
| 2 | 40.7 | 0.030 | 0.050 | 3.5 | 0.57 | 0.0563 | 0.78 |
| 3 | 28.6 | 0.028 | 0.034 | 18.7 | 1.36 | 0.0515 | 1.28 |
| 4 | 33.2 | 0.030 | 0.037 | 20.3 | 1.46 | 0.0740 | 1.00 |
| 5 | 11.9 | 0.038 | 0.048 | 15.6 | 1.98 | 0.0332 | 1.63 |
| 6 | 19.0 | 0.0003 | 0.0004 | 72.0 | 2.95 | 0.0531 | 1.69 |
| 7 | 15.8 | 0.00042 | 0.0005 | 24.87 | 1.95 | 0.0336 | 1.84 |
| 8 | 31.3 | 0.008 | 0.017 | 2.97 | 0.55 | 0.049 | 0.77 |
| 9 | 23.8 | 0.0025 | 0.0065 | 3.44 | 0.71 | 0.1392 | 0.28 |
| 10 | 24.7 | 0.0045 | 0.008 | 10.08 | 1.0 | 0.0288 | 1.73 |
| 11 | 42.4 | 0.035 | 0.050 | 7.07 | 0.74 | 0.0439 | 1.21 |
| 12 | 41.0 | 0.062 | 0.065 | 5.02 | 0.67 | 0.028 | 1.75 |
| 13 | 20.8 | 0.027 | 0.031 | 12.99 | 1.42 | 0.0695 | 0.83 |
| 14 | 24.2 | 0.050 | 0.051 | 34.07 | 1.88 | 0.0507 | 1.48 |
| 15 | 25.4 | 0.014 | 0.016 | 21.4 | 1.83 | 0.0388 | 1.94 |
| 16 | 23.9 | 0.0018 | 0.0085 | 16.99 | 1.56 | 0.0304 | 2.17 |
| 17 | 4.5 | 0.00008 | 0.00012 | 26.44 | 2.18 | 0.0332 | 1.08 |
| 18 | 8.7 | 0.002 | 0.005 | 21.22 | 2.17 | 0.0583 | 0.85 |
| 19 | 8.7 | 0.002 | 0.005 | 20.23 | 1.98 | 0.0393 | 1.18 |
| 20 | 8.6 | 0.058 | 0.070 | 28.33 | 1.49 | 0.0239 | 1.6 |
| 21 | 18.5 | 0.015 | 0.022 | 7.84 | 0.94 | 0.0220 | 1.88 |
| 22 | 30.0 | 0.075 | 0.080 | 10.81 | 1.19 | 0.0414 | 1.49 |
| 23 | 41.2 | 0.0005 | 0.0009 | 12.3 | 0.94 | 0.0303 | 2.03 |
| 24 | 41.2 | 0.0005 | 0.0009 | 12.23 | 0.93 | 0.0202 | 3.01 |
| 25 | 31.6 | 0.012 | 0.020 | 17.67 | 1.67 | 0.0785 | 1.01 |
| 26 | 7.8 | 0.0002 | 0.0005 | 24.02 | 1.77 | 0.0305 | 1.34 |
| 27 | 31.4 | 0.020 | 0.036 | 3.58 | 0.86 | 0.0428 | 3.76 |
| 28 | 24.7 | 0.028 | 0.054 | 10.71 | 1.01 | 0.0315 | 1.59 |
| 29 | 38.0 | 0.020 | 0.032 | 15.79 | 1.37 | 0.0287 | 2.65 |
| 30 | 31.3 | 0.020 | 0.032 | 18.09 | 1.76 | 0.0456 | 1.79 |
| 31 | 27.5 | 0.0010 | 0.0015 | 51.49 | 3.00 | 0.0677 | 1.61 |
| 32 | 43.4 | 0.0006 | 0.0012 | 27.75 | 1.17 | 0.0352 | 2.07 |
| 33 | 11.7 | 0.051 | 0.053 | 30.53 | 2.06 | 0.0315 | 1.76 |
| 34 | 5.4 | 0.020 | 0.023 | 2.8 | 0.60 | 0.0519 | 1.01 |

| No | $\times 10^4$ Slope | $D_{35}$ m | $D_{50}$ m | Surface Width m | R m | 'n' observed | Vel m/s |
|---|---|---|---|---|---|---|---|
| 35 | 7.0 | 0.031 | 0.037 | 60.99 | 4.03 | 0.0772 | 0.86 |
| 36 | 8.2 | 0.0001 | 0.00015 | 20.83 | 2.04 | 0.0475 | 0.97 |
| 37 | 31.6 | 0.031 | 0.042 | 22.86 | 1.97 | 0.0528 | 1.67 |
| 38 | 55.1 | 0.054 | 0.063 | 5.43 | 0.82 | 0.0312 | 2.09 |
| 39 | 117.0 | 0.020 | 0.035 | 17.38 | 1.01 | 0.1117 | 0.97 |
| 40 | .13.6 | 0.055 | 0.065 | 7.51 | 1.28 | 0.0289 | 1.50 |
| 41 | 38.0 | 0.010 | 0.026 | 16.74 | 1.49 | 0.0476 | 1.69 |
| 42 | 13.6 | 0.022 | 0.040 | 17.35 | 2.26 | 0.0256 | 2.47 |
| 43 | 25.5 | 0.032 | 0.046 | 30.69 | 2.65 | 0.0961 | 1.01 |
| 44 | 34.8 | 0.030 | 0.043 | 13.17 | 1.07 | 0.0775 | 0.80 |
| 45 | 90.0 | 0.017 | 0.026 | 24.56 | 2.07 | 0.0736 | 2.09 |
| 46 | 135.0 | 0.035 | 0.060 | 3.37 | 0.64 | 0.1106 | 0.78 |
| 47 | 3.1 | 0.003 | 0.006 | 31.34 | 1.67 | 0.0307 | 0.81 |
| 48 | 12.1 | 0.090 | 0.120 | 14.43 | 1.21 | 0.0445 | 0.89 |
| 49 | 49.3 | 0.090 | 0.120 | 13.09 | 1.14 | 0.0367 | 2.09 |
| 50 | 6.1 | 0.015 | 0.024 | 23.74 | 2.82 | 0.0151 | 3.26 |
| 51 | 19.0 | 0.020 | 0.029 | 9.25 | 0.87 | 0.0223 | 1.78 |
| 52 | 29.6 | 0.038 | 0.048 | 44.21 | 2.53 | 0.0379 | 2.67 |
| 53 | 3.6 | 0.0002 | 0.0006 | 20.16 | 1.98 | 0.0170 | 1.76 |
| 54 | 59.7 | 0.002 | 0.019 | 53.18 | 3.29 | 0.0679 | 2.51 |
| 55 | 39.0 | 0.0045 | 0.007 | 11.62 | 1.58 | 0.0274 | 3.09 |
| 56 | 15.8 | 0.120 | 0.130 | 30.81 | 2.28 | 0.0284 | 2.43 |
| 57 | 73.0 | 0.038 | 0.060 | 6.98 | 0.74 | 0.0295 | 2.37 |
| 58 | 26.4 | 0.039 | 0.044 | 9.69 | 0.78 | 0.0168 | 2.59 |
| 59 | 293.0 | 0.30 | 0.40 | 11.18 | 0.83 | 0.0358 | 4.20 |
| 60 | 6.3 | 0.0011 | 0.002 | 46.7 | 3.27 | 0.0444 | 1.24 |
| 61 | 12.4 | 0.174 | 0.200 | 12.43 | 1.66 | 0.0557 | 0.89 |
| 62 | 52.0 | 0.020 | 0.030 | 24.71 | 1.72 | 0.059 | 1.76 |
| 63 | 52.0 | 0.020 | 0.030 | 34.11 | 2.19 | 0.0541 | 2.25 |
| 64 | 28.0 | 0.016 | 0.026 | 8.19 | 1.17 | 0.0307 | 1.91 |
| 65 | 28.0 | 0.040 | 0.055 | 9.27 | 1.39 | 0.0587 | 1.13 |
| 66 | 23.8 | 0.006 | 0.012 | 8.08 | 0.88 | 0.0593 | 0.75 |
| 67 | 47.2 | 0.010 | 0.050 | 3.39 | 0.59 | 0.0259 | 1.87 |
| 68 | 26.4 | 0.036 | 0.045 | 11.57 | 1.32 | 0.0260 | 2.38 |

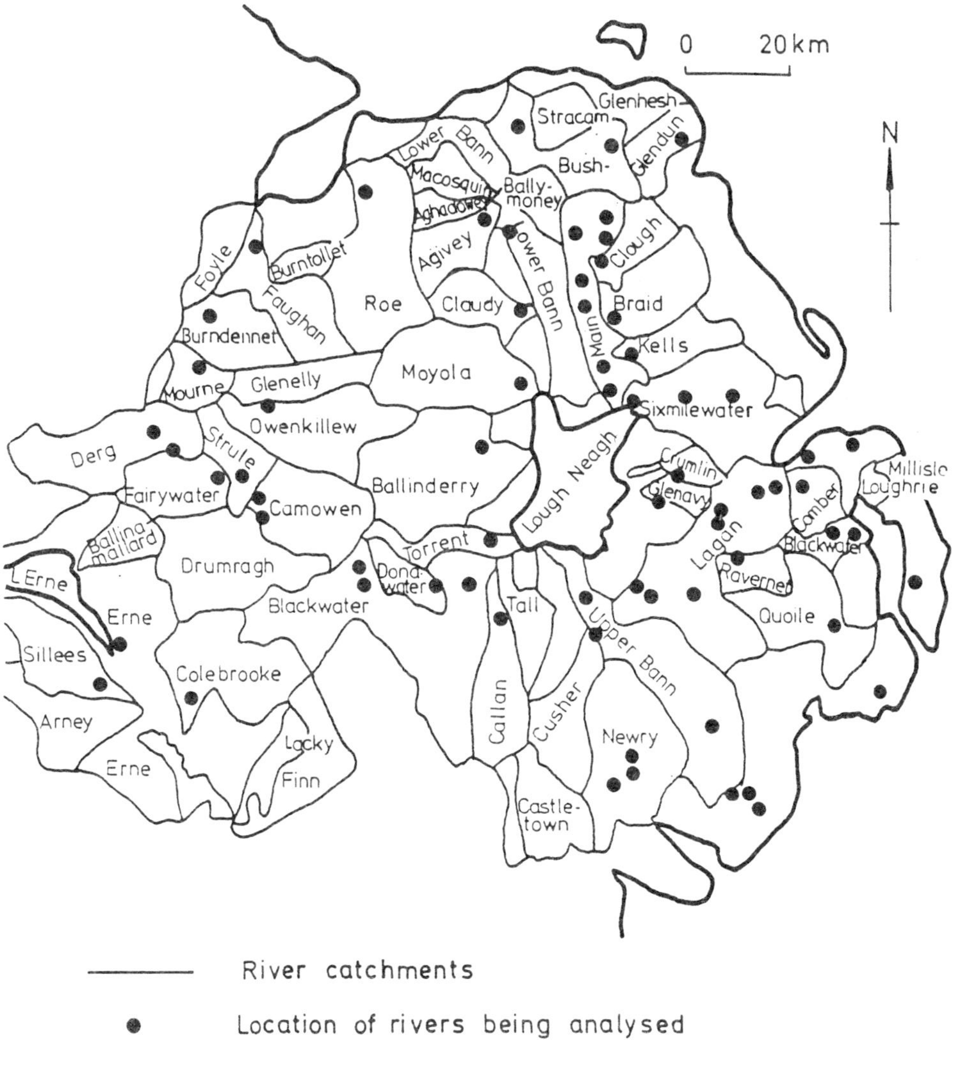


Fig 1 River catchments in Northern Ireland and locations of gauging stations

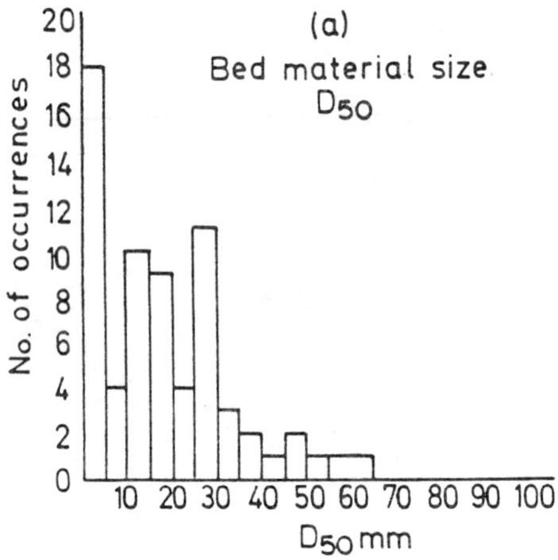


Fig 2a Histogram of $D_{50}$

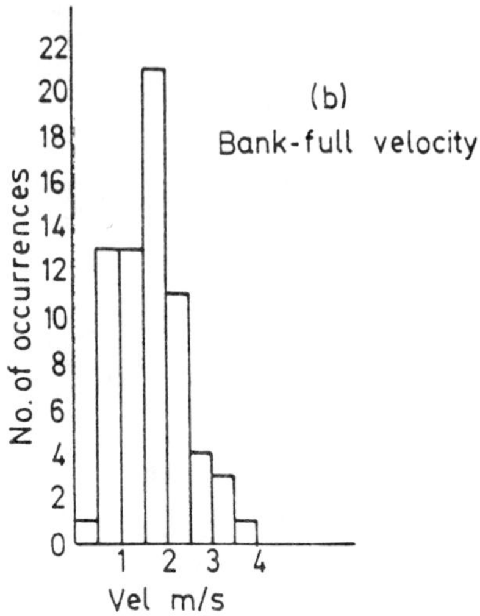


Fig 2b Histogram of Bank-full mean velocities

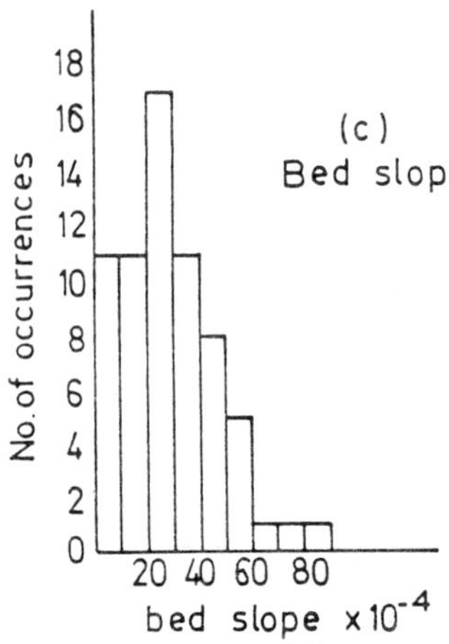


Fig 2c Histogram of channel slopes

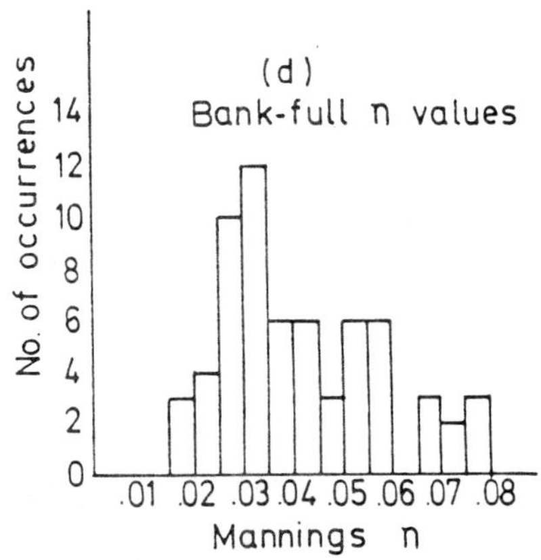


Fig 2d Histogram of bank-full n values

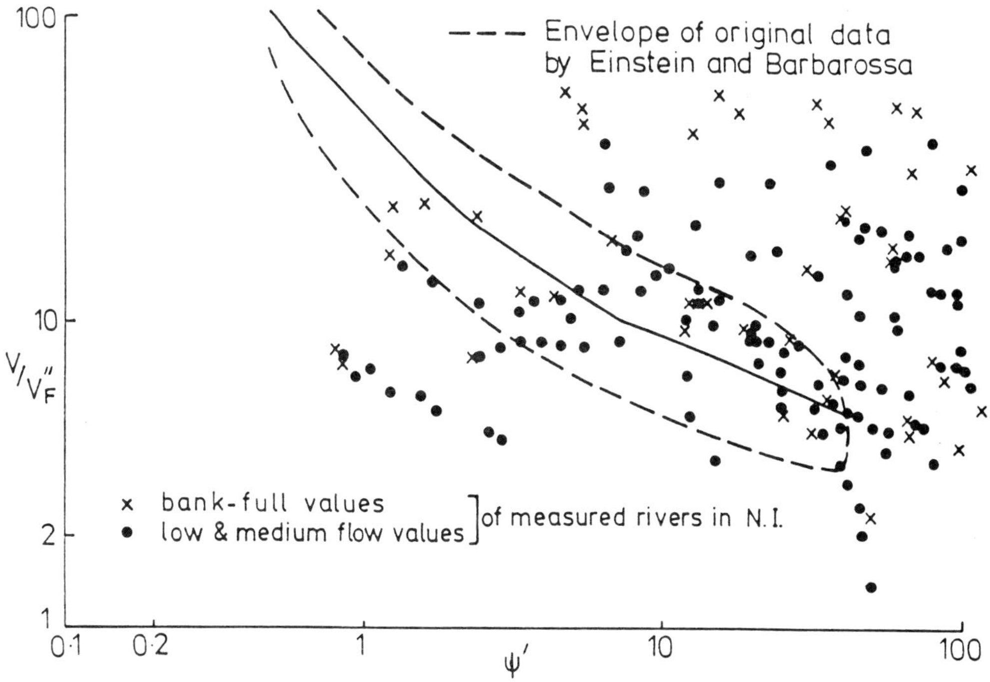


Fig 3 Einstein and Barbarossa diagram

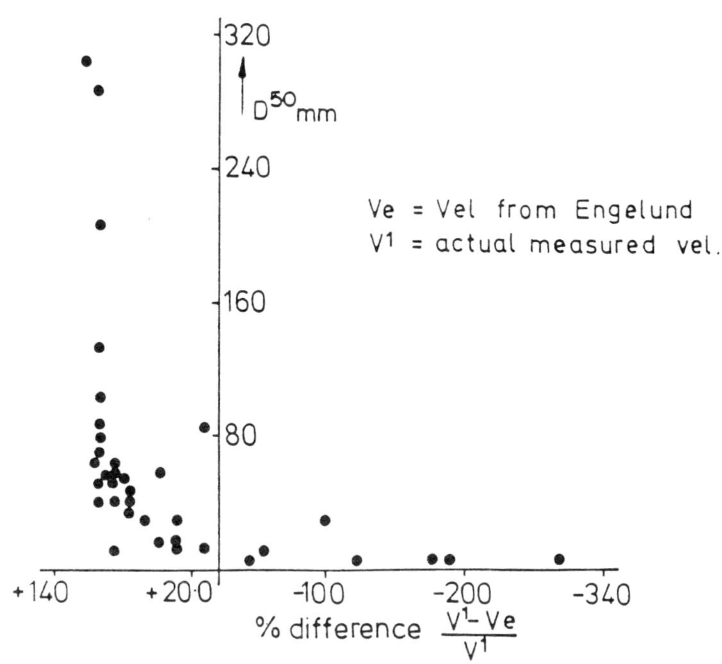


Fig 4 Results from Engelund method

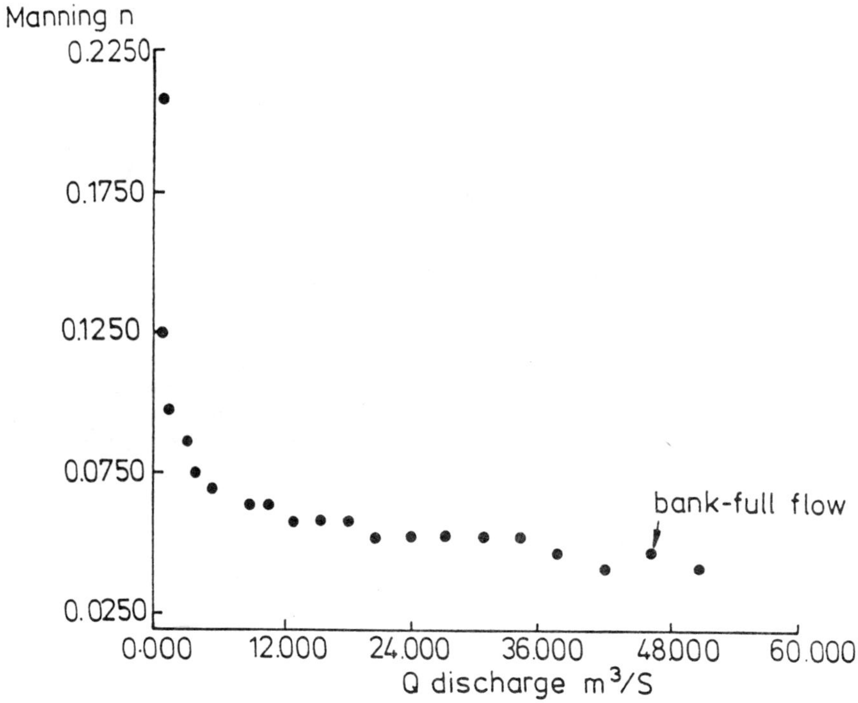


Fig 5 Variation of Mannings n with discharge

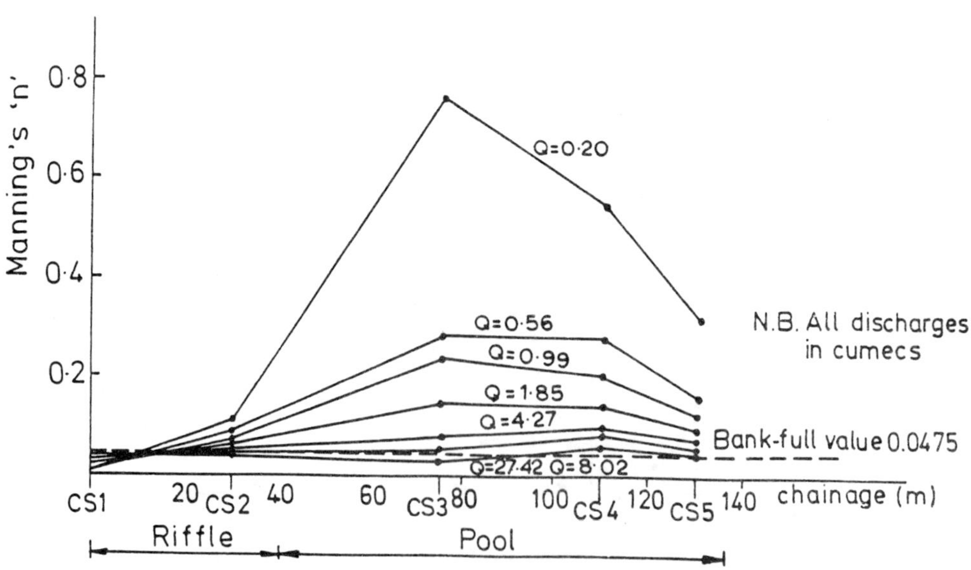


Fig 6 Variation in n over a pool-riffle sequence

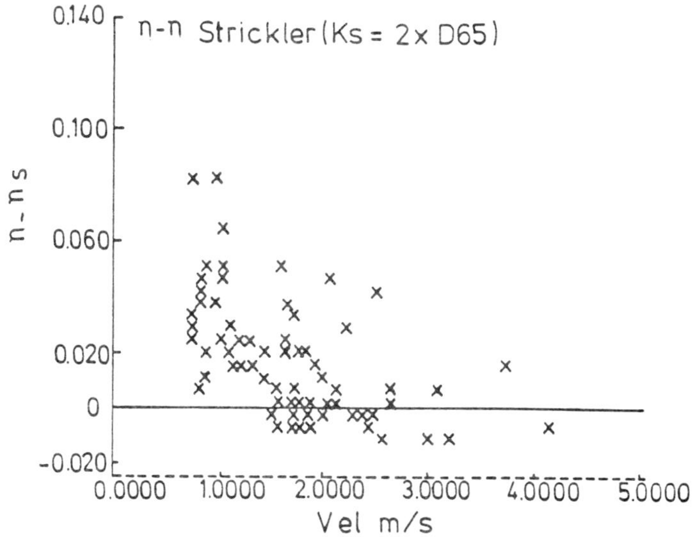


Fig 7 Residual plot of 'n' and velocity

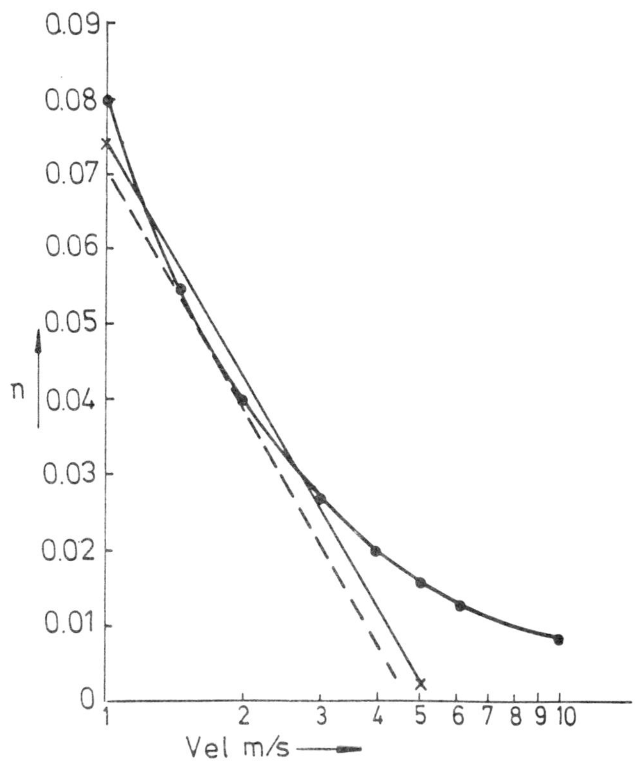


Fig 8 n verses velocity (sample calculation)

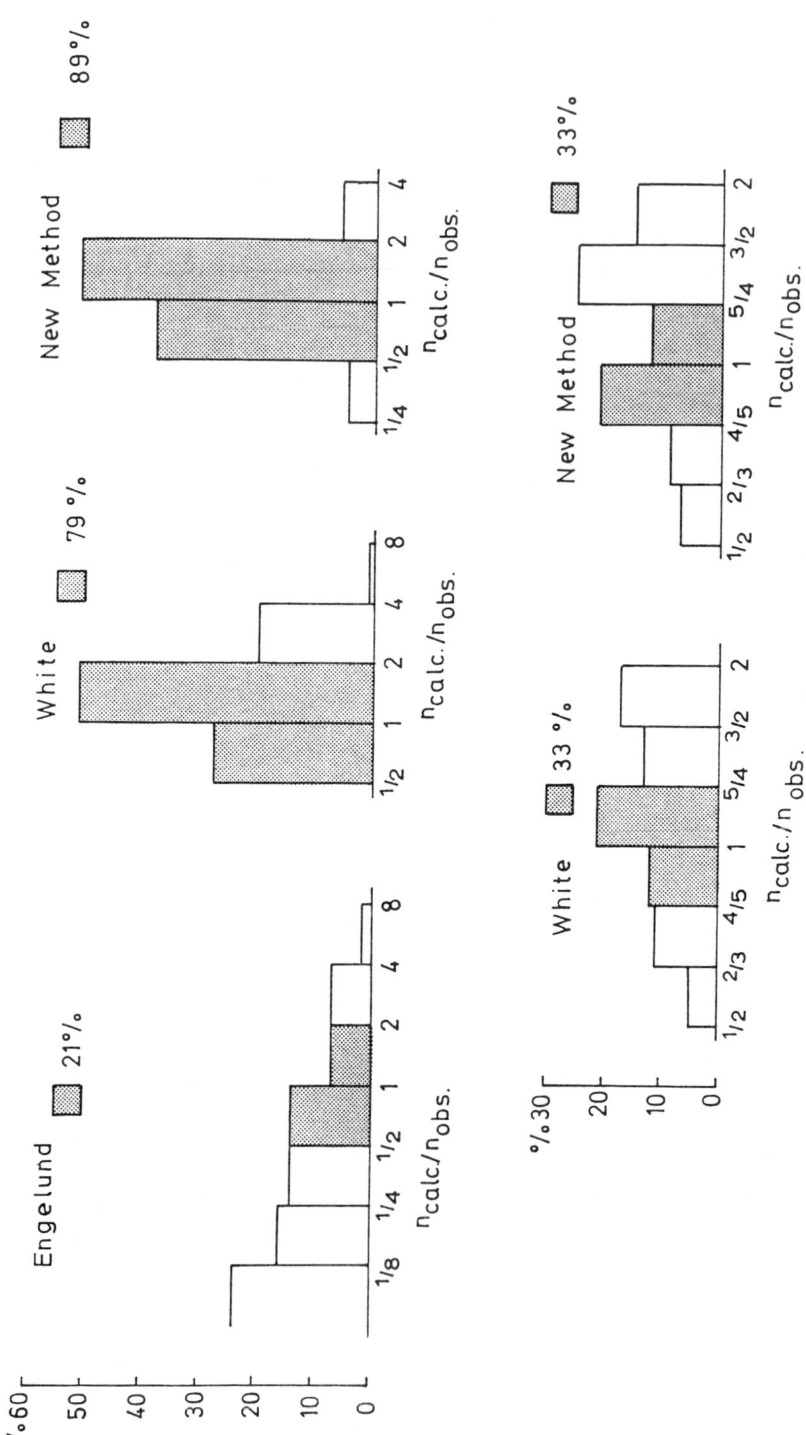


Fig 9  Distribution of discrepancy ratios

International Conference on
**RIVER REGIME**
18-20 May, 1988

PAPER G2

# THE PREDICTION OF MEAN BED AND WALL BOUNDARY SHEAR IN UNIFORM AND COMPOSITELY ROUGH CHANNELS.

T.P. Flintham[1] and P.A. Carling[2]
Dept. of Engineering, University of Leicester, U.K.[1]
Freshwater Biological Association, Ambleside, U.K.[2]

SUMMARY : Experimental procedures and data compilations are presented which relate to the prediction of mean bed and wall boundary shear in uniform and compositely roughened channels. Equations are derived to determine the mean bed and wall shear stress in straight, symmetrical-trapezoidal and rectangular channels. The implications for further work are noted.

## 1. INTRODUCTION

Maximum and mean boundary shear stress are extensively used in sediment transport and river engineering computations. However, the complexities arising from secondary circulations (eg Refs 1, 2, 3) and boundary roughness non-uniformity (Ref 4) makes the theoretical prediction of boundary shear stress distribution extremely difficult, even for straight prismatic open channels. Consequently a variety of empirical techniques have been developed to determine local shear stress values through either direct or indirect measurement (Refs 5 &6). When suitable empirical data are not available only the mean boundary shear stress may be determined from the energy gradient

$$\tau_e = \rho g R S_f \qquad (1)$$

where $\rho$ = fluid density, $g$ = acceleration due to gravity, $R$ = hydraulic radius and $S_f$ = energy gradient. Unfortunately, $\tau_e$ is only of limited use when considering the physical processes involved in flow resistance and sediment mechanics.

A useful empirical method for predicting the mean bed and wall shear stresses in smooth rectangular channels was originally proposed by Knight (Ref 7), using the concept of shear force. The general form of the model was confirmed by Noutsopoulos and Hadjipanos (Ref 8) and the empirical coefficients further modified by Knight et al (Ref 9). In this paper a tentative empirical model is proposed (based on Knight's equation for smooth rectangular channels) to predict the mean bed and side-wall shear stresses for straight symmetrical trapezoidal and rectangular channels having uniform or composite (differential) roughness.

### 1.1 Perspective

The shear force acting on a boundary, per unit length of channel is equal to the mean boundary shear stress ($\bar{\tau}_n$) multiplied by the boundary cross-sectional length ($P_n$). The shear force acting on

International Conference on River Regime, edited by W.R.White.

a channel's side walls ($SF_w$) and bed ($SF_b$) are therefore equal to

$$SF_w = \bar{\tau}_w P_w \tag{2}$$

and

$$SF_b = \bar{\tau}_b P_b \tag{3}$$

The shear force carried by the side-walls or bed can be expressed as a percentage of the total shear force (%SF), for example

$$\%SF_w = \frac{\bar{\tau}_w P_w}{\bar{\tau}_e P} \times 100 \tag{4}$$

Knight found that the relationship between the percentage shear force carried by the side-walls ($\%SF_w$) and the width to depth ratio ($^B/H$) for smooth rectangular channels was of the generalised form

$$\log_{10} \%SF_w = C_1 \log_{10}(\frac{B}{H} + tr) + C_2 \tag{5}$$

where tr = a threshold, introduced to remove curvature in the trend of the experimental data evident when plotted on $\log_{10}$ scales. Knight et al (Ref 9) gave the constants $C_1$, $C_2$ and tr the numerical values of -1.4026, 2.6692, and 3.0 respectively. When expressed in an exponential form, the relationship between $\%SF_w$ and $^B/H$ is given by

$$\%SF_w = \exp^{\alpha} \tag{6}$$

where $\alpha = -3.230 \log_{10}(\frac{B}{H} + 3) + 6.146$.

Once the $\%SF_w$ has been determined the mean side-wall ($\bar{\tau}_w$) and bed ($\bar{\tau}_b$) shear stresses (non-dimensionalised by the two-dimensional shear stress $\rho g H S_f$) can be obtained from the following equations

$$\frac{\bar{\tau}_w}{\rho g H S_f} = 0.01 \%SF_w (\frac{B}{2H}) \tag{7}$$

and

$$\frac{\bar{\tau}_b}{\rho g H S_f} = 1 - 0.01 \%SF_w \tag{8}$$

as

$$\bar{\tau}_e P = \bar{\tau}_w P_w + \bar{\tau}_b P_b \tag{9}$$

To adapt this simple procedure to account for the influence of channel shape and boundary roughness on the mean side-wall and bed shear stress prediction, a series of experiments was conducted in a trapezoidal flume lined with various bed to wall roughness combinations.

## 2. EXPERIMENTAL PROCEDURE

The experimental programme was carried out in a tilting 13 m long trapezoidal channel, constructed out of marine plywood and having the dimensions shown in Fig. 1. The discharge was measured using an orifice plate installed to British Standard specifications and the flow depth controlled by a louvre-type tailgate. The flume bed slope was varied during the experimental programme, to a maximum gradient of $1.429 \times 10^{-2}$. Flow depths were measured using a standard point-gauge attached to a carriage mounted on rails running parallel with the flume bed.

The roughness elements used in this study consisted of 'single sized' rounded river gravels, glued a single layer deep to plywood lining boards. These lining boards were attached to the flume walls using screws located at 300 mm intervals. The sieved gravel samples were very well sorted and the particle size population had a log normal distribution. The standard deviation ($\delta$) of each gravel population, as defined by

$$\delta = 0.5 \log_{10}\left(\frac{d_{84}}{d_{16}}\right) \tag{10}$$

where $d_n$ is the median particle axis for which n per cent (by weight) of the median axes are smaller, varied between 0.08 and 0.19. To maximise the model's versatility the roughness of the laboratory gravels was expressed in terms of Nikuradse's equivalent grain size ($k_s$), the values of which were obtained from velocity profile data (see below). Three gravel sizes were used, and six bed/wall roughness combinations examined (Table 1).

Local boundary shear stresses were measured indirectly, using the logarithmic velocity profile technique.

Within close proximity of a rough boundary the velocity profile distribution is given by

$$\frac{u}{u_*} = \frac{2.303}{\kappa} \log_{10}\left(\frac{y}{k_s}\right) + 8.5 \tag{11}$$

where u = the point velocity at a distance y from the boundary, $u_*$ local mean shear velocity and $\kappa$ = von Kàrmàn's constant. By measuring the velocity at two points on the same velocity profile perpendicular to the boundary, the shear velocity can be calculated from the following equation, assuming $\kappa$ = 0.40

$$u_* = \frac{u_1 - u_2}{5.75 \log\left(\frac{y_1}{y_2}\right)} \tag{12}$$

The local shear stress at the base of the profile can then be obtained from

$$\tau = \rho\, u_*^2 \tag{13}$$

By substituting the $u_*$ value obtained from equation (12) into equation (11) the appropriate $k_s$ value for the velocity profile can be calculated.

A major difficulty in applying boundary layer theory to a hydraulically rough surface concerns the location of the virtual zero bed-datum, which is required to standardise the point velocity distances (y). The elevation of the zero bed-datum is dependent upon the size, shape and spacing of the roughness elements. For densely packed gravel surfaces it is often suppressed below the top of the roughness particles (eg Ref 10).

To ascertain the zero bed-datum height in this study, the following experimental procedure was repeated for each gravel surface. The gravel-bed lining boards were installed and the flume was set to a gradient of $7.17 \times 10^{-3}$. A low discharge (confirmed by direct weighing) was then passed down the flume and uniform flow established. At three cross-sections extensive current-metering was undertaken at 20 equidistant points across the bed (approximately 15 mm intervals), using a miniature impeller-type current meter. At each point the velocity was measured at 5 mm intervals up through the flow to the water surface.

Using the velocity-area technique (integrating the point velocities over the cross-sectional area) the discharge was initially calculated assuming that no roughness elements were present ie the zero bed-datum was located on the lining board surface. Due to the over-estimation of flow area, the velocity-area discharge for all three cross-sections exceeded the true discharge. The datum (line AB in Fig. 2) was then sequentially raised until the velocity-area discharge equalled the true discharge. With each displacement of the datum the new cross-sectional flow area was assumed equal to smooth trapezium $A_nB_nCD$, where the subscript n denotes the number of the datum. The shallow flow depth and relatively large roughness elements ensured that the flow area between the elements represented a significant portion of the total area (approximately 10%). Consequently any inaccuracies in flow area calculation introduced by the roughness elements were accentuated.

Providing the velocity measurements are accurate, when the velocity-area discharge and true discharge are equal the area blocked by the elements above the datum is equivalent to the flow area between the elements below the datum. The mean datum height was averaged over the three cross-sections, which for all the gravel surfaces was approximately equal to $0.5d_{84}$. By adjusting the distance y in the velocity profiles relative to the zero bed-datum, good logarithmic plots were obtained, in accordance with boundary layer theory.

The zero bed-datum was used to define the wetted perimeter, which avoided any complications regarding the effective boundary length of a rough surface. The cross-sectional area was defined as the area enclosed by the zero bed-datum and the free surface. An extensive flow resistance coefficient investigation (Ref 11) conducted simultaneously with the shear stress study confirmed that the empirical $k_s$ values assigned to each surface, together with the hydraulic radius term defined using the zero bed-datum concept, were appropriate.

For all six channel roughness combinations a series of discharges were passed down the flume and uniform flow established in each case. During each flume run the local shear stress was measured at seven points across the bed. The number of side-wall shear stress measurements was dependent upon the flow depth, although for each flume run the same boundary points were taken on both the left and right wall. Each velocity profile consisted of five point velocities taken at regular 5 mm or 10 mm intervals normal to the channel boundary. The five near-boundary velocities invariably followed the logarithmic velocity law, and the slope of the fitted linear regression curve was used to calculate the shear velocity.

The local shear stress values were integrated over the relevant boundary length to obtain mean side-wall and bed values. As a check on the local shear stress measurements the total integrated mean shear stress ($\bar{\tau}_i$), where

$$\bar{\tau}_i = \frac{\bar{\tau}_w P_w + \bar{\tau}_b P_b}{P} \qquad (14)$$

was compared with the mean value computed from the energy gradient ($\bar{\tau}_e$). For all the shear stress experiments $\bar{\tau}_i$ was within ± 11% of $\bar{\tau}_e$. To ensure that the total percentage force carried by the side-walls and bed equalled 100%, $\bar{\tau}_i$ was used in preference to $\tau_e$ in equation (4) to calculate $\%SF_w$.

For each roughness type, the representative $k_s$ value was obtained by averaging data from 40-60 profiles.

## 3. RESULTS

Correlating $\%SF_w$ with the ratio bed/side-wall perimeter length ($P_b/P_w$), rather than with aspect ratio ($B/H$), collapses onto a single curve the data collected from the unroughened plywood trapezoidal channel and the 'smooth-rectangular' data (Fig. 3) listed by Knight et al (Ref 9). As $P_b = B$ and $P_w = 2H$ for a rectangular channel, equation (5) can be readily modified to relate $\%SF_w$ with $P_b/P_w$

$$\log_{10} \%SF_w = -1.4026 \log_{10}\left(\frac{P_b}{P_w} + 1.5\right) + 2.247 \qquad (15)$$

Equation (15) is shown in Fig. 3, together with rectangular and trapezoidal data.

To determine the influence of uniform boundary roughening on the percentage shear force carried by the walls, the variation in $\%SF_w$ with $P_b/P_w$ is shown in Fig. 4 for the trapezoidal channel uniformly lined with 6 mm gravel, together with equation (15). Also shown in Fig. 4 are the equivalent data for uniformly roughened rectangular and trapezoidal (side-angle $\theta = 45°$) channels given by Ghosh and Roy (Ref 5). The exact mean bed and side-wall shear stresses were not stated by Ghosh and Roy, so the values were estimated from graphical representations of shear stress distributions given by the authors.

From $\%SF_w$ the mean side-wall and bed shear stress for symmetrical trapezoidal ($\theta > 45°$) and rectangular channels can be calculated using modified forms of equations (7) and (8),

$$\frac{\bar{\tau}_w}{\rho g H S_f} = 0.01\ \%SF_w \left[\frac{B + P_b}{2P_w}\right] = 0.01\ \%SF_w \left[\frac{(B + P_b)\sin\theta}{4H}\right] \qquad (16)$$

whilst

$$\frac{\bar{\tau}_b}{\rho g H S_f} = 1 - 0.01\ \%SF_w \left[\frac{B}{2P_b} + 0.5\right] \qquad (17)$$

Figs 5 and 6 show the non-dimensionalised mean side-wall and bed shear stress against $P_b/P_w$ respectively, for the plywood and 6 mm gravel lined trapezoidal channel ($\theta = 68°$) together with the predicted curves. Figs 7 and 8 show the predicted variation of $\bar{\tau}_w/\rho g H S_f$ and $\bar{\tau}_b/\rho g H S_f$ with $P_b/P_w$ for uniformly roughened channels having side-wall angles ($\theta$) of 45°, 68° and 90°. The predicted non-dimensionalised bed shear stress for the three channel cross-sections is also shown as a function of aspect ratio in Fig. 9.

Equation (15) appears to adequately predict the influence of cross-sectional shape on the mean side-wall and bed shear stress in uniformly roughened rectangular and trapezoidal ($\theta > 45°$) channels. For these shaped channels having composite roughness, where the bed is differentially roughened relative to the side-walls, a further variable parameter is required to account for the effect of boundary roughness heterogeneity on the shear stress distribution. A suitable dimensionless roughness parameter is represented by the bed to side-wall roughness ratio expressed in terms of Nikuradse's equivalent grain size ($^{k}sb/k_{sw}$).

For any channel in which $^{P}b/P_w$ is held constant, the continual increase in bed roughening relative to the side-walls will systematically reduce the percentage shear force carried by the walls. Conversely, if the side-walls are roughened relative to the bed, the percentage shear force carried by the walls will increase. However, irrespective of the ($^{k}sb/k_{sw}$) ratio the percentage shear force carried by the walls in extremely narrow channels will tend towards 100%, until when $^{P}b/P_w = 0$, $\%SF_w = 100\%$. For an infinitely wide channel $\%SF_w$ will approach zero. Thus, when correlated with $^{P}b/P_w$, the $\%SF_w$ will plot as a series of exponential-type curves, dependent upon the ($^{k}sb/k_{sw}$) ratio.

When $^{P}b/P_w = 0$, the generalised form of equation (15) becomes

$$2 = C_1 \log (tr) + C_3 \qquad (18)$$

from which

$$C_3 = 2.0 - C_1 \log (tr)$$

where $C_1 = -1.4026$, $C_3 = 2.247$, and $tr = 1.5$. Substituting equation (18) into equation (15) gives

$$\log \%SF_w = C_1 \log \left(\frac{P_b}{P_w tr} + 1\right) + 2.0 \qquad (19)$$

or alternatively

$$\%SF_w = \exp^{\alpha *}$$

where $\alpha_* = -3.230 \log (^{P}b/P_w tr + 1) + 4.606$. From equation (19) the influence of composite roughness ($^{k}sb/k_{sw} \neq 1$) on $\%SF_w$ may be either a function of $C_1$ or tr or both. When constraining the predictive curves to pass through $\%SF_w = 100\%$ when $^{P}b/P_w = 0$, the closest approximation to the experimental data is achieved by maintaining $C_1$ constant and varying tr. Fig. 10 shows the logarithmic relationship between tr and ($^{k}sb/k_{sw}$) for the experimental trapezoidal data. When $^{k}sb/k_{sw} = 1$, $tr = 1.5$ (equation 15) and therefore constraining the linear regression curve to pass through this point gives the functional relationship

$$\log tr = -0.203 \log \left(\frac{k_{sb}}{k_{sw}}\right) + 0.176 \qquad (20)$$

Having equation (19) in conjunction with equation (20) the predicted $\%SF_w$ curves are shown together with the experimental data in Fig. 11. In Fig. 12 the predicted mean bed shear stresses for various ($^{k}sb/k_{sw}$) ratios are shown as a function of aspect ratio ($^{B}/H$) for a trapezoidal channel having a side-wall angle ($\theta$) of 68°.

Unfortunately little published low aspect ratio ($^{B}/H < 15$) field data are available to test the model, probably as a consequence of the high flows which are often associated with lower aspect ratios and the physical difficulties in measuring boundary shear stress under such conditions. As a spot check on the predictive accuracy of the model the mean bed shear stress obtained by the numerical integration of local measured values was compared with the predicted mean bed shear stress for a small gravel-bed river having an aspect ratio of 13.6 (Fig. 13). Each bed velocity profile was measured instantaneously using a suite of 6 Ott-type current meters. The predicted mean bed shear stress was approximately 6% smaller than the measured value.

## 4. DISCUSSION

The empirically derived equations (19) and (20) together provide a tentative method of estimating the mean bed and wall shear stress in small straight rectangular and trapezoidal channels (where $\theta > 45°$). The respective influences of channel shape and boundary roughness are correlated with the simple parameters $P_b/P_w$ and $k_{sb}/k_{sw}$. The resulting form of equations (19) and (20) ensures that the procedure is both quick and easy to use.

Although the model is based on data collected over a limited $P_b/P_w$ range (0.371-4.113) the variation in $\%SF_w$ is quite wide. As the relationship between $\%SF_w$ and $P_b/P_w$ appears to be of an exponential nature, equation (19) can be extrapolated beyond the experimental range with a reasonable degree of confidence. The general form of equations (19) and (20) would seem appropriate, although as additional experimental data become available the empirical coefficient values may require slight modification.

### 4.1 Smooth and Uniformly Roughened Channels

The single expression relating the shape parameter ($P_b/P_w$) with $\%SF_w$ implies that the influence of side-wall angle ($\theta > 45°$) on the mean bed and side-wall shear stress may be adequately predicted. The experimental evidence indicates that this relationship derived from flow in smooth channels is also applicable to uniformly roughened channels, thus inferring that uniform roughening does not significantly alter the relative proportioning of shear force carried by the bed and side-walls.

When non-dimensionalised by the two-dimensional shear stress ($\rho gHS_f$), the mean wall shear stress ($\bar{\tau}_w$) varies with $P_b/P_w$ as a complex function, which can be separated into an ascending and descending phase (Fig. 7). The complexity of this relationship is a result of the combined variation of $\bar{\tau}_w$ and H with ($P_b/P_w$). At high values, the ratio $P_b/P_w$ is particularly sensitive to small changes in the flow depth (H). With increasing flow depth (decreasing $P_b/P_w$) this sensitivity diminishes, until at low $P_b/P_w$ values (3<) relatively large variations in H are required to significantly alter $P_b/P_w$. For a trapezoidal channel having a constant energy gradient the predicted mean wall shear stress is negatively correlated with $P_b/P_w$.

In the ascending limb of the curve shown in Fig. 7 the decrease in mean wall shear stress is outweighed by the reduction in H required to lower $P_b/P_w$, and consequently $\tau_w/\rho gHS_f$ increases with $P_b/P_w$. At higher $P_b/P_w$ ratios the decrease in $\bar{\tau}_w$ exceeds the reduction in H, and hence accounts for the descending limb. Lowering the side-wall angle ($\theta$) accentuates these effects.

The relationship between the non-dimensionalised mean bed shear stress and ${}^{P}b/P_w$ (Fig. 8) is positively correlated, indicating that the relative rate of reduction in $\bar{\tau}_b$ with ${}^{P}b/P_w$ exceeds that of H. This difference in the predicted non-dimensionalised shear stress curves implies that at high ${}^{P}b/P_w$ values $\bar{\tau}_w$, although smaller than $\bar{\tau}_b$, is more sensitive to changes in flow depth (H). For all channels $\bar{\tau}_b$ equals $\bar{\tau}_w$ when ${}^{P}b/P_w = 0.55$. At higher ${}^{P}b/P_w$ values $\bar{\tau}_b > \bar{\tau}_w$, and when ${}^{P}b/P_w < 0.55$, $\bar{\tau}_b < \bar{\tau}_w$.

### 4.2 Composite Roughened Channels

The $\%SF_w$ data collected from the differentially roughened channels (${}^{k}sb/k_{sw} > 1$) are more scattered than the equivalent smooth channel data, although the systematic variation in $\%SF_w$ with ${}^{k}sb/k_{sw}$ is clearly apparent (Fig. 12). This scatter is due to the inherent difficulties associated in measuring the shear stress over a hydraulically rough boundary. However, as all the integrated mean shear stresses ($\bar{\tau}_i$) are within ± 10% of the mean shear stress obtained from the energy gradient ($\bar{\tau}_e$) gross errors in the data have been excluded.

The empirical model is suitable for use in laboratory flume studies where the bed has been roughened relative to the smooth side-walls. Under these boundary conditions equations (19, 20) provide a simple side-wall correction procedure to remove the influence of the side-walls on the flow-resistance.

The percentage shear force carried by the walls is sensitive to differential roughening of the bed relative to the side-walls. The logarithmic nature of equation (20) implies that this sensitivity is accentuated over the roughness range $1 < {}^{k}sb/k_{sw} < 10$ and consequently the accurate estimation of side-wall and bed $k_s$ values under these conditions is of prime importance.

### 4.3 Limit of a 'Wide' Channel

The concept of a 'wide' channel is extensively used by river engineers to simplify flow resistance problems in open channels. In a 'wide' channel flow resistance is primarily attributed to bed roughness and the flow is normally considered two-dimensional. At low aspect ratios the flow becomes fully three-dimensional owing to the increased strength in secondary circulation induced by the relatively close proximity of the side-walls.

When the flow is two-dimensional the mean bed shear stress ($\bar{\tau}_b$) equals the two-dimensional shear stress ($\rho gHS_f$). Knight (Ref 7) used the limits ${}^{\tau}b/\rho gHS_f = 0.90$ and ${}^{\tau}b/\rho gHS_f = 0.95$ to define 90% and 95% two-dimensional flow respectively. For uniformly roughened rectangular channels Knight quotes limiting aspect ratios of 12.7 (90%) and 22.6 (95%). Referring to Fig. 9 the approximate limiting aspect ratios for the 68° side-wall angle trapezoidal channel are 12.2 (90%) and 19.5 (95%). For the 45° angle trapezoidal channel the equivalent values are 11.5 (90%) and 19.0 (95%). These aspect ratios defining 90% two-dimensional flow are comparable with limiting aspect ratios used by field workers eg ${}^{B}/H = 15$, (Ref. 12); ${}^{B}/H = 11$, (Ref 13).

Roughening the bed relative to the side-walls (${}^{k}sb/k_{sw} > 1$) reduces the aspect ratio at which ${}^{\tau}b/\rho gHS_f = 0.90$, as the bed now offers a relatively greater resistance to flow. This is apparent in Fig. 12, where the limiting aspect ratio defining 90% two-dimensional for the 68° trapezoidal channel drops from 12.2 to 5 when $^{k}sb/k_{sw}$ changes from 1 to $10^2$.

### 4.4 Limits of Model Application

The experimental data covered the following range of conditions: Reynolds' number ($4R\bar{U}/\nu$) $3.8 \times 10^4 - 5.1 \times 10^5$; Froude number ($\bar{U}/\sqrt{gh}$) 0.22-1.30; aspect ratio ($^B/H$) 1.5-9.7; $^{k}sb/k_{sw}$ 1-91.1; where $\bar{U}$ = mean velocity and $\nu$ = kinematic viscosity.

The model should be applied with caution to larger-scale channels as Reynolds' numbers will probably exceed the upper limit of the experimental range, although no systematic variation in $\%SF_w$ with Reynolds' number is apparent in the experimental data. The model is suitable for stable sub-critical and super-critical flows. The majority of flows in natural channels are sub-critical, with Froude numbers typically $< 0.5$ (Ref 14). Equation (20) is strictly only applicable to composite roughened channels in which the bed roughness exceeds the side-wall roughness ie $^{k}sb/k_{sw} > 1$. Further experimental data is required to check whether equation (20) is still appropriate when $^{k}sb/k_{sw} < 1$.

It should be remembered that the experimental data were collected under uniform flow conditions and that similar conditions should be approached in the full scale channel. The model is suitable for use on straight trapezoidal and rectangular channels ($45° < \theta < 90°$) where skew-induced secondary current effects on the shear stress distribution are minimal. Appropriate field channels might include canalised waterways, and riffle sections located in straight gravel-bed rivers. In gravel-bed rivers, bed forms are often of low amplitude or absent and $k_s$ may be adequately expressed as a multiple of a particular gravel percentile (eg Refs 12, 15, 16). During high discharges in gravel-bed rivers the longitudinal variation in mean bed shear stress, induced by the relative flow depth differences in pool and riffle reaches (Ref 17), may become suppressed, at which point equation (19) might apply to the whole channel section length. The good agreement between the predicted and measured mean bed shear stress in the small gravel-bedded river example is encouraging, and further field trials in other gravel-bedded rivers are currently being undertaken.

## 5. CONCLUSIONS

Equations (16, 17, 19, 20) provide a quick and simple method of determining the mean bed and side-wall shear stress in straight symmetrical trapezoidal and rectangular channels ($45° < \theta < 90°$), in which the bed roughness is equal or greater than the side-wall roughness. Further data on differentially roughened channels of various cross-sectional shape are required to substantiate equations (19) and (20), and extend their use to channels having side-walls rougher than the bed. A series of field trials are also required to test the validity of equations (19) and (20) on larger scale channels.

## 6. ACKNOWLEDGEMENTS

This investigation was conducted whilst T.P. Flintham was in receipt of a SERC research studentship. Dr D.J. Cockrell, University of Leicester, supervised laboratory investigations and is thanked for useful discussion. Mrs M. Hurley (FBA) gave valuable statistical advice and a computational service.

7. REFERENCES


1. Tracy, H J: "Turbulent flow in a three-dimensional channel". Journal of the Hydraulic Division, American Society of Civil Engineers, 91, No HY6, Proc. Paper 4530, November 1965, pp 9-35.

2. Bathurst, J C: "Distribution of boundary shear stress in rivers". In: Adjustments of the Fluvial System (D D Rhodes and G P Williams, eds), Dubuque Iowa, Kendall/Hunt Publishing Co., 1979, pp 95-116.

3. Knight, D W and Patel, H K S: "Boundary shear in smooth rectangular ducts". Journal of Hydraulic Engineering, American Society of Civil Engineers, 110, pp 29-47.

4. Ghosh, S N: "Boundary shear distribution in channels with varying wall roughness". In: Proc. Institute Civil Engineers (London), 53, Part 2, Proc. Paper 7572, December 1971, pp 529-544.

5. Ghosh, S N and Roy, N: "Boundary shear distribution in open channel flow". Journal of the Hydraulics Division, American Society of Civil Engineers, 96, No HY4, Proc. Paper 7241. April 1970, pp 967-994.

6. Srinivasam, K: "Techniques in determining tractive force in open channels". Irrigation and Power, 31, Part 1, 1974, pp 43-54.

7. Knight, D W: "Boundary shear in smooth and rough channels". Journal of the Hydraulics Division, American Society of Civil Engineers, 107, No HY7, Proc. Paper 16364, July 1981, pp 839-851.

8. Noutsopoulos, G C and Hadjipanos, P A: Discussion of "Boundary shear in smooth and rough channels". by D W Knight, 108, No HY6, June 1982, pp 809-812.

9. Knight, D W, Demetrious, J D and Hamed, M E: "Boundary shear in smooth rectangular channels". Journal of Hydraulic Engineering, American Society of Civil Engineers, 110, No 4, Proc. Paper 18744, April 1984, pp 405-422.

10. Kamphius, J W: "Determination of sand roughness for fixed beds". Journal of Hydraulic Research, 12, No 2, 1974, pp 193-203.

11. Flintham, T P: "Flow resistance in moorland and forestry drains: narrow channels having composite roughness". Thesis presented to University of Leicester, England in 1988, in fulfillment of the requirements for the degree of Doctor of Philosophy.

12. Hey, R D: "Flow resistance in gravel-bed rivers". Journal of the Hydraulics Division, American Society of Civil Engineers, 105, No HY4, Proc. Paper 14500, April 1979, pp 365-379.

13. Carling, P A: "Threshold of coarse sediment transport in broad and narrow natural streams". Earth Surface Processes and Landforms, 8, 1983, pp 1-18.

14. Richards, K: "Rivers: Form and Process in Alluvial Channels". London, Methuen, 1982, pp 358

15. Charlton, F G, Brown, P M and Benson, R W: "The hydraulic geometry of some gravel rivers in Britain". Report IT 180, Hydraulic Research Station, Wallingford, England, July 1978, 48 pp.

16. Bray, D I: "Flow resistance in gravel-bed rivers". In: Gravel-Bed Rivers: Fluvial Processes, Engineering and Management (R D Hey, J C Bathurst and C R Thorne, eds), John Wiley & Sons, Chichester, 1982, pp 109-138.

17. Bathurst, J C: "Theoretical aspects of flow resistance". In : Gravel-Bed Rivers: Fluvial Processes, Engineering and Management (R D Hey, J C Bathurst and C R Thorne, eds), John Wiley & Sons, Chichester, 1982, pp 83-108.

Table 1

| Trapazoidal $\theta = 68°$ | $\frac{k_{sb}}{k_{sw}}$ | $\frac{P_b}{P_w}$ | $\%SF_w$ | Trapazoidal $\theta = 68°$ | $\frac{k_{sb}}{k_{sw}}$ | $\frac{P_b}{P_w}$ | $\%SF_w$ |
|---|---|---|---|---|---|---|---|
| Plywood walls | 1 | 2.952 | 23.6 | 6 mm gravel walls & | 4.2 | 2.481 | 19.1 |
| & bed | 1 | 1.723 | 35.2 | 20 mm gravel bed | 4.2 | 2.366 | 21.2 |
| " | 1 | 1.105 | 44.0 | " | 4.2 | 1.867 | 23.2 |
| " | 1 | 0.546 | 64.8 | " | 4.2 | 1.435 | 28.8 |
| " | 1 | 0.427 | 71.1 | " | 4.2 | 1.206 | 35.6 |
| 6 mm gravel | 1 | 2.154 | 29.0 | " | 4.2 | 0.660 | 49.0 |
| walls & bed | 1 | 1.801 | 33.7 | " | 4.2 | 0.524 | 58.6 |
| " | 1 | 1.757 | 34.9 | Plywood walls & | 43.5 | 3.342 | 15.2 |
| " | 1 | 1.550 | 38.1 | 6 mm gravel bed | 43.5 | 1.562 | 21.5 |
| " | 1 | 1.335 | 41.0 | " | 43.5 | 0.930 | 29.2 |
| " | 1 | 1.113 | 44.2 | " | 43.5 | 0.608 | 41.0 |
| " | 1 | 1.071 | 47.0 | " | 43.5 | 0.488 | 45.6 |
| " | 1 | 1.027 | 49.0 | Plywood walls & | 91.1 | 4.113 | 5.5 |
| " | 1 | 0.699 | 61.0 | 14 mm gravel bed | 91.1 | 2.604 | 10.0 |
| " | 1 | 0.455 | 73.4 | " | 91.1 | 1.944 | 14.5 |
| 6 mm gravel | 2.1 | 1.687 | 30.8 | " | 91.1 | 1.218 | 19.1 |
| walls & 14 mm | 2.1 | 1.030 | 43.5 | " | 91.1 | 1.203 | 21.3 |
| gravel bed | 2.1 | 0.657 | 55.9 | " | 91.1 | 0.825 | 28.4 |
| " | 2.1 | 0.440 | 67.7 | " | 91.1 | 0.804 | 29.0 |
| " | 2.1 | 0.340 | 75.0 | " | 91.1 | 0.546 | 37.3 |
| | | | | " | 91.1 | 0.371 | 51.2 |

NOMENCLATURE

$B$ = flow width

$C_{1-3}$ = empirical coefficients

$d_n$ = intermediate axis of gravel particles at which n percent of the median axes in a sample are smaller

$g$ = acceleration due to gravity

$H$ = flow depth

$k_s$ = Nikuradse's equivalent grain size

$k_{sb}$ = Nikuradse's equivalent grain size for the bed

$k_{sw}$ = " " " " " " side-wall

$P$ = total cross-sectional wetted perimeter

$P_b$ = bed sub-perimeter

$P_w$ = side-wall sub-perimeter

$R$ = hydraulic radius

$S_f$ = energy gradient

$SF_b$ = bed shear force

$SF_w$ = side-wall shear force

$\%SF_w$ = percentage shear force carried by the side-walls

tr = a threshold

$\bar{U}$ = mean velocity at cross-section

$u$ = point mean velocity

$u_*$ = local mean shear velocity

$y$ = distance normal to the boundary

$\kappa$ = von Karman's constant

$\nu$ = kinematic viscosity

$\rho$ = fluid density

$\delta$ = standard deviation of gravel particle size distribution

$\tau$ = local shear stress

$\bar{\tau}_e$ = mean boundary shear stress from energy gradient

$\bar{\tau}_i$ = mean boundary shear stress from integrating local values

$\bar{\tau}_b$ = mean bed shear stress

$\bar{\tau}_w$ = mean wall shear stress

$\theta$ = side-wall angle

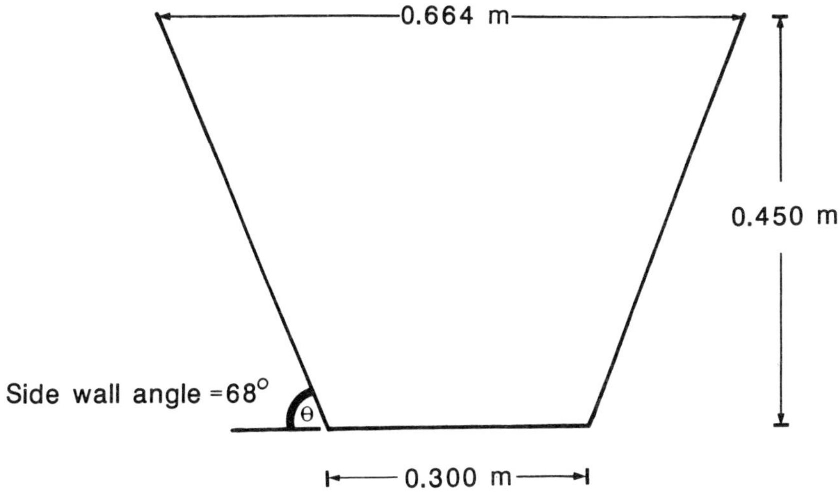


Fig. 1. Flume dimensions.

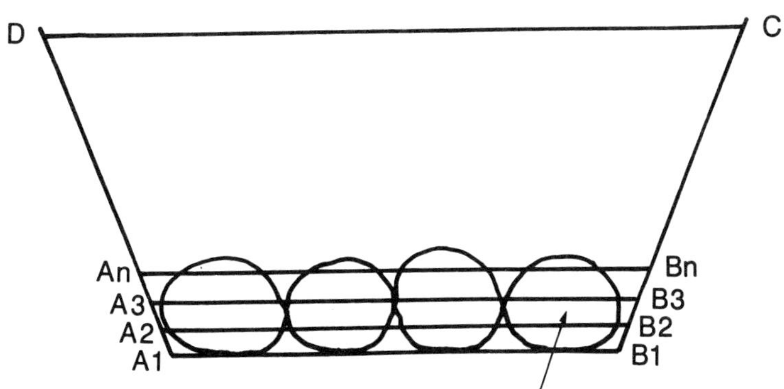


Fig. 2. Location of the virtual zero bed-datum for a hydraulically rough surface.

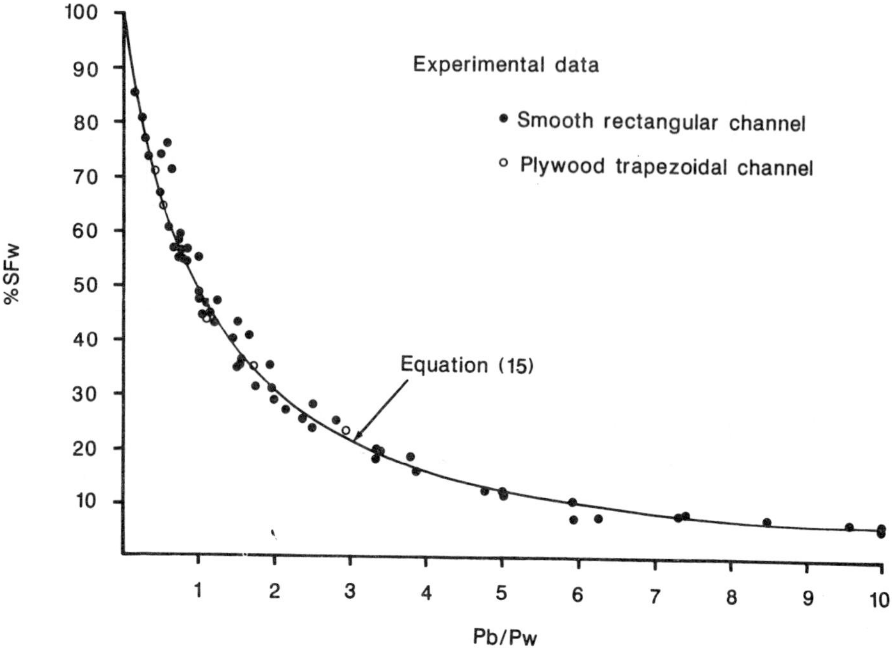


Fig. 3. Variation in percentage shear force carried by the walls ($\%SF_w$) with the ratio bed perimeter/wall perimeter ($P_b/P_w$).

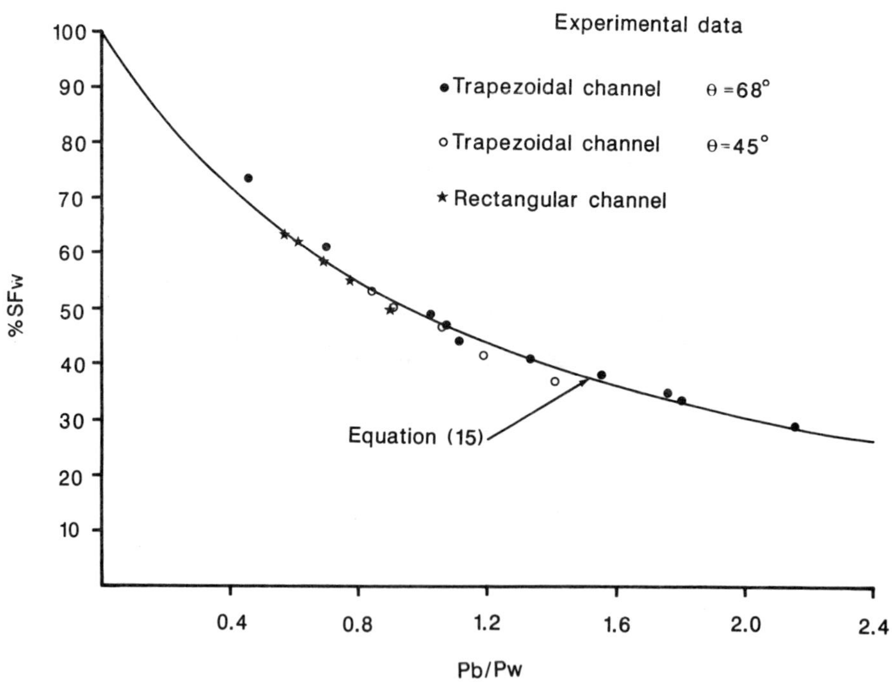


Fig. 4. Relationship between percentage shear force carried by the walls ($\%SF_w$) and the ratio bed perimeter/wall perimeter ($P_b/P_w$).

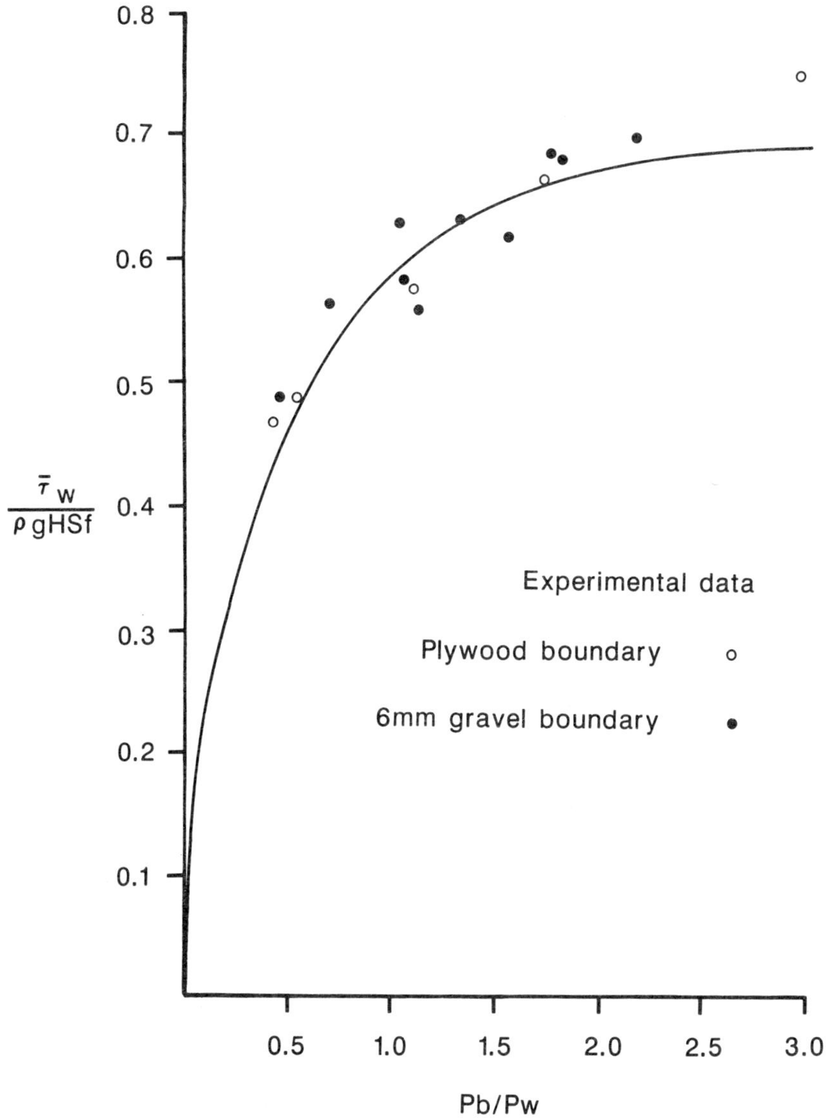


Fig. 5. Relationship between non-dimensionalised mean wall shear stress ($\bar{\tau}_w/\rho gHS_f$) and the ratio bed perimeter/wall perimeter ($P_b/P_w$).

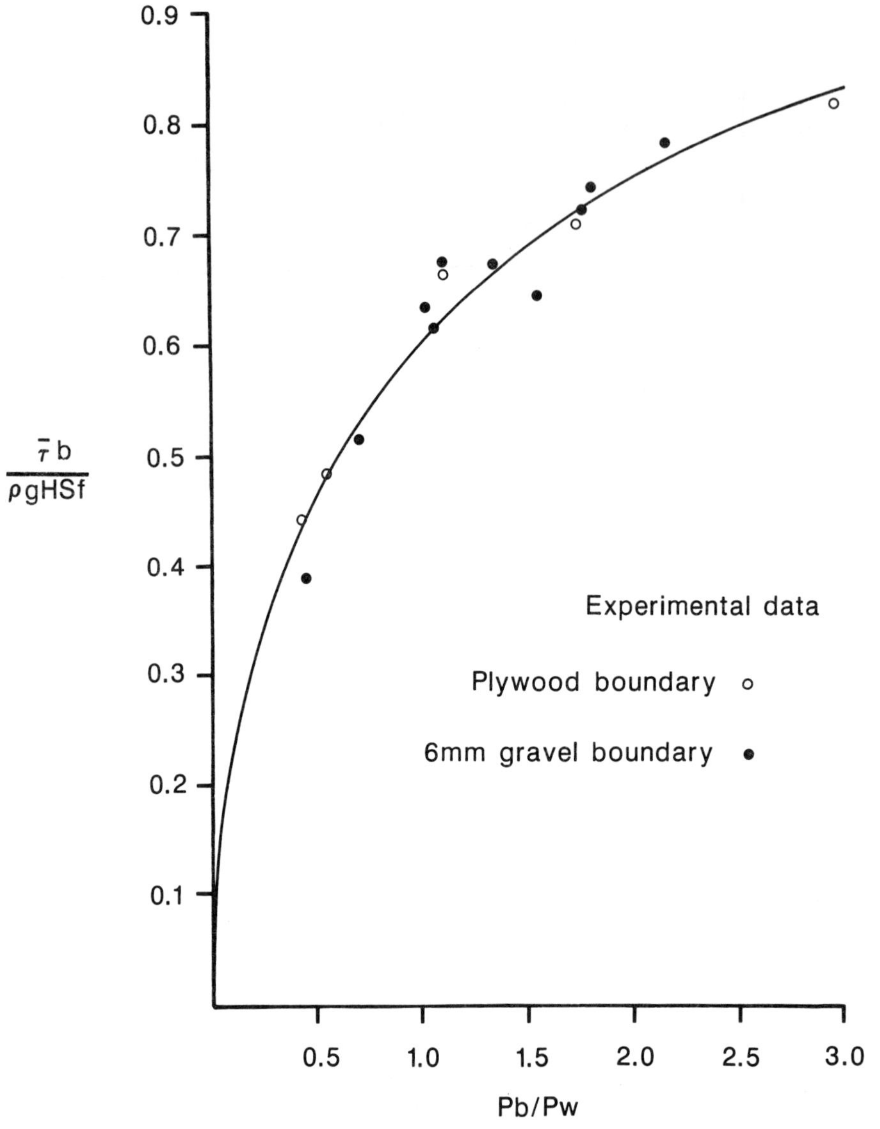


Fig. 6. Relationship between non-dimensionalised mean bed shear stress ($\bar{\tau}_b/\rho gHS_f$) and the ratio bed perimeter/wall perimeter ($P_b/P_w$).

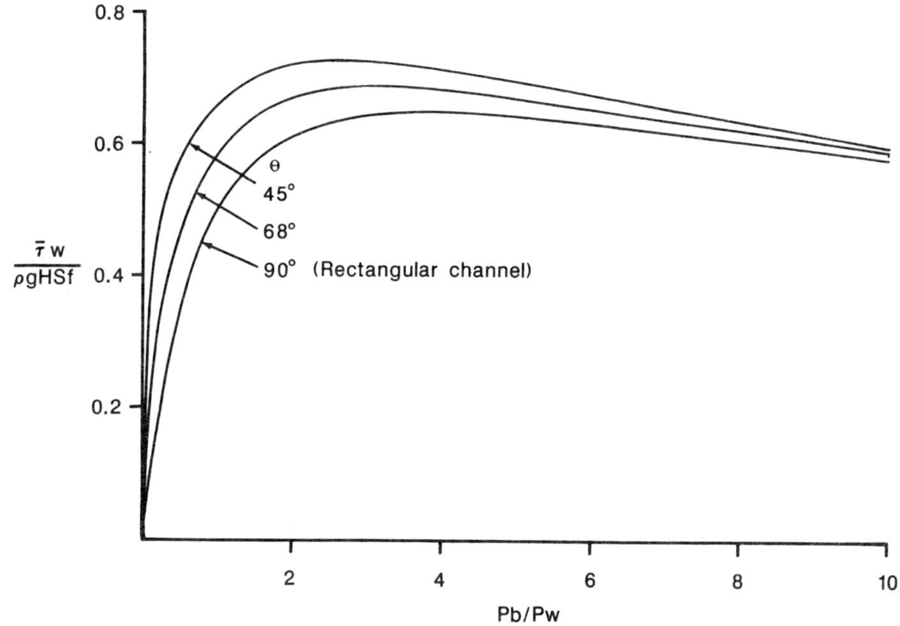


Fig. 7. Predicted variation in non-dimensionalised mean wall shear stress ($\bar{\tau}_w/\rho g H S_f$) with the ratio bed perimeter/ wall perimeter ($P_b/P_w$) for uniformly roughened trapezoidal and rectangular channels.

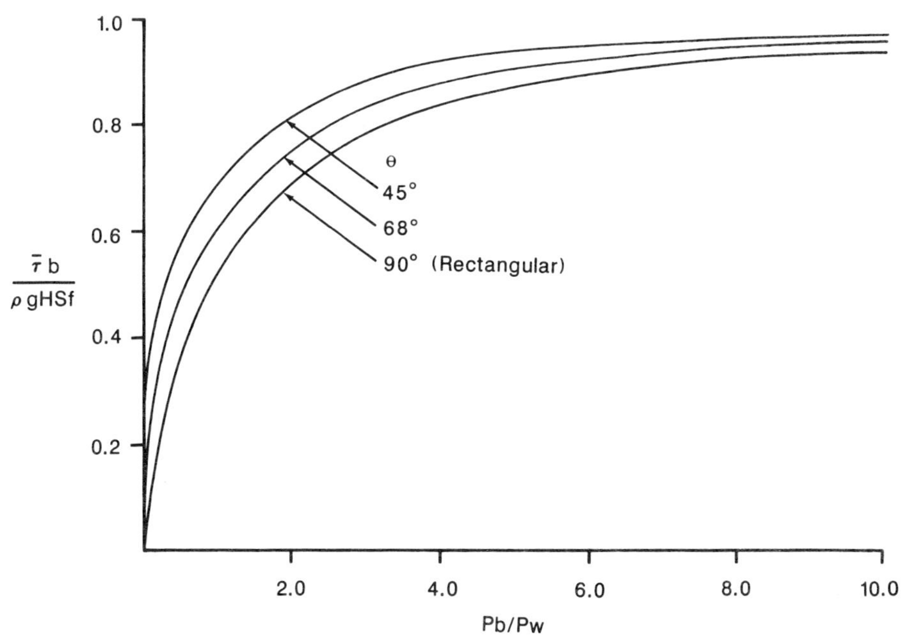


Fig. 8. Predicted variation in non-dimensionalised mean bed shear stress ($\bar{\tau}_b/\rho g H S_f$) with the ratio bed perimeter/ wall perimeter ($P_b/P_w$) for uniformly roughened trapezoidal and rectangular channels.

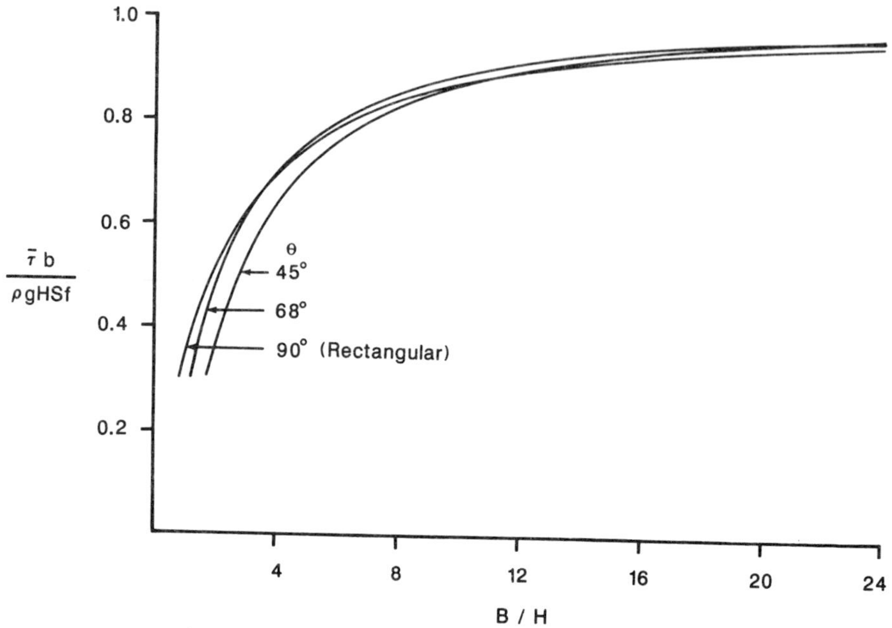


Fig. 9. Predicted variation in non-dimensionalised mean bed shear stress ($\bar{\tau}_b/\rho g H S_f$) with width/depth ratio for uniformly roughened trapezoidal and rectangular channels.

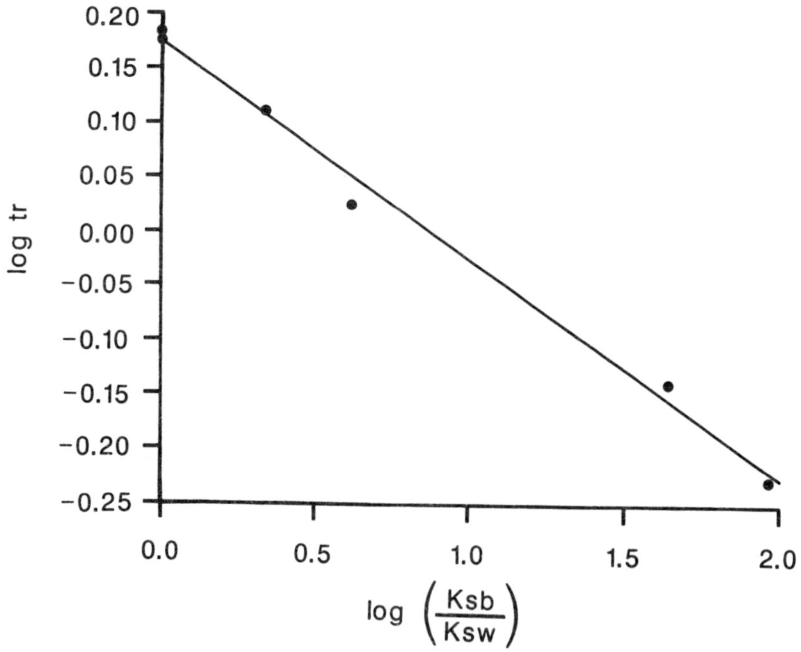


Fig. 10. Logarithmic relationship between the threshold (tr) and roughness ratio ($k_{sb}/k_{sw}$).

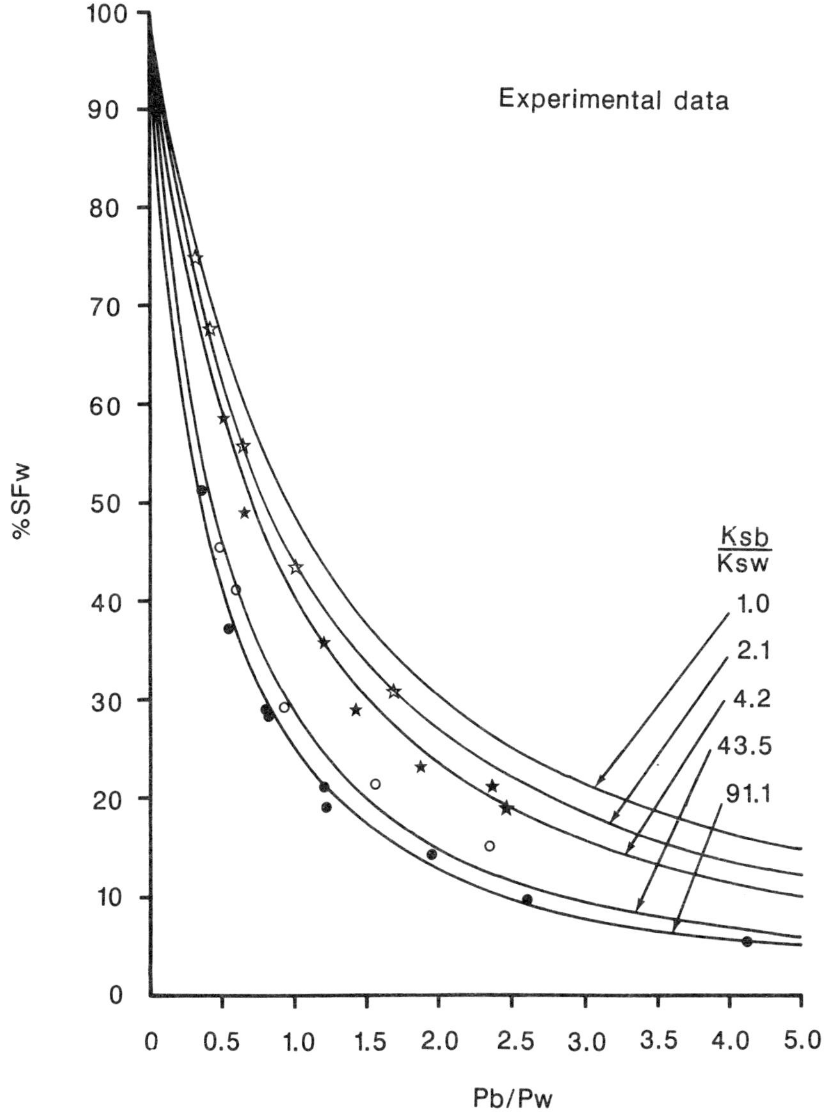


Fig. 11. Relationship between the percentage shear force carried by the walls ($\%SF_w$) and the ratio bed perimeter/wall perimeter for a differentially roughened trapezoidal channel ($\theta = 68°$).

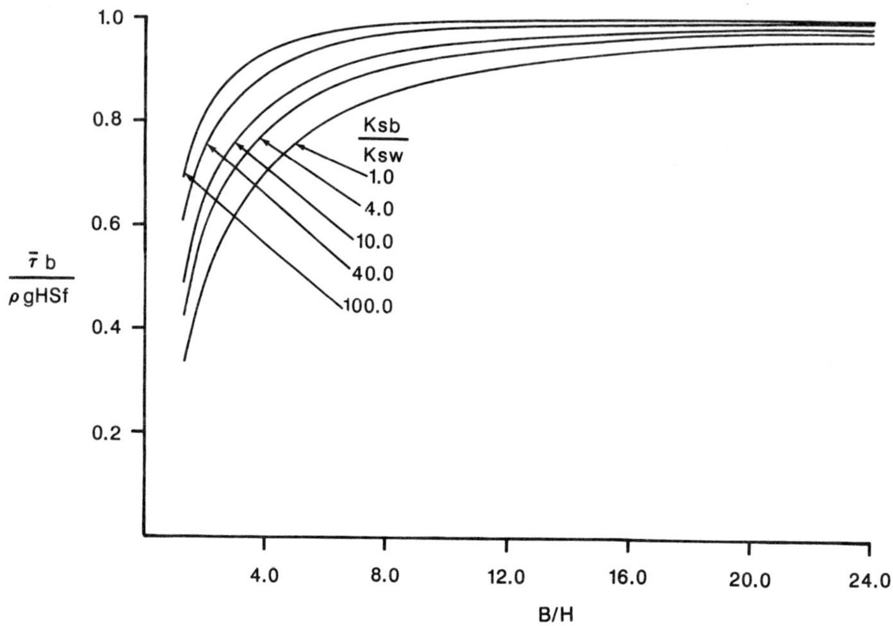


Fig. 12. Predicted variation in non-dimensionalised mean bed shear stress ($\bar{\tau}_b/\rho gHS_f$) with width/depth ratio (B/H) for differentially roughened trapezoidal channels having side-wall angles of 68°.

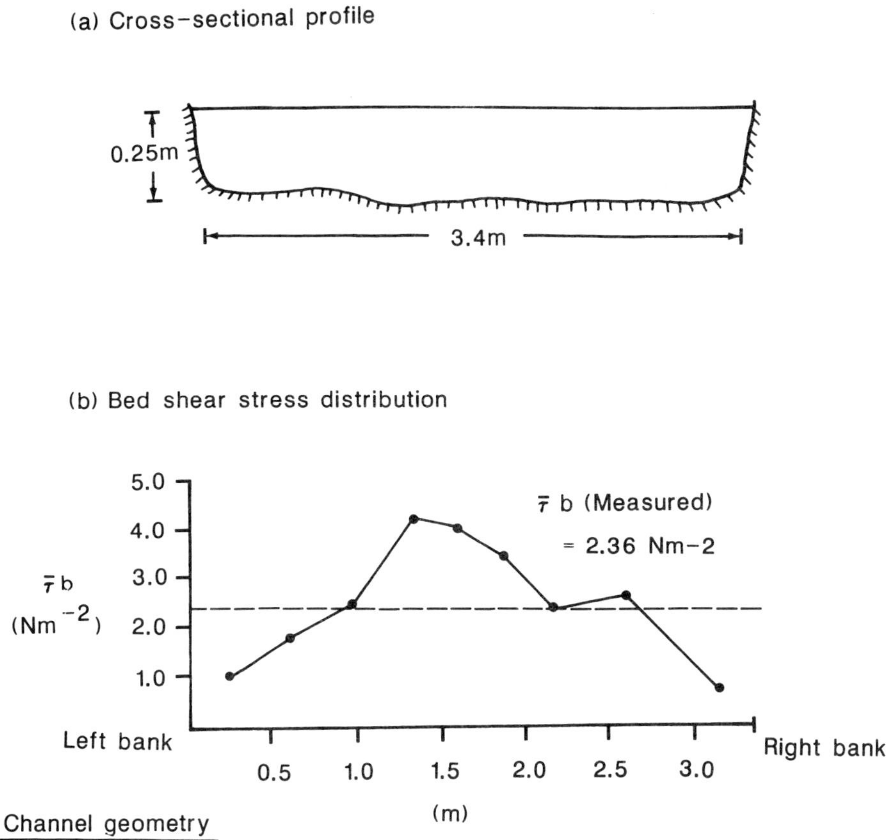


Channel geometry

bed = 3.4m

depth = 0.25m

Pb/Pw = 6.8

$\frac{Ksb}{Ksw} = 1$

Sf = $1 \times 10^{-3}$

%SFw = 9.1%

$\bar{\tau}$b (predicted)= 2.23 Nm$^{-2}$

$\bar{\tau}$w (predicted)=1.51 Nm$^{-2}$

$\bar{\tau}$e = 2.14 Nm$^{-2}$

$\bar{\tau}$b (measured)= 2.36 Nm$^{-2}$

$$\text{Error} = \frac{\bar{\tau}b(\text{pred}) - \bar{\tau}b(\text{meas})}{\bar{\tau}b(\text{meas})} = 6\%$$

Fig. 13. Prototype test data for a small gravel-bed river.

International Conference on
RIVER REGIME
18-20 May, 1988



# MULTIPLE BARS AND STREAM BRAIDING

Yuichiro FUJITA, Assoc.Prof., Dr.Eng. and Yoshio MURAMOTO, Prof., Dr.Eng.
Disaster Prevention Research Institute
Kyoto University, Gokanosho, Uji, Kyoto 611, JAPAN

SUMMARY : Laboratory experiments were carried out on the formation of sand bars leading to braided streams. The number of bars across the width of channel is referred to as the mode; higher modes implying division of the stream into many parallel sections. In the experiments, it was found that the higher modes turned into lower modes and then braided patterns appeared in cases with small depths. Individual bars in high modes were found to behave like alternate bars. Their formations is predictable by a tentative criterion which allows the co-existence of several modes.

## 1. introduction

Braided streams and multiple row bars, in other words, bars in high mode (mode means the number of symmetrical repeat of rows of alternate bars as shown in Fig 5) which are regarded as a general form of incipient stream meandering. Their complicated features have attracted many geomorphologist who have attempted to find out governing factors in their formation. Braided streams also present a difficult problem on river regulation to hydraulic engineers because they change rapidly and radically even during small floods and they sometimes cause severe local scour of river bed and bank. This difficulty is ascribable to little knowledge of their hydraulic characteristics.

Behaviour of multiple row bars and formation of braided streams were investigated by experiments conducted in three flumes with different widths. Results are discussed on changes of bed configuration, in order to clarify hydraulic characteristics of braided streams. Changes of the bar mode were deduced from bed features and surface measurement. Specific properties, such as geometrical characteristics and development time, of multiple bars were compared with those of alternate bars. Formative conditions of braided streams and multiple bars mode were derived.

## 2. Experiments on braided stream formation

### 2.1 Apparatus and procedures

Experiments were carried out in three flumes in the Ujigawa Hydraulic Laboratory, Disaster Prevention Research Institute, Kyoto University.

In all of runs except for re-conducted ones, a servo type water gauge and a sand surface detector of optical type were used to measure water stages just after the beginning and bed topographies just before the end of the water supply respectively. At the stoppage, continuous plan pictures of the bedforms were taken from camera perches attached to the carriage.

Runs of A-series were carried out in 50.2 cm wide, 20.5 m long, slope changeable steel flume. A carriage for grading and measurement could travel at

International Conference on River Regime, edited by W.R.White.

various speed, on which measuring devices were installed, The experimental reach was an 18 m long. Sand was laid 6 cm thick to make an alluvial bed for the experiment. Sand was not supplied at the inlet during experiments, because local deposit of the sand supplied is inevitable in such experiment with very small depths and it has a great influence on the occurrence of sand bars. Coordinates for the measurement were defined as: x-axis is taken in the downstream direction and x=0 m at the upstream end of the flume; y-axis from the left-hand side wall to the right and y=0 m at the left-hand side wall. The outputs of these measurements were recorded continuously in the downstream direction. Calibrations for the absolute difference between water stage and bed level measurements were found to be insufficient. Therefore, almost all runs of A-series were re-conducted till the first stoppage in order to obtain the accurate hydraulic quantities and to confirm the reproducibility of the experiments. In this case, measurement of the elevations were made by a point gauge at 1 m intervals. Because the measurement of the water stages took whole flow duration, bed levels were interpolated to coincide to the time of water level measurement by using measured values at the initial and final state.

Runs of B-series were carried out in a 3 m wide, 43 m long channel which was set up by partition walls of tin plates in a 7.5 m wide, 243 m long concrete flume (Ref 1). The stream bed was smoothed by a 3 m blade attached to a shovel device of an existing sand scraper which travels along rails laid on the concrete sidewalls being self-propelled. A little sand was manually fed in run B-3 to prevent the formation of armor coats. As for the measurement coordinates, x-axis is taken in the downstream direction and x=0 m at the upstream end of the experimental reach, and y-axis from the right-hand side wall to the left and y=4.0 m at the center of the stream channel. The measurement was done continuously in the y-direction by the same devices mounted on a traverser which was installed on a measurement carriage self-propelled.

Runs of C series were conducted in a 1.8 m wide, 15 m long, 0.2 m deep flume, which was made by widening the middle 15 m reach of the flume used in the A series. A mobile carriage was used for measurement and initial arrangement of the bed. A traverser with the same measuring devices and a stand for photographing were mounted on the carriage. Sand was laid between 0.7 m from the upstream end and 0.2 m from the downstream end in 6.5 cm thickness. In a 0.9 m reach from the upstream end, a steel plate with full width was buried and pulled up automatically in order to maintain the water stage constant. The buried wall was returned down on the channel bottom at stoppage after the bed surface measurement and sand was supplied as same amount as to fulfill this hole. When this sand supply was short, a suitable amount of sediment was injected manually. x-axis is taken in the downstream direction and x=0 m at the upstream end of the expanded reach and y-axis from the left-hand side wall to the right and y=100 cm at the center of the flume. Controls of the measurement carriage and data collection were automated as far as possible, by using a personal computer attached with a PIO board and a AD conversion board.

### 2.2 Experimental conditions and hydraulic quantities measured

Bed materials used for the A and C series experiment were selected so that grain motions could be observed easily and move actively even under small depths, hence almost same uniform sands of $d_m$=0.99 mm and of $d_{50}$=0.92 mm used in the A series and that of $d_m$=1.05 and $d_{50}$=0.92 mm in the C. In the B series, a rather widely graded sand that been laid previously was used, $d_m$ and $d_{50}$ of which were 0.88 and 0.61 mm respectively. Experimental conditions listed in Table 1 were determined to cover broadly the formative conditions of braided streams (Ref 2).

$$(h/d)/(B/d)^{2/3}<0.15 \qquad \text{for} \quad 1<U_*^2/U_{*c}^2<12 \qquad (1)$$

In the first A series, the above inequality was covered from the critical condition between alternate bar and braided streams in order to observe overall aspects of the formative process of braided streams. In B series, values of the parameter were selected almost same as the mean values in groups of runs of A series where the similar bed configuration appeared. In C series, the experimental reach was so short compared with the width as in the previous ones on the river bed configuration of meso scale. In cases of bar formation in higher modes than several order, this length was concluded to be sufficient because the wave length decreases in inverse proportion to the bar mode. Referring the results of A and B series, the experimental conditions were chosen to mainly cover a region that $(h/d)/(B/d)^{2/3}<0.07$.

In all experiments, profiles of cross-sectionally averaged water stage and bed elevation showed degradations in the most upstream reach, reflecting insufficient sand supply, but changes in the mean slopes were negligible because of generally steep slopes. Longitudinal changes of the water stage followed the bed profiles well, hence those changes in the mean depth also were small and hydraulic quantities averaged in the longitudinal direction could represent experimental conditions. These quantities also are listed in Table 1.

In A-series, though the hydraulic resistance in a few runs somewhat differed from the prediction because of very small depths, all of values of $F_r$ are greater than unity and are consistent with the upper regime observed. The smallest value of the depth was 0.28 cm in run A-10 and the mean aspect ratio was about 180. In the other five runs, the depths were smaller than 0.5 cm and then the aspect ratios exceeded 100. $U_*$ were 2.7 to 3.7 cm/sec, 15 to 60 percent greater than the critical value of $U_{*c}$=2.33 cm/sec.

In runs of B series, because the depths were relatively large, measuring errors became small, which produced reasonable hydraulic resistance that was almost the same as the sand roughness. In run B-1, the shear velocity was nearly equal to the critical one of $U_{*c}$=2.19 cm/sec, whereas it was factor of 1.6 in run B-2 and 1.2 in run B-3. Their aspect ratios of 180 and 120 agreed with those in corresponding runs in A series. The Froude numbers of 0.7 to 0.9 were consistent with the observed transition regime.

In the C series, though scatter of the mean depth tend to increase with slope, its extent was within 30 % for that of the depth of 0.6 to 1.5 cm. This small scatter was considered as an effect of detailed measurement in the transverse direction. $F_r$ changed within an extent of $F_r$=0.8 to 1.0 for all experiments and corresponded to transition regime observed in runs with small discharge. In contrast, in runs C-7 and C-9 with large discharge, variations of $F_r$ became large according to the bar development. Fluctuations of $U_*$ also were small and within 0.5 cm/sec. Absolute values of $U_*$ were $U_*$=2.5 to 5.5 cm/sec and were one to two times of the critical value that $U_{*c}$=2.4 cm/sec. Flow resistance also varied within a reasonable extent in spite of the large bar formation. Averaged quantities also had no trend with time.

No systematic changes in the mean hydraulic quantities were discerned in longitudinal direction or with time, but various bedforms developed and in some runs braided streams appeared with clearly emerged bars. Specific features in such variations are described in the following section.

## 3. Transformation of sand bars

### 3.1 Specific features of bars in high modes and braided streams

In runs of A and C series carried out under fairly steep slopes, it was common to every experiment that regular, symmetrical bars in very high mode appeared in the early stage and that they changed into lower mode as time elapsed. In late to final stage, large differences due to prescribed conditions appeared in the states of streams, bars and bar mode. In cases with relatively large depths and/or discharge, bars became similar to alternate ones as in runs A-6 and C-9. In cases with not so small depth, as runs A-5, A-7, C-4 C-5 and C-7, fairly symmetrical regular patterns of double row bar appeared to keep submerged. For example, in run C-5 (Photo 1 and Fig 1), bedforms in early stages were bars in mode 6 to 7 forming a scale-like pattern and they lasted throughout the run in the most upstream reach. Except for this reach, symmetrical, regular bars of mode 4 appeared in the whole reach and they remained from middle to lower stream reach while They turned into mode 3 from the most downstream reach and this alternation propagated upstream.

In cases with small enough depth, such as runs A-9, A-10, C-2, C-3, C-8 and C-10, occurrence of emerged bars divided by small streams and of main flow pattern in combination of 8-letter shaped meanders presented features of the braided streams when the bar mode was reduced to 2 to 3. In run C-8 (Photo 2 and Fig 1), the bed at T=5' was covered with scale-like bars of the highest mode of 8 to 9 which show a beautiful, geometrical pattern. Then their edges became round, showing the development. According to the decrease in the bar mode to 5-7 at T=10', and to 3-4 at T=25', bars became large and irregular, concentrating the flow into scour hollows. When some part was occupied by bars in mode 2 the bed turned into a braided stream, remaining small bars on the large emerged bars.

In contrast, runs of B series with relatively mild slope were characterized by longitudinal helicoid flow in the early stage and the bar mode higher than 4 were difficult to see. As shown in Fig 2, double or triple row bars were observed from the upstream reach in quite early stages and they propagated downstream orderly, in developing from the upstream. In run B-2, after a upstream reach 5 to 10 m long was covered with small water surface waves which followed disappearance of helicoid flows, a line like as bar edge was recognized there. This reach migrated downstream, leaving thin bar edges as if the turbulence caused by the surface wave would have produced bars. The first bar edge had a alternate bar like pattern from left to right but it was divided into two part near the center bifurcating a new edge to the left-hand side wall.

### 3.2 Bed variation during the development of multiple row bars and braided streams

In case of runs of C series, cross-sectional variation of the bed could be divided into two categories, that is, large-scale, simple changes and small-scale, complicated changes. Though these changes apparently seemed to occur simultaneously in many cross-sections because of bar's migration, the former changes were found often in runs with relatively large depths where the bar surfaces were kept smooth being covered with sufficient flow to move sand grains even at the final state, such as run C-4 shown in Fig 3, where very clear changes in cross-sectional shapes are seen at a middle reach of x=4.5-9.0 m. Scoured trenches decrease in their number from 4-5 to 2-4 at T=30' and from 3-4 to 2 at T=70'. Almost all bars present smooth surfaces and intervals of the scoured trenches are nearly same. Such features implies that bars formed in neat order reducing their mode and superiority of lower modes bars in cases with sufficient depths that development of lower modes bars were not affected by the previous bars of higher mode.

In contrast, the latter changes corresponded partly to notched cross-sectional shapes of large bars due to small bars in higher mode remained on them. For instance, in run C-8, though changes in cross-sectional shapes are complicated as a whole and minute in height, looking at the cross-sectional shapes at each stoppage carefully, it can be noticed that crests and troughs corresponding to regular bars in higher mode at T=7' decreased in their number and make their intervals uneven at T=27'. Along with such decrease in bar mode to 2-4, scour holes became large while bed varied a little on large bars, such as a left half of the cross-section at x=9 m.

Final bed profiles in run B-2 are depicted in Fig 4 along three longitudinal lines, channel center and proximate to both sidewalls. The profiles along both sidewalls resemble each other very much, presenting periodical variation of three wave lengths. Similarly, the center profile shows almost the same variation with a phase lag of one wave length. It suggest that double row bars developed regularly not only in plan but also vertical profile. Moreover, sand surface level on bars aggraded slightly above the initial level while scour holes became deep, which coincide with the bed variation characteristics in the alternate bar development.

## 4. Changes of bar mode and formative conditions of each mode bar and braided streams

### 4.1 Determination of bar mode

The bar mode is judged easily in cases with regular arrangement, such as in run B-2 and early stages in runs of C-series, but the judgment becomes difficult and inevitably arbitrary when bar geometry is uneven and bar edges are entangled. In order to determine the bar mode objectively as possible, the following two methods were considered separately.

In the first method, because the number and arrangement of scour ditches correspond to the bar mode, as shown in Fig 5, the mode of braided stream was determined from longitudinal changes in numbers of ditches scoured in a cross section. Namely, when this number is unity in any sections the mode is unity, if it repeats 1 and 2 alternatively the mode is 2, and it is always 2 then mode 3. In general, the mode is regarded to be 2m-1 when the number of scour holes is always m, and to be 2m if the number changes alternatively m, m+1, m, m+1,.... More over, mode 1.5 is defined as a state that mode 1 bars are accompanied with small bars at opposite side alternatively.

The second method is based upon a fact that the cross-sections shows a bell-shape in cases of the alternate bar bed when they are averaged longitudinally in a reach including even number of bars and that the bell-shape increases in its height corresponding to the development of alternate bar (Ref 3). In other word, the stream bed becomes Quonset-hut shape as a mean trend when alternate bars appear in it. Accordingly, it was expected that the averaged cross section shows double bell-shape in case of mode 2, triple in mode 3, ....., that is, number of peaks that appeared in an averaged cross-section presumably is as same as the order of the bar mode.

The bar mode referred in the foregoing section was judged by the first method. The results by the second are mentioned in the next paragraph.

Longitudinally averaged cross-sectional shapes in some runs of A series depicted in Fig 6. Cross-sections in formerly conducted and in re-conducting cases agree well with each other in the number and height of peaks and their shapes to affirms reproducibility of the experiments. A comparison of the modes determined here with those from the plane pictures presented a good agreement. Superior bars masking the existence of small bars make it difficult to judge the intermediate configuration from the averaged cross-section. In run A-7, however, where bars of mode 2 and 1.5 were taken into pictures at early and final stages respectively, the averaged cross-sections show that a regular shape with two peaks in the early turned into a mode 1 form with a depressed crest, which is regarded to be typical and specific form in cases of the definite intermediate configuration.

Fig 7 shows a clear double bell-shape in run B-2 as if two rows of alternate bar series were put in parallel. This agrees completely with double row bars found quite regular. The figure in run B-3 consists of large bell-shape with depressed crest and small incomplete one. The number of peaks counted simply is three and is consistent with the mode from the picture. Such a cross-sectional shape was probably caused by the main flow slightly meandering.

In run C series, longitudinally averaged cross-sections were calculated not only for the whole reach but also for three sub-reaches, upper (x=2-6 m), middle (x=6-10 m) and lower (x=10-14.5 m), because it was observed clearly in some runs that bedforms varied spatially, reducing the bar mode in the downstream direction. When large bars appeared in low mode with the wave lengths comparable to the sub-reach lengths, the judgment becomes incorrect or impossible. Fig 8 shows the averaged shapes in runs C-3 and C-8, as examples of braided stream formation, and those in runs C-4 and C-5 where no bar emerged.

The former shapes are notched and overall difference in height are small. It is possible to read the bar mode from these figures though they are entangled. The averaged shapes in individual sub-reaches resembled to those in the whole reach and neither definite difference nor systematic trend was recognized among these figures. It implies that bed variations had almost same properties in the longitudinal direction in cases of braided stream formation. In contrast, the averaged shapes generally have simple, smooth forms with large difference in height in the latter cases, and therefore, the bar mode was judged more easily and agreed better with those determined from pictures. The cross-sectional shapes averaged for the whole reach are quite different from those for the sub-reaches and it suggests a spatial change in mechanism of bed variations in cases without emerged bars.

Characteristics of changes in the bar mode with time elapsed are depicted in Fig 9 by using the mode judged from pictures in runs of C series. Bars in quite high mode more than 8 in runs of C-7 and C-8 with slope of 1/50 while those of mode about 5 in runs C-3 and C-5 with slope of 1/100 appeared in early stages and they both altered to bars in lower modes with time. Such changes were definitely rapid in runs with steep slopes rather than with mild slopes. In cases without braided streams, the bar mode is clearly higher in the upstream reach than in the downstream reach except for early stages, whereas any systematic change in the longitudinal direction was not discerned in cases with occurrence of braided streams. Such difference of changes in the bar mode is inferred that under a hydraulic conditions in cases with larger depths, the upstream boundary conditions of uniform levels of both the bed and water surface seemed to propagate to the upper reach so strongly that the bed tend to maintain the same conditions as initial state. In contrast, with small depths, flow states seemed to be easily controlled by local conditions of the bed surface, which is considered to result in the formation of a similar bedforms for the whole reach.

### 4.2 Geometrical characteristics of double row bar

The wave lengths and the wave height, normalized both by B and by $B_B$, in run B-2 were compared with previous data of alternate bars in Fig 10 (Ref 2). Data normalized by $B_B$ in the same range of the previous suggest that individual bars have the same geometrical feature as the alternate bar. Similarly, the development time $T_e$ of these bars was compared with an equation (2) for the prediction of that of alternate bars in Fig 11 (Ref 3).

$$T_e = 0.57\ l_B Z_B / q_B \qquad (2)$$

$T_e$ in run B-2 is considered to be in the same range of the previous data of the alternate bar. The developing time of these bars also could be estimated by the same equation for alternate bars. Furthermore, a bar height predicted for mode 2 by our equation (3) (Ref 4) is about 7 cm, replacing B by $B_B$, and it agrees well with the observed value.

$$Z_B/B = 0.0051(B/h)^{2/3}(h/d)^{-1/3}/(1-u_g') \qquad (3)$$

$(1-u_g')$ approach to $U_{*c}/U_*$ when the effect of the transverse slope is negligible. Similarly, supposing that bars in mode 1 (alternate bars) would develop fully, the heights are predicted to exceed 23 cm, more than 8 times of the mean depth. It is incredible that bars in mode 1 keep developing to such a height.

### 4.3 Formative conditions of bars in high mode and braided streams

Development of bars in various mode under the same hydraulic conditions, demonstrated in Fig 9, indicates an overlap of their formative conditions. them. In addition, even in cases of high modes, individual bars have very similar plan to alternate bars, and the cross-sectional shapes averaged longitudinally were confirmed to have peaks of the same number of modes of regularly formed bars at each phase in the formative process. It can be concluded that individual bars in the multiple row have the same mechanism of bed variation as develops the alternate bar. This conclusion is confirmed by an inspection for bar characteristics in run B-2. Accordingly, even for such bars in quite high modes, the formative conditions presumably become predictable, by paying attention to unit bars which compose the scale-like pattern and by an application of the alternate bar's formative condition.

Formative conditions of alternate bar found empirically is (Ref 2),

$$0.15<(h/d)/(B/d)^{2/3}<0.45 \quad \text{for} \quad 1<U_*^2/U_{*c}^2<12 \tag{4}$$

Assuming that the bar mode is m and B is replaced by the mean bar width of mode m, $B_B=B/m$, the formative condition bars in mode m becomes, as shown in Fig 12,

$$2.2m^{2/3} < (B/d)^{2/3}/(h/d) < 6.7m^{2/3} \tag{5}$$

Formative regions for each bar mode overlap fairly wide range, corresponding to the observations. Bar modes are also shown in Fig 13; experimental data are also plotted for all runs. The highest modes in the experiment do not exceed the upper line which expresses the higher boundary of an inequality (5). They are systematically below a broken line in Fig 13, expressed as:

$$2.8m^{2/3} = (B/d)^{2/3}/(h/d) \tag{6}$$

Since photographs were not necessarily taken immediately after the beginning in all the experiments, it cannot be guaranteed that the highest modes that appeared in the experiment are depicted in Fig 13. Even so, the bar mode that becomes visible first seems still to be lower than the higher boundary of the inequality (5).

As for the final modes, in runs where bars were submerged till the end of experiment, all data are plotted above the lower line. In contrast, in cases where the occurrence of emerged bars, data at the last phases plot below the lower line. They corresponded to the formation of braided streams, as shown by closed circles. Additionally, almost all open circles below the lower line are data on bars which emerged at later stages. Therefore, the lower line in Fig 13 can give both the lowest modes of bars covered with sufficient flow depths after a long flow duration and a measure for the highest modes of braided streams.

## 5. Conclusion

Detailed hydraulic experiments were carried out on the behaviour of multiple row bars and the formation of braided streams in three flumes with different widths of 0.5, 1.8, and 3.0 m. Characteristic features of stream bed variations were described. The higher modes of bars were observed to alter to lower ones and braided patterns appeared in cases with small depths. Bars in high modes proved to have the same hydraulic properties as alternate bars. Several results of studies on alternate bars could be applied to predict behaviour of multi-mode bars. It was concluded that the formations of bars in various mode and braided streams can be predicted by using diagrams such as Fig 14 which allows the co-existence of several modes. However, it is necessary to examine the formative process quantitatively from the viewpoint of sediment transport in order to predict the time required for the alternation of the bar mode and the appearance of braided streams.

## Nomenclature

- $B$ the full width of the channel
- $B_B$ individual widths of the bars in high modes
- $d$ the representative size of the bed material (the mean diameter)
- $d_m$ the mean diameter of the bed material

$d_{50}$ the median diameter of the bed material
$F_r$ the Froude number
$h$ the depth averaged for B
$l_B$ the wave length of bars
$m$ the bar mode
$q_B$ the bedload per unit width
$T$ the experimental time elapsed
$T_e$ the developing time of bars
$u_g'$ the non-dimensional grain velocity of bed load.
$U_*$ the mean shear stress for h
$U_{*c}$ the critical shear stress for d
$x$ the longitudinal coordinate
$y$ the transverse coordinate
$z$ the vertical coordinate
$Z_B$ the wave height of bars

## Acknowledgment

We are grateful to Mr. H. Akamatsu for his effort in the execution of all the experiments. The present research was partly supported by Joint US-Japan Research promoted by NSF and JSPS on "Development and Application of the Theory of River Meander" organized by Prof. S. Ikeda and Prof. G. Parker.

## References

1) Fujita, Y and Muramoto, Y: "Experimental study on stream channel processes in alluvial Rivers". Bull. Disast. Prev. Res. Inst., Kyoto Univ., Vol. 32, Part 1, 1982, pp 49-96.
2) Muramoto, Y and Fujita, Y: "A classification and formative conditions of river bed configuration of meso scale". In: Proc. 22nd Japanese Conf. on Hydraulics, Hydraulic committee, JSCE, 1978, pp 275-282 (in Japanese).
3) Fujita, Y and Muramoto, Y: "Study on the process of development of alternate bars". Bull. Disast. Prev. Res. Inst., Kyoto Univ., Vol. 32, Part 1, 1982, pp 49-96.
4) Fujita, Y and Muramoto, Y: "On the mechanisms of alternate bar development". In: Proc. 26th Japanese Conf. on Hydraulics, Hydraulic committee, JSCE, 1982, pp 25-30 (in Japanese).

## Tables

Prescribed conditions and hydraulic quantities measured in the experiments

## Illustrations and Photographs

### List of figures and Photographs

Fig 9. Changes in bar modes in runs C-4 and C-7 without emerged bars (right) and runs C-3 and C-8 with braided streams (left).

Fig 10. Geometrical characteristics of double row bars in run B-2.

Fig 11. Comparison of the development time between double row bar in run B-2 and alternate bars in previous studies.

Fig 12. A tentative criterion for bar formation in various modes.

Fig 13. Comparison of the criterion with the bar mode observed (open circles, submerged bars; closed circle, braided streams).

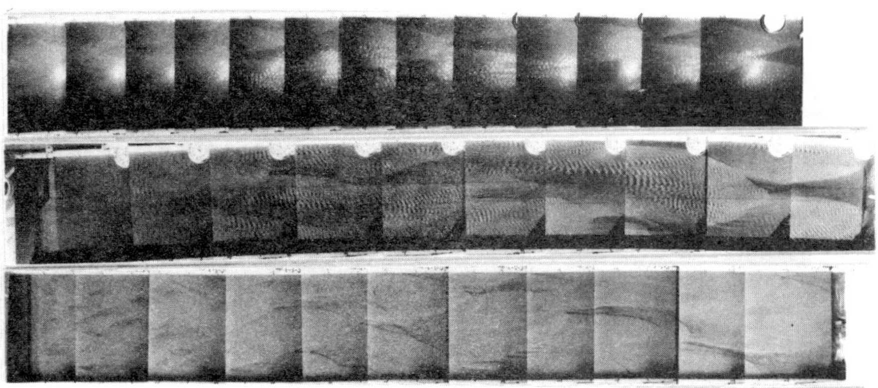


Photo 1. Bedform variations in run C-5 where bars did not emerge.

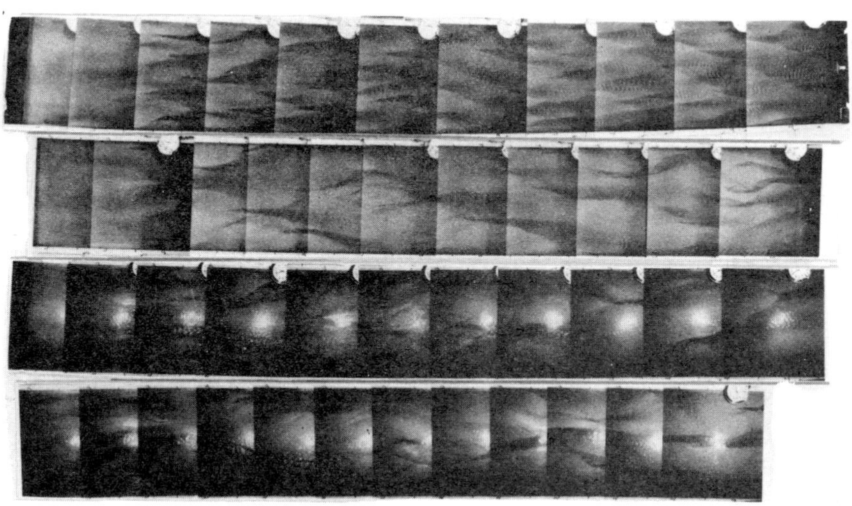


Photo 2. Bedform variations in run C-8 where a braided stream appeared.

## Tables

Prescribed conditions and hydraulic quantities measured in the experiments

| | Time | Flume width | Bed slope | Dis-charge | Mean depth | Mean velocity | Froude No. | Energy slope | Shear velocity | Velocity coeff | Manning's roughness |
|---|---|---|---|---|---|---|---|---|---|---|---|
| Run No. | T | $B$ (cm) | $I_0$ | $Q$ (l/sec) | $h$ (cm) | $V$ (cm/s) | $F_r$ | $I_e$ | $U_*$ (cm/s) | $U/U^*$ | $n$ |
| A- 1 | 30′00″ | 50.2 | 1/50 | 0.65 | — | — | — | — | — | — | — |
| A- 2 | 45′00″ | 50.2 | 1/50 | 0.45 | — | — | — | — | — | — | — |
| A- 3 | 15′00″ | 50.2 | 1/50 | 0.45 | 0.37 | 24.23 | 1.27 | 19.87 | 2.67 | 9.09 | 0.0138 |
| A- 4 | 18′00″ | 50.2 | 1/50 | 0.30 | 0.30 | 19.92 | 1.16 | 19.78 | 3.40 | 8.32 | 0.0145 |
| A- 5 | 16′53″ | 50.2 | 1/50 | 1.00 | 0.54 | 36.89 | 1.60 | 19.61 | 3.19 | 11.57 | 0.0115 |
| A- 6 | 15′00″ | 50.2 | 1/50 | 1.20 | 0.72 | 33.20 | 1.25 | 19.61 | 3.67 | 9.05 | 0.0154 |
| A- 7 | 20′00″ | 50.2 | 1/50 | 0.75 | 0.44 | 33.96 | 1.64 | 19.39 | 2.87 | 11.85 | 0.0109 |
| A- 8 | 13′30″ | 50.2 | 1/50 | 2.50 | 1.37 | 36.35 | 0.99 | 19.45 | 4.98 | 7.31 | 0.0212 |
| A- 9 | 18′15″ | 50.2 | 1/25 | 0.50 | 0.35 | 28.46 | 1.54 | 37.85 | 3.58 | 7.96 | 0.0156 |
| A-10 | 8′02″ | 50.2 | 1/25 | 0.30 | 0.28 | 21.34 | 1.29 | 37.83 | 3.21 | 6.66 | 0.0178 |
| A-11 | 10′05″ | 50.2 | 1/25 | 0.75 | 0.37 | 40.38 | 2.12 | 37.76 | 3.68 | 10.99 | 0.0114 |
| B- 1 | 4hr40′ | 301. | 1/200 | 9.90 | 1.24 | 26.52 | 0.83 | 5.01 | 2.43 | 10.90 | 0.0144 |
| B- 2 | 3hr40′ | 301. | 1/200 | 30.75 | 2.62 | 38.99 | 0.77 | 4.90 | 3.52 | 13.74 | 0.0157 |
| B- 3 | 11hr00′ | 301. | 1/200 | 14.95 | 1.41 | 35.23 | 0.95 | 4.79 | 3.56 | 11.09 | 0.0114 |
| C- 1-4 | 1hr41′ | 180 | 1/100 | 5.0 | 1.0 | 28.77 | 0.93 | 9.55 | 3.00 | 9.94 | 0.0157 |
| 5 | 2hr10′ | | | | 1.12 | 25.80 | 0.79 | 9.64 | 3.20 | 8.35 | 0.0188 |
| C- 2-1 | 31′ | 180 | 1/100 | 4.5 | 0.86 | 29.15 | 1.01 | 10.27 | 2.91 | 10.18 | 0.0145 |
| 2 | 1hr07′ | | | | 0.70 | 36.05 | 1.39 | 9.90 | 2.57 | 14.37 | 0.0100 |
| C- 3-1 | 23′ | 180 | 1/100 | 4.0 | 0.92 | 24.10 | 0.80 | 10.15 | 2.99 | 8.16 | 0.0182 |
| 2 | 1hr11′ | | | | 0.86 | 26.24 | 0.92 | 10.48 | 2.93 | 9.10 | 0.0162 |
| 3 | 2hr14′ | | | | 0.92 | 24.48 | 0.82 | 9.61 | 2.91 | 8.51 | 0.0175 |
| 4 | 3hr15′ | | | | 0.92 | 24.71 | 0.84 | 10.18 | 2.97 | 8.51 | 0.0178 |
| C- 4-1 | 30′ | 180 | 1/100 | 7.5 | 1.42 | 29.04 | 0.78 | 10.43 | 3.76 | 7.84 | 0.0204 |
| 2 | 1hr09′ | | | | 1.54 | 26.77 | 0.68 | 9.93 | 3.82 | 7.15 | 0.0228 |
| C- 5-2 | 1hr15′ | 180 | 1/100 | 10.0 | 1.68 | 33.19 | 0.82 | 10.04 | 3.99 | 8.47 | 0.0196 |
| 3 | 1hr53′ | | | | 1.46 | 38.01 | 1.00 | 9.14 | 3.57 | 10.72 | 0.0149 |
| C- 6-2 | 36′ | 180 | 1/50 | 4.5 | 0.98 | 25.88 | 0.85 | 19.37 | 4.28 | 6.12 | 0.0245 |
| 3 | 58′ | | | | 0.95 | 26.85 | 0.90 | 19.64 | 4.22 | 6.42 | 0.0234 |
| C- 7-1 | 11′ | 180 | 1/50 | 9.0 | 1.06 | 46.18 | 1.44 | 24.34 | 4.90 | 9.86 | 0.0162 |
| 3 | 1hr15′ | | | | 0.99 | 51.59 | 1.72 | 19.82 | 4.25 | 13.20 | 0.0125 |
| C- 8-3 | 46′ | 180 | 1/50 | 4.0 | 0.82 | 27.30 | 0.97 | 19.54 | 3.94 | 6.98 | 0.0208 |
| 4 | 1hr20′ | | | | 0.78 | 28.82 | 1.06 | 18.97 | 3.79 | 7.73 | 0.0188 |
| C- 9-1 | 14′ | 180 | 1/50 | 10.5 | 1.50 | 39.33 | 1.03 | 19.53 | 5.30 | 7.47 | 0.0213 |
| 2 | 45′ | | | | 1.28 | 46.64 | 1.34 | 18.26 | 4.70 | 10.16 | 0.0157 |
| 3 | 1hr12′ | | | | 1.33 | 44.52 | 1.25 | 17.92 | 4.76 | 9.53 | 0.0167 |
| C-10-2 | 18′ | 180 | 1/33 | 3.5 | 0.70 | 28.16 | 1.09 | 30.92 | 4.58 | 6.23 | 0.0229 |
| 3 | 40′ | | | | 0.84 | 23.68 | 0.84 | 30.30 | 4.94 | 4.87 | 0.0301 |

## Illustrations and Photographs

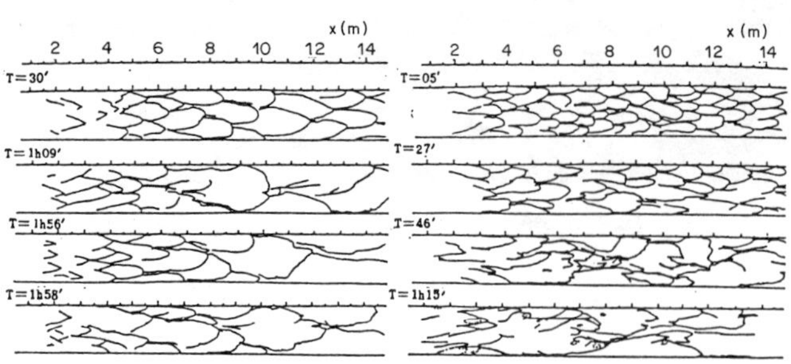


Fig 1.

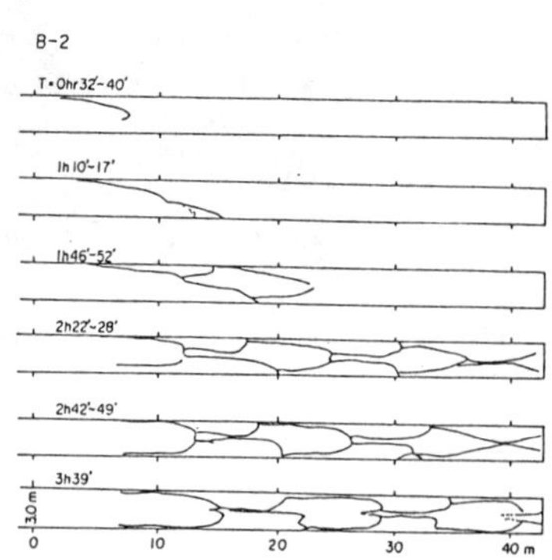


Fig 2.

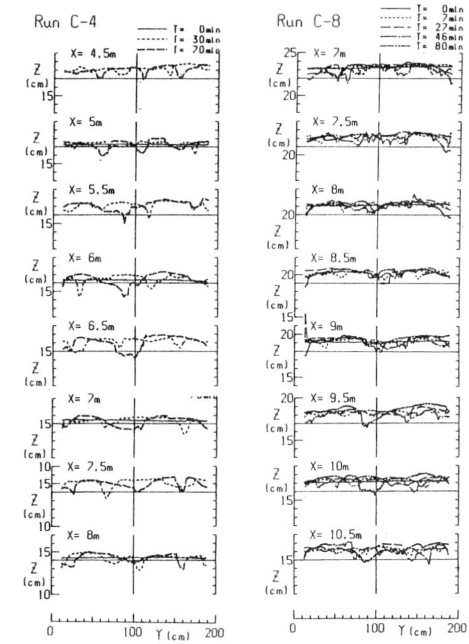


Fig 3.

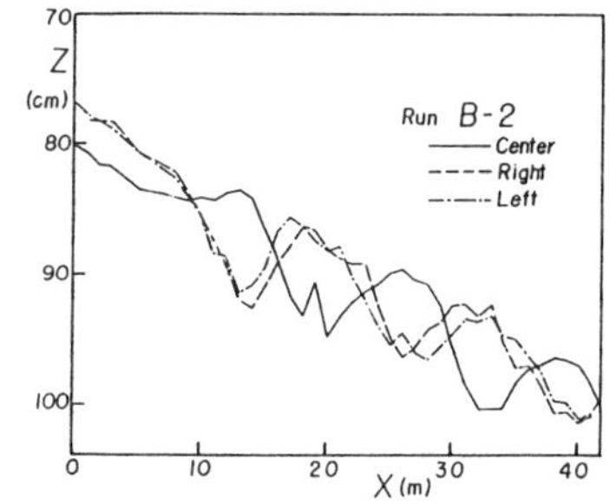


Fig 4.

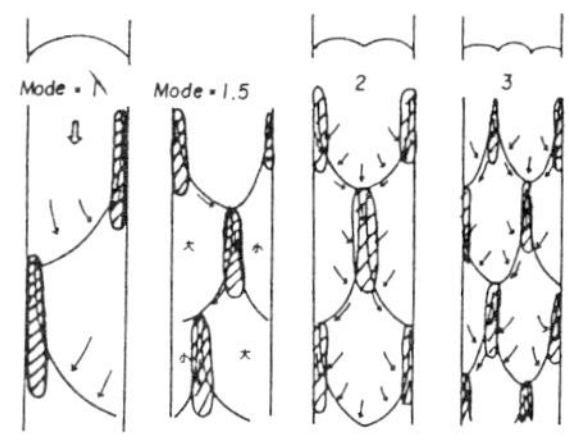


Fig 5.

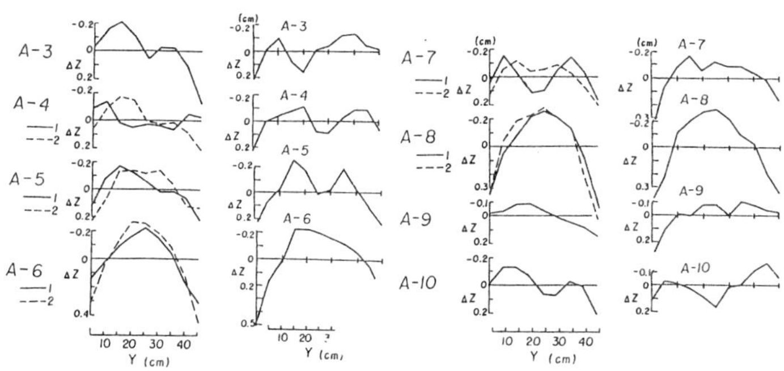


Fig 6.

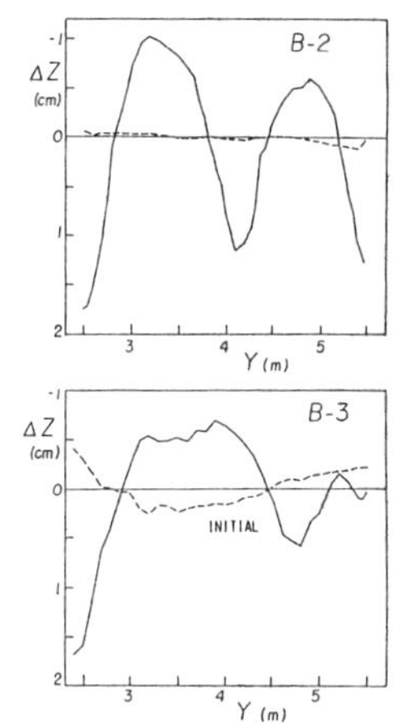


Fig 7.

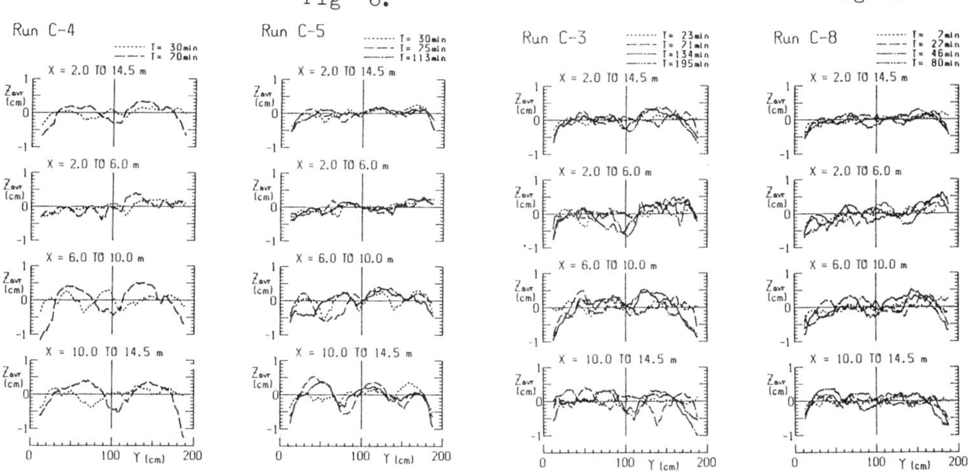


Fig 8.

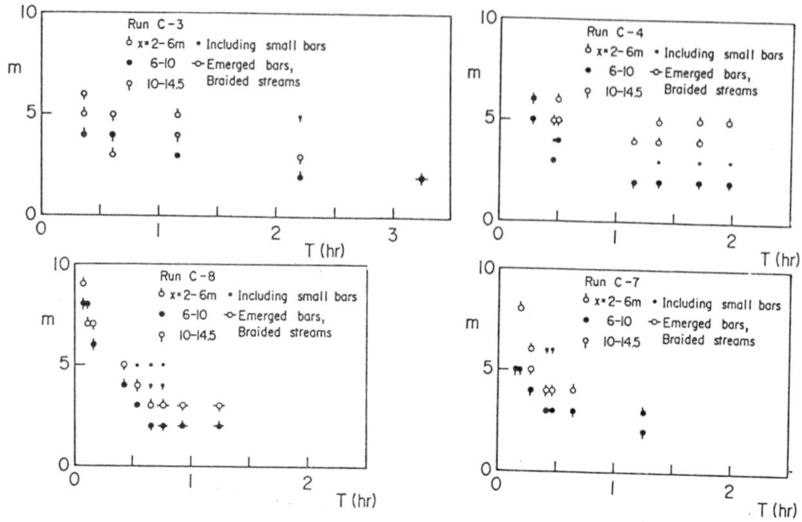


Fig 9.

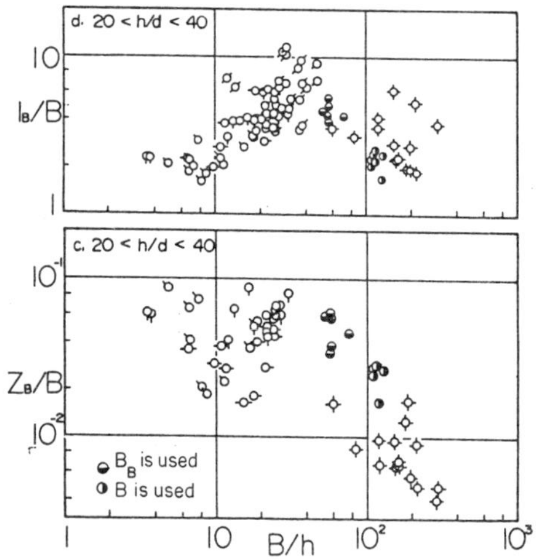


Fig 10.

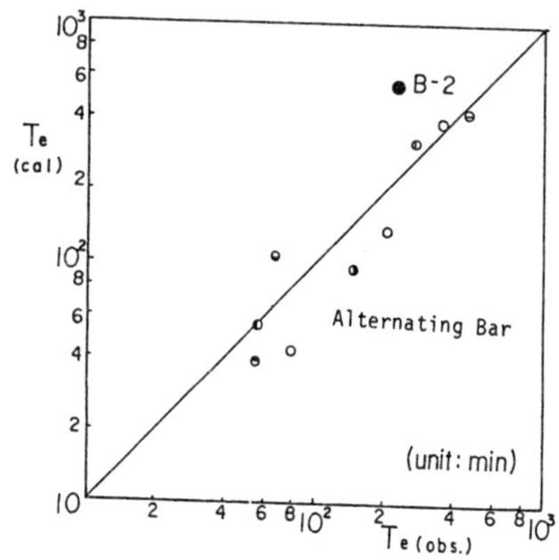


Fig 11.

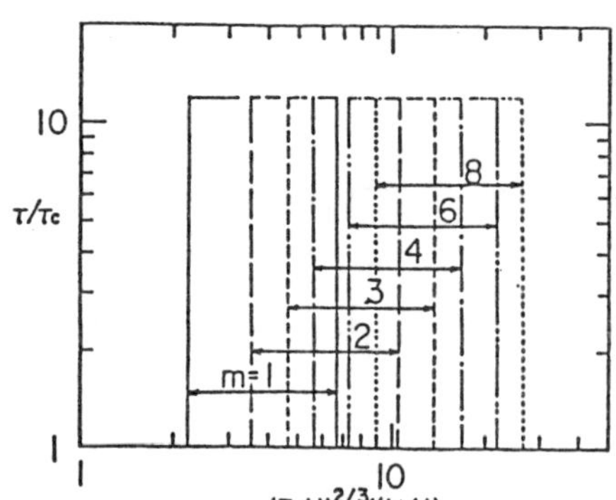


Fig 12.

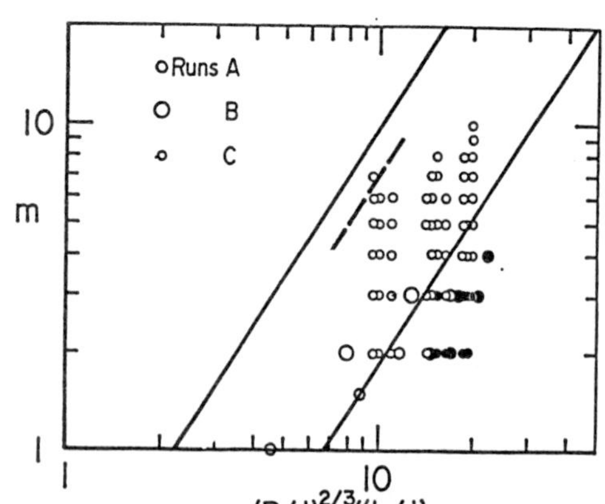


Fig 13.

International Conference on
RIVER REGIME
18-20 May, 1988

PAPER G4

# On Bed Form Configurations of a River Bend

Hong-Yuan Lee and Hsin-Ya Ho

Department of Civil Engineering, National Taiwan University

Taipei, Taiwan, R.O.C.

## ABSTRACT

Series of model tests are conducted to investigate the bed form characteristics, including height, wave length and orientation of a river bend. It is found that there are significant differences between the bed form characteristics of a river bend and those of a straight channel. Hence, traditional bed form prediction criteria can not be applied to a curved channel.

In a river bend, significant bed forms are observed, even before the flow reaches incipient conditions. The following phenomena are also observed: (1) The bed form height and wave length are strongly correlated to the water depth, local densimetric Froude number and strength of secondary current. (2) The dimensions of the bed forms at the outside bend are larger than those of the bed forms at the inside bend. Also, the dimensions of the bed forms grow in streamwise direction. (3) The orientions of the bed forms are strongly correlated to the strength of the secondary current and the transverse bed slope. In the early developing stage, tremendous amount of sediments are carried from outside toward inside bend by secondary current, hence, a significant deviation angle is formed between the bed form crestline and the line perpendicular to the mean flow direction. As time proceeds, inward movement of sediments reduces, the deviation angle decreases accordingly.

## INTRODUCTION

Bed form dimensions are very important in the design of crosswater structures crossing a stream. Hence, the accuracy of bed form prediction is very critical. Bed form configurations of a straight channel have been extensively studied and the criteria of bed form prediction have been developed with a substantial degree of success. However, the bed form patterns of a river bend have seldom been discussed. In the previous studies, for example, Yalin (1964) [8] analyzed the relationships between the water depth and the height and length of the bed forms, Fredsoe (1975) [4] and Yalin and Karahan (1979) [9] investigated the steepness of the dunes etc. All these studies are

International Conference on River Regime, edited by W.R.White.

concentrated on the straight channels. The effects of the river bend hydraulic characteristics upon the bed form configurations were not discussed in these studies. Bridge and Jarvis (1982) [2], analyzed the field data of River South Esk, Scotland, found that the length of the dunes in a river bend is shorter than that of a straight channel. Also, the dune height in a river bend was not found to be related to the water depth.

In this study, series of model tests were conducted to investigate the characteristics of the bed form configurations of a curved channel. It is observed that there are significant differences between the bed form configurations of a river bend and a straight channel. In a curved channel, significant bed forms can be seen, even before the flow reaches incipient condition. Hence, the criteria valid for a straight channel [4,8,9] can not be applied to a curved channel. In a river bend, the length and height of the bed forms are found to be strongly correlated to the water depth, strength of secondary current and the local densimetric Froude number. The bed form dimensions of the channel bed close to the inside bank are smaller than those of the channel bed close to the outside bank. The bed form dimensions grow in streamwise direction. Also, the orientations of the bed forms in a curved channel are correlated to the strength of the secondary current and the transverse bed slope. In the case of constant discharge, during the early developing stages, because of the effects of secondary current, tremendous amount of sediment are transported from outside bank toward inside bank. Hence, a significant deviation angle is formed between the bed forms crestline and the line perpendicular to the mean flow direction. As time proceeds it approaches to an equilibrium condition, the inward movement of sediment reduces, the deviation angle also reduces.

## EXPERIMENTS AND FIELD DATA

The flume used in this study is 31.5 M long, 1 M wide. The radius of curvature of the center of the flume is 3 M and the total angle in the meander part is 210° . The layout of the flume is shown in Figure 1. Twenty working sections, 3 at the approaching straight reach, 15 at the bend and 2 at the downstream straight reach, were chosen to take data. The median diameter of the bed material used is 0.25 MM and standard deviation is 1.44. The discharge chosen is 0.025 CMS.

At the beginning of the experiment, the river bed was leveled flat, and constant discharge was released from upstream. The average depth and velocity at the approaching straight reach were 0.127 M and 0.2 M/S respectively. The flow was stopped at 2, 4, 7, 13, 25 and 40 hours to measure the dimensions and orientations of the bed forms.

The longitudinal variations of the bed forms at hours 4, 13 and 40 are shown in Figures 2(A), 2(B) and 2(C) respectively. Since the approaching flow field did not reach the incipient condition, the dimensionless shear stress $\theta$ = 0.029, there were no significant changes at the approaching straight channel and the beginning part of the bend. This phenomena can be seen very clear in Figure 3. The bed forms were observed starting from section 6 and the dimensions of the bed forms increases in the streamwise direction. Sizes of the bed forms close to the outside bank are larger than those of inside bank (shown in Figure 4). Becauses of the effects of secondary current, sediments are transported from outside bank toward inside bank. There is a deviation angle between the bed form crestline and the line perpendicular to the mean flow direction. Between sections 6 and 16, the deviation angle increases in streamwise direction. After section 16, it starts decreasing. The variations of the deviation angles are shown in Figures 5 and 6. The temporal developments of the bed form heights, lengths and deviation angles are shown in Figures 7(A), 7(B) and 7(C) respectively. The bed form height and length increase with time while the deviation angle decreases as time proceeds. The variation tendency is close to an exponential function.

The dune heights and lengths of River South Esk collected by Bridge and Jarvis [2] and experiment data of this study are shown in Figures 8(A) and 8(B). The relative height and length of bed forms of field data and experiment data are of same order. Both data indicate that the bed form lengths in a river bend is shorter than that of a straight channel.

## THEORETICAL ANALYSIS

Secondary current is a very important phenomenon of a curved channel, and has significant impacts on the bed form geometry of the river bend (Bridge and Jarvis [2]). According to Rozovskii [7], the development of the secondary current in a rigid flat bed curved channel can be expressed in the following equations:

$$u_n = u_{n\infty}[1 - \exp(-bs)] \qquad s < s_o \tag{1}$$

$$u_n = u_{n\infty}[\exp(-b(s - s_o))] \qquad s > s_o \tag{2}$$

where $b = 2k\sqrt{g}/Cd$;
- $d$ : water depth;
- $u_n$ : radial velocity component;
- $u_{n\infty}$: fully developed radial velocity component;
- $k$ : Karman constant;
- $g$ : acceleration of gravity;
- $C$ : Chezy's constant;
- $s$ : longitudinal distance originated at the starting point of the bend;

$s_o$ : distance from the starting point to the end of the bend.

According to Ali [1] and Falcon [3], the strength of the secondary current close to the channel bed can be expressed as

$$\left(\frac{u_n}{u_s}\right)_\delta = 10\,\frac{d}{r} \tag{3}$$

where, $\delta$ : reference elevation close the channel bed;
$u_s$ : longitudinal velocity;
$r$ : radius of curvature.

Based on the radial force balance of the sediment particle, Odgaard [6], obtained the fully developed transverse bed slope $S_T$, reads

$$S_T = 4.8\sqrt{\theta}\; F_D\,\frac{d}{r} \tag{4}$$

where, $S_T$ : transverse bed slope;
$\theta$ : dimensionless shear stress ( $\theta = \tau_o / \gamma_s' D_s$ );
$F_D$ : densimetric Froude number ($F_D = u/\sqrt{\Delta g D_s}$ );
$\tau_o$ : bed shear stress;
$D_s$ : particle size;
$\Delta$ : $(\rho_s - \rho)/\rho$ ;
$\gamma_s'$ : $(\rho_s - \rho)$ .

According to equations (3) and (4), it is evident that the transverse bed slope is proportional to the multiplication of the strength of secondary current and densimetric Froude number. Their relationship is shown as follows:

$$S_T = 0.48\sqrt{\theta}\; T_n\; F_D \tag{5}$$

where $T_n = \left(\frac{u_n}{u_s}\right)_\delta$

The streamwise variations of $T_n$, $S_T$ and $T_n F_D$ are shown in Figure 9. Since $T_n$ is based on the result of fixed bed model, some deviations between the streamwise variations of $T_n F_D$ and transverse bed slopes are observed. Comparing Figures 2(A), 2(B) and 9, it can be seen that variations of the heights and lengths of the bed forms in a curved channel is similar to that of $T_n F_D$ .

## REGRESSION RESULTS

The expression forms of bed form height and length obtained by Yalin [8,9] are

$$\frac{H}{d} = \frac{1}{6} \left(1 - \frac{d_{cr}}{d}\right) \tag{6}$$

for R. > 25

$$\frac{L}{d} = 2\pi \tag{7}$$

for R. < 25

$$\frac{L}{d} = 5 , \qquad \text{when } D_s > 0.38 \text{ MM} \tag{8}$$

$$\frac{L}{D_s} = 1000 , \qquad \text{when } D_s < 0.18 \text{ MM} \tag{9}$$

where, $d_{cr}$ : water depth of incipient condition;
$H$ : bed form height;
$L$ : bed form length;
$D_s$ : sediment size;
R. : particle Reynolds number.

In this study, the flow field did not reach a incipient condition. Hence, equation (6) can not be applied here to predict the bed form height. Also, according to Bridge and Jarvis' field data and experiment data presented herein, we can see that the bed form length of the curved channel is shorter than that of the straight channel.

Figures 2(A), 2(B) and 2(C) show that, bed form sizes of the outside bank are larger than those of the inside bank. The transverse distributions of the water depth have similar tendency. Also, the tendency of the streamwise variation of the bed form sizes is similar to that of $T_n F_D$ water depth and $T_n F_D$.

$$\frac{H}{d_o} = F_1\left(\frac{d}{d_o}, T_n F_D\right) \tag{10}$$

$$\frac{L}{d_o} = F_2\left(\frac{d}{d_o}, T_n F_D\right) \tag{11}$$

where $d_o$: water depth of the approaching straight channel.

Using the data (t = 40 hr.) presented herein, the following regression formulas are obtained:

$$\frac{H}{d_o} = 0.13 \left(\frac{d}{d_o}\right)^{0.91} \left( T_n \ F_D \right)^{0.06} \qquad (12)$$

for the whole curved reach from section 6 to 20.

$$\frac{H}{d_o} = 0.13 \left(\frac{d}{d_o}\right)^{0.48} \left( T_n \ F_D \right)^{0.29} \qquad (13)$$

for section 11 to 18, i.e. for fully developed reach only.

$$\frac{L}{d_o} = 1.9 \left(\frac{d}{d_o}\right)^{1.2} \left( T_n \ F_D \right)^{0.17} \qquad (14)$$

for the whole curved reach.

$$\frac{L}{d_o} = 1.7 \left(\frac{d}{d_o}\right)^{0.41} \left( T_n \ F_D \right)^{0.80} \qquad (15)$$

for the fully developed reach only.

The comparisons of the calculated bed form dimensions using equations (12), (13), (14), (15) and the data measured herein are shown in Figures 10(A), 10(B), 11(A) and 11(B). The variations of the calculated values are smaller than the variations of the experiment data. Comparing Figures 8(A), 8(B), 10(A), 10(B), 11(A) and 11(B), it can be seen that the correlation between bed form sizes and water depth alone is not good. Adding the term $T_n \ F_D$, can improve the correlation. Hence, the strength of the secondary current and densimetric Froude number are the dominant factors affecting the bed form sizes in a curved channel.

The direction of the particle movement is related to the direction of the bed shear stress and transverse bed slope. (Kikkawa et. al, [5]). The experiment data also show that the deviation angle of the bed form crestline increases in a streamwise direction, and similar to the tendency of the change of the strength of secondary current. Moreover, during the initial developing stage of the transverse bed slope, the deviation angle is larger. As time proceeds, the path of sediment particle moves toward center of the channel, and the deviation angle decreases. From these phenomena, it is evident that the strength of secondary current and the transverse bed slope are the dominant factors affecting the deviation angles of the bed form crestline. Their relationships can be expresses in the following form:

$$\varphi = F_3(S_T \ , \ T_n \ ) \qquad (16)$$

where $\varphi$ : deviation angle of the bed form crestline.

According to regression analysis, the following formula is obtained:

$$\varphi = 30.9 \left(S_T \frac{B}{d_o}\right)^{-0.27} (T_n)^{0.65} \tag{17}$$

for the whole curved reach.

$$\varphi = 9.9 \left(S_T \frac{B}{d_o}\right)^{-0.58} (T_n)^{-0.32} \tag{18}$$

for the fully developed reach only.

where B : channel width.

The comparisons between the calculated value using equations (17), (18) and measured data are shown in Figures 12(A) and 12(B). The results are quite satisfactory.

## CONCLUSION

1. Because of the effect of secondary current, the bed form characteristics of a curved channel are quite different from those of a straight channel. The differences are listed as follows :

(a) Significant bed forms in a curved channel are observed even when $\theta < \theta_{cr}$ in the approaching straight channel. The traditional bed form prediction formula for a straight channel may not be appliable.

(b) Because of the effect of secondary current, there are some deviations between the direction of sediment movement and mean flow.

(c) The bed form length is smaller than that in a straight channel, and the bed form height is correlated to the strength of the secondary current, densimetric Froude number and water depth.

2. The bed form sizes close to outside bank are larger than those close to the inside bank. The size increases in a streamwise direction and the variation behavior is close to an exponential function.

3. The deviation angle is correlated to the strength of the secondary current and transverse bed slope. As time proceeds the deviation angle decreases as an exponential function.

ACKNOWLEDGMENTS

Authors like to thank Professor C. L. Yen at the National Taiwan University for his support and helpful comments, and the staff of the Hydraulic Laboratory, National Taiwan University for their support in conducting experiments.

References

1. Ali, A. M., "Longitudinal and Transverse Variations of Secondary and Depth-Averaged Flows in Meandering Channels," Thesis presented to Univ. of Tokyo in partial fulfillment of the requirements for the degree of Doctor of Philosophy, July, 1985.
2. Bridge, I. S. and Jarvis, J., "The Dynamics of a River Bend : A Study in Flow and Sedimentary Processes," Sedimentology, 1982, pp. 499-541.
3. Falcon Ascanio, M., "Analysis of Flow in Alluvial Channel Bends," Thesis presented to Univ. of Iowa in partial fulfilment of the requirements for the degree of Doctor of Philosophy, 1979.
4. Fredsoe, J., "The Friction Factor and Height-Length Relations in Flow over a Dune-Covered Bed," Progress Report 37, Institute of Hydrodynamics, Technical University of Denmark, Copenhagen, Denmark, Dec., 1975.
5. Kikkawa, H., Ikeda, S. and Kitagawa, A., "Flow and Bed Topography in Curved Open Channels," ASCE, Vol. 102, No. HY9, September, 1976, pp. 1327-1342.
6. Odgaard, A. J., "Flow and Bed Topography in Alluvial Channel Bend," Journal of Hydraulic Eng., Vol. 110, No. 4, April, 1984, pp. 521-536.
7. Rozovskii, I. L., Flow of Water in Bends of Open Channels, Academy of Sciences of Ukranian SSR., Kiev, 1957, Translated by Prushansky, Y., The Israel Program for Scientific Translations, 1961.
8. Yalin, M. S., "Geometrical Properties of Sand Waves," Journal of the Hydraulics Division, ASCE, Vol. 90, No. HY5, Proc. Paper 4055, Sept., 1964, pp. 105-119.
9. Yalin, M. S. and Karahan, E., "Steepness of Sedimentary Dunes," Journal of the Hydraulics Division, Proceeding of the ASCE, Vol. 105, No. HY4, April, 1979, pp. 381-392.

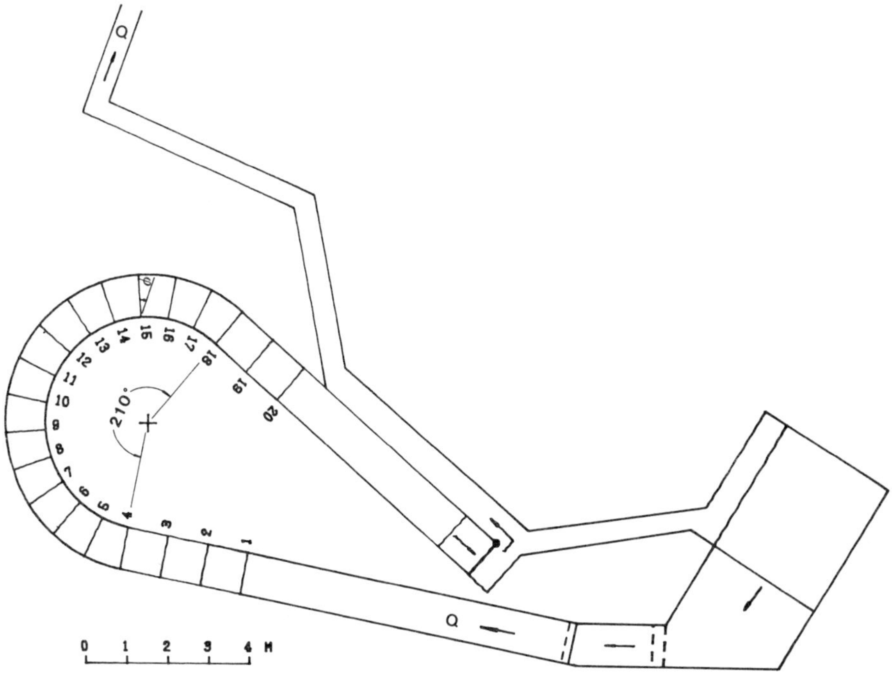


Fig. 1. Layout of the Flume

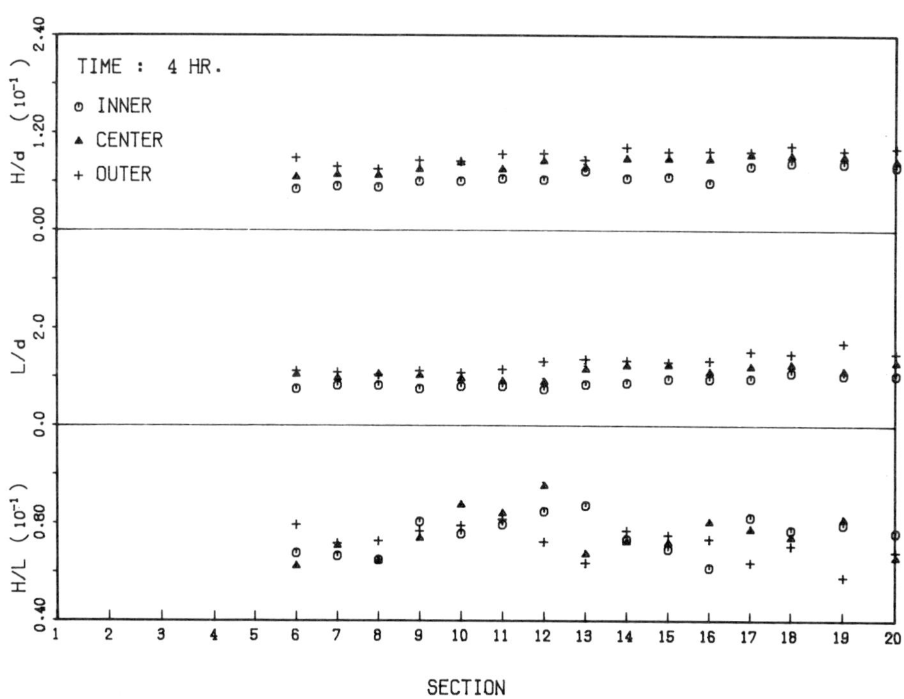


Fig. 2(a). Spatial Variations of Dune Size, t = 4 hr.

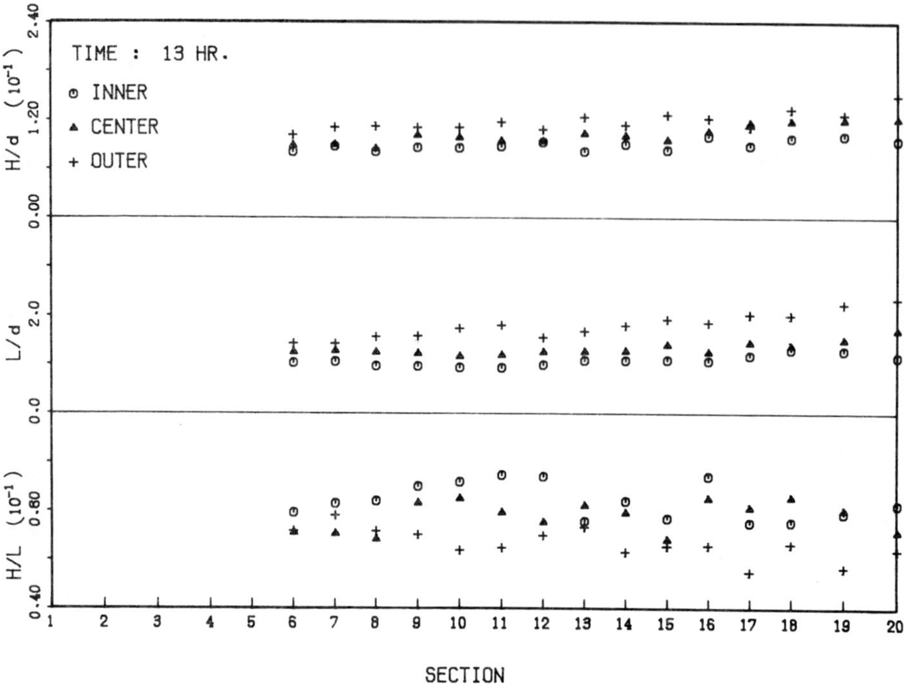


Fig. 2(b). Spatial Variations of Dune Size, t = 13 hr.

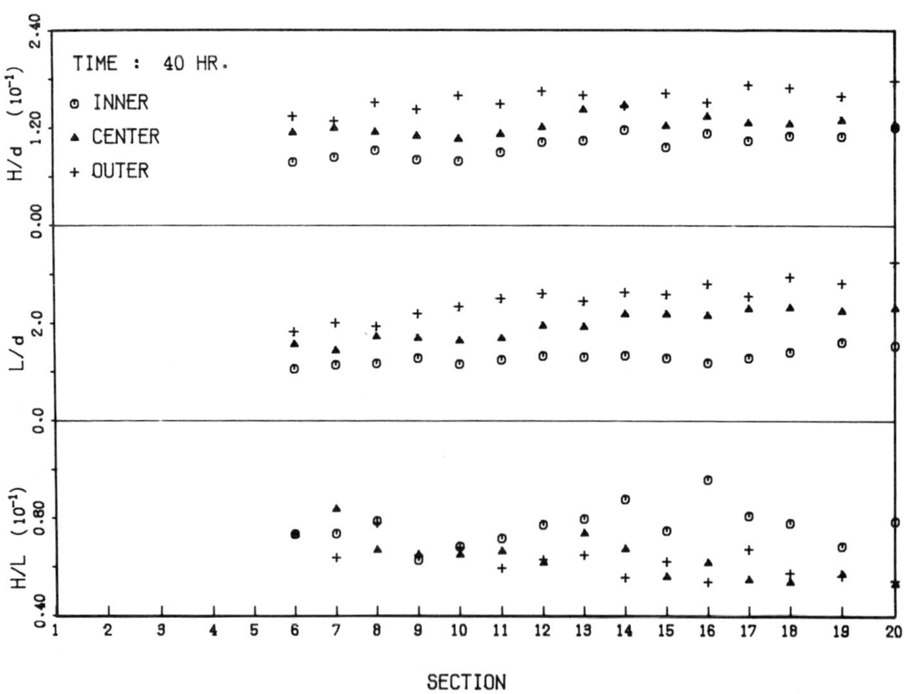


Fig. 2(c) Spatial Variations of Dune Size, t = 40 hr.

Fig. 3. Comparsion of the Bed form Characteristics of the Approaching Straight Reach and the Curved Reach ( $\theta < \theta_{cr}$ )

Fig. 4. Transverse Variations of Dune Sizes (Fully Developed, Section 15)

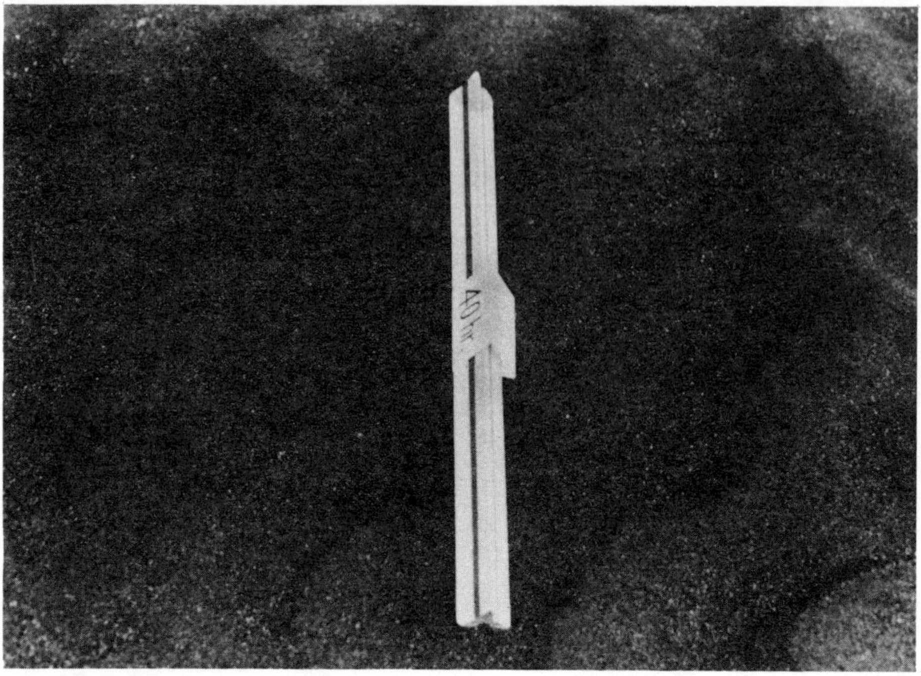


Fig. 5. Orientation of the Dunes (Fully Developed, Section 15)

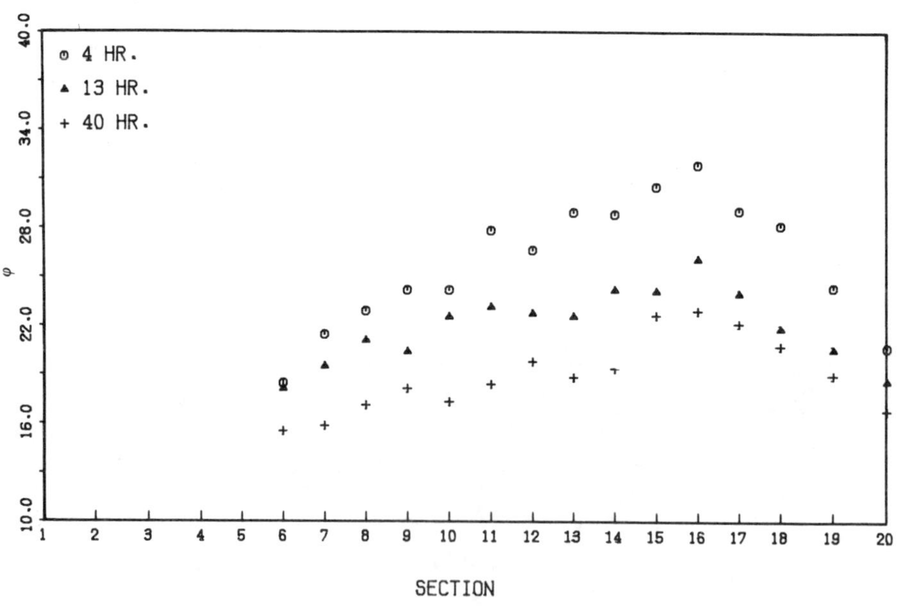


Fig. 6. Variations of the Deviation Angle

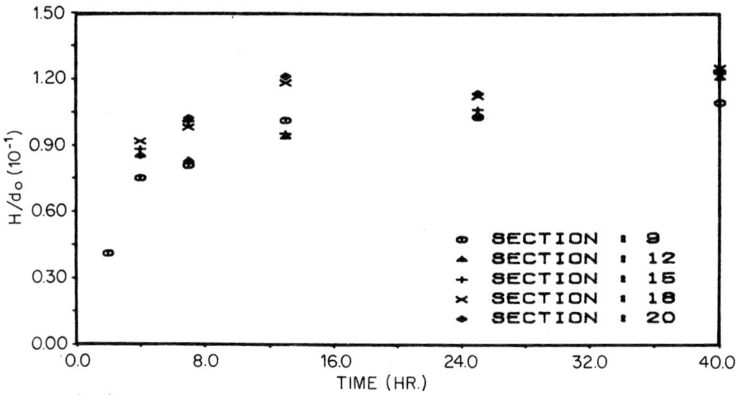


Fig. 7(a). Temporal Developments of the Dune Height

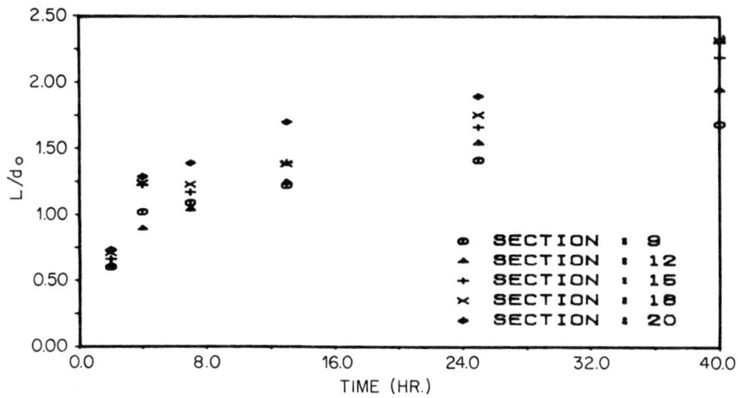


Fig. 7(b). Temporal Developments of the Dune Length

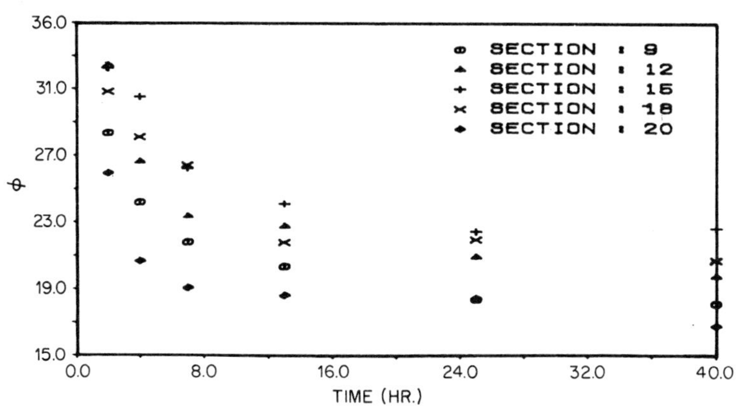


Fig. 7(c). Temporal Developments of the Deviation Angle

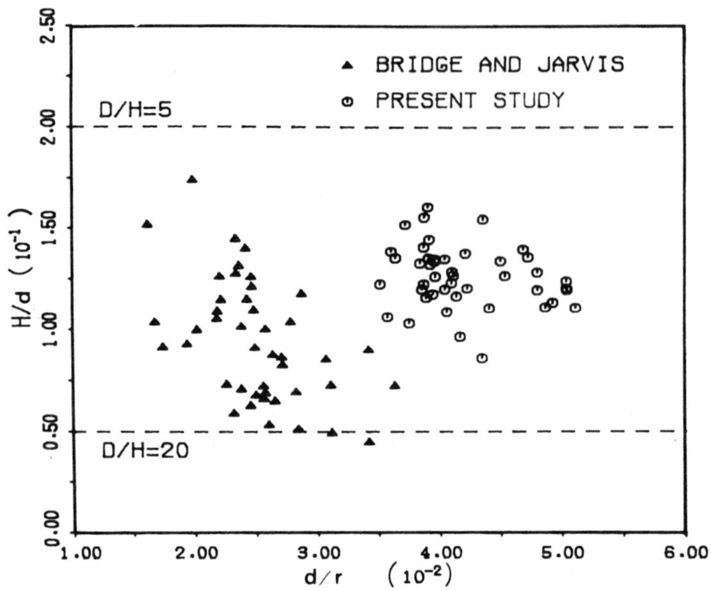


Fig. 8 (a). Correlation between Dune Height and Water Depth

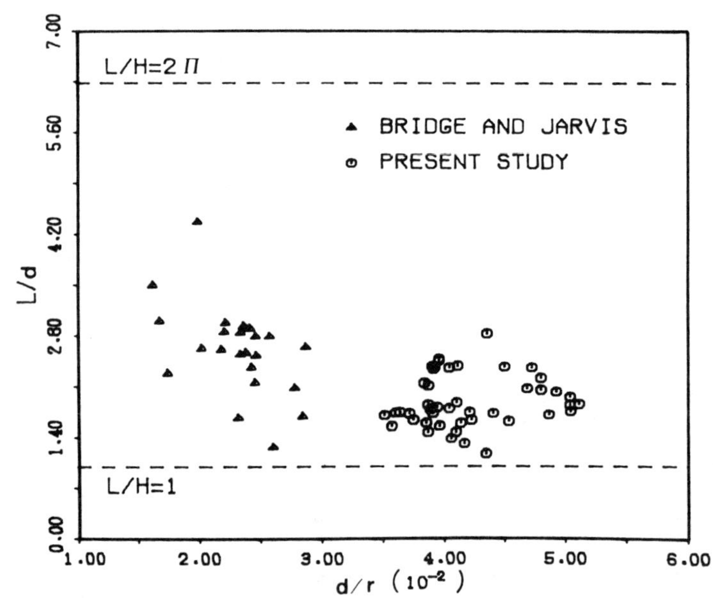


Fig. 8 (b). Correlation between Dune Length and Water Depth

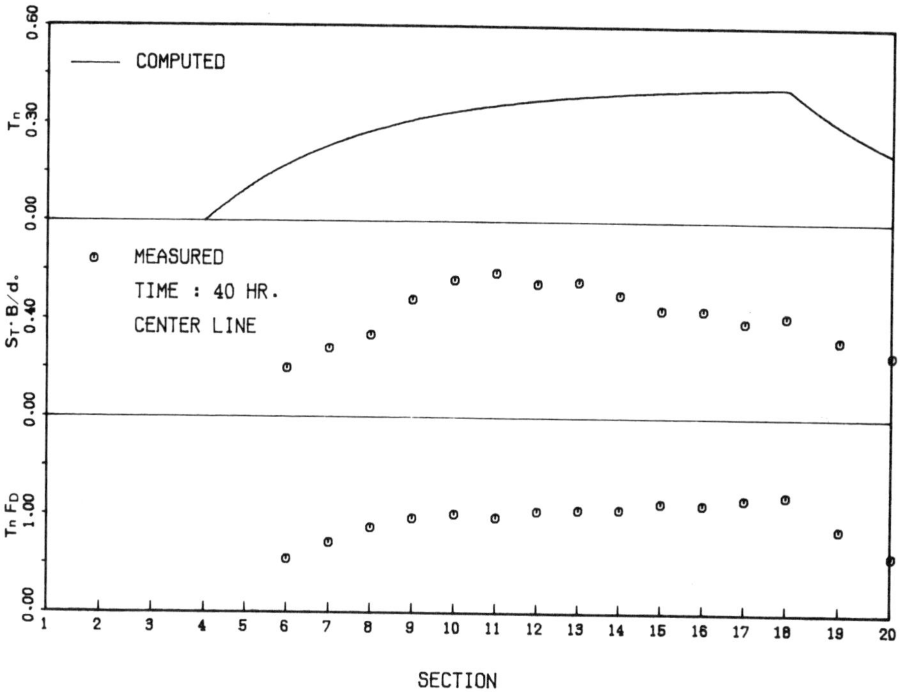


Fig. 9. Streamwise Variations of $T_n$ , $S_T$ , and $T_n F_D$

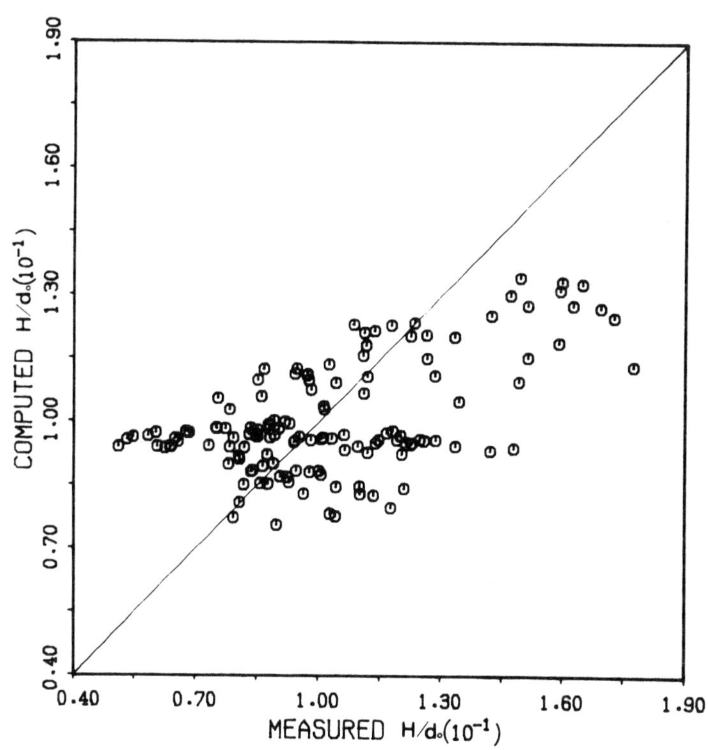


Fig. 10. Comparsion between Calculated Dune Height (Eq.12) and Experiment Data

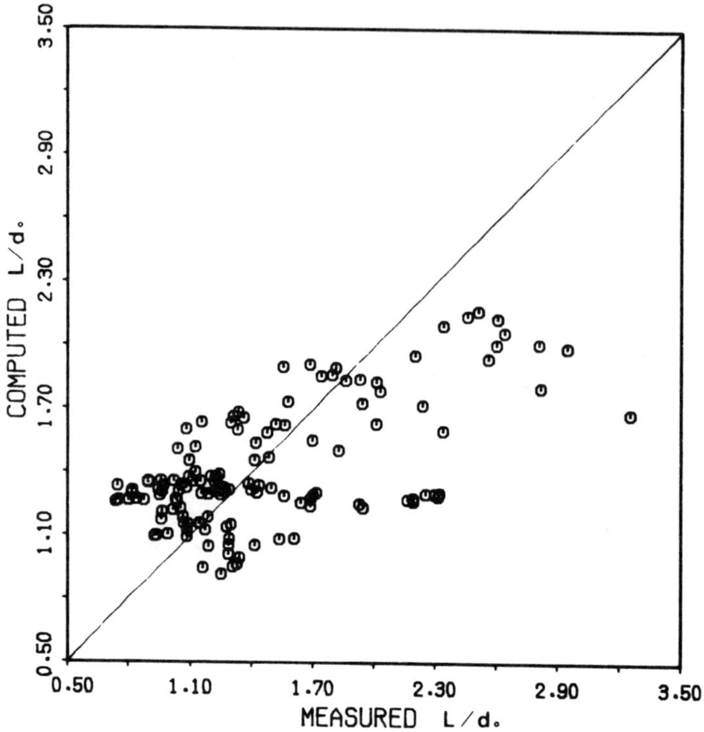


Fig. 11. Comparsion between Calculated Dune Length (Eq. 13) and Experiment Data

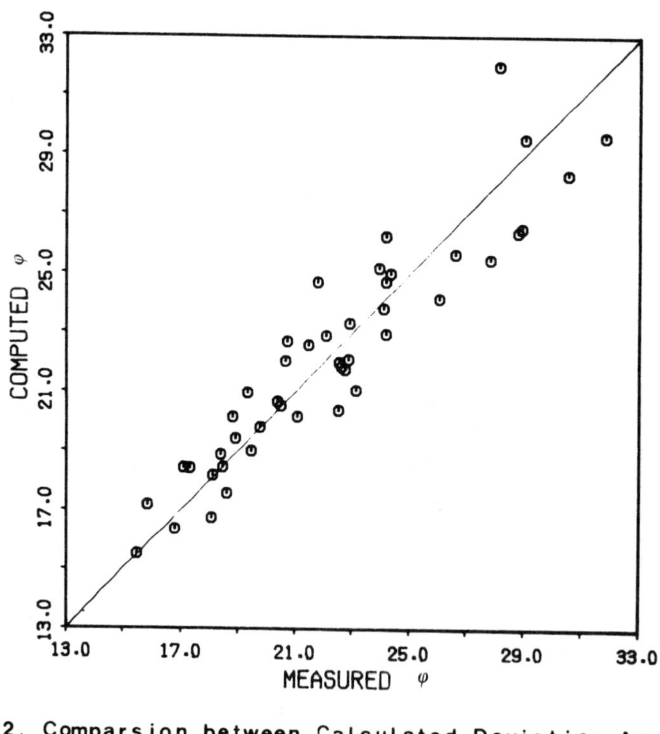


Fig. 12. Comparsion between Calculated Deviation Angle (Eq.15) and Experiment Data

International Conference on
**RIVER REGIME**
**18-20 May, 1988**

**PAPER H1**

# REGIME ASPECTS OF FLOOD CONTROL CHANNELIZATION

Charles R. Neill and Eugene K. Yaremko, principal engineers, Northwest Hydraulic Consultants, Edmonton, Alberta, Canada.

Summary. Channelization of natural streams causes various hydrologic and morphologic responses that are difficult to predict and treat. These include flood peak augmentation, profile degradation and aggradation, widening and deepening of cross-sections, and initiation of meandering or acceleration of meander shifting. A regime viewpoint is useful especially at the preliminary study stage for assessing the predominant characteristics of the stream and identifying likely responses, but quantitative application of the regime approach raises various problems especially with regard to channel planforms and bed-sediment transport.

Problems in channel response and in applying the regime approach are illustrated with reference to embanked rivers in Bangladesh, a gravel river in Alaska, a channelized hill stream in Mississippi, a desert stream in New Mexico, and a channelized river in northern Alberta.

## 1. INTRODUCTION

The conversion of natural streams into flood control channels by clearing, straightening, enlarging or diking is a common practice. Channels tend to respond with various forms of erosional and depositional instability, ranging from mild to severe depending on the hydraulic geometry, the nature of the boundary materials, and the basin's hydrologic and sediment characteristics. Problems include headcutting, lateral migration and meandering, and sedimentation with loss of flood capacity. Remedial measures include bank protection, slope control structures, reservoirs and sediment basins. These measures are usually expensive and not always successful.

Quantitative prediction of channel response is difficult even where a mass of data is available. At a preliminary study stage data are often scarce, time and resources are limited, and an economical method of approximate response assessment is desirable. The regime approach, with its basic premise that a channel to carry given inflows of water and sediment in a given environment will adjust to an equilibrium condition of cross-section and slope, appears to offer a rational approach to preliminary assessment and at least conceptual design. Although the approach originated from experience with irrigation canals, it has proved useful for understanding river behaviour and for designing river works (Refs 1-3). Application to flood control conversions, however, raises problems that are not so acute in other applications.

The paper discusses problems in applying the regime approach and uses some actual cases to illustrate various points.

## 2. KEY FEATURES OF REGIME APPROACH

The regime approach has been reviewed at length by various authors (Refs 4-7). It postulates a unique combination of width, depth and slope that will provide

International Conference on River Regime, edited by W.R.White.

long-term equilibrium for given water and sediment inflows. Design relationships express width, depth and slope as power functions of design discharge and sediment properties. Some key features are as follows.

(i) Width is given explicitly as an empirical function of discharge and sometimes of other parameters, usually varying as the square root of discharge. This form of relationship has been confirmed for various sets of natural streams (Refs 8-12, 16,17) but is difficult to derive from fluid-mechanics considerations.

(ii) Slope for channels with low bed transport is given as an empirical function of discharge and bed material, usually varying inversely with a small power of discharge. If an empirical width equation is accepted, normal hydraulic relationships enable the derivation of a slope equation if a roughness is assumed; however, the usually-quoted slope relationship for sand channels is essentially empirical. In the case of gravel channels, adoption of a relationship between roughness and grainsize enables derivation of slope-discharge-grainsize relationships.

(iii) Channel planform - straight, meandering, braided, etc -is not explicitly provided for in most regime relationships, except that meander wavelength is usually given as a square root function of discharge. Many meandering sand-bed rivers seem to follow the same type of width and slope relationships as straight canals, but with higher slopes. Braided channels tend to have much higher slopes.

(iv) Bed-sediment transport was not explicitly provided for in the original relationships. A practical difficulty is that actual rates are seldom known. Channels with high inflows of bed sediment may maintain long-term equilibrium on much steeper slopes than regime canals. White et al (Ref 13) presented modified regime-type relationships incorporating bed-sediment concentrations.

(v) Although slope is treated as dependent and transport as imposed, in the short term slope may appear imposed and transport dependent. In the long term, however, the slope of a reach is assumed to adjust to be in balance with the external sediment supply.

(vi) For natural streams, discharge fluctuations are represented by a "dominant" discharge assumed to be responsible for channel formation, normally taken as the bankfull discharge. Sometimes it is defined in terms of flood return period, usually in the range of 1.5 to 5 years.

## 3. TYPICAL RESPONSES TO FLOOD CONTROL CONVERSIONS

The prime objective of converting a natural stream for flood control is to increase within-bank flows and reduce or eliminate overbank flows. Methods include:

(i) constructing levees along the top of the banks or set back some distance;

(ii) enlarging the cross-section by widening or deepening or both;

(iii) improving hydraulic capacity by clearing vegetation and snags;

(iv) shortening the channel and augmenting its slope by straightening or by cutting off severe bends;

(v) providing high flow by-passes or short-cuts.

The effect of these various measures on discharges can be quite complex. Flow that formerly passed outside the channel is now confined within it, and the flood routing characteristics of the system are altered by elimination of floodplain storage. In the case of "perched" channels, overbank flow that formerly escaped from the basin to other stream systems is now retained.

The way that cross-section and slope will ultimately respond to increased flood peaks is fairly clearly indicated by regime relationships: widths and depths tend to increase, and slope to reduce by degradation upstream and aggradation downstream. Effects on planform, channel shifting and sediment transport are less clear. If the natural stream has a shifting meandering planform, shifting will tend to accelerate and meander wavelengths will tend to lengthen; also, meanders will tend to widen and cause the stream to impinge on setback levees. Because these changes result in substantial inputs of erosion products, slope flattening may not be initially apparent. In certain cases there may be a change in planform type: a stable straight or sinuous planform may start to meander, or a meandering channel may tend to become braided.

Clearing vegetation can have dramatic effects on lateral stability. Discharges are increased by improved hydraulic capacity, and the erosional resistance of the banks may be severely reduced. Widths may increase substantially and meander shifting may accelerate. Removal of timber snags that formerly acted to stabilize the bed may result in active bed movement and slope flattening, reinforcing the effect of increased discharges.

Straightening increases the hydraulic capacity and imposes an immediate increase of slope. The most immediate response is often slope flattening by upstream-progressing degradation (headcutting). Erosion products from bed degradation tend to deposit downstream, reducing or negating the planned increase in flood capacity. Accelerated meandering and bank erosion may also develop downstream and in the straightened reach.

## 4. PARTICULAR DIFFICULTIES WITH REGIME APPROACH

The first difficulty that arises in flood control applications is the question of dominant discharge: for what discharge is the channel to meet regime dimensions? If the project flood protection criterion is a fairly low return period, say up to 5 years or so, there is perhaps no great difficulty is using that value. For longer return period criteria, however, use of the design capacity will tend to indicate inordinately large channels. One solution is a compound cross-section consisting of a channel for the dominant discharge plus overbank areas or berms within levees forming a compound channel for higher discharges.

A second problem relates to planform. Straightening of a meandering stream may invoke serious problems of slope response. Yet if augmented discharges are put in a meandering channel, meander shifting may greatly accelerate. Nature (and environmentalists) appear to prefer a meandering planform, but in many environments meandering channels are subject to continual shifting and migration.

A third problem is transport of bed sediment. If transport is high, a channel designed to regime canal criteria will have too flat a slope to transport the sediment through. At the conceptual design stage, the use of equations that explicitly provide for transport rates is unlikely to be satisfactory, as rates are unlikely to be known.

A final problem with coarse-bed channels is that the bed grainsize constitution observed in the natural channel may not represent the final constitution of the converted channel, because of armouring phenomena. If bed transport is relatively low, armouring may reduce the expected slope response. Generally, severe problems of slope response are characteristic of channels with fine bed materials.

## 5. CASE ILLUSTRATIONS

A number of cases of response to flood control conversion or of the application of regime concepts to such channels are discussed briefly in the following sections. Principal characteristics of the rivers described are summarized in Table 1.

5.1 Khowai and Manu Rivers, northeast Bangladesh. The main point of this case is the effect of flood control embankments on discharges and on retention of sediment.

The Khowai and Manu rivers flow into the flat deltaic lands of Bangladesh from steep hill country in northeast India, and historically have been subject to severe floods. Because their drainage basins lie mostly within India, Bangladesh has little control over water and sediment inflows. Deposition and overspill of sediment in historical times have perched the river channels above the surrounding landscape (Fig 1) so that natural overflows do not return to the rivers but reach the sea by other routes.

Bangladesh has attempted to provide flood protection by diking, to the point that substantial lengths of river inside Bangladesh are now enclosed between high embankments (Photo 1). When the most recent embankments were planned in the mid-1970s, the design criterion was considered to be a 50-year return period. Within a few years of construction, however, embankments have been overtopped or their freeboard has been severely encroached upon on several occasions. The main reason is a substantial increase in discharge peaks resulting from cutting off flood escapes and accelerating floodwave propagation.

The morphologic and stability consequences of the conversions are of serious concern. Sediment that formerly escaped with flood water now deposits overbank inside the embankments, reducing designed conveyance. There are indications that bed aggradation is occurring in the lower reaches along with degradation in the upper reaches, as the rivers attempt to flatten slope to adjust to larger discharges. Initially, most embankments were well set back from the natural meanders, but meander encroachment and erosional attack are now developing.

The bulk of the sediment transport is fine sand and silt, transported mainly as suspended sediment in floods. Bed-sediment loads of medium and coarse sand appear to be modest and the river slopes are not greatly in excess of expected regime values for low bed-material transport (Fig 4). Channel widths are within the expected regime range in relation to bankfull discharges (Fig 3). Reliable evidence of substantial width adjustment to larger flood peaks is lacking, but the embankments have mostly been in place for 10 years or less.

5.2 Tanana River at Fairbanks, Alaska. This case illustrates regime associations between planform, slope and bed-material. Long-term response to limited flood-control measures is presently unclear. A detailed report is available (Ref 14).

The river at Fairbanks is multi-channel with a broad floodplain and islands, partly wooded. Upstream of the city it is highly braided with a slope of about 1 per 1000 and a mainly gravel bed with a 75% passing size of 40 mm. Downstream it has a meandering planform, a slope of 0.3 per 1000, and a mainly sandy bed with a 75% passing size of 10 mm. Between is a transitional length of only about 16 km, in which planform, slope and bed-material grainsizes merge from the upstream to the downstream regimes (Fig 2). Geomorphologically, the transition zone apparently represents the toe of an advancing wedge of gravel derived from glacial outwash far upstream.

The bankfull discharge is about 2000 $m^3/s$, return period about 2 years. Widths in multi-channel reaches range up to 1000 m or more, but where all the flow passes in a single channel the bankfull width is approximately 210 m, within the usual regime range for that discharge (Fig 3).

The slope at Fairbanks in the transition can be taken as about 0.5 per 1000 and the representative bed-material size as about 20 mm. This is high in regime terms for a single channel of 2000 $m^3/s$ dominant discharge (Fig 4). But as the transitional reach has generally at least 2 channels, the dominant discharge might be taken as 1000 $m^3/s$ or less. On this basis the slope is only slightly higher than a low-transport regime slope.

Bedload and suspended sediment transport at Fairbanks have been measured systematically near the middle of the transitional reach. Bedload concentrations are from 10 to 40 parts per million by weight under most high-flow and flood conditions: these figures would be generally classed as low. Suspended concentrations are roughly 50 times greater under flood conditions.

Between 1973 and 1981, about 30 km of the right floodplain upstream of and through Fairbanks was closed off by a setback levee system with a few protective groins. This construction caused some shifting of bars and channels, but no significant effect on the overall regime of the river. What would be the slope response if the river were to be diked on both sides and secondary channels closed off to confine bankfull flows to a single channel? The effective dominant discharge for slope would then increase to

2000 $m^3/s$ or more. Reference to Fig 4 suggests that the slope at Fairbanks would tend to reduce, perhaps setting off a longterm trend to upstream degradation and downstream aggradation.

5.3 Big Sand Creek, Mississippi. This case illustrates slope response to straightening and diking, and the sensitivity of regime slopes to bed-sediment transport.

Big Sand Creek is one of many "bluff line" streams that drain hill country in northern Mississippi and emerge abruptly into the Mississippi Valley. Since early in the present century, the final 11 km in the valley has been subjected to a sequence of diversions and dikings and is now a straight flood control channel with levees of varying setback (Photo 2). Geomorphic aspects of other streams in this region have been analyzed extensively by Schumm et al (Ref 15).

Fig 5 shows a sequence of slope profiles extending from the first straightening of unknown date to a survey of 1973. Before the first straightening ("original ditch") the stream had a longer course. The response has been a flattening of the imposed slope through upstream degradation and downstream aggradation. An original drop of 14.3 m in a 6.4 km straightened length had reduced to 8.2 m by 1932 and to 6.1 m by 1947. In 1973 the final 4.7 km, which until then had followed a natural course, was straightened and enclosed as a short steep reach followed by a long flat reach. This diversion filled in a few weeks to produce a uniform slope of about 0.9 per 1000 over the whole 11 km.

The slope flattening represents a response not only to straightening but to increase in channel discharges as a result of levee construction over most of the length and incision at the upper end. The bankfull capacity of the "original ditch" was about 100 $m^3/s$. The final 4.7 km diversion of 1973 was designed for a bankfull capacity of about 250 $m^3/s$ and a total capacity between levees of about 500 $m^3/s$, later revised to 900 $m^3/s$. Return periods are in the order of 5 years for 500 $m^3/s$ and 50 years for 900 $m^3/s$.

The slope of 0.9 per 1000 appears to represent more or less an equilibrium condition for a dominant discharge of about 250 $m^3/s$. This is much higher than a regime low-transport slope for a medium sand channel, which would be about 0.15 per 1000 (Fig 4). As the channel is straight, the difference must be due mainly to bed-sediment transport. Sediment transport has been estimated from non-hydraulic considerations as about 600 000 tonnes per year, mostly of sand. Application of the Ackers-White sediment transport procedure indicates relatively high bed-sediment concentrations of 3000 to 4000 ppm by weight for flows between 250 and 500 $m^3/s$, and produces total load estimates that are of the same order as the non-hydraulic estimate.

The tables of White et al (Ref 13) for stable alluvial channels indicate an even greater sensitivity of slope to transport than demonstrated by Big Sand Creek. For a dominant discharge of 200 $m^3/s$ and a bed-sediment size of 0.3 mm, they show equilibrium slope increasing from 0.16 to 1.3 per 1000 for an increase in sediment concentration from 100 to 2000 ppm.

5.4 Puerco River at Gallup, New Mexico. A brief comment is made on this case to illustrate a more extreme deviation of slope from low-transport regime values. (The Puerco River is not the Rio Puerco that has featured in several geomorphologic publications.) The Puerco River in its natural state is an ephemeral stream or 'arroyo' with a single meandering channel, a wide floodplain, a bed of fine to medium sand, and compact silt banks. Through the town of Gallup, New Mexico it has been partly channelized and diked for flood control.

The stable slope of the Puerco River is about 2 per 1000. The natural bankfull discharge is about 140 $m^3/s$. A regime low-transport slope for these conditions would be about 0.13 per 1000, therefore the actual slope is about 15 times greater. Concentrations of bed-sediment under flood conditions are probably 10 000 ppm or greater. It is often supposed that channels with such high bed-sediment transport and slope would adopt a braided planform, even when the material is sandy. Lack of braiding in the Puerco River may be related to wash load, the nature of the banks, and the intermittent flashy nature of the flow regime.

5.5 <u>East Prairie River, northwest Alberta</u>. This case illustrates slope flattening in response to straightening and diking of a meandering stream in fine-grained glaciolacustrine soils.

In its natural state the river had a serpentine meander pattern and dense bank vegetation with channel debris, and frequently flooded adjacent agricultural land. The bed was of medium sand overlying lacustrine clay or glacial till at shallow depths. The slope was 0.5 per 1000 or less and the bankfull discharge about 60 $m^3/s$. In 1962 a length of about 14 km above the mouth was replaced with a straight excavated channel and low setback dikes for a design capacity of about 140 $m^3/s$, on a slope of 1 per 1000 over the upper 11 km and 0.5 per 1000 over the lower 3 km.

The slope response between 1968 and 1984 is illustrated in Fig 5. By 1984 the lower half of the channelized length where sand is depositing had adjusted to a slope of about 0.3 per 1000, and the upper degradational half to about 0.5 per 1000. A steeper "nick zone" (Schumm et al 1984) had advanced in silt and clay materials to about 5 km upstream of the channelization. Assuming that bed-sediment transport will be low after stability is achieved, a final regime slope of about 0.2 per 1000 or less has been estimated for the channelized reach (Fig 4).

## 6. CONCLUSIONS FROM CASE STUDIES

1. A primary question in contemplating a flood control channelization is the potential for augmentation of flood peaks, both immediately and in the long term.

2. Many instability response problems arise from the fact that channelization measures impose an increased or at best an unaltered slope, where regime considerations require a reduced slope to suit augmented discharges and/or reduced resistance.

3. Lacey-type regime relationships derived from irrigation canals are

not very useful for some classes of stream. Sand-bed streams in some environments may maintain a form of stability with very high sand loads, slopes that exceed regime values by 10 or more times, and cross-sections that are relatively small.

4. Regime relations of the Wallingford type appear promising for high-transport as well as low-transport channels, but more verification of field transport rates is required, as well as consideration of high wash loads and discontinuities in transport relations.

## REFERENCES


1. Lacey, G. 1929-30. Stable channels in alluvium. Min. Proc. Inst. Civ. Engrs. v.229 pt.1.

2. Lacey, G. 1958. Flow in alluvial channels with sandy mobile beds. Proc. Inst. Civ. Engrs. v.9.

3. Blench, T. 1957. Regime behaviour of canals and rivers. Butterworths, London.

4. Bogardi, J. 1974. Sediment transport in alluvial streams. Hungarian Academy of Sciences, Budapest.

5. Henderson, F.M. 1966. Open channel flow. Macmillan.

6. Mahmood, K. and Shen, H.W. 1971. Regime concept of sediment transporting canals and rivers. In River Mechanics (2 vols), H.W.Shen, Fort Collins, Colorado.

7. Raudkivi, A.J. 1976. Loose boundary hydraulics. Pergamon.

8. Nixon, M. 1959. A study of the bankfull discharges of rivers in England and Wales. Proc. Inst. Civ. Engrs. v.12.

9. Bray, D.I. 1975. Representative discharges for gravel-bed rivers in Alberta, Canada. J. Hydrology v.27 no 1/2.

10. Kellerhals, R. 1967. Stable channels with gravel-paved beds. J.Waterways Div, Amer. Soc. Civ. Engrs, February.

11. Leopold, L.B. and Maddock, T. 1953. The hydraulic geometry of stream channels and some physiographic implications. U.S Geol. Survey Prof. Paper 252.

12. Neill, C.R. 1973. Hydraulic geometry of sand rivers in Alberta. Hydrology Symp. 9, Nat. Research Council of Canada.

13. White, W.R, Paris, E, Bettess, R. 1981. (i) River regime based on sediment transport concepts; (ii) Tables for the design of stable alluvial channels. Hydraulics Research Station Reports IT 201 and 208.

14. Corps of Engineers, U.S. Army, 1984. Overview of Tanana River monitoring and research studies near Fairbanks, Alaska. Cold Regions Research and Engineering Laboratory, Hanover N.H.

15. Schumm, S.A, Harvey, M.D, Watson, C.C. 1984. Incised channels - morphology, dynamics and control. Water Resources Publications, Fort Collins, Colorado.

16. Charlton, F.G, Brown, P.M, Benson, R.W. 1978. The hydraulic geometry of some gravel bed rivers in Britain. Hydraulics Research Station Report IT 180.

17. Hey, R.D. and Thorne, C.R. 1986. Stable channels with mobile gravel beds. J.Hydr.Engin, Amer. Soc. Civ. Engrs, August.

Acknowledgements. The paper is based in part on studies for U.S. Army Corps of Engineers, Alberta Environment, and Bangladesh Water Development Board.

Table 1. Representative data for quoted cases

| River | Bankfull discharge $m^3/s$ | Bankfull width m | Bankfull depth m | Slope per 1000 | Bed material |
|---|---|---|---|---|---|
| Khowai | 200 | 50 | 3.5 | 0.25 | Med. sand |
| Manu | 400 | 75 | 4.0 | 0.16 | Med. sand |
| Tanana | 2000 | 210 (where single channel) | 4.3 | 0.5 | Med. gravel |
| Big Sand | 250 | 40 | 3.5 | 0.9 | Med. sand |
| Puerco | 140 | 30 | 1.8 | 2.0 | Fine sand |
| E. Prairie | 140 | 35 | 3.4 | 0.5 | Med. sand |

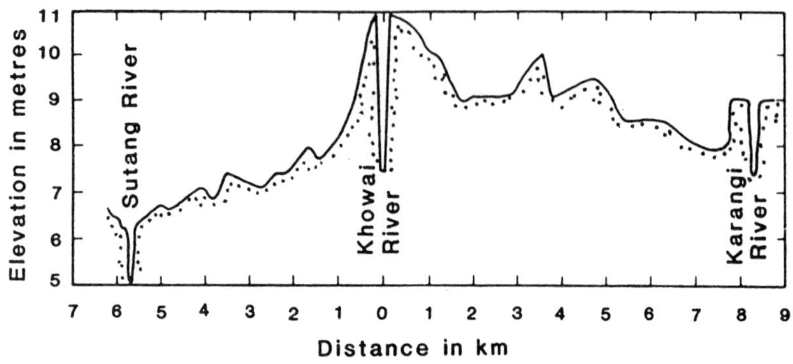


Fig 1. Landscape cross-section showing perched condition of Khowai River, Bangladesh.

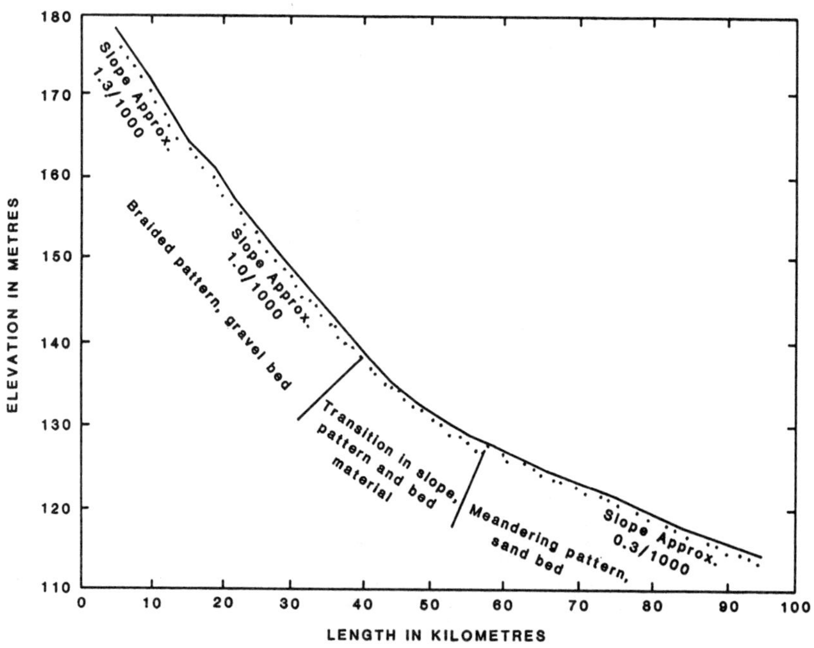


Fig 2. Slope profile of Tanana River near Fairbanks, Alaska.

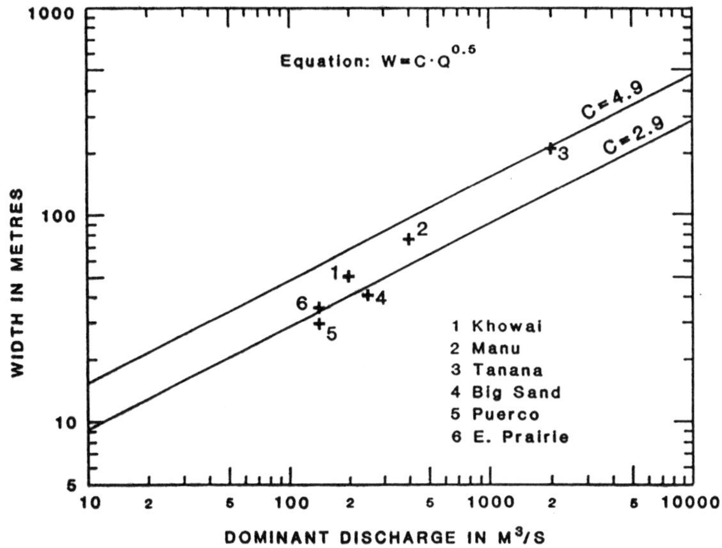


Fig 3. Form of regime width plot showing data for quoted rivers.

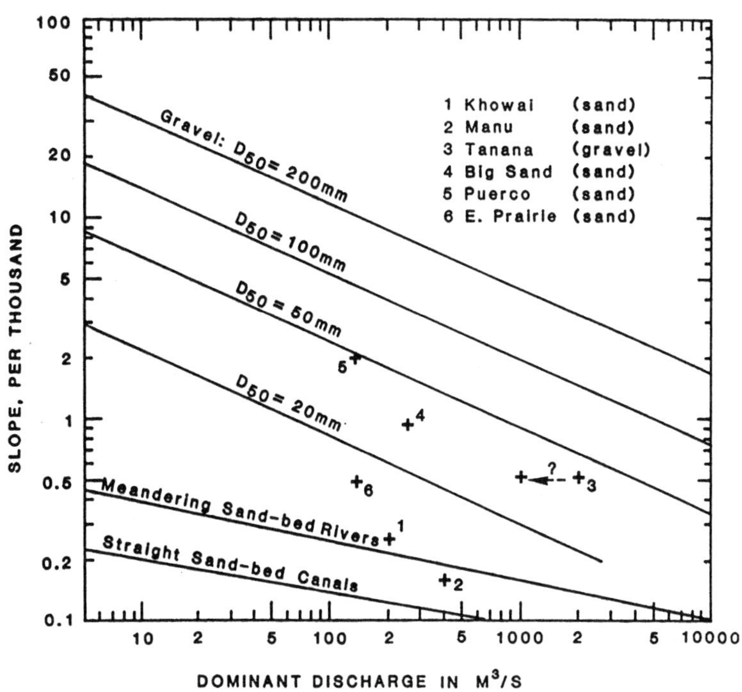


Fig 4. Form of regime slope plot showing data for quoted rivers.

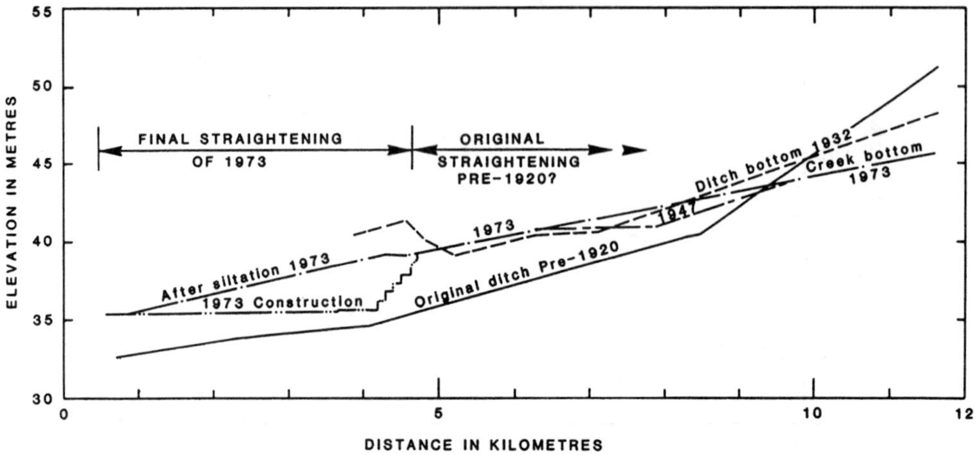


Fig 5. Slopé profiles of Big Sand Creek channelization, Mississippi.

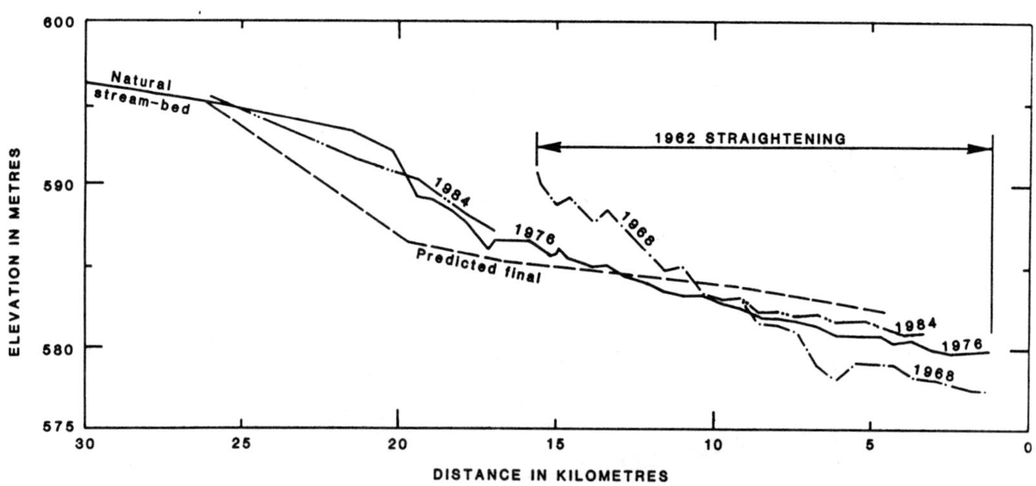


Fig 6. Slope profiles of East Prairie River channelization, Alberta.

PHOTO 1:
Khowai River at Shaistaganj
9 May, 1985
showing road and rail bridges and BWDB embankments.

PHOTO 2:
Big Sand Creek near Greenwood, Mississippi
September, 1986
looking downstream.

PHOTO 3:
Puerco River at Gallup, New Mexico
October, 1987
looking upstream.

International Conference on
RIVER REGIME
18-20 May, 1988

PAPER H2

# THE INFLUENCE OF RIVER REGIME ON THE FLOOD MANAGEMENT OF SUKKUR BARRAGE, PAKISTAN

by

I G G Hogg — Sir M MacDonald & Partners Ltd
K Guganesharajah — Sir M MacDonald & Partners Ltd
P D S Gunn — Sir M MacDonald & Partners Ltd
P Ackers — Consultant

## SUMMARY

Varying combinations of discharge, sediment transport, channel roughness and downstream hydraulic control, make flood water levels on the Indus river difficult to predict with any precision. Some historic floods caused inundation of agricultural lands and scour at diversion structures such as Sukkur Barrage (irrigated command area 3.2 million hectares). As a result there has been extensive bund development throughout the lower Indus basin. Nevertheless, some recent floods have caused considerable damage. Paradoxically, others of similar magnitude, have not made any serious impact. Because the agricultural lands served by the Indus barrages are crucial to the economy of Pakistan, it is important to ensure that adequate flood management procedures are in place.

This paper considers the hydraulic and sediment characteristics of the river during the 1976 and 1986 floods; the two largest on record. Although peak flow rate in each case was reported to be about 1.2 million $ft^3/s$ (34 000 cumecs), maximum water levels in 1976 were about 3.5 ft (1.1 m) higher than in 1986. Hydraulic calculations indicated that the channel roughness coefficient (Manning's 'n') varied between 0.011 and 0.029 during the 1976 flood and between 0.011 and 0.020 during the 1986 flood. Because the bed material is almost homogeneous ($D_{50}$ = 0.2 mm), these variations of roughness were interpreted in terms of regime changes. Low values corresponded to a smoothed longitudinal bed profile (upper regime) while higher values represented pronounced bed forms such as ripples and dunes (lower regime).

The Ackers-White sediment transport function was found to provide a reasonable predictor for a given hydraulic condition. The observed behaviour was simulated successfully using the mathematical model HYDSED which couples the hydraulic equations and the Ackers-White sediment transport function and solves them simultaneously. The model was used to route the 200-year design flood (1.5 million $ft^3/s$ - 4 2500 cumecs) given the probable combinations of regime, sediment inflow and hydraulic control. This allowed an assessment of the flood hazard and scour with existing and proposed channel and barrage geometry.

International Conference on River Regime, edited by W.R.White.

## NOMENCLATURE

| | |
|---|---|
| $A$ | total flow area ($m^2$) |
| $A_{gr}$ | value of sediment mobility $F_{gr}$ at initial motion |
| $c$ | coefficient of the dimensionless sediment transport parameter $G_{gr}$ |
| $C$ | suspended sediment load |
| $d$ | mean depth of flow (m) |
| $D$ | median sediment diameter (m) |
| $Dgr$ | dimensionless grain size |
| $F_{cg}$ | sediment mobility, coarse grains |
| $F_{fg}$ | sediment mobility, fine grains |
| $F_{gr}$ | sediment mobility, general |
| $g$ | acceleration due to gravity in metres/per second per second |
| $n$ | transition exponent dependent on particle size |
| $m$ | exponent of the dimensionless sediment transport parameter $G_{gr}$ |
| $q$ | lateral inflow into the channel per unit length in $m^3/s/m$ |
| $q_s$ | lateral sediment load, weight per unit time per unit length of channel |
| $Q$ | discharge in the channel in $m^3/s$ |
| $s$ | channel sediment transport rate |
| $S_f$ | friction slope |
| $V$ | mean velocity in metres/second |
| $V_v$ | velocity of wind along the channel relative to the channel flow |
| $W$ | relative density of sediment (dimensionless) |
| $\nu$ | kinematic viscosity of fluid in $m^2/s$ |
| $\gamma$ | weight density of sediment |
| $\rho$ | mass density of fluid |

## 1.0 INTRODUCTION

The Indus river rises in the Karakorum mountains and flows to the Arabian Sea through the plains of Punjab and Sind Provinces, Pakistan. From a number of major barrages on the river, irrigation waters are widely distributed. In the plains, the river flows above the general elevation of the surrounding lands. Natural levees have been artificially heightened and an extensive system of flood bunds has been progressively developed to contain flood flows within the river channels. As a result, the sporadic flooding that was a historical feature of the river has largely been eliminated. Nevertheless, some recent major floods have caused loss of life, damage to croplands, infrastructure and property as well as serious scour in the vicinity of the barrages.

Sukkur Barrage was constructed between 1927 and 1932 to provide irrigation waters to much of the lower Indus plain. The area now served is approximately 3.2 million hectares (Figure 1). The barrage was originally designed to pass 1 500 000 $ft^3/s$, (42 500 cumecs). Early sedimentation problems led to a remodelling of the approach channels in the late 1930s, and the design flood was reduced to 900 000 $ft^3/s$ (25 500 cumecs). The implementation of the sediment exclusion works involved closure of 10 of the 66 gates originally in operation.

Since 1973, the Indus has experienced a series of very large floods. According to direct measurements at Sukkur, five of these had peak flows in the range 900 000 to 1 200 000 $ft^3/s$, (25 500 to 34 000 cumecs). Some, but not all, caused inundation in the region and scour at Sukkur. As a result, there was growing concern for the safety of the barrage. In 1979, Pakistani authorities commissioned studies of the options for the future flood management of the Barrage. These studies were based mainly on physical modelling of the barrage and the immediate upstream river reach. They resulted in recommendations for modifications to the sediment exclusion works and the re-opening of eight of the ten closed gates.

In 1986, Sir M MacDonald and Partners, already acting as Consultant to the Sukkur Barrage Gates Rehabilitation Project, was invited to review the findings of the study. As there was no scope for further physical modelling in the terms of reference, the review was based on the analysis of available hydraulic data and mathematical modelling of the hydraulic and sediment transport characteristics of the river. This paper discusses the results of the review.

## 2.0 DESIGN FLOODS

Discharge measurements have been made close to Sukkur Barrage since 1901. Annual maximum flow rates in the range 300 000 to 1 200 000 $ft^3/s$, (8 500 to 34 000 cumecs) have been reported. Although the detailed measurement procedures and gauging sites have changed somewhat over the years, it is considered (Ref 1) that the accuracy of the measurements lies in the range - 6% to +14% owing to a combination of random and systematic errors; the latter introduced by the weighted line depth sounding procedure that is employed.

Annual maximum flow rates extracted from the 86 year record (Figure 2) show an apparent inhomogeneity. It is noticeable that major floods (greater than 900 000 $ft^3/s$, (25 500 cumecs) say) have become more frequent since the 1950s. Available hydrological data from the Indus basin, do not suggest that this pattern is the result of an intensification of flows due to flood bund development, because much of the development took place prior to 1950. The record of seasonal and annual runoff does not show any significant increase in runoff due to catchment denudation. Nor can it be shown that the operation of the main dams in the basin (Mangla, Chasma and Tarbela) exert any appreciable influence on peak flow rates measured at Sukkur.

An attempt was made to explain the inhomogeneity in the series of annual maximum flow rates in the following way. The eight greatest annual flow rates since 1950 were reduced by a conservative 10% each. A Gumbel probability distribution was fitted to the adjusted data set (Figure 3). The distribution does not exhibit an obvious discontinuity and indicates that, with some adjustment to those reported peak flow rates, where the discharge may have been overestimated because of drag on the sounding line, reasonably consistent design floods may be derived. Both corrected and uncorrected distributions were considered in deciding the design flood.

Although somewhat low by contemporary international design standards, barrages in Pakistan are generally designed against 100-year flood events and flood embankments are provided with a freeboard of 6 ft (1.8 m) above the highest recorded water level. However, in view of the great economic importance of Sukkur and neighbouring barrages, and the severe consequences of failure of the barrages or the bunds, it has been suggested to the Authorities that higher return period flood events should be considered for design purposes.

The 1979 study considered the possibility that future bund developments could well eliminate flooding altogether. A corollary of this of course would be the intensification of peak flow rates in the main channels. The potential impact of this was quantified by using a mathematical model to route major historical floods (during which breaching of bunds took place) in order to simulate breachless conditions at Sukkur Barrage (Ref 2). The WMO model used by WAPDA (Water and Power Development Authority) for flood forecasting purposes was employed for the routing exercise. As a result, it was predicted that peak flow rates would be about 18% greater than those observed, if no breaching had occurred. This factor was therefore applied to the preliminary flood estimates suggested here to produce design floods at Sukkur Barrage (Table 1).

## TABLE 1

### Design Flood Magnitude

| Return Period (years) | Peak Discharge (million $ft^3/s$) | (cumecs) |
|---|---|---|
| 100 | 1.2 | 34 000 |
| 200 | 1.5 | 42 500 |
| 1000 | 1.7 | 48 200 |

## 3.0 SEDIMENT TRANSPORT

In order to gain an understanding of the hydraulic and sediment transport characteristics of the river, a study was made of the 1976 and 1986 floods. These are the two highest floods on record. Although they are reported to have had similar peak flows (1.2 million $ft^3/s$, (34 000 cumecs)) flood water levels were significantly different. Thus the two floods may demonstrate quite well the range of hydraulic behaviour to be found even at similar flow rates.

Daily discharge and velocity measurements are available from the Outfall section, 6 000 ft (1 800 m) upstream of the barrage, and daily water levels are reported from five sites in the vicinity of the barrage. Sediment transport rates were estimated for each flood with the Ackers-White function (Ref 3). This relates sediment transport to a sediment mobility number ($F_{gr}$) where:

$$F_{gr} = \frac{V_*^{n}}{\sqrt{gD(w-1)}} \left[\frac{V}{32 \log(10\ d/D)}\right]^{1-n} \qquad .......1$$

The constant n can be evaluated in terms of a dimensionless grain size ($D_{gr}$) where:

$$n = 1.00 - 0.56 \log (D_{gr}) \quad \text{for} \quad 1 \leq D_{gr} < 60 \qquad .......2$$

$$n = 0 \quad \text{for} \quad D_{gr} \geq 60 \qquad .......3$$

and

$$D_{gr} = \left[\frac{g(w-1)}{\nu^2}\right]^{1/3} D$$

The sediment transport $G_{gr}$ a non-dimensional parameter, is expressed in terms of the mobility number $F_{gr}$ such that:

$$G_{gr} = c \quad \frac{F_{gr}}{A_{gr}} - 1 \quad ^{m} \qquad .......4$$

$$m = \frac{9.66}{D_{gr}} + 1.34 \qquad .......5$$

$$A_{gr} = 0.17 \qquad \text{for} \qquad D_{gr} \geq 60 \qquad .......6$$

$$A_{gr} = (0.23/\sqrt{D_{gr}}) + 0.14 \qquad \text{for} \qquad D_{gr} \leq 60 \qquad .......7$$

Each of the parameters $A_{gr}$, m and c is related to the dimensionless grain size $D_{gr}$ as follows:

$$m = \frac{9.66}{D_{gr}} + 1.34$$

The rate of sediment transport X is related to $G_{gr}$ by the following equation:

$$G_{gr} = \frac{X\,d}{wD} \left(\frac{V*}{V}\right)^n \qquad .......8$$

Thus, from observed water slopes, velocities and depths, values of $F_{gr}$ may be calculated. Further calculations lead directly to values of sediment concentration and transport. A median bed material grain size D of 0.2 mm was adopted on the basis of particle size analyses of local bed samples. The estimated transport rates at peak flow rate during the 1976 and 1986 floods are shown in Table 2 together with alternative estimates made from other sources. In the course of the Lower Indus Project (LIP) representative sediment concentrations from the Indus basin were compiled into a rating (Ref 4). It should be noted that this rating predates the construction of Mangla and Tarbela dams, both of which might now influence sediment concentrations in the lower Indus. In addition, the rating refers to total sediment concentration (suspended load as well as bed load) rather than bed load alone. However, there is reasonable agreement between the estimates produced by the Ackers-White function and the LIP rating. Estimates produced from water samples taken in the left-hand pocket of Sukkur Barrage during the 1976 and 1986 floods are significantly less than those produced by both the LIP rating or the Ackers-White function, although this is by no means clear cut in the case of 1976, as the reported sample was not collected at the time of peak flow rate.

TABLE 2

ESTIMATED SEDIMENT CONCENTRATION IN PPM AT PEAK FLOW RATE

| | 1976 | 1986 |
|---|---|---|
| LIP rating | 7 600 | 6 300 |
| Ackers-White function | 8 800 | 9 600 |
| Samples from barrage | 3 000* | 1 500 |

Note: * not sampled at flood peak
Flow rate = 1 200 000 $ft^3/s$, (34 000 cumecs)

The variation of sediment mobility during the 1976 and 1986 floods is demonstrated in Figure 5. The calculations for 1976 demonstrate considerable hysteresis, with the sediment mobility number ($F_{gr}$) on the rising limb being higher than on the falling limb. This feature indicates considerable variation in the hydraulic characteristics and in the inflow of sediment during the course of the flood hydrograph and serves warning that simulation exercise must be subject to uncertainty.

It is generally accepted that Manning's coefficient 'n' is a reasonable measure of channel roughness. Values of 'n' were therefore calculated at five day intervals during the two flood events under study (Figure 4). During the course of the 1976 flood there was considerable variation. The period covered by the calculations begins at a discharge of 300 000 $ft^3/s$, (8 500 cumecs) with a calculated value of 'n' = 0.014. Such a low value indicates a relatively smooth channel, presumably because most bed features had already been washed out and the river was flowing in upper regime. As the discharge increased, 'n' fell to a minimum of 0.012 before rising again as peak flow was reached. Around the time of peak flow, there was a sudden increase in 'n' to about 0.024 and then a further, more gradual increase to about 0.029 as flow receded and lower regime flow was established. Records show that considerable scour took place in the vicinity of the barrage at about this time.

The 1986 flood displayed similar characteristics on the rising limb of the hydrograph. At 300 000 $ft^3/s$, (8 500 cumecs) 'n' had a value of 0.015 and this fell to about 0.011 at peak flow. On the receding limb, 'n' rose gradually to reach about 0.020 when flow had fallen again to 300 000 $ft^3/s$, (8 500 cumecs). Sounding data, taken at the time, do not indicate that major scour took place during this flood even though its magnitude was similar to that in 1976.

## 4.0 CHANNEL RESISTANCE

The analysis of observed flood behaviour using sediment transport theory revealed large fluctuations in sediment mobility and in channel roughness. It showed that a knowledge of sediment concentrations and discharge alone, would be an inadequate tool with which to simulate design flood hydraulics close to Sukkur Barrage. It is necessary to introduce additional variables to account for changes in channel roughness.

An alternative approach to the examination of the hydraulics of the river is afforded by resistance functions of the type put forward by Hydraulics Research Ltd (Ref 4). It is instructive to examine how well these functions represent the conditions encountered during the two floods under study. The White-Paris-Bettess function expresses the sediment mobility number as:

$$1 - \frac{F_{gr} - A_{gr}}{F_{fg} - A_{gr}} = 0.76 \left[ 1 - \frac{1}{\exp[(\log D_{gr})^{1.7}]} \right] \quad .......9$$

where

$$F_{fg} = \frac{V_*}{\sqrt{gD(w-1)}} \quad ......10$$

$$A_{gr} = (0.23/\sqrt{D_{gr}}) + 0.14 \quad ......11$$

Once $D_{gr}$ is determined from laboratory analyses of bed material, the function represents a single valued relationship between $F_{gr}$ and $F_{fg}$.

The actual values of $F_{gr}$ during the study floods are readily calculated using the Ackers-White function and these may be compared with corresponding values predicted by the White-Paris-Bettess relation in Figure 4. Clearly, there is poor agreement on the rising limbs of the hydrographs when channel resistance was low and flow was evidently in upper regime. However there is reasonable agreement on the hydrograph recession when channel resistance is higher and the transition to lower regime flow has taken place. On the basis of this comparison, it would seem that lower regime flow in the Indus river close to Sukkur Barrage can be represented by the White-Paris-Bettess function. Furthermore, it is noticeable from Figure 4 that upper regime flow on the rising hydrograph can be reasonably represented by a simplified relationship where $F_{gr}$ is equated to $F_{fg}$ i.e. the bed resistance comprises grain resistance only. These results offer hope that design flood hydraulics may be simulated to the extent that both upper and lower regime conditions can be reproduced mathematically although transitional conditions cannot yet be satisfactorily treated.

## 5.0 MATHEMATICAL MODEL

There are a number of key characteristics of the hydraulics and sediment transportation close to Sukkur Barrage which present great demands on any modelling exercise. Among these are:

- flood water velocities which reach 20 ft/s (6.1 m/s) in places thus pronouncing the velocity heads in the hydraulic equations;

- a mobile river bed comprising almost homogeneous sand (D50 = 0.2 mm) and having a tendency to change bed form according to the level of sediment inflow;

- a wide river and correspondingly low hydraulic radius;

- changes of bed profile take place rapidly relative to the hydrograph and thus the system may be regarded as being in steady state at any particular moment;

- there is a division of flow around Bukkur island, 10 000 ft, (3 000 m) upstream of the barrage;

- there is a possibility of critical flow at the barrage under low tailwater conditions.

The hydraulic mathematical model HYDSED, developed by Sir M MacDonald & Partners Ltd, has been used successfully on a number of complex flood problems, notably in the fenlands of the UK and braided river systems in Australia. It is particularly suited to the treatment of looped flow systems and flow through structures. The model uses the principles of continuity of flow and potential sediment transport at nodes and within channels, and conservation of momentum within channels. The upstream and downstream limits of a channel or structural reach are defined by nodes. The governing equations are illustrated below:

(a) Continuity of flow at node

$$\sum Q_i = 0 \qquad \text{......12}$$

(b) Continuity of sediment transport at node

$$\sum s_i = 0 \qquad \text{......13}$$

(c) Continuity of flow within channels

$$\frac{\partial Q}{\partial x} + \frac{\partial A}{\partial t} - q = 0 \qquad \text{......14}$$

(d) Continuity of sediment transport within channels

$$\frac{\partial s}{\partial x} + \gamma \cdot \frac{\partial A}{\partial t} + \frac{\partial AC}{\partial t} = q_s \qquad \text{......15}$$

(e) Momentum equation

$$\frac{\partial Q}{\partial t} + \frac{2Q}{A}\frac{\partial Q}{\partial x} - (Q^2/A^2)\frac{\partial A}{\partial x} + gA\frac{\partial y}{\partial x} + g\,A\,S_f = 0 \qquad \text{......16}$$

The equations associated with sediment transport were illustrated in Section 3.0. The above equations are transformed to a finite difference implicit scheme using Preissman's weighting procedure. For steady state the solutions are derived by omitting the time related terms.

The model was adapted for use on the Indus river by incorporating routines designed to represent a mobile river bed based on the Ackers-White sediment transport function. Variations in channel resistance, equivalent to different flow regimes were expressed as variations in Manning's 'n'.

## 6.0 CALIBRATION OF SEDIMENT TRANSPORT MODEL

The mathematical model HYDSED with the Ackers-White function built in, was calibrated against the 1976 and 1986 floods. Model inputs were flow rate, sediment inflow, Manning's 'n' and the tailwater level downstream of the barrage. A river reach of approximately 6 miles, (9.6 km) hydraulic length was modelled. Its upstream limit was 1 000 ft, (300 m) above Bukkur island and its downstream limit was about 1 000 ft below the barrage (Figure 6). Observed water levels were available from six gauging stations within this reach. Observed bed levels were also available from the Outfall gauging section, about 6 000 ft, (1 800 m) upstream of the barrage. It was assumed that, at the time of peak discharge, all the operational barrage gates were fully open. Sediment concentrations as sampled in the left-hand barrage pocket were adopted, although it has already been noted that these seem uncharacteristically low when compared with both theoretical estimates and regional empirical estimates.

### 6.1 1976 Flood

Simulated water levels and river bed levels for the 1976 flood are shown in Figures 7 and 8 respectively and compared with observed values. Four simulations are demonstrated in order to show the influence of channel roughness as expessed by Manning's 'n' which is used here as a corollary for flow regime.

The comparison shows that at the time of peak flow, Manning's 'n' lay in the range 0.015 to 0.020 and that the flow was probably in a transitional state between upper and lower regime. As it is known that there was a large abrupt change in bed level at around the time of peak flow, the range of observed bed levels at the Outfall section in the 20 day period centred on the day of the peak, is shown in Figure 8. It can be seen that the simulated bed levels fall within this range. It can also be seen that the transition from upper to lower regime could have led to considerable scour throughout the model reach with bed levels falling by about ten feet immediately upstream of the barrage and by as much as 15 feet, (4.5 m) in the Bukkur Gorge.

### 6.2 1986 Flood

The simulation of conditions at the time of peak flow during the 1986 flood was less successful (Figures 9 and 10). It is not known how representative the measurement of sediment concentration in the left pocket was of concentrations in the main river channel. Accordingly, a hypothetical sediment transport rate of 188 t/s was introduced. This represented a compromise value between the sampled rate of 51 t/s and those suggested by the LIP sediment rating (215 t/s) and the estimates produced by the Ackers-White function (326 t/s).

## 7.0 SIMULATION OF DESIGN FLOOD

Flood frequency has already been discussed and it was argued, in view of standing hydraulic design procedures in Pakistan, that 200 years was a suitable return period for the design flood at Sukkur Barrage although this may be regarded as low in some engineering circles. The 200 year peak flow rate was estimated as 1.5 million $ft^3/s$, (42 500 cumecs). The mathematical model HYDSED, was therefore run at this flow rate. As there is doubt about the design sediment concentration three values were chosen, 977, 255 and 175 t/s. These were derived mainly from the LIP sediment rating and in the case of the upper and middle values, correspond to extrapolations of the upper envelope of data points and the mean rating respectively. The low value represents an extrapolation of the sediment concentrations observed at the barrage during the 1976 and 1986 floods. The three chosen sediment transport rates represent the probable range that might be encountered during the design flood.

The results of the simulation are shown in Figures 11 and 12. On the basis of an analysis of the variations of stage/discharge ratings downstream of the Barrage (Ref 1), tailwater levels at design flow might lie in the range 200 to 205 ft above mean sea level. The upper limit was chosen for present purposes so as to represent worst-case conditions in terms of high water levels, although not necessarily in terms of drawdown and scour at the Barrage.

On the basis of the analysis of the 1976 and 1986 floods, a Manning's 'n' of 0.015 was chosen. The results show that flooding is possible at all likely levels of sediment inflow. If there were a change to the lower regime flow, then of course water levels would be significantly higher and flooding more pronounced. Such a regime change would scour out the channel; if this coincided with a fall in sediment inflow, other excessive scour would take place at the barrage.

## CONCLUSION

For specified flood discharges and tailwater levels in the Indus river close to Sukkur Barrage, the risk of flooding and scour depends upon sediment inflow and channel roughness. On the basis of observed flood behaviour, there is no single valued relationship between either sediment concentration, or channel roughness, and discharge. In general, however, flow appears to be in upper regime on a rising flood hydrograph. Under these conditions, channel resistance is relatively low and water velocities are correspondingly high. At about the time of peak flow, the regime becomes transitional; channel resistance increases on the falling hydrograph and lower regime conditions are re-established.

Significant changes in sediment inflow may modify this pattern, as happened in 1976; a sudden fall in sediment concentration led to abrupt scour and rapid transition to lower regime flow. Design procedures must take account of the possibility of regime changes, particularly around peak discharge. The modelling of design floods should therefore be carried out with a combination of representative sediment concentrations and channel roughnesses. Channel roughness may be expressed in terms of channel resistance; the White-Paris-Bettes equation may be used in lower regime flow and a plane bed function in upper regime.

Design calculations for the Indus river must allow for uncertainty, but where there is serious risk of loss of life, agricultural production and infrastructure, a worst case scenario is advisable.

**Acknowledgement**

The authors are grateful to the Irrigation and Power Department of Sind Province, Islamic Republic of Pakistan, for permission to present this paper.

**References**

1 Sukkur Barrage Flood Management Review, Sir M MacDonald & Partnrs Limited, Cambridge, UK, Sept 1987.

2 NESPAK - Feasibility Studies for Proper Management of Flood Flows of Sukkur Barrage. National Engineering Services (Pakistan) Ltd.

3 Ackers, P and White, W R, Sediment Transport new approach and analysis. J Hydrauli Dir ASCE, Nov 1973.

4 White, W R, Paris, E and Bettes, R, The Frictional Characteristics of Alluvial Streams: a New Approach, Proc Instr Civ Engrs, Sept 1980.

5 Lower Indus Report, Physical Resources, Hunting Technical Services Limited & Sir M MacDonald & Partners, 1965.

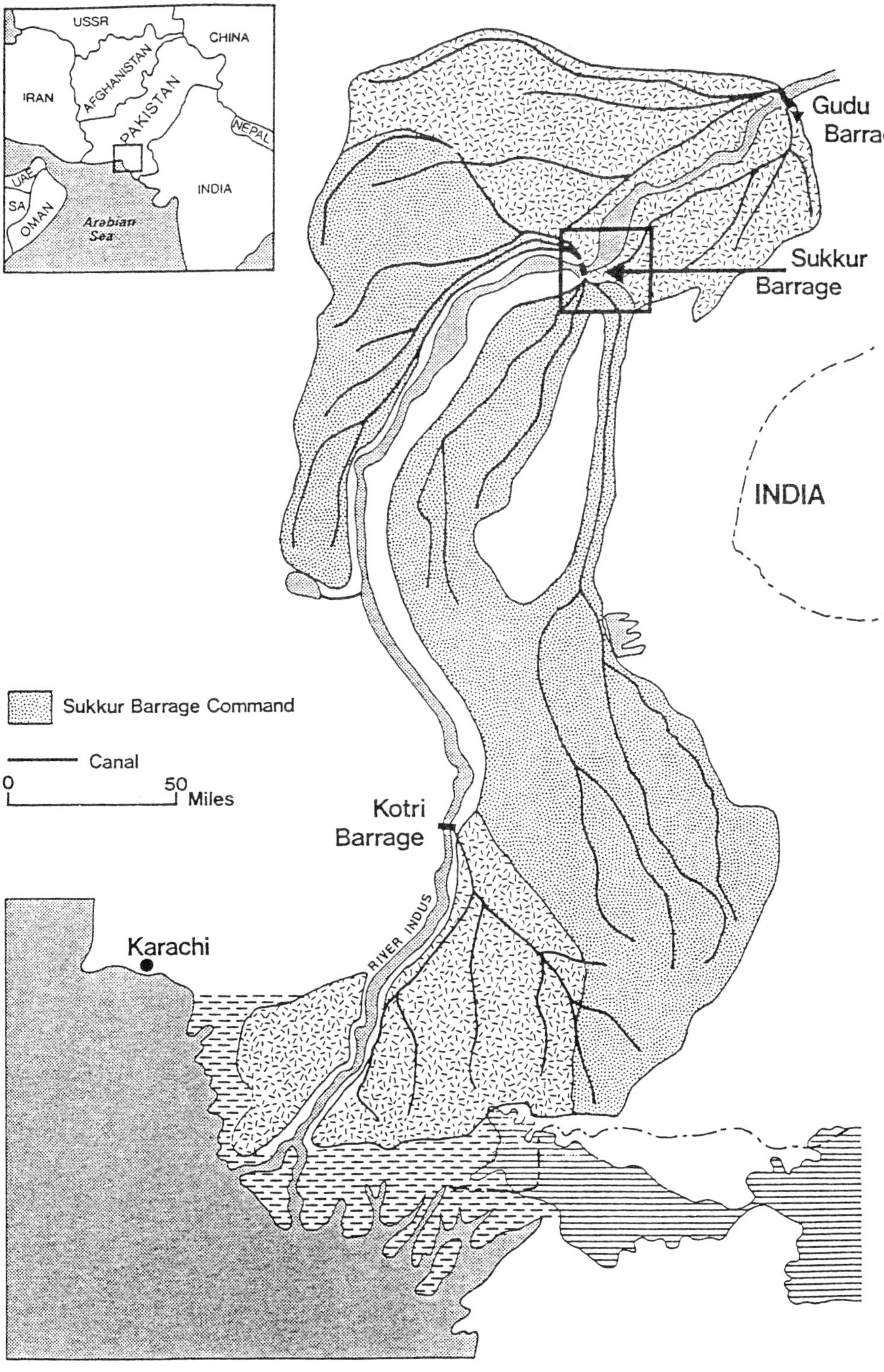


Figure 1 Location Plan

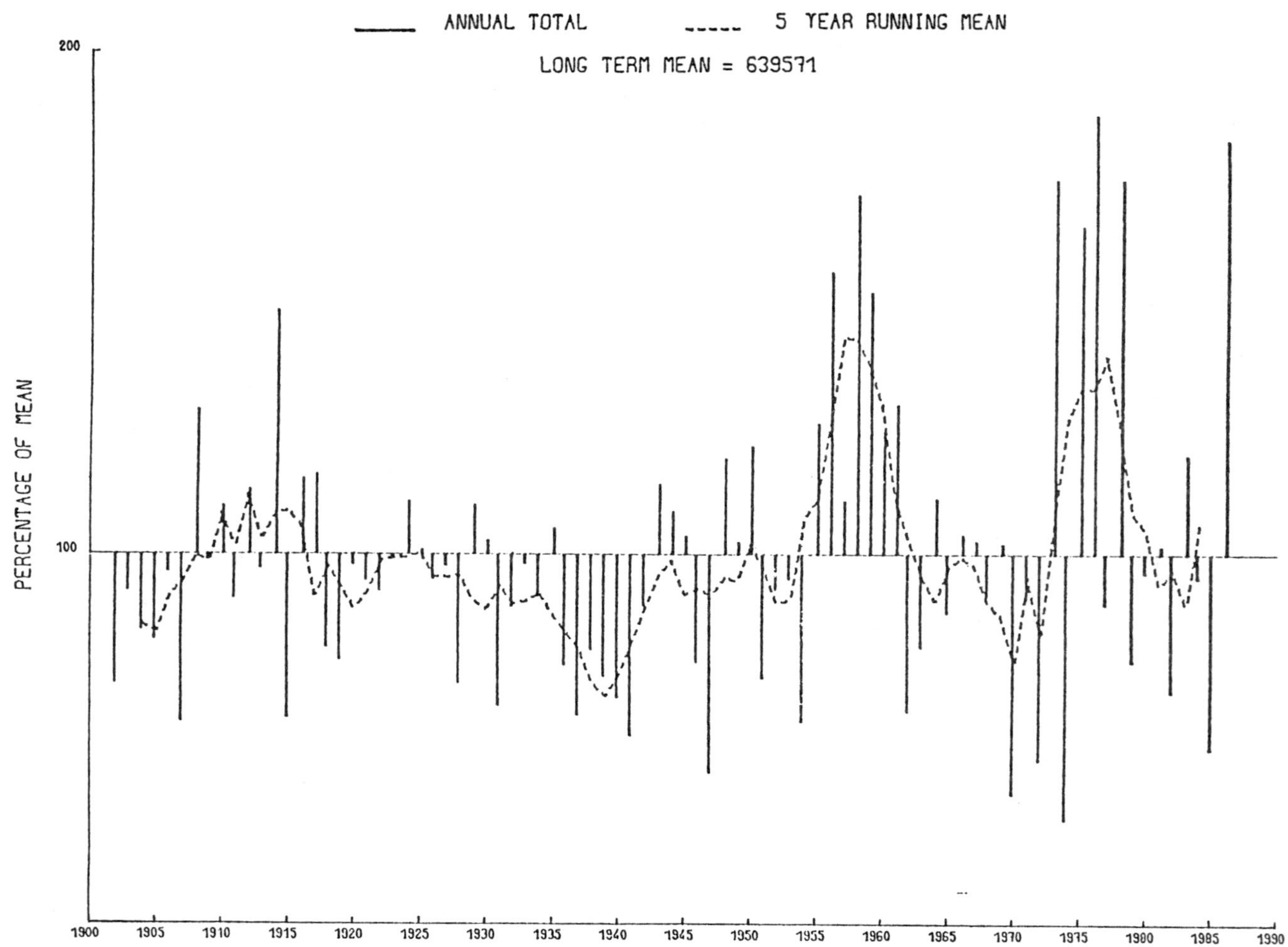


**Figure 2** Annual Maximum Discharges at Sukkur Barrage

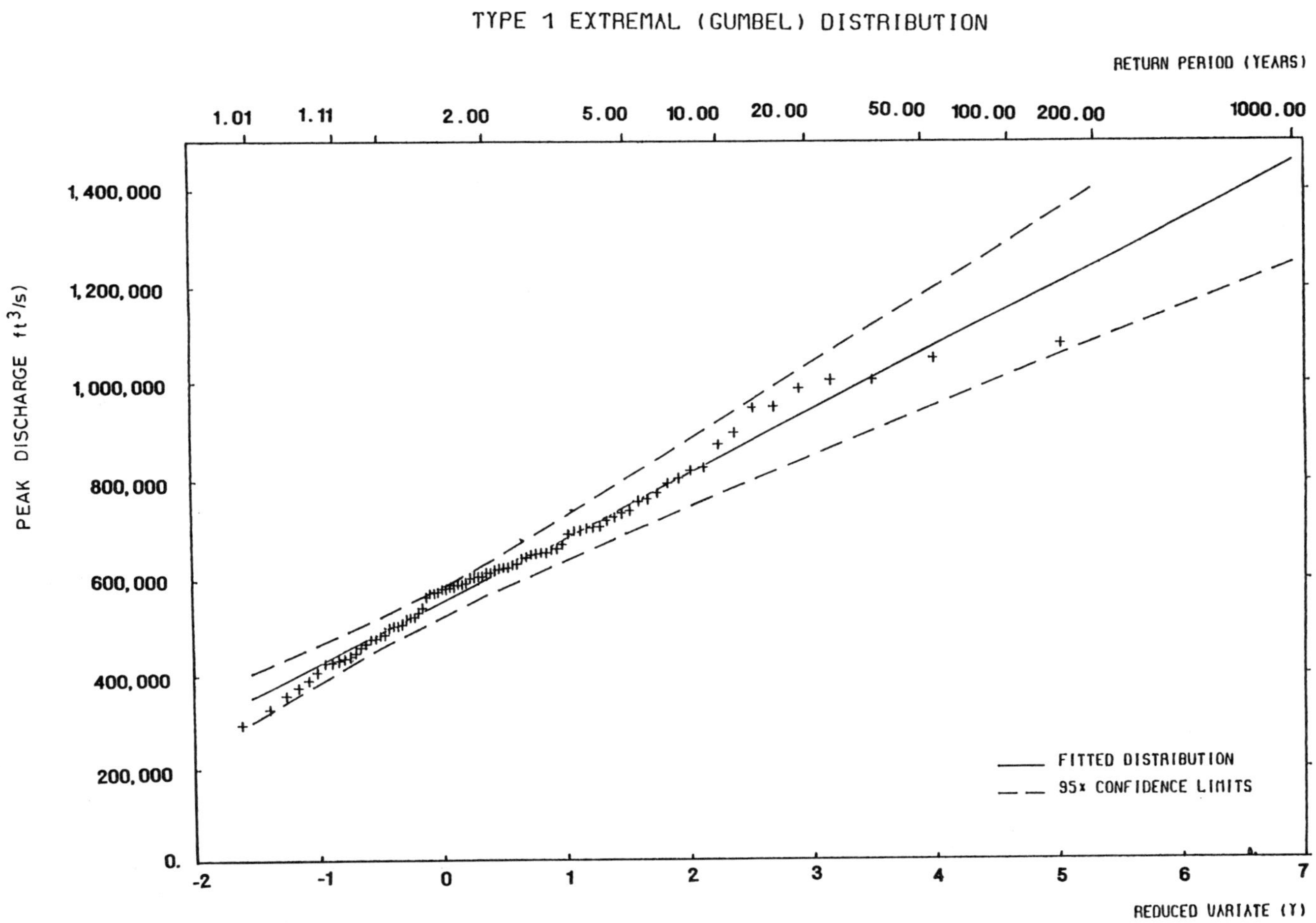


**Figure 3** Flood Frequency at Sukkur Barrage - Adjusted Record

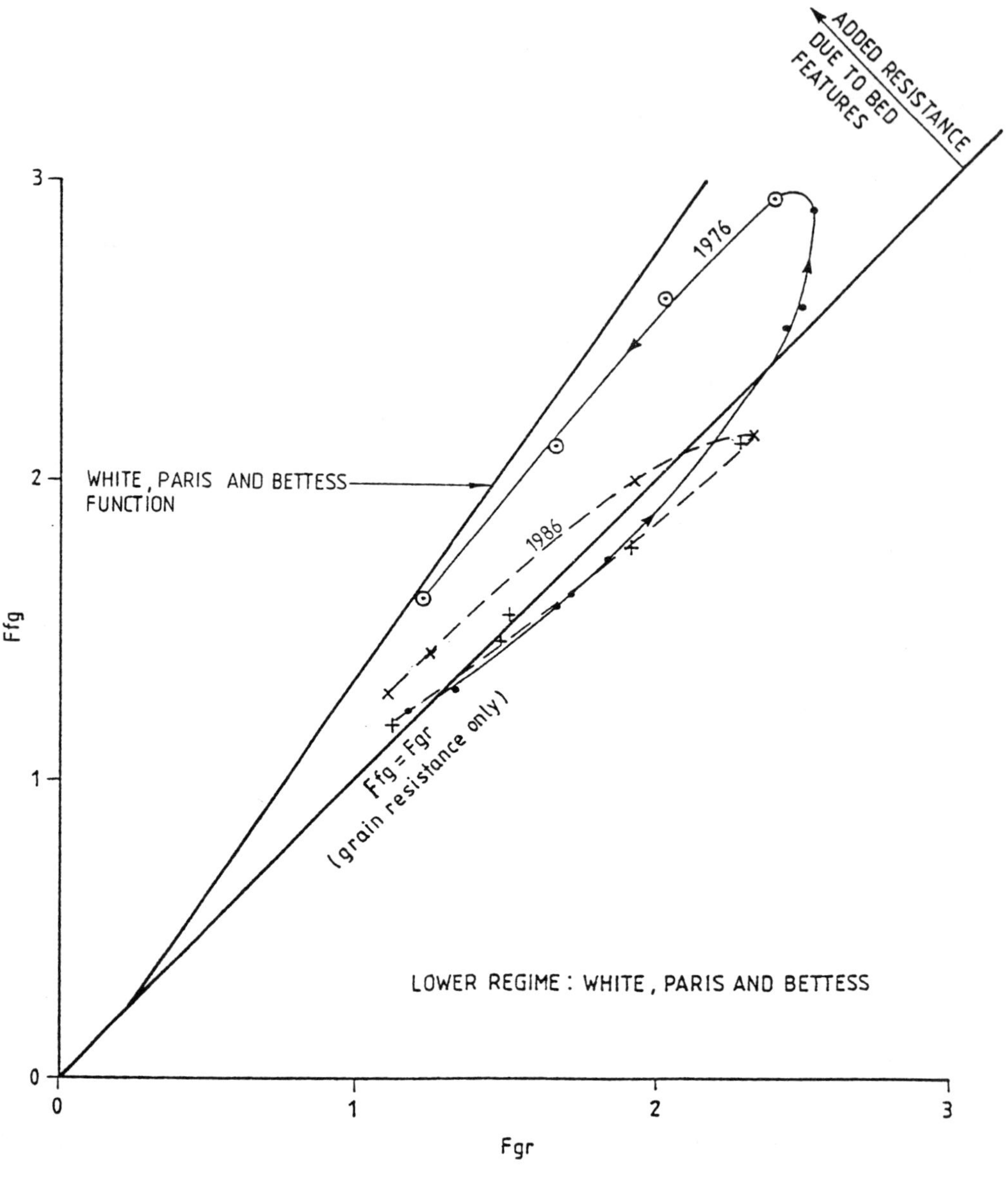


LEGEND

1976 • Before peak

⊙ After peak

1986 + Before peak

× After peak

Dgr. = 4.63

Agr. = 0.247

LOWER REGIME: $\frac{Fgr - A}{Ffg - A} = 0.7007$

**Figure 4** **Alluvial Resistance During 1976 and 1986 Floods**

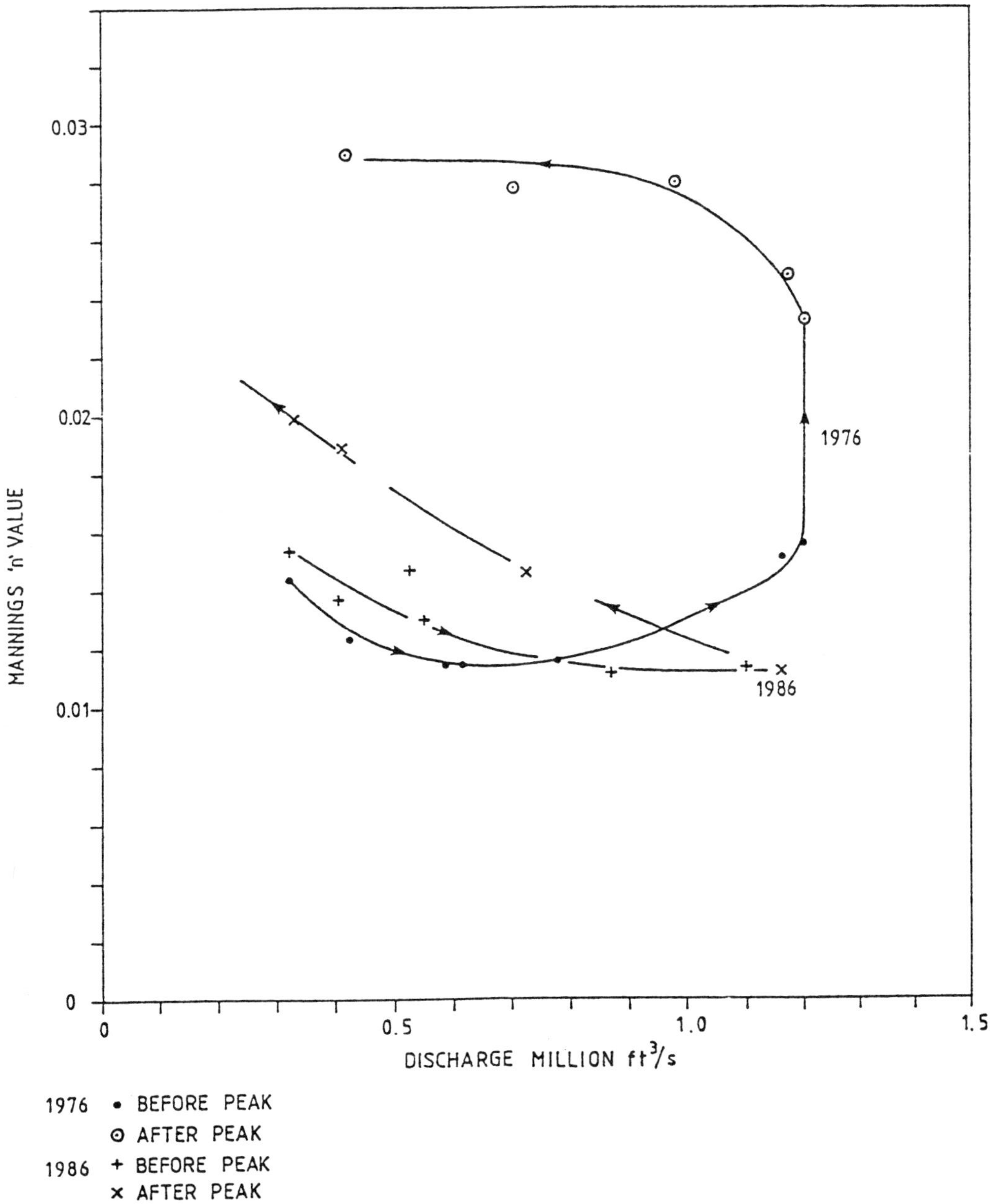


**Figure 5** Manning's 'n' at Outfall During 1976 and 1986 Floods

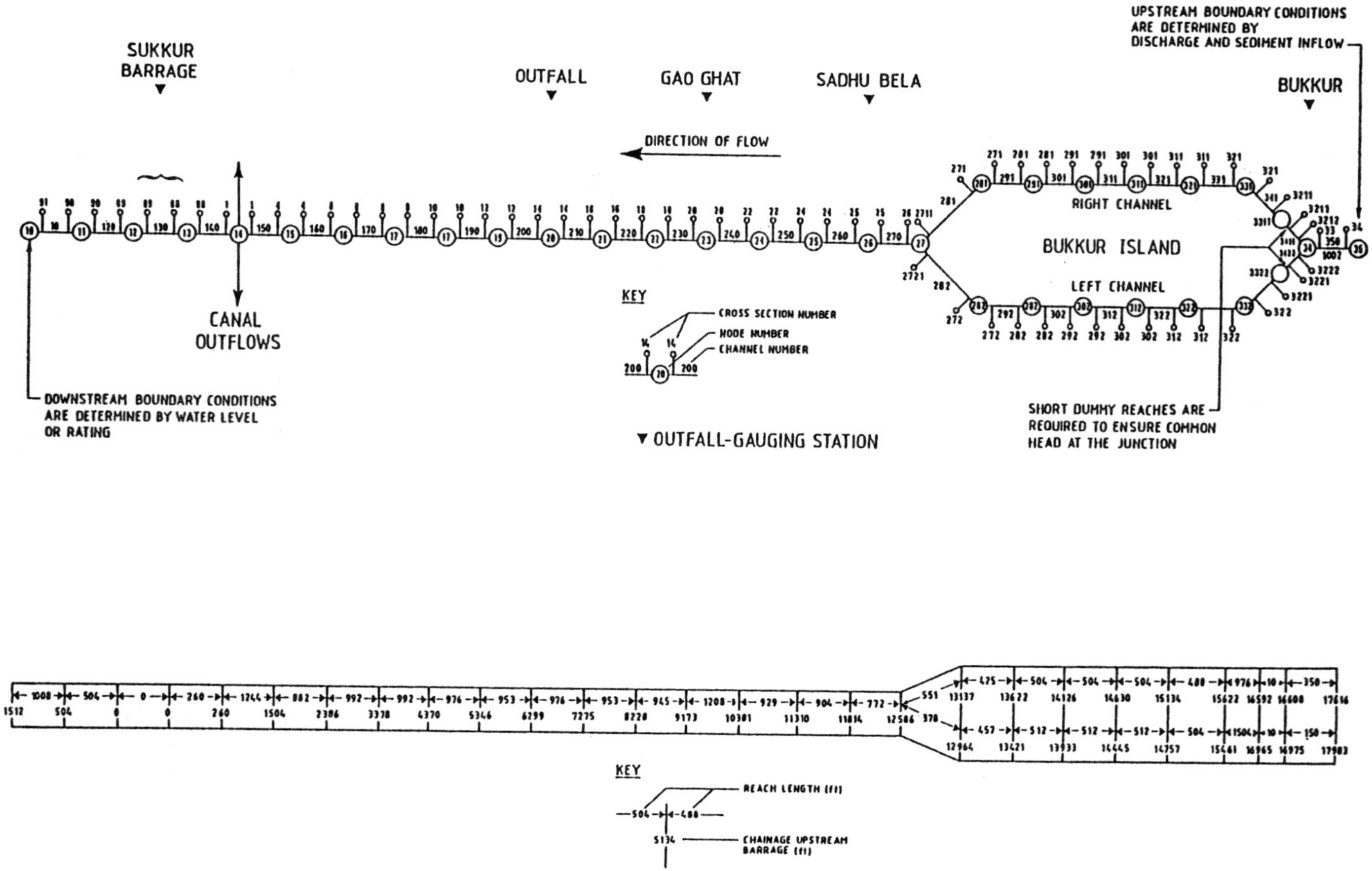


Figure 6 Schematic View of Hydraulic Network

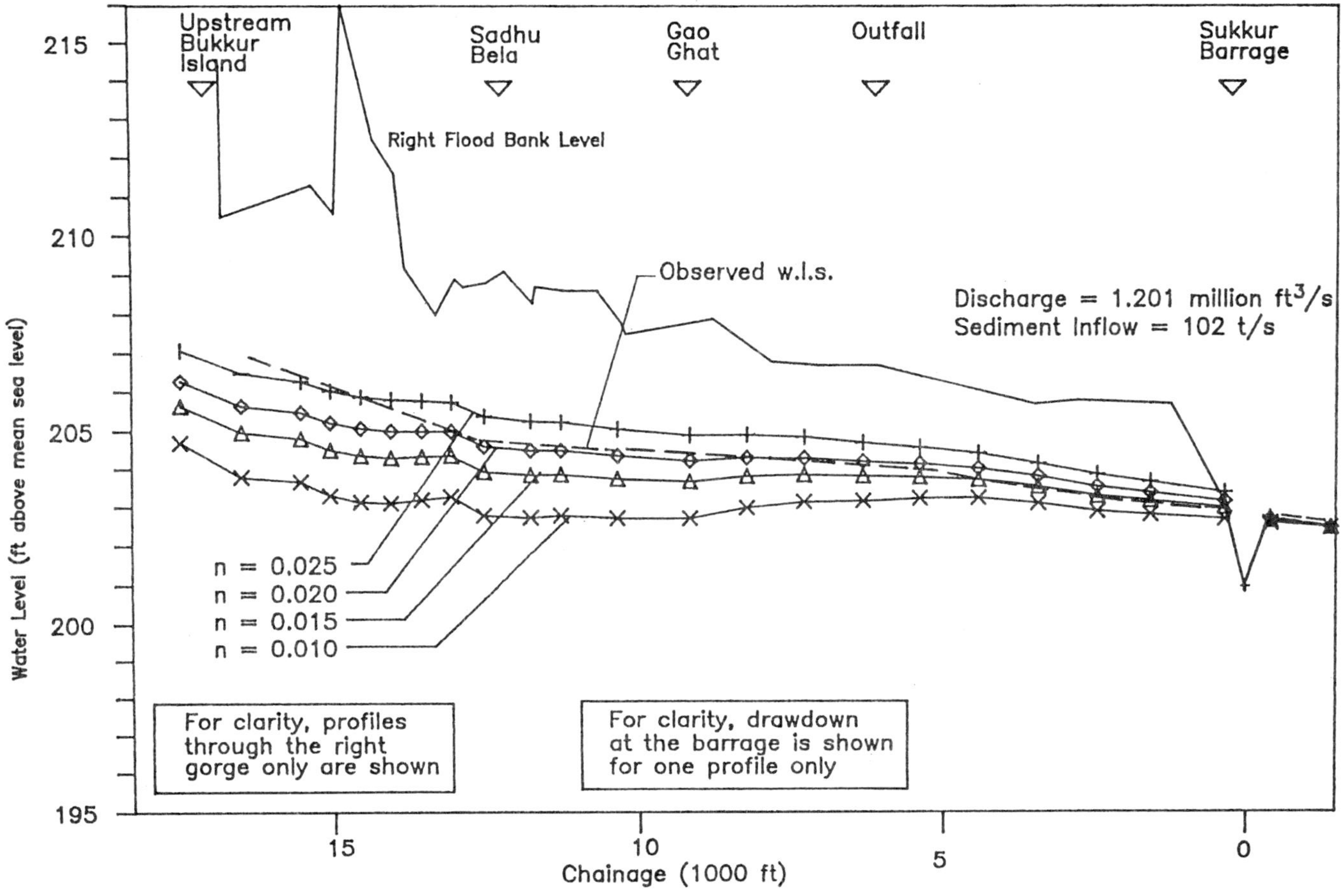


Figure 7 1976 Simulated Water Levels

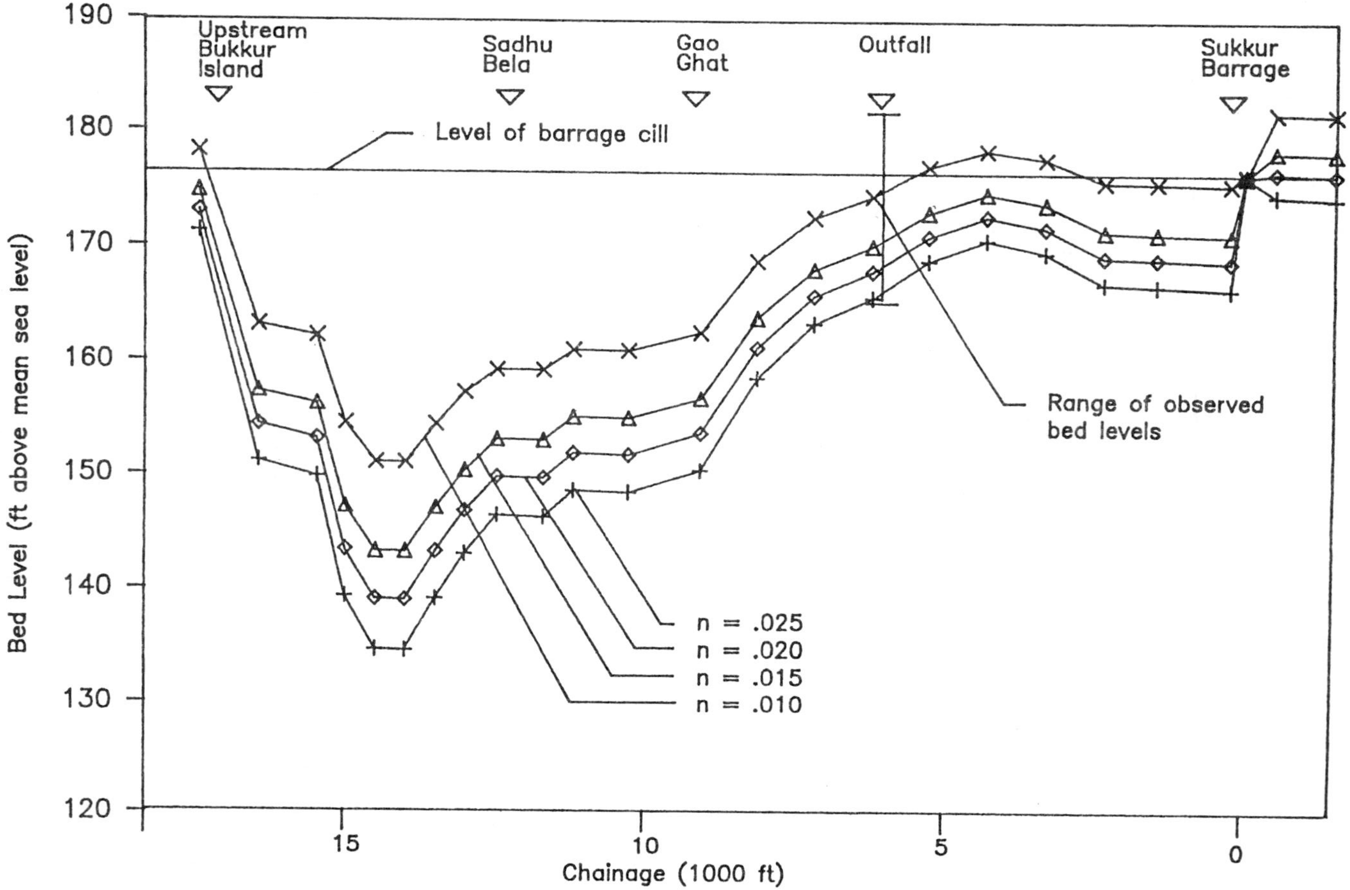


Figure 8 1976 Simulated Bed Levels

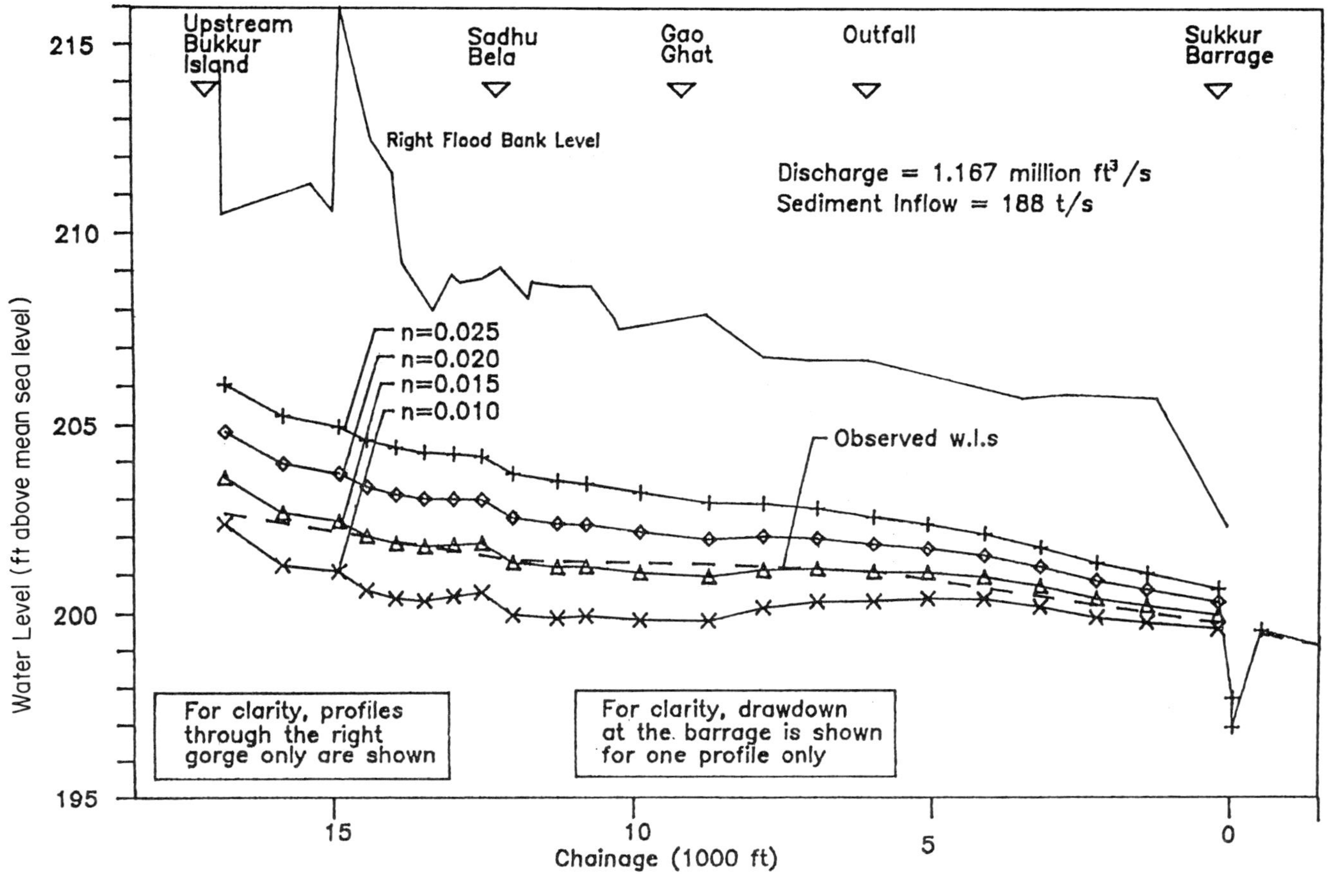


Figure 9 1986 Simulated Water Levels

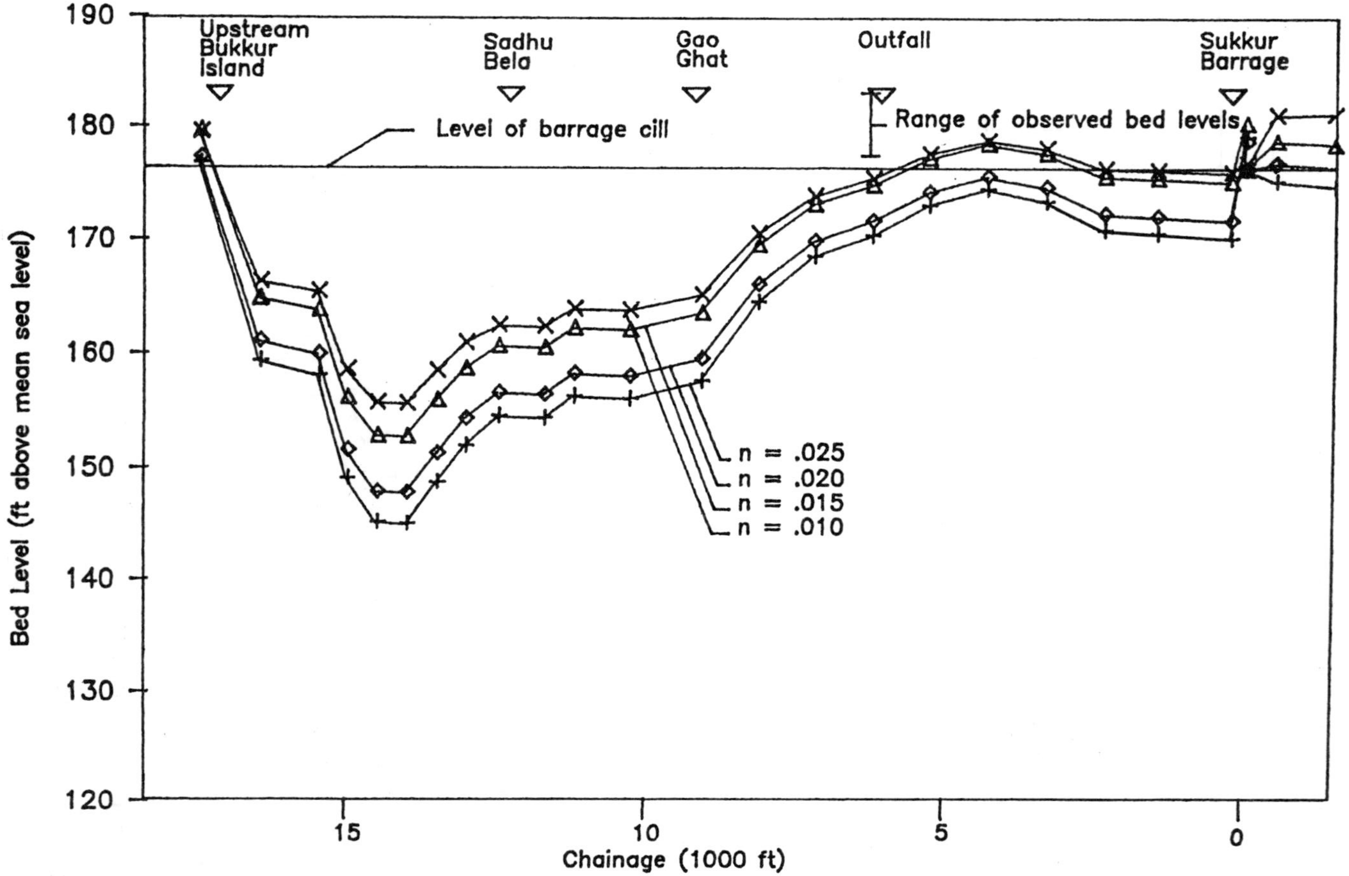


**Figure 10** 1986 Simulated Bed Levels

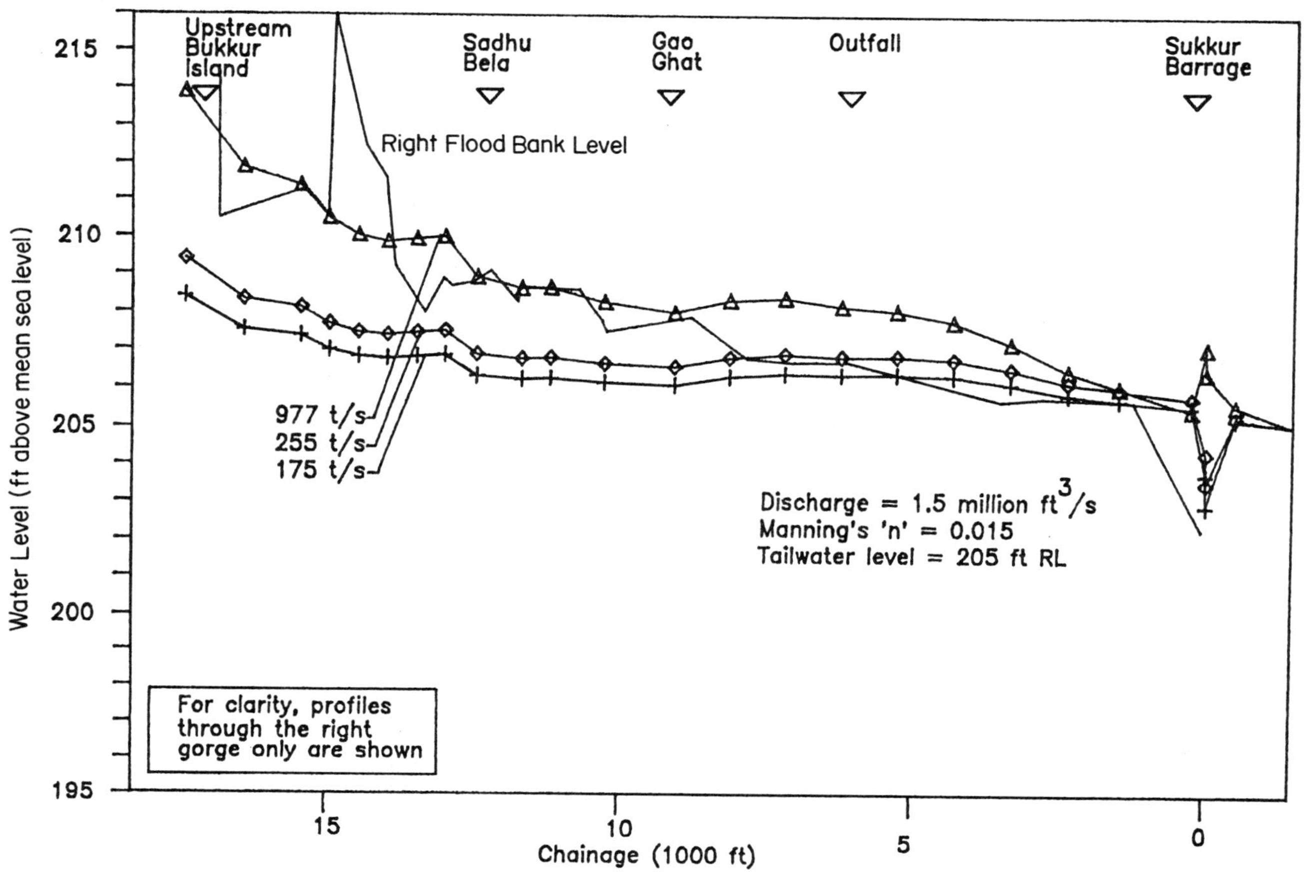


**Figure 11** Simulated Water Levels at Design Discharge (Return Period 200 Years)

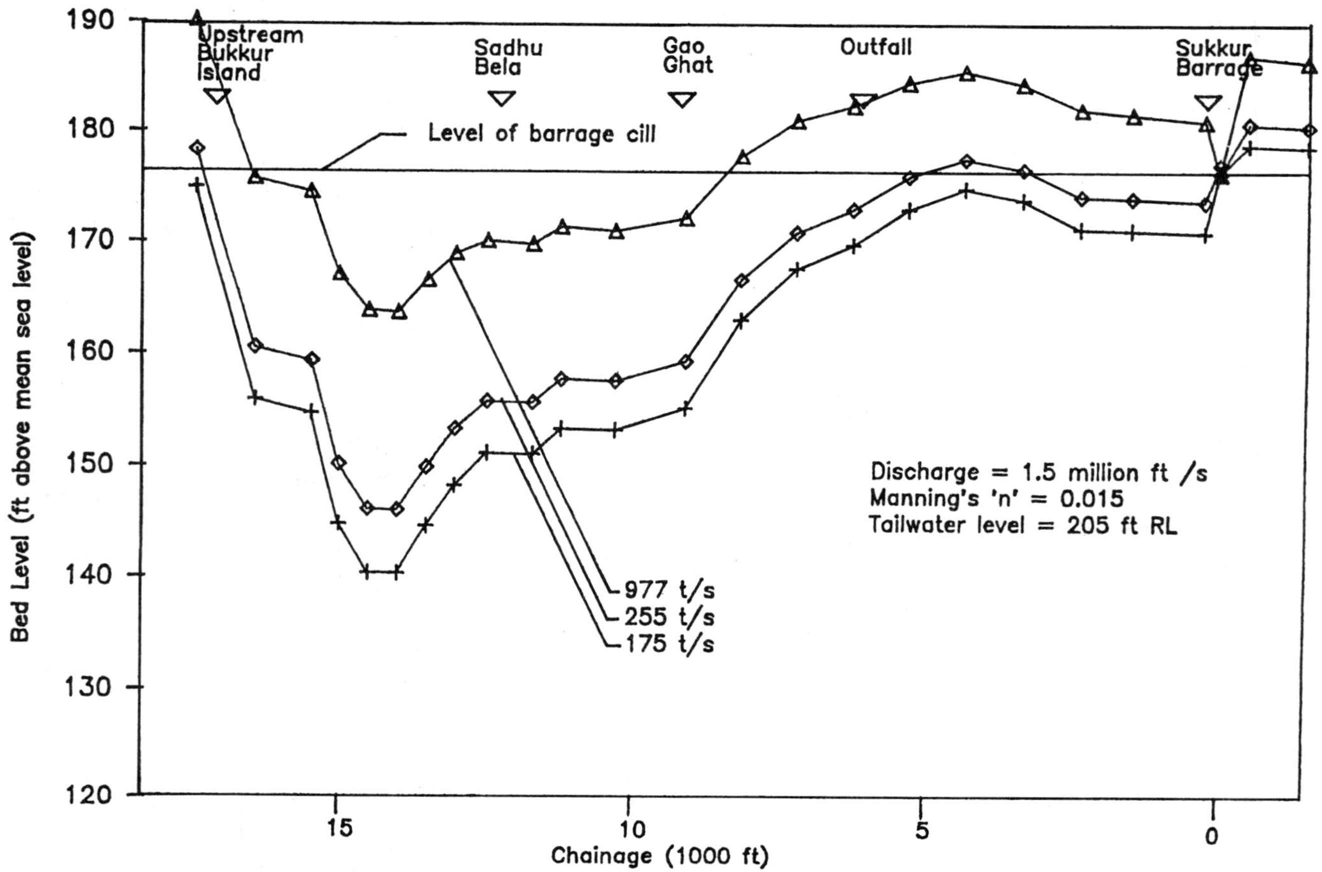


**Figure 12** **Simulated Bed Levels at Design Discharge (Return Period 200 Years)**

International Conference on
RIVER REGIME
18-20 May, 1988

PAPER H3

# THE APPLICATION OF BED STABILITY CRITERIA IN RIVERS WITH CONSECUTIVE RUN-OF-RIVER POWER PLANTS

Helmut Scheuerlein

Obernach Hydraulics Laboratory

Technical University Munich

SUMMARY: Using the Lech River in Bavaria as an example, the flow phenomena and the sediment behaviour in an almost completely controlled river are explained. For one selected range between two consecutive barrages the stability of the river bed is investigated by means of five different approaches (HJULSTRÖM, NEILL, MEYER-PETER, SHIELDS, ZANKE). The results are compared and discussed.

## 1 INTRODUCTION

Whenever a dam or a barrage is built in a river, degradation of the river bed downstream of the structure is the consequence. Degradation must not be mistaken as scouring due to insufficient energy dissipation downstream of the structure. Degradation occurs because the sediment balance of the river is disturbed by backing up the water upstream of the barrage causing settling down of a substantial portion of the sediment transported by the river. When the flow conditions downstream of the barrage are not changed significantly, as common in the case of run-of-river power plants, the deficiency concerning the sediment load of the flow is compensated by picking up sediment from the bed. This process continues until the slope of the bed and in consequence the tractive forces of the flow have decreased sufficiently to be balanced by the resistance of the bed material, and/or the river bed has experienced some armouring through accumulation of coarse material which could not be carried away by the flow.

Very often the development of a new equilibrium cannot be awaited, or the degradation corresponding to the future equilibrium cannot be tolerated. In this case the construction of a further barrage downstream is usually considered the only solution to the problem. However, downstream of the new barrage the conditions are the same as described previously, i.e. the problem has not been solved for good, it just has been shifted further downstream. Consequent adherence to the above mentioned solution finally leads to a more or less uninterrupted chain of barrages in the river. As long as the exploitation of the water power potential of the river proves to be economic, the construction of one barrage (respectively run-of-river power plant) after the other is an acceptable if not desirable solution. However, when the profitability of further power plants becomes questionable the readiness to build further barrages diminishes.

In the Federal Republic of Germany there are some rivers where exactly this has happened. In various cases progressive degradation downstream of run-of-river power plants has forced to add further barrages although from economic point of view the investment could not be expected to pay off. The power plants have been built mainly in order to stop further degradation (Lech, Wertach, Isar, Inn).

International Conference on River Regime, edited by W.R.White.

The installation of turbines and the production of hydroelectric energy was decided upon merely in order to get at least a certain benefit out of the investment.

It is generally taken for granted that a chain of consecutive barrages presents the final answer to all sediment problems. It is true that an uninterrupted chain of barrages contributes substantially to the stability of a river, however, it is advisable not to believe unlimitedly in the idea that now all sediment problems were definitely put to an end. Even in a supposedly well controlled river, for correct evaluation of the situation it is inevitable to investigate carefully what might happen at varying flow conditions.

## 2 BED STABILITY IN A RIVER CONTROLLED WITH CONSECUTIVE BARRAGES

Let us consider the hypothetical situation that the hydrostatic upstream water level of one barrage is identical with the downstream water level of the next barrage upstream. This means that the backing up effect covers the whole range between the two barrages. There is no river portion left with undisturbed flow conditions. Strictly speaking, however, this is only true for purely hydrostatic conditions, i.e. for discharge $Q = 0$. At discharges $Q > 0$ the range of the backing up effect decreases, leaving a certain river portion with undisturbed flow. The length of this river portion depends on the discharge. The higher the discharge the longer the reach of uninfluenced flow conditions, i.e. the shorter the extension of the backing up effect.

It is obvious that bed protection is restricted to the range where backing up is effective. This means that the river portion characterized by undisturbed flow conditions is again subject to erosion and degradation. For proper evaluation of the morphological condition in a river with consecutive barrages it is necessary to investigate the real situation by means of some bed stability criteria.

## 3 BED STABILITY CRITERIA

Erosion starts when the flow-induced forces acting on the river bed exceed the resisting forces of the bed material. The process depends on the intensity flow and on the properties of the bed material. There are numerous approaches to an analytical formulation of the equilibrium describing the critical or threshold situation. The various approaches can be classified in two general groups, namely

- stability criteria based on critical flow velocity
- stability criteria based on critical shear stress.

Out of a large number of analytical approaches, five were selected to be applied to the practical case this paper is dealing with. Two of them can be seen as exponents of the first group (critical flow velocity: HJULSTRÖM, NEILL), The remaining three belong to the second group (critical shear stress: MEYER-PETER, SHIELDS, ZANKE).

HJULSTRÖM (in GRAF, Ref.1) presents the critical condition in the form of a graphically given relation

$$v_c = f(d_m). \qquad (1)$$

NEILL (Ref. 3) describes the critical condition with the equation

$$\frac{v_c^2}{(\frac{\rho_s}{\rho} - 1) \cdot g \cdot d_m} = 2 \cdot (\frac{d_m}{h_c})^{-1/3} \qquad (2)$$

According to MEYER-PETER (Ref. 2) the critical condition is represented by the equation

$$\frac{\tau_c}{(\rho_s - \rho) \cdot g} = \frac{0,047 \cdot d_m}{\frac{Q_s}{Q} \cdot \left(\frac{k_s}{k_r}\right)^{3/2}} \tag{3}$$

In SHIELDS' well-known diagram (Ref. 4) the critical condition can be seen as being represented by

$$\frac{\tau_c}{(\rho_s - \rho) \cdot g \cdot d_m} = f\left(\frac{v_{*c} \cdot d_m}{\nu}\right) \tag{4}$$

ZANKE (Ref. 5) presents two equations describing critical conditions

$$\frac{\tau_c}{(\rho_s - \rho) \cdot g \cdot d_m} = 0,13 \cdot \frac{\nu}{v_{*c} \cdot d_m} \quad \text{for} \quad \frac{v_{*c} \cdot d_m}{\nu} \leqq 3,32 \tag{5a}$$

and

$$\frac{\tau_c}{(\rho_s - \rho) \cdot g \cdot d_m} = 0,432 \left(\frac{\nu}{v_{*c} \cdot d_m}\right)^2 + 0,04\left(1 - 3,32 \frac{\nu}{v_{*c} \cdot d_m}\right)$$

$$\text{for} \quad \frac{v_{*c} \cdot d_m}{\nu} > 3,32 \tag{5b}$$

In the following the application of the above mentioned five bed stability criteria by means of a practical example will be demonstrated.

## 4 INVESTIGATION OF BED STABILITY BETWEEN TWO CONSECUTIVE BARRAGES AT THE RIVER LECH

The Lech river in Southern Bavaria is an interesting example of a river with an almost uninterrupted chain of consecutive barrages resp. run-of-river power plants (Fig. 1). 20 barrages are existing already, 7 more are projected. One of them is Barrage 16 at Landsberg. The river portion between the existing Barrage 15 and the planned Barrage 16 is subject to the investigation which shall be reported here. The plan of the river range in question is shown in Fig. 2, the appertaining cross-sections are given in Fig. 3.

The hydrostatic water level of the planned Barrage 16 is identical with the minimum downstream water level of Barrage 15. The water levels corresponding to four representative discharges (80, 440, 900 and 1500 $m^3/s$) could be taken from a hydraulic model of Barrage 16 (scale 1:35) which covered the range from Barrage 16 to km 85,4. The water levels of the river portion between km 85,4 and Barrage 15, not covered by the hydraulic model, were caluculated using the Manning-Strickler equation. The water levels are shown in Fig. 4.

Based on these water levels the corresponding flow velocities and shear stresses were calculated. The results are given in Fig. 5 and 6. Additionally in Fig. 5 and 6, critical values of flow velocity (after HJULSTRÖM) and shear stress (after SHIELDS) are indicated. These values are based on the assumption that the bed material consists of quartz sand ($\rho_s$ = 2650 $kg/m^3$) with a mean grain size of $d_m$ = 28 mm.

From Fig. 4, 5 and 6 various conclusions can be drawn:

1. Although the hydrostatic water level reaches up to the neighbouring barrage upstream the effect of bed protection is rather limited (the critical values are exceeded already at rather low discharges, see Fig. 6)
2. The flow conditions and in consequence also the forces acting upon the river bed are significantly influenced by the local topography of the river (affinity of the water levels, even those corresponding to low discharges, see Fig. 4).
3. There is evidence that the properties of the material forming the bed must have some discontinuity (significant peak of flow velocity respectively shear stress at km 85,6; see Fig. 5).
4. In the vicinity of Barrage 15 no further degradation must be expected (the erosion of the river bed between Barrage 15 and km 85,8 has already changed the flow conditions towards a stable state).

With respect to the results of the analysis further information from the field was collected. Consultation of old surveys indeed showed that considerable degradation has occured downstream of Barrage 15. Other checks at the site proved the evidence of some very local rock formations at km 85,6 which apparently have some armouring effect. Additional analyses of bed material showed that the grain sizes varied considerably, not only from one intake to the other but also within one sample.

In order to make a more realistic evaluation possible, in a further step of the investigation the grain size was considered an independent variable. With the definition of the critical value $d_c$ as the grain size which under given flow conditions just starts moving, it was possible to calculate for each cross-section corresponding values of critical grain size $d_c$ for a given discharge Q (or vice versa, i.e. the critical discharge $Q_c$ for a given grain size d). In Fig. 7 some results of this calculation are presented. By means of three representative cross-sections the correlation between grain size and discharge at threshold conditions is shown.

Fig. 7 can also be used to compare the performance of the five selected stability criteria. It is obvious, that there is no consistency in terms of a certain tendency. One cannot say that one criteria generally produces high values and another one low values of $d_c$. Fig. 7 shows the whole complexity of the problem. The same method of analysis can at one time be conservative and at another time be optimistic. Therefore it is hard to judge which one of the approaches might be the most reliable one. If the lack of producing extreme values is considered an indication of reliability, then the SHIELDS method can be addressed as being the one that deserves the most confidence.

## 5 CONCLUSION

The results which have been found through the investigation this paper is dealing with can be summarized as follows:

1. The existence of an uninterrupted chain of barrages does not necessarily mean that sediment motion comes to a complete standstill. The barrages simply function as fix-points in the longitudinal section of the river. Between these fix-points sediment motion will go on and substantial rearrangement of the bed configuration will take place. The final slope representing stable stage will not be significantly different from that one that would develop without barrages, however, due to the bed fixation at the barrages the length of the degradation wedge is

restricted to the distance between two barrages, hence the degradation is limited, too.

2. The actual behaviour of the river bed between two consecutive barrages depends on the topography of the river course rather than on the backing up effect of the downstream barrage. The backing up effect diminishes with increasing discharge and under certain circumstances disappears completely. Consequently the range of undisturbed flow and therefore the range of unhindered erosion of the bed increases.

3. For verification of the morphological changes between two consecutive barrages a thorough sediment transport analysis is necessary (analysis of bed stability, development of stable slope).

4. Simultaneous application of five different approaches concerning verification of bed stability shows that criteria based on critical velocity (HJULSTRÖM, NEILL) have the tendency to produce optimistic results for low flow velocities ($v < 2$ m/s) and conservative results for high flow velocities ($v > 2$ m/s), compared with stability criteria based on critical shear stress (MEYER-PETER, SHIELDS, ZANKE). Out of all methods that have been applied the one from SHIELDS proved to be the most balanced one, neither producing too conservative nor too optimistic results.

## NOMENCLATURE

| | | |
|---|---|---|
| $d_m$ | [m] | mean grain size of bed material |
| $d_c$ | [m] | critical grain size related to bed stability |
| $g$ | [m/s$^2$] | acceleration due to gravity |
| $h_c$ | [m] | flow depth corresponding to $v_c$ |
| $k_s$ | [m$^{1/3}$/s] | Strickler's roughness coefficient ($k_s = \frac{1}{n}$; Manning's n) |
| $k_r$ | [m$^{1/3}$/s] | grain roughness (Meyer-Peter) |
| $Q$ | [m$^3$/s] | total discharge |
| $Q_s$ | [m$^3$/s] | partial discharge related to sediment transport (Meyer-Peter) |
| $v$ | [m/s] | flow velocity |
| $v_c$ | [m/s] | critical flow velocity related to bed stability |
| $v_{*c}$ | [m/s] | critical shear velocity related to bed stability |
| $\rho$ | [kg/m$^3$] | density of water |
| $\rho_s$ | [kg/m$^3$] | density of sediment |
| $\nu$ | [m$^2$/s] | kinematic viscosity |
| $\tau$ | [N/m$^2$] | shear stress |
| $\tau_c$ | [N/m$^2$] | critical shear stress related to bed stability |

REFERENCES

1 GRAF. W.: "Hydraulics of Sediment Transport" New York, McGraw Hill, 1971, pp. 88-89.

2 MEYER-PETER, E. and MÜLLER, R.: "Formulas for bed-load transport". In: Proc. second IAHR Conference (Stockholm Sweden: June 7-9, 1948) Appendix 2, pp. 39-64.

3 NEILL, C.R.: "Note on initial movement of coarse uniform bed-material". Journal of Hydraulic Research, Vol. 6, No. 2, 1968, pp. 173-176.

4 SHIELDS, A.: "Anwendung der Ähnlichkeitsmechanik an der Turbulenzforschung auf die Geschiebebewegung". Mitteilungen der Preußischen Versuchsanstalt für Wasserbau und Schiffbau, Heft 26, 1936, pp. 26.

5 ZANKE, U.: "Grundlagen der Sedimentbewegung". Berlin, Springer, 1982, pp. 151-153.

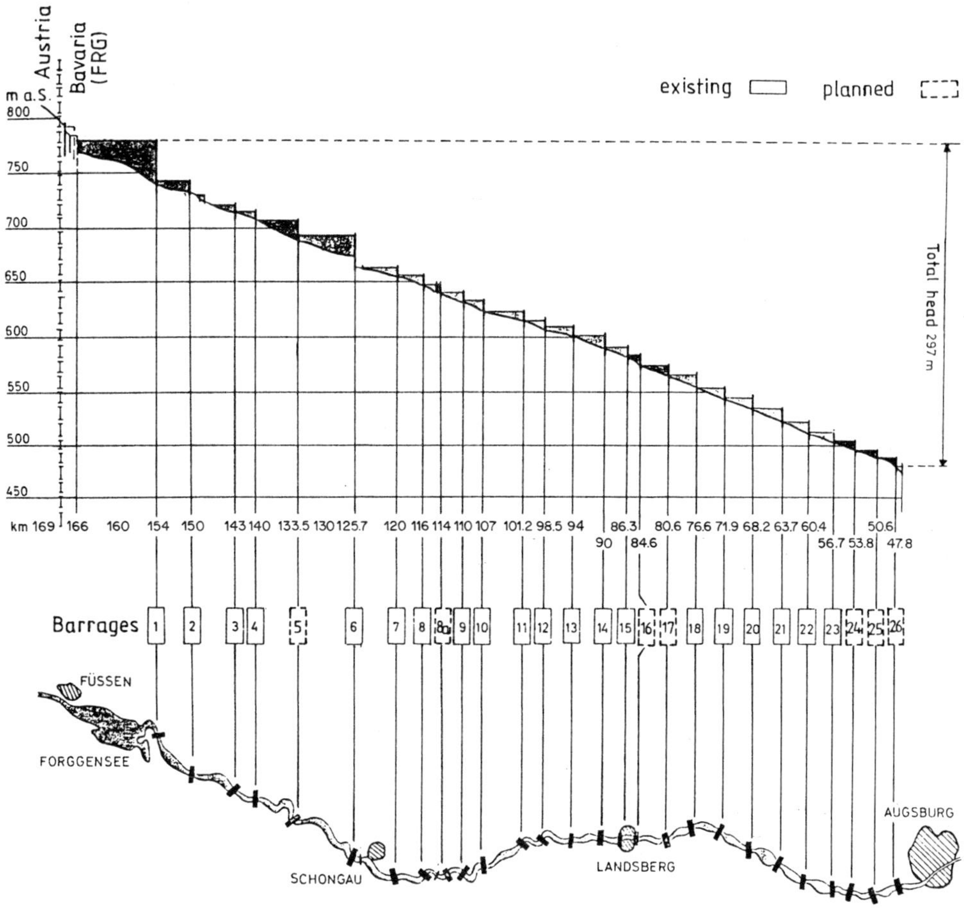


Fig.1: Consecutive chain of run-of-river power plants at the Lech river (state 1984)

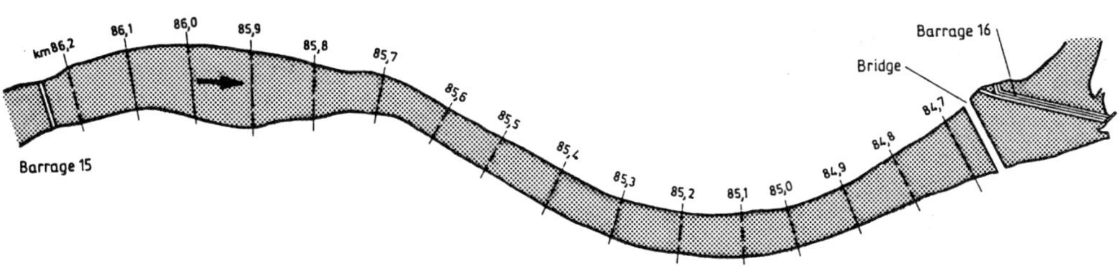


Fig.2: Plan of Lech river between Barrages 15 and 16

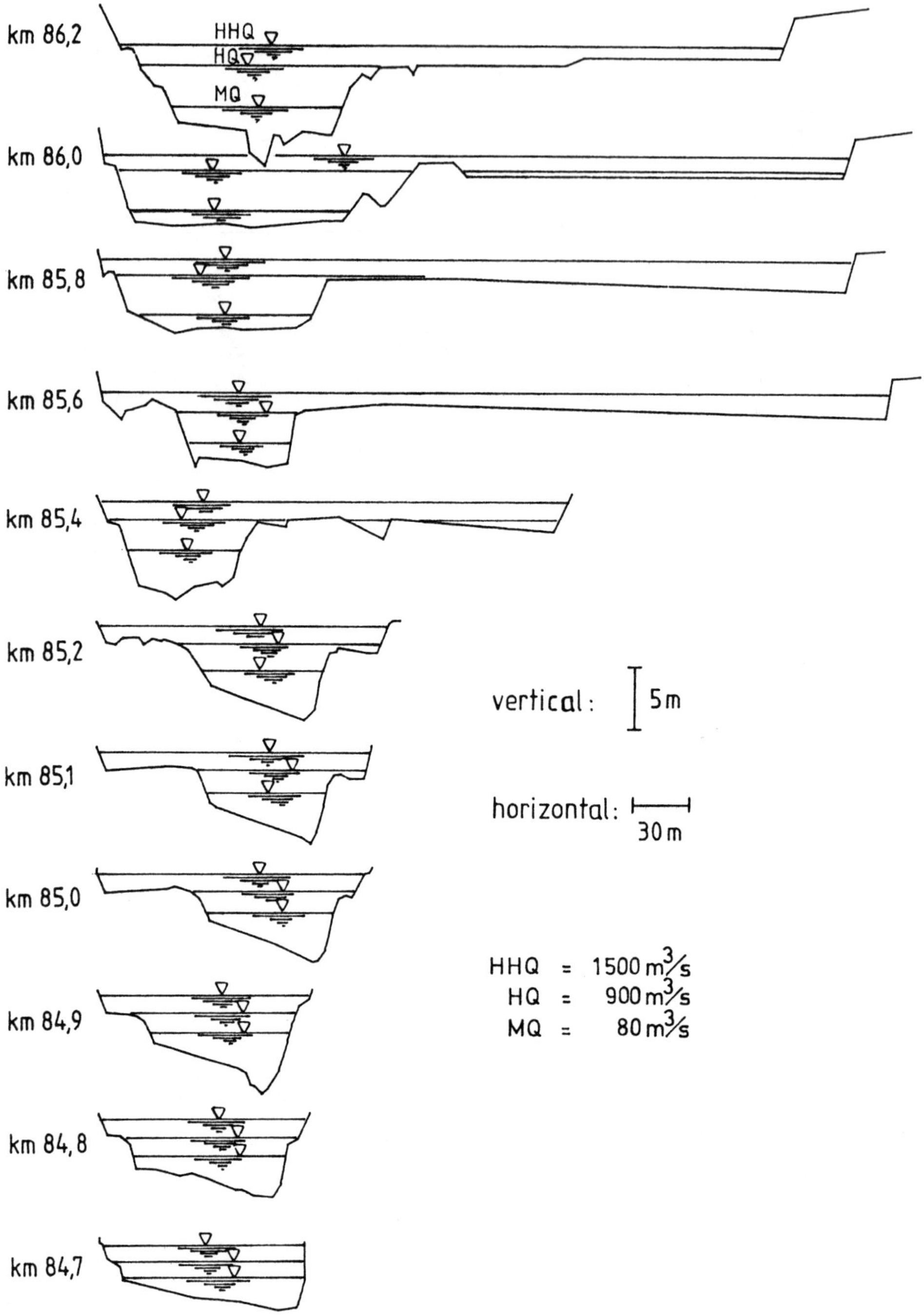


Fig.3: Cross-sections of Lech river between Barrages 15 and 16

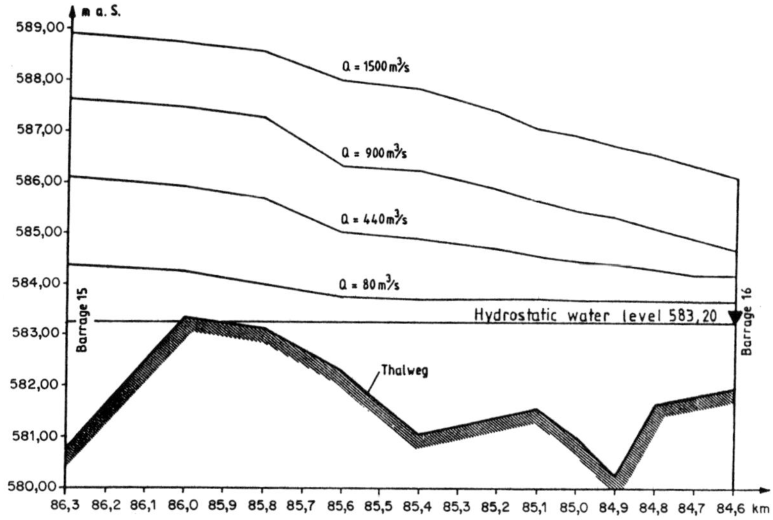


Fig.4: Water level in the Lech river between Barrages 15 and 16

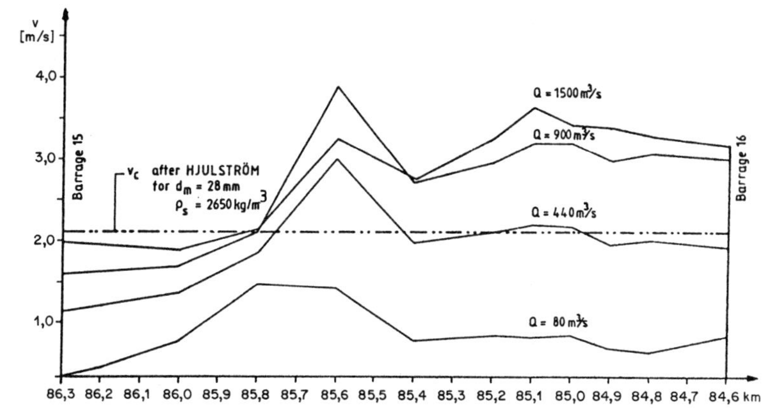


Fig.5: Flow velocity in the Lech river between Barrages 15 and 16

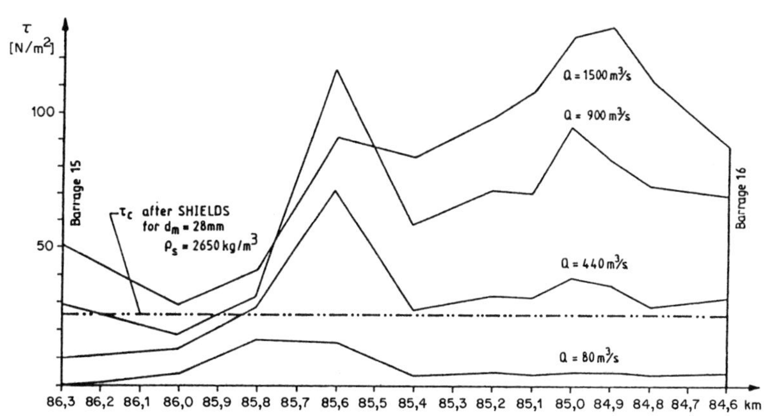


Fig.6: Shear stress in the Lech river between Barrages 15 and 16

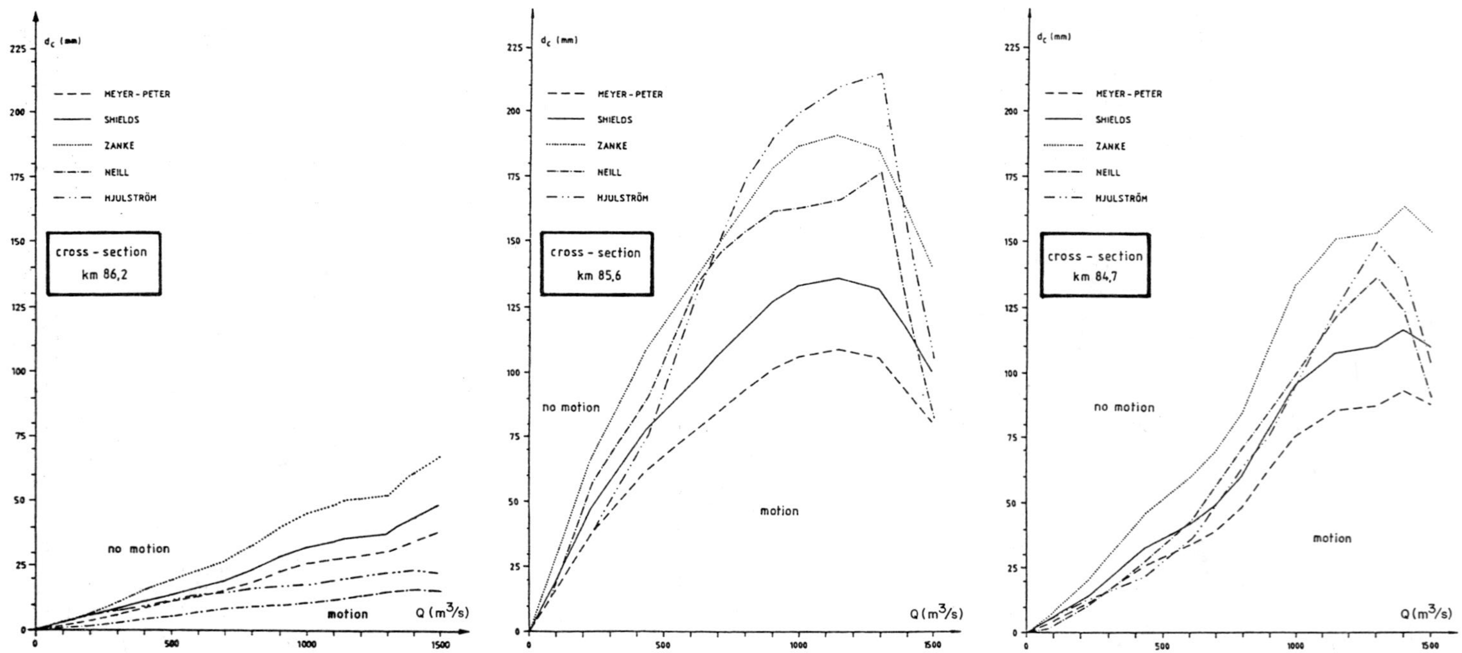


Fig.7: Critical grain size according to various stability criteria at three selected cross-sections

International Conference on
RIVER REGIME
18-20 May, 1988

PAPER I1

# REGIME CHANGES IN THE CALEDON RIVER ASSOCIATED WITH SEDIMENT DEPOSITION UPSTREAM OF WELBEDACHT BARRAGE

A Rooseboom and S J van Vuuren
Department of Civil Engineering
University of Pretoria, Pretoria, South Africa

SUMMARY : Different modelling techniques have been used to predict further deposition upstream of Welbedacht Reservoir, which has lost two-thirds of its capacity within fourteen years. Historical records, empirical extrapolation, mathematical modelling and a physical model have been used in the predictions. This paper deals with the historical record and empirical and mathematical analyses.

## 1. INTRODUCTION

Dams and weirs which are constructed across rivers which carry heavy sediment loads cause major changes in the river regimes over considerable distances. Insufficient provision for sediment build-up in the reaches upstream of a dam may necessitate costly remedial measures especially where developed areas become affected due to increasing flood levels.

Analyses of sediment build-up in these cases form important components of dam engineering and also provide insight into the relationships which govern river forming processes.

In the case of Welbedacht Reservoir physical and mathematical models as well as empirical relationships based on recorded sedimentation patterns were used to predict the rate and extent of further sediment build-up.

### Historical Background

Welbedacht dam was completed near Wepener on the Caledon River, South Africa, in 1973. It was created with the primary aim of providing Bloemfontein with water and it was predicted that the reservoir would have an assured yield of $198 \times 10^6$ m$^3$ per annum (Ref 1). According to the White Paper the average sediment load of the Caledon River at Jammersdrift was determined at $14,4 \times 10^6$ tonne per annum and it was envisaged that no more $13 \times 10^6$ m$^3$ of the storage capacity would be lost during the first 30 years of the dam's existence. Thus it was expected that only 11% of the reservoir's original gross storage capacity of $115 \times 10^6$ m$^3$ would have been lost through sedimentation by the year 2003.

It soon became evident that sedimentation was taking place at a much faster rate than originally envisaged. Surveys of the reservoir basin presented the following picture :

| Year | Percentage of storage capacity filled by sediment | Remaining capacity ($10^6$ $m^3$) |
|---|---|---|
| 1973 | 0 | 115,2 |
| 1976 | 32,5 | 77,7 |
| 1978 | 45,5 | 62,8 |
| 1981 | 51,7 | 55,7 |
| 1982 | 50,6 | 56,9 |
| 1986 | 65,6 | 39,6 |
| 1987 | 66,7 | 38,3 |

Different internal reports by the Department of Water Affairs indicated that a serious situation was developing and that urgent and intensive action was required.

It became necessary to predict both the storage volume that would remain as well as the sediment levels at different sections as functions of time.

## 2. SEDIMENT DISCHARGE PATTERNS

Regular sampling of suspended sediment on a daily basis was undertaken just upstream of Welbedacht by the Department of Water Affairs during the period 1931 to 1942. Annual loads were calculated (Ref 2) and these are as follows :

| Hydrological year | Sediment Load ($10^6$ t) | Runoff ($10^6$ $m^3$) |
|---|---|---|
| 1931 - 1932 | 3,9 | 610 |
| 1932 - 1933 | 2,8 | 240 |
| 1933 - 1934 | 53,6 | 3 370 |
| 1934 - 1935 | 11,5 | 1 410 |
| 1935 - 1936 | 3,9 | 400 |
| 1936 - 1937 | 16,8 | 1 720 |
| 1937 - 1938 | 4,9 | 630 |
| 1938 - 1939 | 9,3 | 1 070 |
| 1939 - 1940 | 5,7 | 890 |
| 1940 - 1941 | 9,2 | 990 |
| 1941 - 1942 | 8,1 | 890 |
| AVERAGES : | 11,8 x $10^6$ tonne/annum | 1 110x $10^6$ $m^3$/annum |

TABLE 1 : ANNUAL SEDIMENT LOADS AND RUNOFF VALUES RECORDED AT JAMMERSDRIFT (D2M01)

The high degree of variability in annual loads is typical of rivers which carry heavy loads of fine sediment in suspension. Generally speaking the loads are availability-limited and the carrying capacity of the river tends to be far in excess of the available load of fine sediments. Under these conditions no direct relationship exists between sediment concentration and discharge (Fig 1) and it is only when the stream is retarded to the extent that the carrying capacity becomes limiting that a singular relationship exists between sediment concentration and flow characteristics (Fig 2). In this case recorded concentration values were plotted against average stream power.

## 3. EQUILIBRIUM CONDITIONS

Sediment deposition within a reservoir continues until the sediment transporting capacities of inflows remain high enough to carry incoming sediment loads right through. Some sedimentation tends to take place during periods of lower inflows and scour may occur under high flood conditions. A stage is however reached when the average storage capacity decreases very slowly with time.

Welbedacht reservoir should soon be approaching such a condition. The remaining storage capacity when the sediment profile approaches equilibrium conditions will be small and once again, depends on the velocities which are maintained through the reservoir.

Equilibrium conditions are very important as they represent the average long term balance between the transporting capacity of flows and the transportability of both transported and deposited materials. As such they also determine the shape and volume of the reservoir basin which can be maintained by means of a set of operating rules.

Whereas the conditions under which cohesionless material will be entrained can readily be quantified it is more difficult to quantify critical conditions for materials which are held together by cohesive forces. Cohesive forces play a dominant role in determining the transportability or resistance to entrainment of deposited fine sediments as in the case of Welbedacht Reservoir. The (critical) stream transporting capacity which is required to remove consolidated materials as well as that which is required to keep discrete particles in suspension can be expressed in terms of $(g\ R\ S_f)^{\frac{1}{2}}$, the so-called shear velocity (Ref 3).

A useful model for determining longer term equilibrium conditions in Welbedacht Reservoir is provided by an analogous example, Buchuberg Reservoir, a small sediment filled reservoir downstream of Welbedacht and the Caledon-Orange confluence on the Orange River.

Buchuberg dam was completed in 1931 and since 1959 its storage capacity has remained nearly constant. Upstream of the dam the bed material is so cohesive and compacted that it cannot be flushed out. It may be assumed therefore that especially under flood conditions the carrying capacity right through Buchuberg Reservoir is sufficient to pass all incoming sediments through and perhaps pick up some recently-deposited sediments. Sediments which were deposited in Buchuberg in the past are similar to those which have been entering Welbedacht reservoir. Critical flow conditions in Buchuberg should therefore be similar to those which will eventually prevail in Welbedacht reservoir. Fig 3 contains shear velocities which were calculated for 1 in 2 year and 1 in 10 year floods upstream of the dam. The minimum, relatively constant values observed up to 9 km from the dam represent the equilibrium conditions as neither scour nor deposition takes place on a significant scale.

## 4. EXTRAPOLATED SEDIMENTATION RATES FOR WELBEDACHT RESERVOIR

The historical progression of deposited volume in Welbedacht reservoir has been extrapolated by means of an empirical relationship which has been developed from records of older reservoirs (Ref 4). This curve is shown in Fig 4. Confidence in this curve is strengthened by the fact that the curve passes through both the 1982 and 1986 recorded values.

Independent (modelled) values of future sediment volumes have also been obtained from mathematical analyses based on :

- constant average sediment inflows
- trap efficiencies varying according to the Churchill relationship
- density functions as calibrated by Braune (Ref 5) for similar sediment deposits.

As can be seen in Fig 4, the "modelled" and "empirical" values agree well from 1983 onwards. The deviation prior to 1983 is attributed to exceptionally high sediment loads during 1974. All the indications are therefore that without special sediment flushing operating rules very little storage space will remain in Welbedacht reservoir by the year 1998.

The equilibrium shape of the bed profile (Fig 5) was calculated by means of a special backwater programme which not only balances the energy equation but also fills in the bed up to the level where equilibrium exists. In this way it is possible to predict what the longitudinal bed slope of the lower basin region will be. Typical cross-sections through deep deposits in South African reservoirs contain near-horizontal beds.

Further upstream, the very high sediment levels lead to very wide beds and ill-defined flow patterns. As these surfaces are above the full supply level of the reservoir, conditions here do not affect the storage capacity of the reservoir. The equilibrium bed slope in this region is important however for flood level calculations and was assumed to be equal to that which has been observed in the upper delta region under similar flow conditions.

The amount of storage space which could be maintained in future will depend upon the magnitude of minimum shear velocities which can be maintained within the reservoir under flood conditions.

If the reservoir level is kept at Full Supply Level in future, according to the mathematical models, less than $5 \times 10^6$ $m^3$ of storage space will be maintained. These analyses indicate that $10 \times 10^6$ $m^3$ of storage space could be retained if all incoming floods are to be passed through with the gates fully opened. Physical model studies have confirmed the results of the mathematical models.

## 5. CONCLUSIONS

(i) In the case of Buchuberg Reservoir very little storage space has been lost over the past 30 years and it has been possible to calculate critical flow conditions under which no further build-up of the bed takes place i.e. equilibrium regime conditions.

(ii) The critical flow conditions which were established for consolidated sediments within Buchuberg Reservoir have been used to predict the equilibrium storage capacities which can be maintained in another reservoir (Welbedacht) under different operating rules.

## 6. ACKNOWLEDGMENTS

The authors wishto thank the South African Department of Water Affairs for permission to publish the information contained in this paper. It is believed that this information will be useful to others who have to do prediction on equilibrium conditions upstream of smaller reservoirs on major rivers.

## 7. REFERENCES


(1) Government Printer (1968), Secretary for Water Affairs, Pretoria South Africa. Second Supplementary Report on the Orange River Development Project. White Paper LL-68.

(2) Rooseboom A, and Maas N F (1974: "Sedimentafvoergegewens vir die Oranje-, Tugela- en Pongolariviere". Technical Report No 59, Department Water Affairs, South Africa.

(3) Rooseboom, A and Mülke, F J (1982): "Erosion Initiation", Proc IAHS Conf Exeter, July 1982. IAHS Publ No 137.

(4) Rooseboom, A (1976): "Reservoir Sediment Deposition Rates", Proc ICOLD Congress Mexico Q47 R21.

(5) Braune, E (1984): "Density of sediment in South African reservoirs". Proc First S Afr Ntl Hydrol Symp. Pretoria. Technical Report No 119, Department Water Affairs, Pretoria.

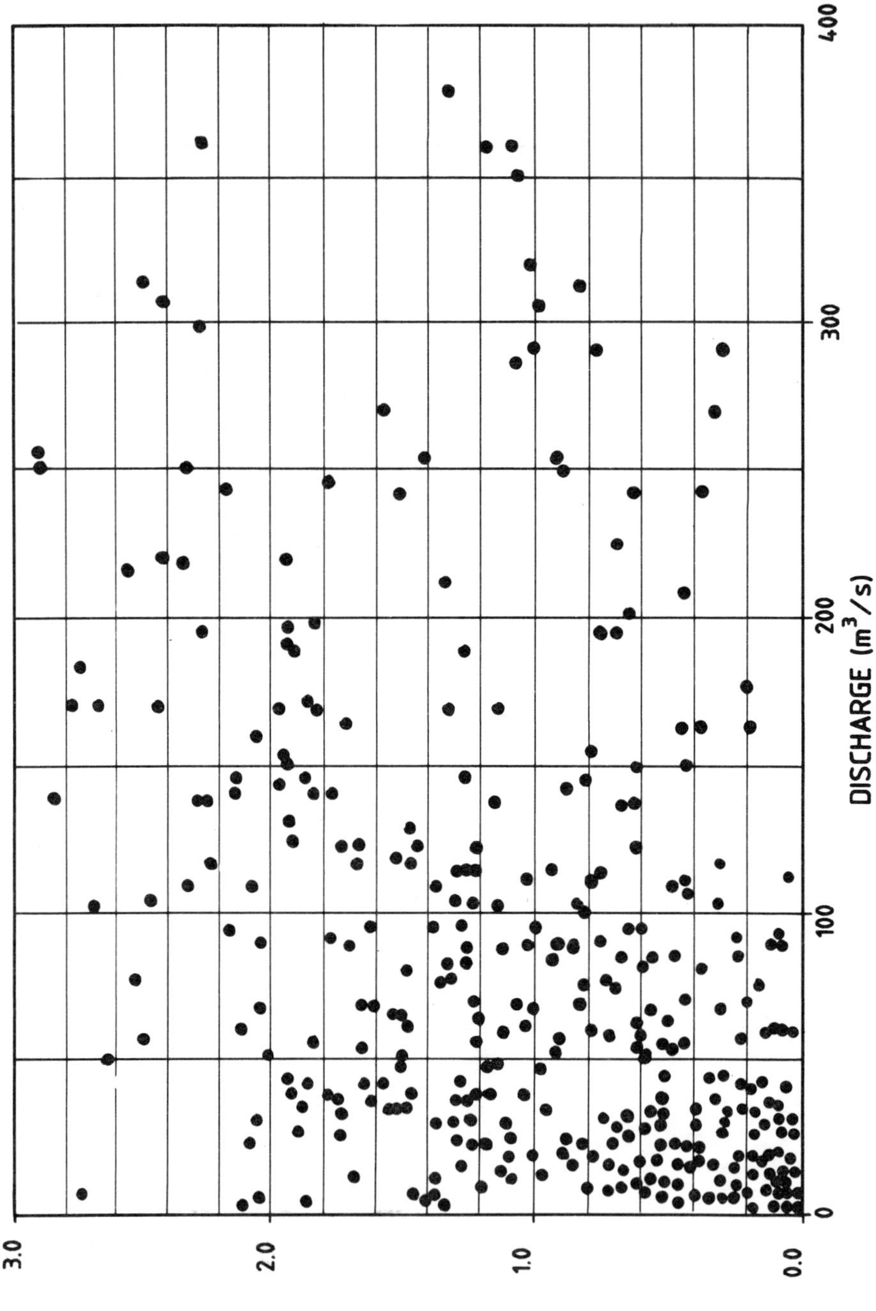


FIGURE I : CALEDON - JAMMERSBERGDRIFT SEDIMENT CONCENTRATION VS DISCHARGE
MAY 1969 - MAY 1976

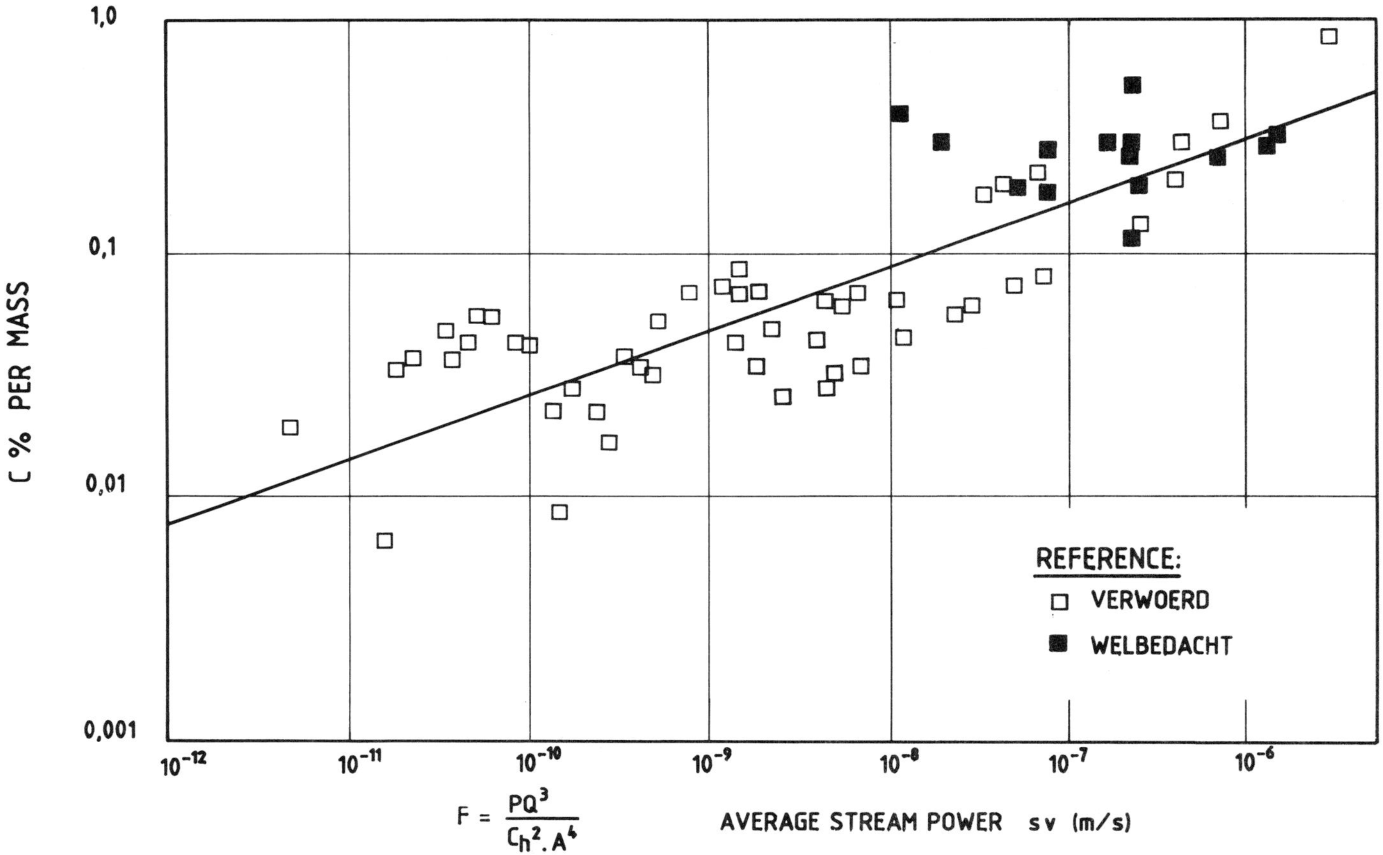


FIGURE II : SEDIMENT CONCENTRATION VS STREAM POWER

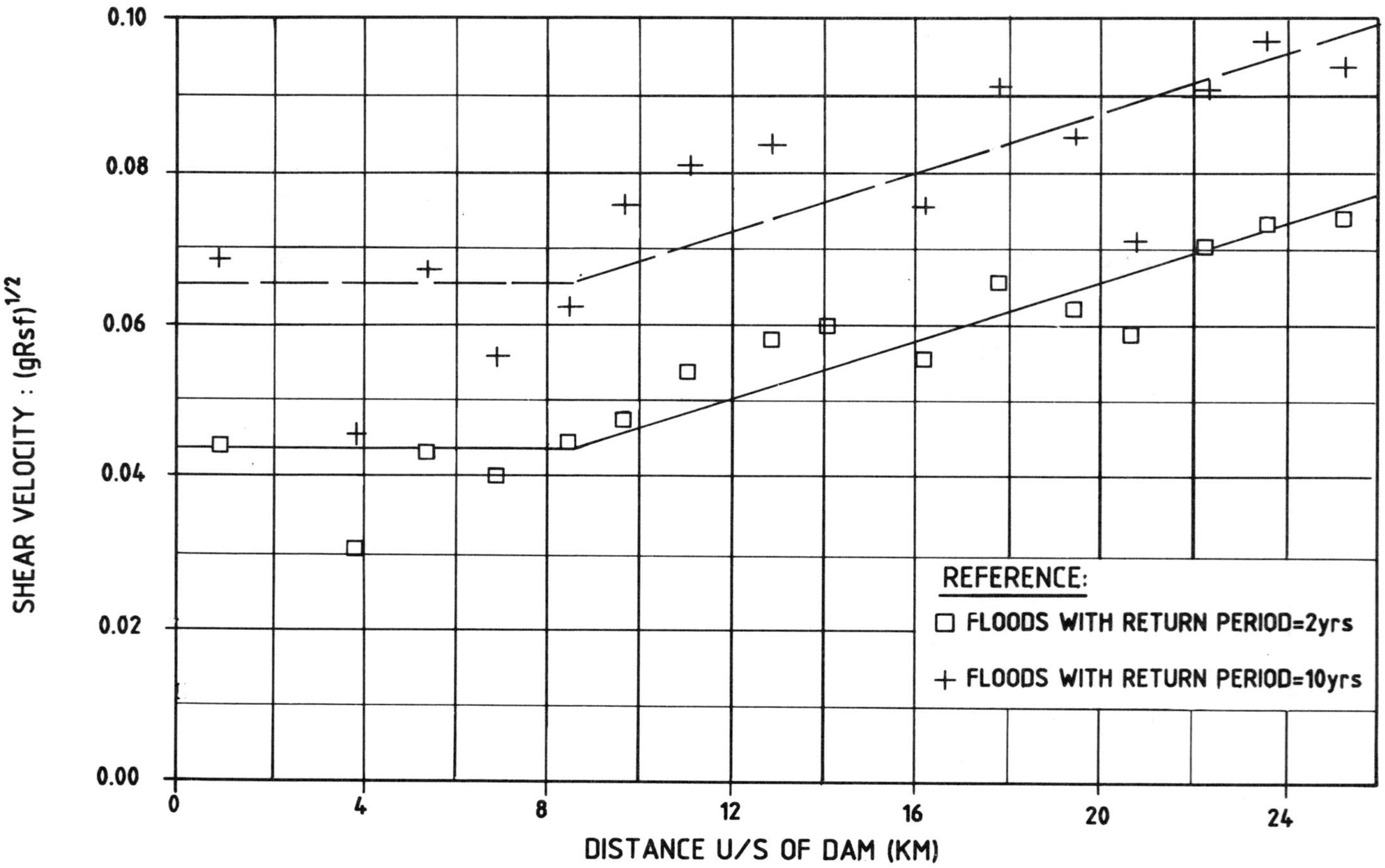


FIGURE III: BUCHUBERG RESERVOIR: SHEAR VELOCITY VS DISTANCE

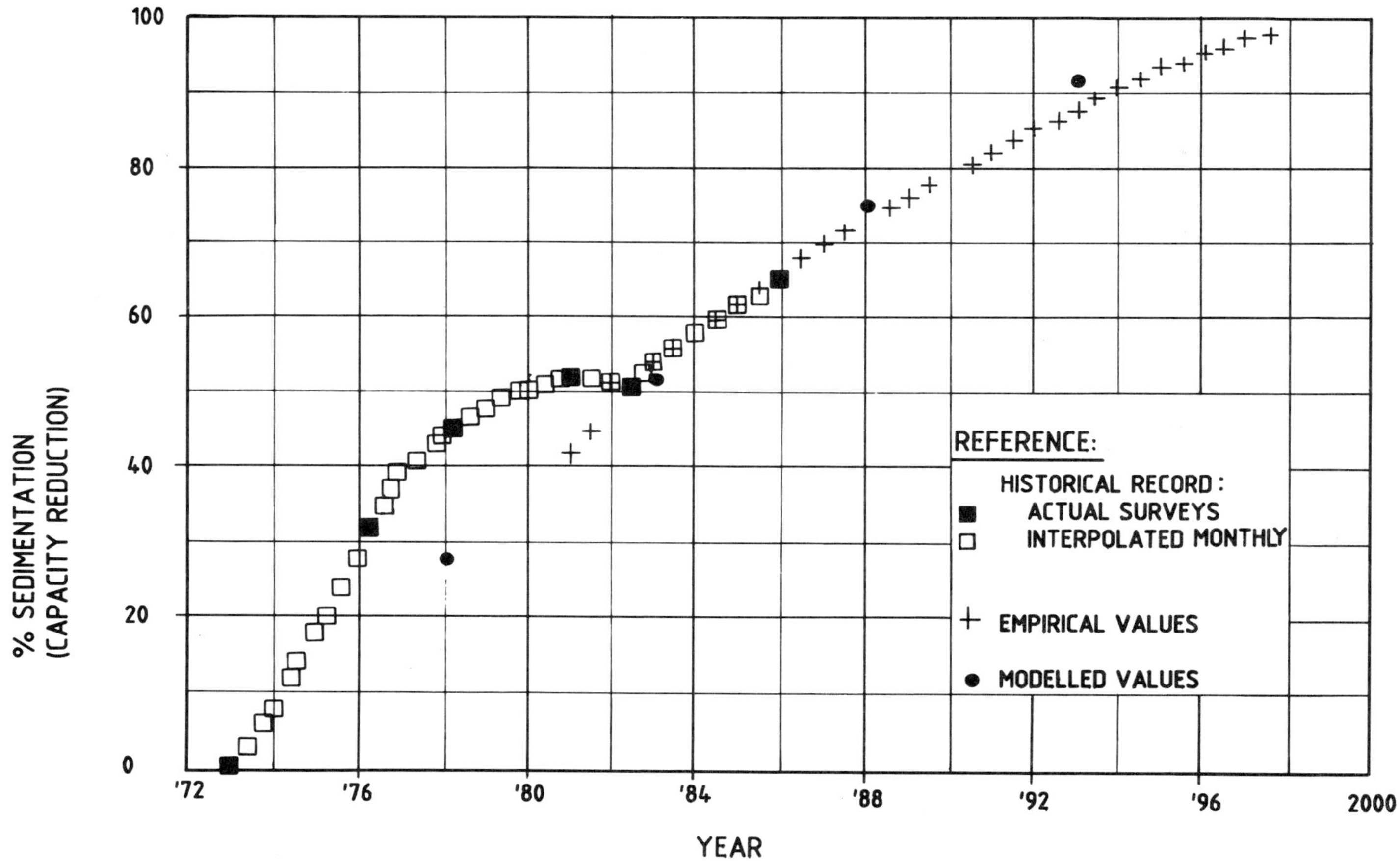


FIGURE IV : WELBEDACHT RESERVOIR : SEDIMENT BUILD-UP (CAPACITY REDUCTION)

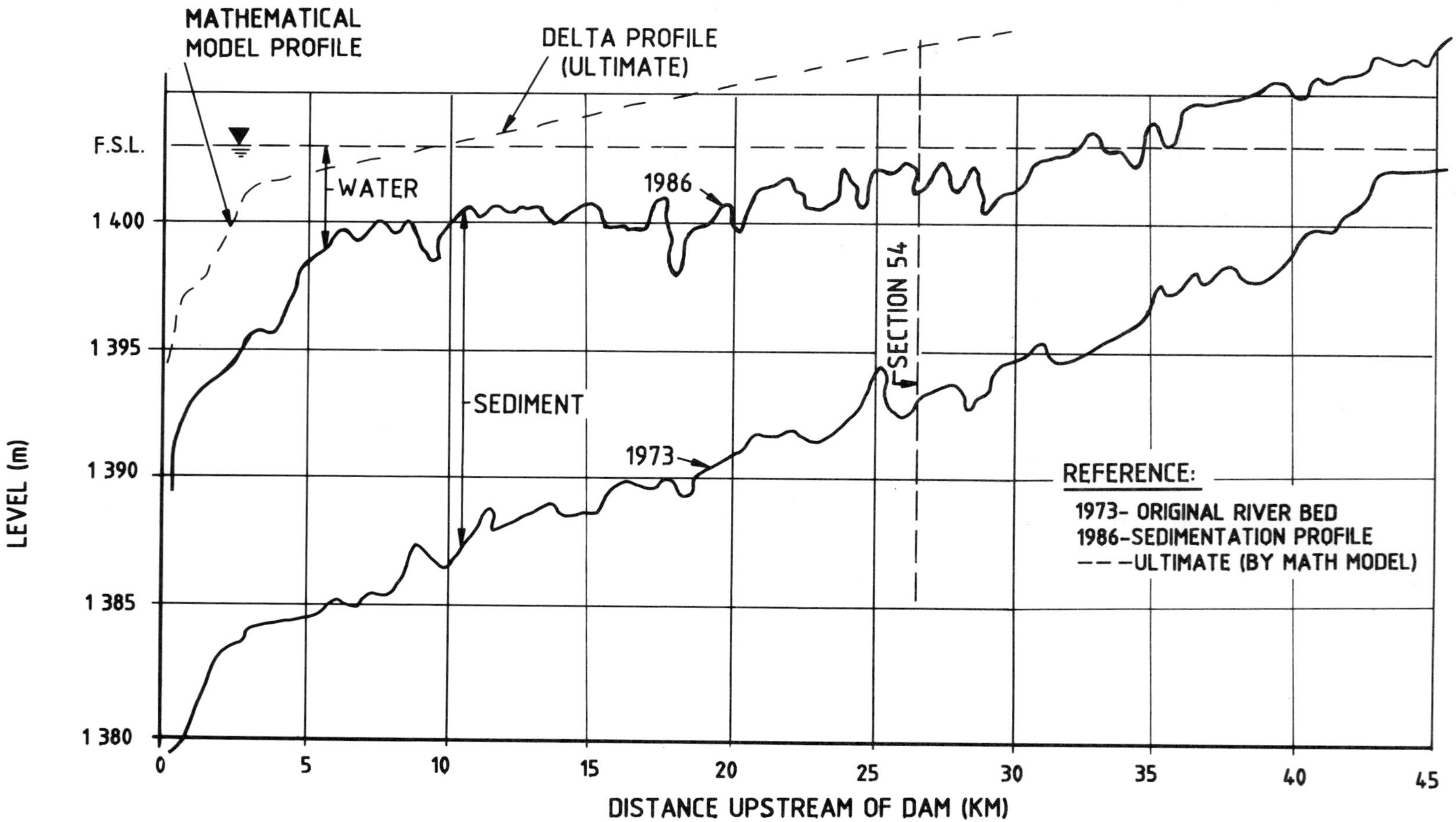


FIGURE V : WELBEDACHT RESERVOIR : RECORDED AND PREDICTED BED PROFILES

International Conference on
RIVER REGIME
18-20 May, 1988

PAPER I2

# REGIME CHANGES IN THE CALEDON RIVER ASSOCIATED WITH WELBEDACHT BARRAGE: PHYSICAL MODEL, PROTOTYPE AND THEORETICAL CORRELATIONS

J.M. Jordaan and D. Clark
Chief Engineer (Specialist: Hydraulics) and Engineer
Department of Water Affairs, Private Bag X313, Pretoria, 0001 South Africa

SUMMARY: Historical sedimentological and hydrological records are compared with experimental observations obtained by means of a scale model of the reservoir (Ref 1) and with theoretical and empirical relationships reported in the paper by Rooseboom and Van Vuuren (Ref 2). Excellent correlations and reliable predictions as to the future fate of the reservoir were obtained.

## 1. INTRODUCTION

A physical, movable bed hydraulic scale model, 1:1 000 (horizontal), 1:100 (vertical), was built in order to simulate present behaviour and predict the ultimate fate of the reservoir. It was necessary to know how long useful remaining storage could be extended, and by what operating means, while an additional off-channel storage/pumping scheme was being designed and constructed. The model was built and verified employing pulverized fuel ash as movable bed material to correctly reproduce observed sedimentation trends up to 1986. Useful trends were observed and reliable predictions made as to the reservoir's future behaviour (Ref 1).

## 2. DESCRIPTION OF REGIME CHANGES BROUGHT ABOUT BY THE RESERVOIR

The reservoir planform and longitudinal section are shown in Fig Ia and b. A typical cross-section, section 10, in the wider region of the reservoir is shown in Fig IIa and the longitudinal distribution of sediment in 1986 in Fig IIb.

Studying these figures reveals that the stream power for transporting sediment would be greatly reduced downstream of section 40 where the reservoir starts to widen out (Fig Ia). The average longitudinal slope, originally 0,0005 in 1973 has at present reduced to 0,00025 (Fig Ib) and the impediment by the sediment causes a considerable increase in the backwater effect and a reduction of stream transportive power. The typical cross-section is characterised by asymmetry: a narrow river channel and unequal flood plains separated from the main channel by low berms. Sedimentation progressively reduces the contrast in relief as shown by the hydraulic model results past 1986, (Fig IIa). The longitudinal distribution of sediment in the lower region of the reservoir is shown in Fig IIb, up to segment 27, correlating model tests with prototype observations. An equilibrium condition has been reached in the narrower sections upstream of Segment 13 while additional sedimentation can be expected to occur in the wider reaches further down.

3. HYDROLOGY

Fig III indicates the distribution of run-off during the 8 wet months and 4 dry months. The Caledon River drains a well-watered highland watershed and carries an average sediment load of nearly 0,5% by volume. Due to the dam's impounding volume being about one-tenth the mean annual run-off, the rate of sedimentation is high, annually about 5% of its volume. The annual hydrograph shown was repeated for each year simulated in the model, with the water carrying an up to a 1% by volume suspension of lightweight very fine pulverized fuel ash (pfa) to represent the natural sediment. The greater transportability of the pfa enabled realistic simulation in the 10 times vertically distorted model. A morphological time scale equal to the hydraulic time scale (1/100) was obtained in the model.
The submerged material density scale in the model was 0,73, the velocity scale 1/10 and the discharge scale 1/10E8.

4. STREAM POWER AND EQUILIBRIUM CONDITIONS

From theoretical considerations a curve was obtained, Fig IV, where the ratio of shear velocity to particle sedimentation velocity is plotted against the stream power as shown by Rooseboom and Mülke (Ref 2, sub 3).

The former is a function of the flow velocity, and the latter of the energy slope of the river. The Grass curve divides the flow regime into a transportation zone to the right and a sedimentation zone to the left. The results for the observed conditions in Welbedacht Reservoir (prototype and model) are given as points identified by section numbers. It is seen that most of the sections No. 12 and below fall into the sedimentation category. With reference to Fig I it is clear that the reservoir would trap most incoming sediment even if it is transported as far down as section 12.

5. SEDIMENTATION RATES - STREAMFLOW INFLUX

Table I presents an analysis of the estimated sediment influx rate as a function of streamflow in the period of existence of the reservoir (1973 - present). The 1931 - 1942 sediment data (Ref 2) was used to arrive at a 1% sediment load (by mass) and a derived 0,54%, by volume, for calculating the cumulative influx values.

The measured sediment retention (from the original survey and the six sedimentological surveys made since) is also given in the last two columns, the nett figure representing the reduction in reservoir storage capacity and the gross figure the total sediment accumulation including that above Full Supply Level (FSL) (due to the backwater effect). It is seen that the gross measured retention lags behind the calculated sediment influx, due to the portion that escapes the system via the turbid water discharged through the gates. Except for the period 1981/82, when the reservoir was kept purposefully low in a largely unsuccessful attempt to scour existing deposits, sediments once deposited tended to remain in place. The marginal effect achieved in 81/82 was completely nullified by the increased sedimentation in the ensuing years. The cross-sections also indicate that sediment deposits on the flood plain tend to remain, while that in the river channel show minor fluctuations. It was concluded that applying gate operation rules, i.e. keeping the reservoir level low, or even resorting to dredging, could have no lasting benefit where flow velocities within the reservoir drop below critical values.

## 6. SEDIMENTATION RETENTION RATES

Table II indicates an average volumetric sediment retention rate of 0,5% during the life of the reservoir, showing marked fluctuations in the observed values. A trap efficiency of approximately 90% of the incoming sediment is indicated as compared to the originally estimated value of 3% reported in Ref 2. With only 33,3% of the reservoir's original capacity remaining in 1987, after 14 years of useful life, it was clear that a complete replacement in the form of a 125 million $m^3$ offshore pumped/storage reservoir was urgently required. This is now being built and is situated on a tributary of the Caledon, having its own annual mean run-off of only 1/8th its volume. It is to be filled with water partly desilted drawn off by a pump station in the present reservoir at sec. 54.

## 7. FUTURE OF WELBEDACHT RESERVOIR

The remaining storage in Welbedacht Reservoir will continue to be utilised for water supply. Ultimately the reservoir, when almost fully sediment filled, would merely form a conveyance from the off-channel reservoir to the present offtakes and treatment works situated at Welbedacht Dam. Model tests showed that if the reservoir is kept at full supply level in future, less than $5 \times 10^6$ $m^3$ of equilibrium storage space will ultimately be maintained (Ref 1). According to the mathematical modelling analyses, Ref 2, however, $10 \times 10^6$ $m^3$ of storage space may be retained ultimately if all incoming floods are passed through without much retardation. Figure V indicates the correlation between prototype records, theoretical calculations and model tests, showing the decline of remaining storage capacity with gross input of sediment and time. An equilibrium sedimented condition is expected to be reached within a decade from the present, stabilising at 10 million $m^3$ or 9% of the original capacity, when some 225 million $m^3$ of sediment influx would have had occurred (twice the original storage volume) and after a reservoir lifetime of some 25 years - when run-off some 250 times its volume would have occurred. At that stage the system would again be considered to be in a natural regime or equilibrium state, similar to the Buchuberg Reservoir cited in Ref 2.

## 8. CONCLUSIONS

1. This study has shown that a dam that is situated in an area where the sediment yield is high should, wherever possible, be of the barrage type and so operated, with the retention water level at any time being kept to the minimum in keeping with the required necessary storage.

2. Numerical and physical modelling techniques, when coupled with adequate field data and surveys for verification and calibration purposes, did show good agreement in the equilibrium profiles and therefore yield valuable prognostications as to future behaviour of rivers and reservoirs undergoing regime changes. The agreement regarding eventual equilibrium conditions between the physical and mathematical models was very good.

3. This study provided support in two ways to decisions to add an off-channel storage reservoir to the system. It showed that continued loss of storage due to imminent sedimentation was unavoidable and also provided the basis for the optimum location of the draw-off point.

9. ACKNOWLEDGEMENT: The permission of the Director-General of Water Affairs to present this paper is acknowledged.

## 10. REFERENCES


1. Jordaan, J.M., Elges, H.F., Clark, D.R. and Skotnicky, P.: "Welbedacht Dam River Model Study. Report No. D200-01-DH03, Department of Water Affairs, Pretoria, Republic of South Africa (1986).

2. Rooseboom, A. and Van Vuuren, S.J.: "Regime changes in the Caledon River associated with sediment deposition upstream of Welbedacht Barrage." In: Proc. of Conf. on RIVER REGIME, Wallingford, England, May 1988.


TABLE I: WELBEDACHT DAM/CALEDON RIVER: ESTIMATED SEDIMENT INFLUX AND VOLUME RETAINED

| Hydrol. year (1 Oct. to 30 Sept) | Measured stream flow | Estimated sediment (1% by mass) | | | Measured sediment retention | |
|---|---|---|---|---|---|---|
| | | (Mass) | (Vol) * | Cumulative (Vol) | Below full supply level | Incl. above full supply level |
| | $m^3x10^6$ | T x $10^6$ | $m^3x10^6$ | $m^3x10^6$ | $m^3x10^6$ (nett) | $m^3x10^6$ (gross) |
| 1971/72 | 1 180 | 11,80 | 6,36 | 6,36 | | |
| 1972/73 | 350 | 3,50 | 1,89 | 8,25 | | |
| 1973/74 | 990 | 9,90 | 5,34 | 13,59 | | |
| 1974/75 | 1 100 | 11,00 | 5,93 | 19,52 | | |
| 1975/76 | 3 420 | 34,20 | 18,40 | 37,92 | 37,46 | 37,46 |
| 1976/77 | 2 190 | 21,90 | 11,81 | 49,73 | | |
| 1977/78 | 2 250 | 22,50 | 12,13 | 61,86 | 52,35 | 54,06 |
| 1978/79 | 990 | 9,90 | 5,34 | 66,84 | | |
| 1979/80 | 690 | 6,90 | 3,72 | 70,92 | | |
| 1980/81 | 1 580 | 15,80 | 8,52 | 79,44 | 59,50 | 62,94 |
| 1981/82 | 722 | 7,22 | 3,89 | 83,33 | 58,29 | 63,08 |
| 1982/83 | 483 | 4,83 | 2,60 | 85,93 | | |
| 1983/84 | 628 | 6,28 | 3,39 | 89,32 | | |
| 1984/85 | 408 | 4,08 | 2,20 | 91,52 | | |
| 1985/86 | 757 | 7,57 | 4,08 | 95,60 | 75,51 | 82,42 |
| 1986/87 | 846 | 8,46 | 4,56 | 100,16 | 76,84 | 85,09 |

* Mass: (1 - v.r.) x 2,65, (v.r. = 0,3) .˙. mass = 1,855 $T/m^3$

TABLE II: PERCENTAGE CHANGE IN RECORDED SEDIMENT VOLUME

| Period | Run-off $m^3x10^6$ | Measured retention in storage space* $m^3x10^6$ | Storage loss as % of run-off | Average run-off per annum $m^3x10^6$ |
|---|---|---|---|---|
| 1973 - 1976 | 5 510 | 37,46 | 0,68 | 1 837 |
| 1976 - 1978 | 4 440 | 16,60 | 0,374 | 2 220 |
| 1978 - 1981 | 3 260 | 8,88 | 0,272 | 1 087 |
| 1981 - 1982 | 720 | -1,14 | -0,160 | 720 |
| 1982 - 1986 | 2 280 | 19,34 | 0,848 | 570 |
| 1986 - 1987 | 850 | 2,67 | 0,155 | 850 |
| Total | 17 060 | 83,81 | Average 0,491% | 1 219 |

* Including above FSL. The average volumetric percentage of sediment retained by the reservoir is 0,49%, compared to the average gravimetric percentage of sediment influx of 1% (or 0,54% volumetric) giving an average 0,9 trap efficiency.

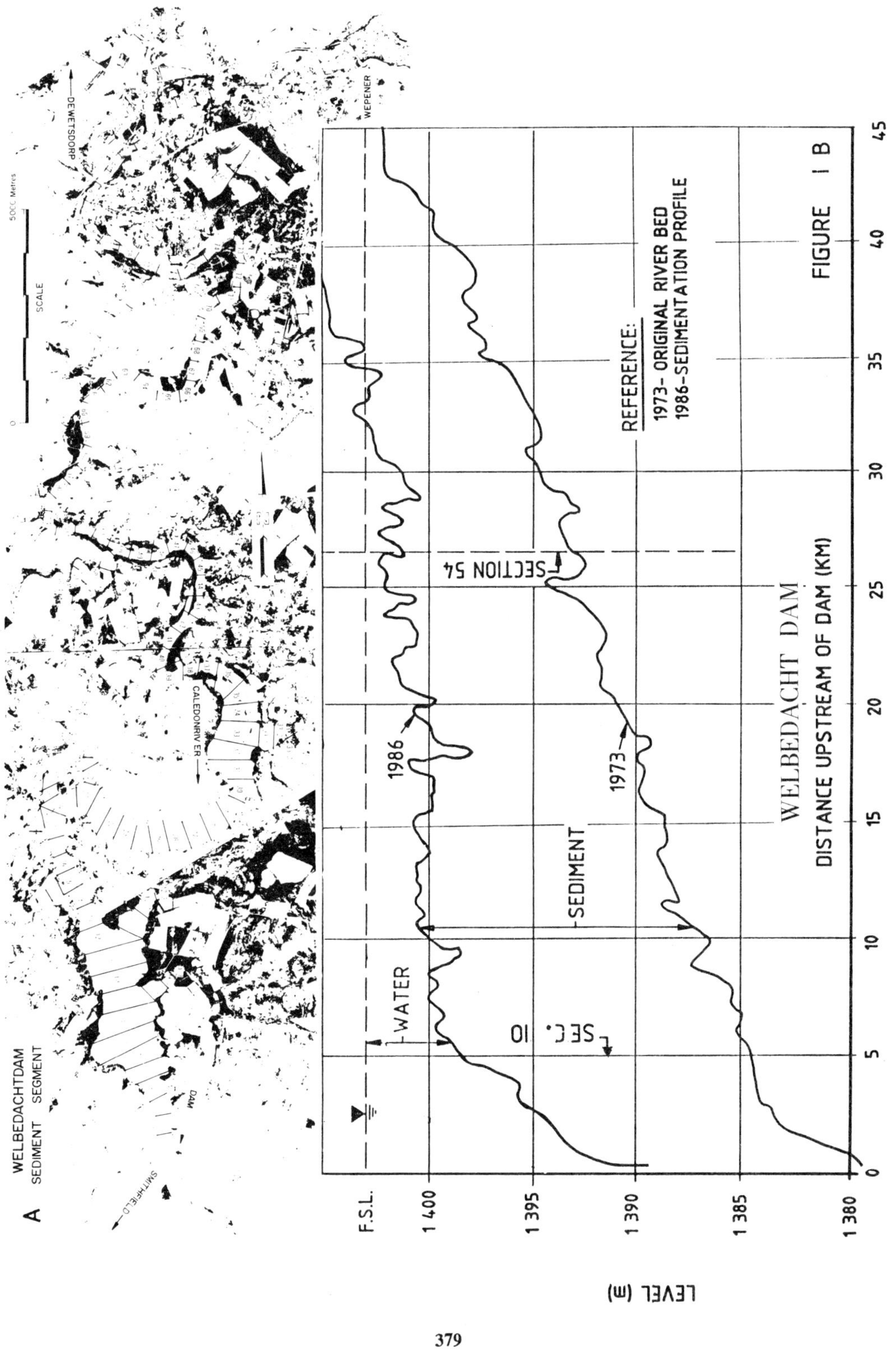

A
WELBEDACHTDAM
SEDIMENT SEGMENT
SCALE
DEWETSDORP
WEPENER
CALEDONRIVER
DAM
SMITHFIELD
F.S.L.
1 400
1 395
1 390
1 385
1 380
LEVEL (m)
WATER
SEC. 10
SEDIMENT
1986
1973
SECTION 54
REFERENCE:
1973- ORIGINAL RIVER BED
1986-SEDIMENTATION PROFILE
FIGURE 1 B
WELBEDACHT DAM
DISTANCE UPSTREAM OF DAM (KM)
0
5
10
15
20
25
30
35
40
45

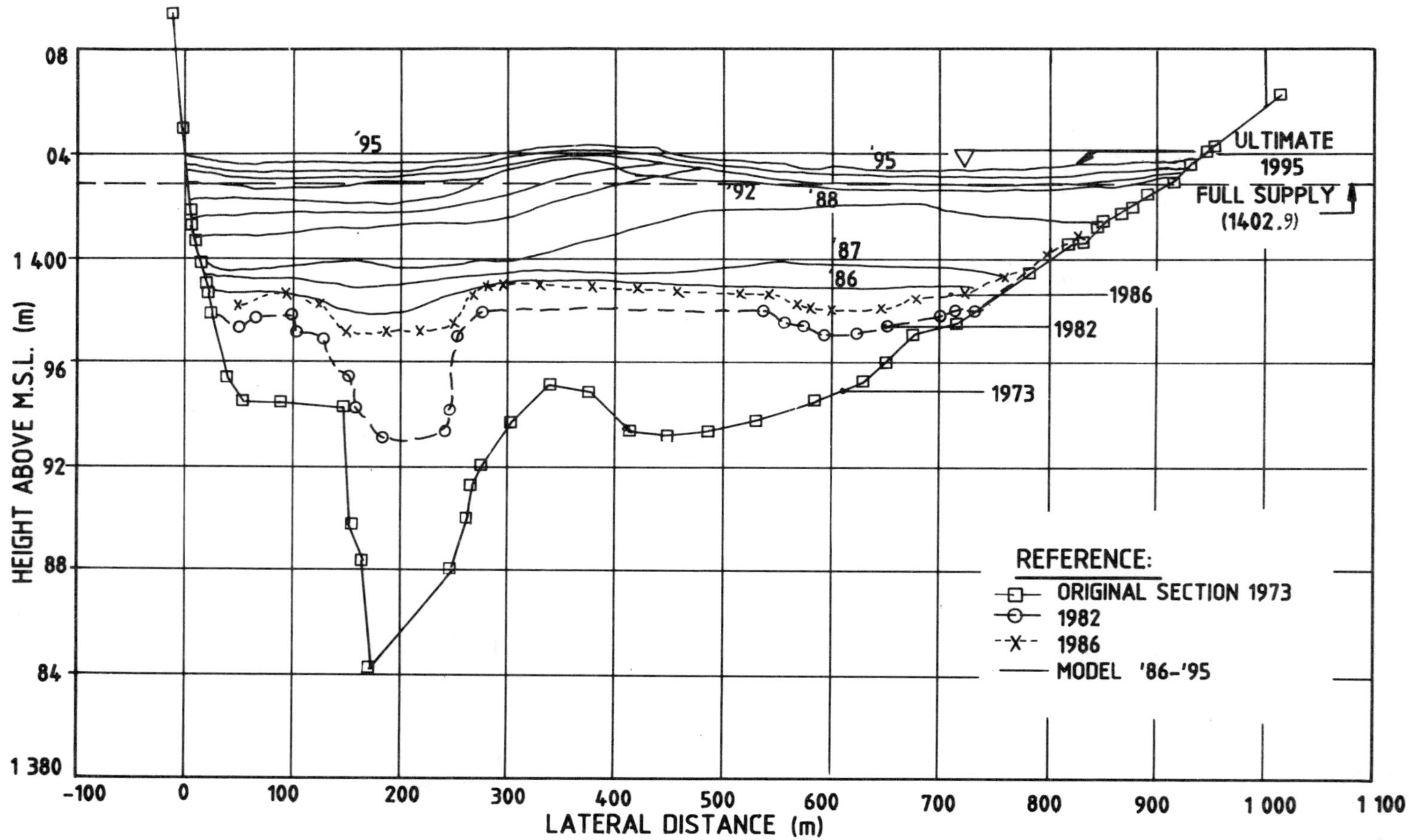


FIGURE IIa : TYPICAL SECTION (No. 10) SHOWING PROTOTYPE OBSERVATIONS (NATURE) AND PHYSICAL PREDICTIONS OF SEDIMENT PROFILES (MODEL)

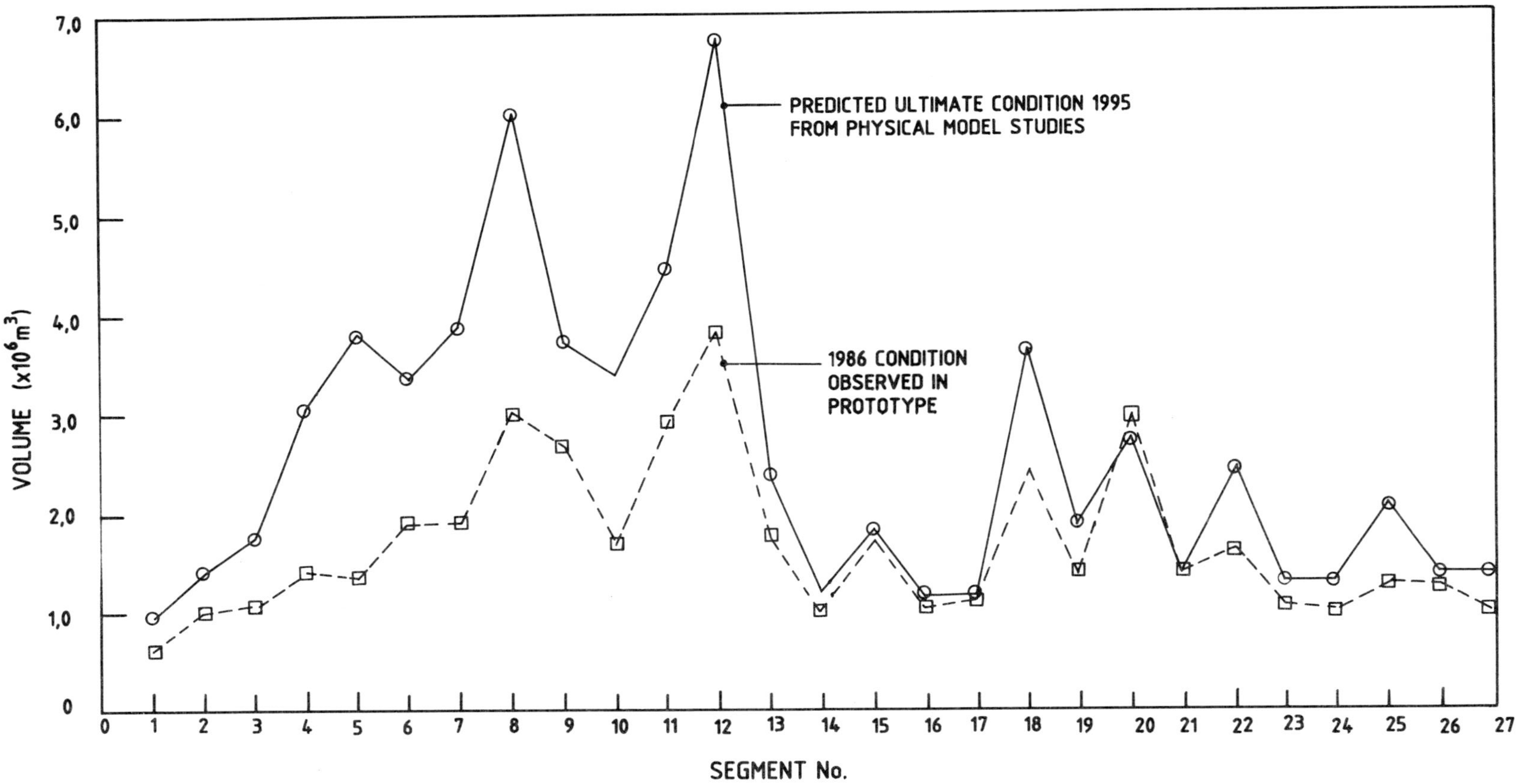


FIGURE IIb : VOLUME OF SEDIMENT PER SEGMENT UP TO F.S.L.

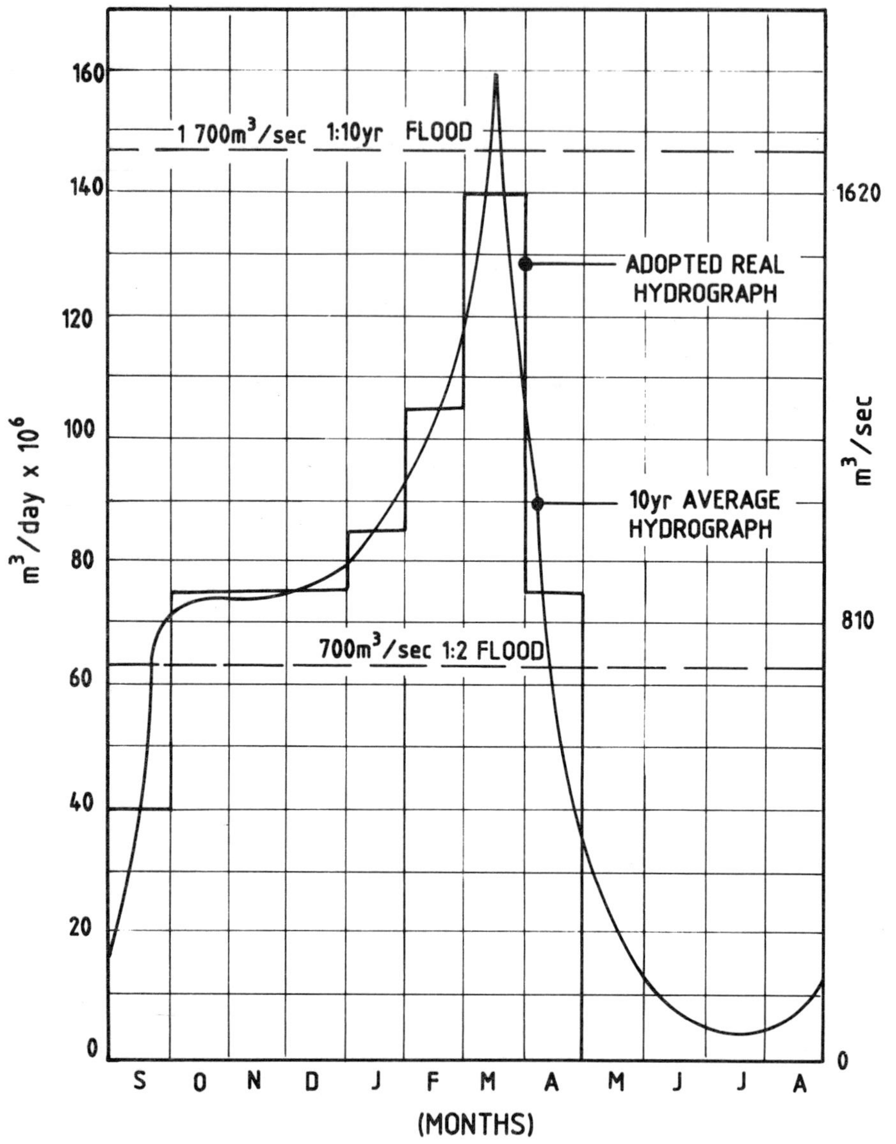


FIGURE III : HYDROGRAPH FOR JAMMERSBERGDRIFT "REAL HYDROGRAPH"

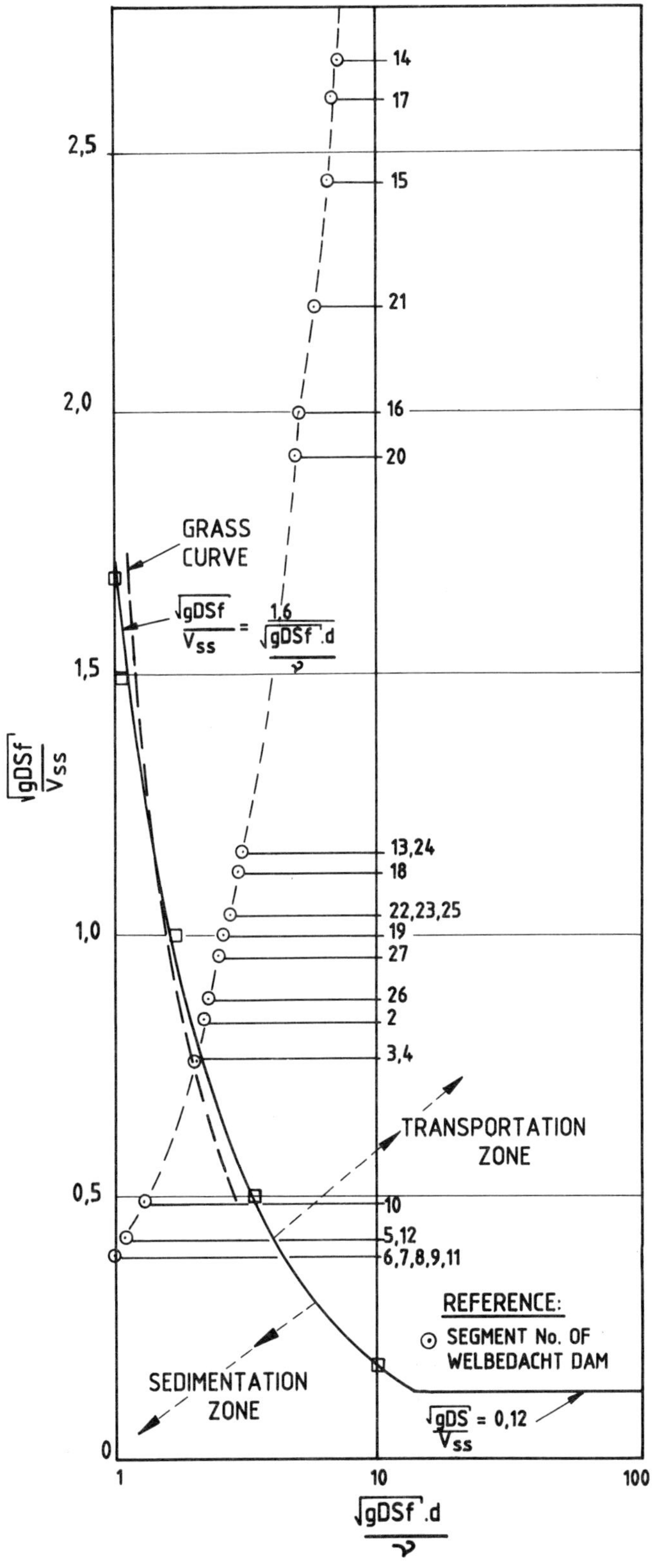


FIGURE IV: EQUILIBRIUM CONDITIONS OF WELBEDACHT DAM IN 1986

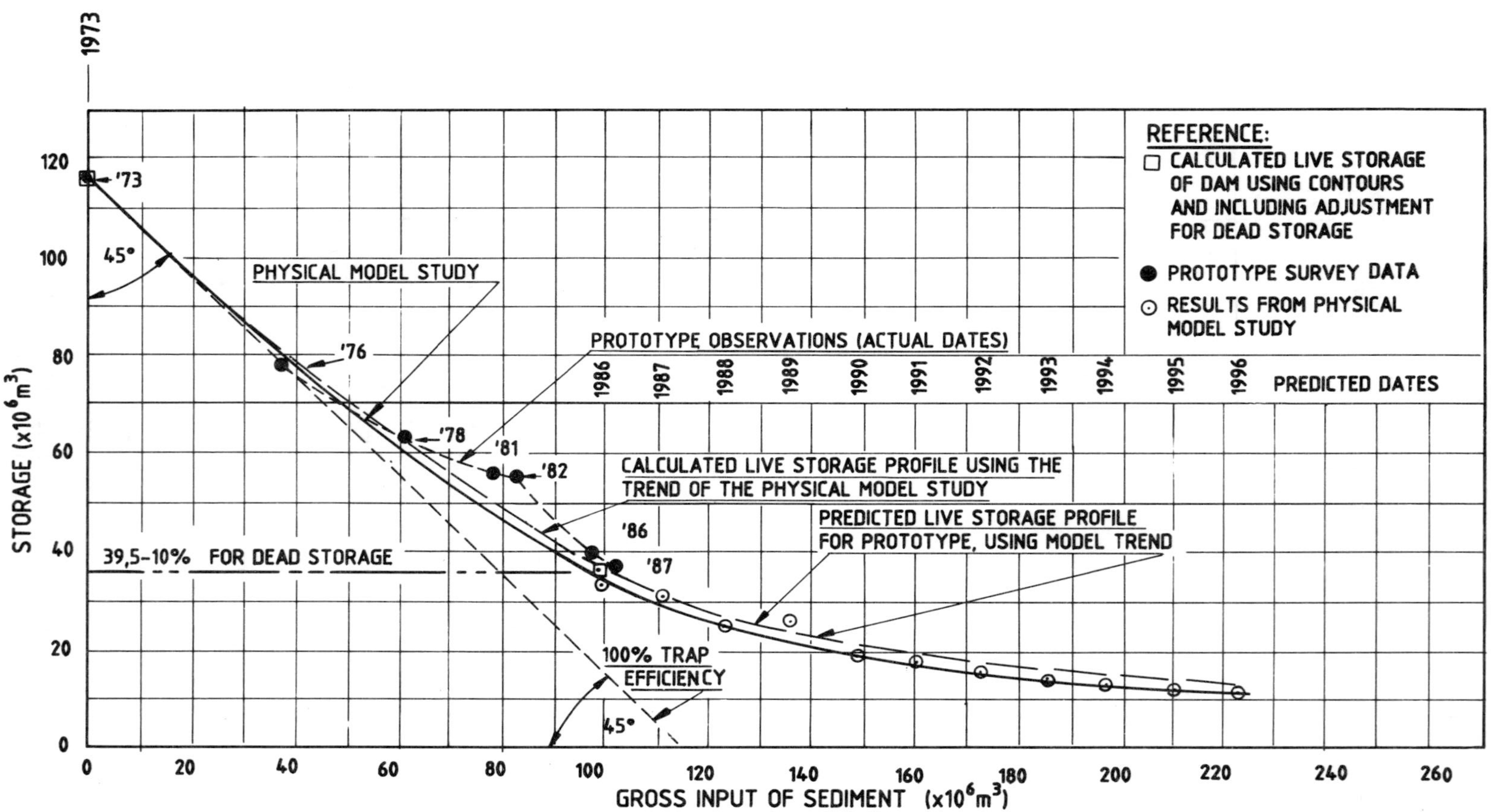


FIGURE V : CHANGE IN LIVE STORAGE VS GROSS INPUT OF SEDIMENT

International Conference on
RIVER REGIME
18-20 May, 1988

PAPER I3

# EFFECTS OF EXTENSIVE DREDGING ON THE RIVER REGIME

S Kornis (Ms) I A Laczay
Research Center for Water Resources Development
"VITUKI", Budapest, Hungary

SUMMARY: Long-term and extensive dredging caused significant changes in the channel-regime and in the low-flow conditions along the Danube river in Hungary. To explore the former and to assess the further consequences of this activity, both hydraulic model tests and field investigations have been performed.

## 1. RIVER REGIME AND HUMAN ACTIVITIES

Rivers or river-reaches when in regime maintain or at least try to maintain the slope-conditions, channel pattern and hydraulic geometry in or around the so called state of dynamic equilibrium. During the process of responding to the water discharge sediment yield and relief imposed from upstream, continuous feedback exist from the river to the basin-regime and from the local hydraulic conditions i.e. the flow velocity, sediment transport rate, bank materials, armouring etc. to the river-regime as well. The system is often called as the most complicated one in the natural environment.

Rivers or river-reaches when in addition influenced by certain kind(s) of human activities try to incorporate the superimposed factors into the system and tend to reach and maintain some "revised" version of regime. The resulting slope/flow/channel etc. conditions may or may *not* meet the demands of all the various kinds of human exploitation. In the latter case conflicting requirements and new problems can occur.

To meet the demands of the construction industry extensive dredging activities have been carried on along certain reaches of the Danube river in Hungary, especially in the vicinity of Budapest the capital (Fig 1). During the last 20 years the material excavated from the channel amounted to the magnitude of $10^7$ $m^3$. On some shorter reaches the channel-bottom degraded by several meters with all the consequences in the water-surface slope conditions, the descending levels of the low-flows, the changing grain-size distribution of the bed material and so on.

By now the need for the row-material came in sharp contradiction with the demands of the rest of the water users such as navigation, bank-filtered water-works, water intakes, river pollution control.

Apart from theoretical considerations, to find the proper balance in fulfilling the various needs and to prepare the necessary measures the following questions had to be answered:

International Conference on River Regime, edited by W.R.White.

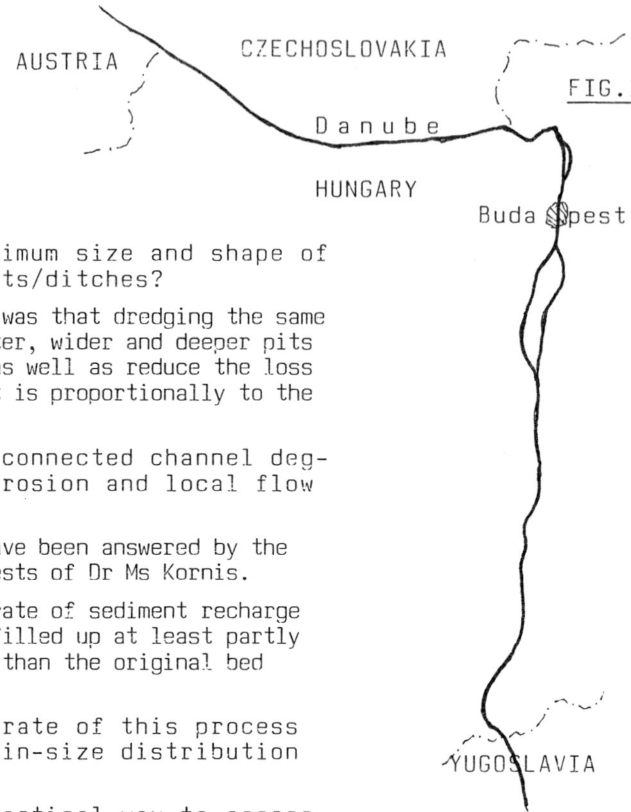


FIG.1.

i - What is the optimum size and shape of the dredging pits/ditches?

The starting idea was that dredging the same amounts from shorter, wider and deeper pits will concentrate as well as reduce the loss of slope, since it is proportionally to the length of the pit.

ii - What about the connected channel degradation, bed-erosion and local flow disturbancies?

These questions have been answered by the hydraulic model tests of Dr Ms Kornis.

iii - Depending on the rate of sediment recharge the pit would be filled up at least partly by finer sediment than the original bed material.

What about the rate of this process and the new grain-size distribution in the pit?

iv - Is there any practical way to assess the would-be sediment/channel balance?

To answer these questions, successive field-surveys and investigations have been performed by Dr Laczay.

## 2. SIZE AND SHAPE OF DREDGING PITS

The possible consequences of "deep-hole" dredging have been investigated on the river-reach upstream of the big bend of the Danube (Fig 1). The governing hydraulic conditions are as follows:

- low/mean/high flow 1000/2300/9000 $m^3$/sec, respectively
- water-surface slope about 7 cm/km at low-flows
- channel width 400 to 500 m
- water-depth at low-flows about 2 to 6 m (riffles and pools)
- bed-material graded sandy-gravel with mean grain-size (by weight) of about 8 mm
- suspended sediment transport about 3 to 5 million $m^3$/yr
- bed-load transport round 10,000 $m^3$/yr.

Dredging activities have been restricted in depth to 5 m below the level of the characteristic low-flow of 94 % duration. Dredging usually caused in the channel 1 to 3 m deep, 100 to 150 m wide and several hundred m long "troughs". Connecting effects of sequences of these troughs have caused the descend of low-flow levels along longer reaches as well. That is why the idea came to restrict dredging in length and allow it deeper down to about 10 m below the low-flow levels.

2.1. Methodology of model tests

The hydraulic model of the Danube-channel of 13 km in length have been built into a model-box in scale 1:400 undistorted by using 52 cross-sectional profiles based on prototype-survey in 1984. The model-box of 50 x 6 x 0.5 m has been filled up with a graded mixture of sand and a special plastic material formerly worked out experimentally by Ms Kornis (Ref 1). According to the hydraulic conditions this mixture provides in the model mobile bed and erodible banks as well.

Bank-protecting and other river-training structures have been built into the model by using coarser gravel.

A "quasi" flow-regime has been simulated during each run by changing lower and higher flows in 10 various steps with 6-hours duration each. Extreme values have been 1050 and 5000 m3/sec in the prototype, the latter being the bankfull discharge.

2.2. Variations investigated

The following experiments have been performed.

RUN 01 - The model was verified in operation by the similarity of characteristic water-surface profiles and surface flow-patterns. After draining the model and having measured the changes of the model cross-sections, similarity of the bed-configuration and the

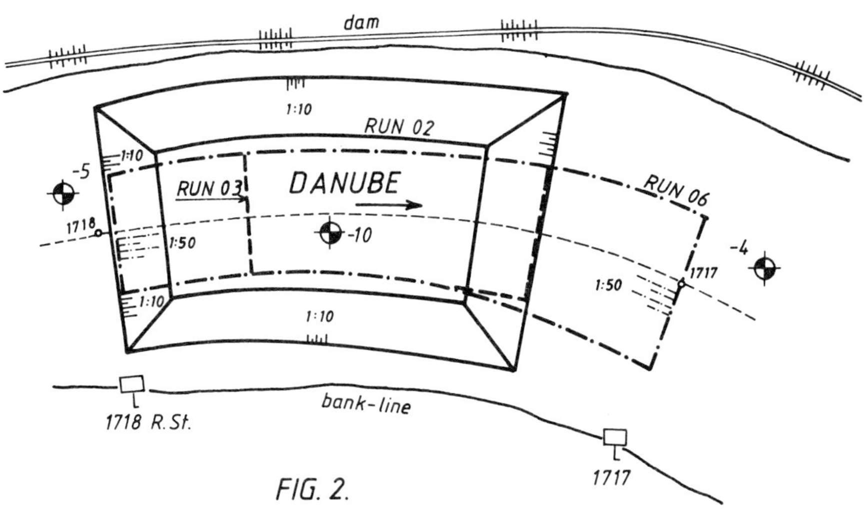


FIG. 2.

overall channel-balance were confirmed. Both practical experience and the model emphasized the importance of a certain crossing 5 km upstream of the planned experimental dredging.

RUN 02 - An experimental pit with dimensions shown in Fig 2 was built into the model-channel between 4 and 5 km from the downstream end of the stretch investigated. Note the slope of 1:10 around the pit.

Observations proved that the current-pattern, velocity distribution in and around the pit were irregular and unfavourable. The entering current-lines could not follow the 1:10 slope therefore a horizontal eddy-roller developed in the pit and caused continuous receding bed-erosion (Fig 3). At the downstream end of the pit a reversed backward flow and vortices with vertical axis developed on both sides and started to erode the banks. As a secondary effect these two vortices by blocking the flow near the banks confined the discharge into the middle of the section causing higher velocities

FIG. 3.
RUN 02.

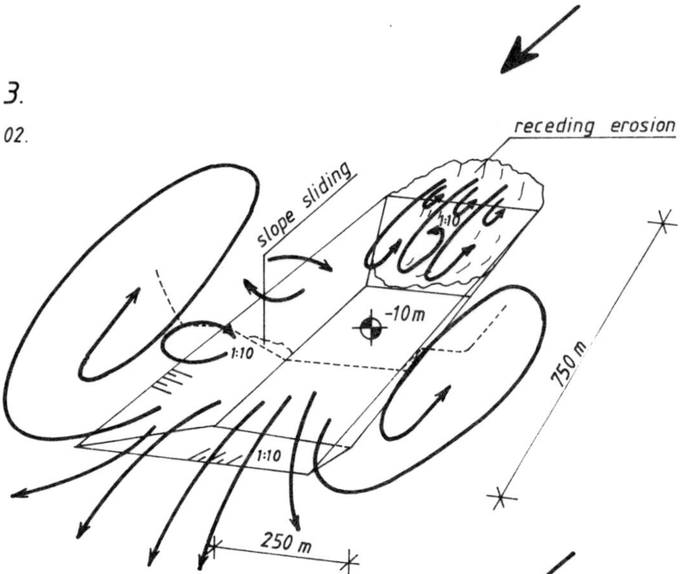


FIG.4.
RUN 03

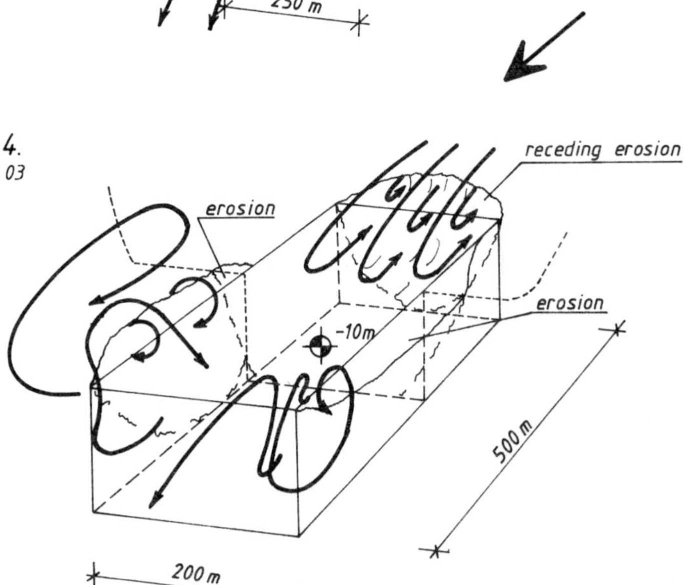


and heavier effectiveness of the horizontal eddy-roller in the pit.

For this and other reasons the pit has drown the current-lines to itself with cross-currents developed at the upstream and of it. However, the head-drop i.e. the high and concentrated local slope occured not at the upper end of the pit but moved about 5 km in the upstream direction to the characteristic crossing mentioned above.

After draining the model bed erosion was observed receding in the upstream direction. Because of the high near-bottom-velocities the material eroded upstream of the pit did not fill in it. Deposits found in the pit originated from the sliding slopes. The downstream slope of the pit was also seriously eroded.

After evaluating these first findings intermediate runs No 03-05 were performed (Fig 4).

RUN 06 - Runs 03-05 proved that the pit ought to be narrower with vertical side-walls. The development of the horizontal eddy-roller in the pit was also to be prevented. The latter has been avoided by excavating very mild slopes of 1:50 on both the upstream and downstream ends of the pit. In addition the downstream slope has been constructed in a diffusor-like pattern.

This configuration has provided the best solution with the possible minimum disturbances of the current-pattern. However, the slope/channel conditions have redeveloped also in this case in such a way that the bed erosion and the head-drop receded back to the upper crossing. The material eroded from this site did not reach the pit but settled in-between. No erosion of the upper slope and/or major filling-up of the pit but minor degradation on and after the downstream slope have been observed.

The actual experimental excavation started in 1986 with size and shape proposed in Run 06. The site is under regular observation.

3. DREDGING VS FLOW- AND CHANNEL-REGIME

The first indications of the unfavourable effects of dredging came some 15 years ago from the rating curves i.e. the discharge vs altitude relationships at the flow-gauging stations. These relations in their lower sections started to turn unstable. The

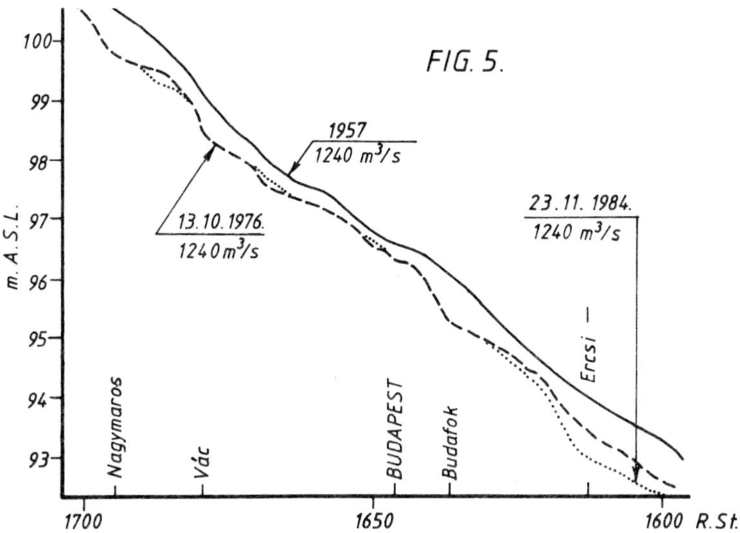


FIG. 5.

levels and also the longitudinal profiles of especially the low--flows begun to descend in very close and confirmed connection with the sites and amounts of actual dredging along the river (Fig 5).

Regular observations and measurements explored fairly well these interconnections and some restrictive measures have also been taken. To clear up the situation concerning changes in the channel-regime, sediment-balance and bed-material field investigations have recently been started.

## 3.1. Channel-changes and sediment-balance

Downstream of Budapest on a 51 km-long reach round 13 million $m^3$ material was excavated from the Danube-channel during the period of 1971-1985. On Fig 1 the left-hand or eastern branch of the river is a canalized side-arm while the western one is the main channel as well as the stretch in question. Along the banks bank-filtered wells belong to the municipal water-works of the capital.

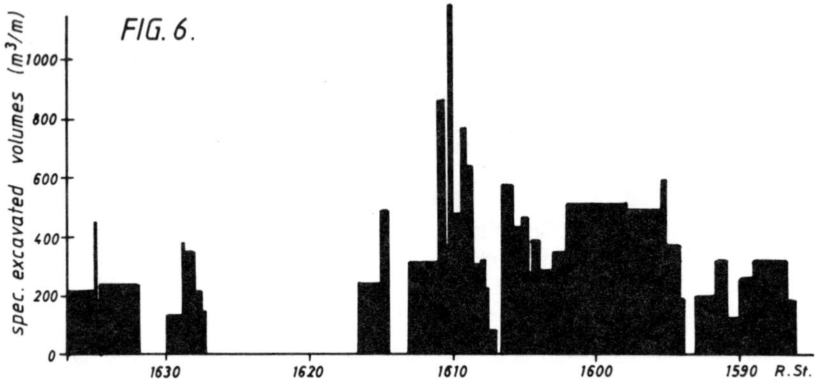


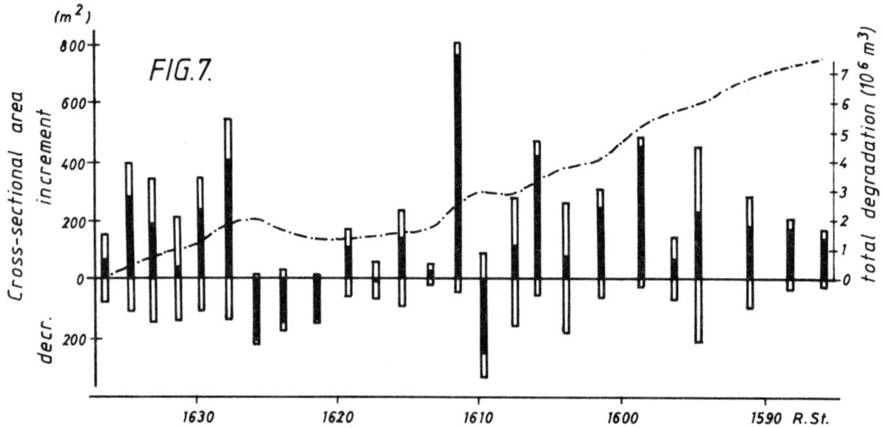


The dredging sites as well as the excavated specific volumes expressed in m3/m are shown in Fig 6. Specific volumes are identical to the averaged cross-sectional area-increments. Upstream of R.St. 1600 where dredging is in operation note the close relation of the area-changes to the level descend in Fig 5. Around R.St. 1630 no excavation-work has been performed since 1977.

Apart from the excavations local bed-interchanges also influence the channel-behaviour (Refs 2 and 3). This natural "shuffling--reshuffling" process as called by Blench in the General Report to Ref 2 is represented in this case by the area-changes of the 25 control cross-sections between the surveys in 1970 and in 1986

(Fig 7). The resulting aggradation/degradation/balance is composed of the area-increments and decrements of the same cross-section(s). Multiplying these area-changes separately by the channel-lengths between the succesive cross-sections, as a result of 13.4 million m3 of local degradation and 6.0 million m3 of aggradation on the reach of excavations, the total amounts to about 7.4 million m3 of degradation.

From the point of the channel/sediment balance that means that during the period of dredging and simultaneously with heavy local changes round 5.6 million m3 or an average of about 0.4 million m3/year of sediment had to be resettled on the reach.

On the other hand, based on regular sediment-sampling at some gauging-stations and on the sediment rating-curves the yearly yields of both suspended sediment and bed-load have been calculated. The bed--load transport has been about 10,000 m3/yr in average therefore it has no importance in the total balance and in the refilling process. However, in the same 15-year period the calculated suspended sediment balance of a 110 km-long reach between two gauging stations and including our stretch has suggested a d e g r a d a t i o n of about 6 million m3 which is roughly 0.2 million m3/yr in average on the stretch investigated.

Comparing the measured channel-changes and the calculated sediment yields and also taking into account all the discrepances and assumptions, these inconsistent results are of not much help in evaluating the situation. This case study has again confirmed that channel processes in themselves and especially when influenced by human activities are much more complex and far less explored than what could be properly handled by means methods and funds available.

### 3.2. Grain-size distribution and dredging

Downstream of Budapest in each of the 25 control cross-sections mentioned above 7 samples of the bed/bottom material have been taken at the same spots both in 1970 and in 1986. Samples No 1 and No 7 represent the left- and the right-hand channel-slopes or the bottom nearby which means in general finer sediment. Samples from No 2 to No 6 from the left to the right provide some kind of five "longitudinal profiles" of the bed-material.

The statistical means of the average grain-size diameters (by weight) in mm of these seven groups of 25 samples each and also the mean of the total of 175 samples taken in 1970 and in 1986 respectively together with the differences are given in Table 1.

| No of samples | 1 | 2 | 3 | 4 | 5 | 6 | 7 | 1-7 |
|---|---|---|---|---|---|---|---|---|
| Mean (1970) | 6.51 | 9.22 | 12.12 | 11.14 | 14.46 | 11.19 | 7.35 | 10.48 |
| Mean (1986) | 4.60 | 12.35 | 11.41 | 16.87 | 17.15 | 12.89 | 4.35 | 12.50 |
| +/- | -1.91 | +3.13 | -0.71 | +5.73 | +2.69 | +1.70 | -3.00 | +2.02 |

The results clearly indicate that in general, as "regime" the bed/bottom material of the reach has not significantly changed.

However, as indicated in Fig 8 (next page), local changes in the grain-size distribution have been common events. In Fig 8 the bed-material on the left-hand side of the channel became significantly finer. These local changes are induced by the dredging activities and by the construction of longitudinal and transverse

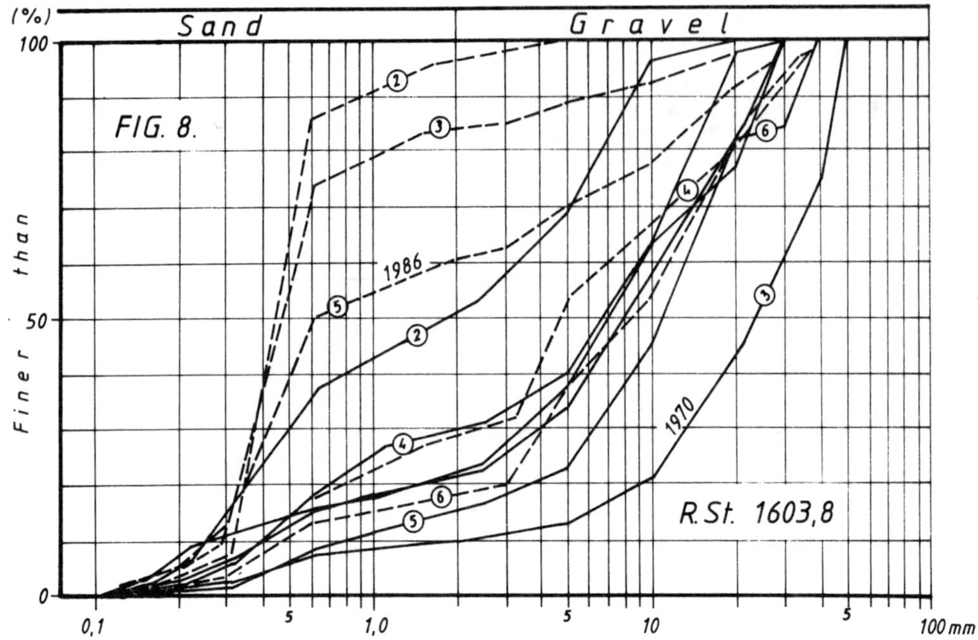


training structures as well. So the next question to be answered is whether a definite relationship between the dredging/training activities and the changes of the grain-size distribution can be found or not.

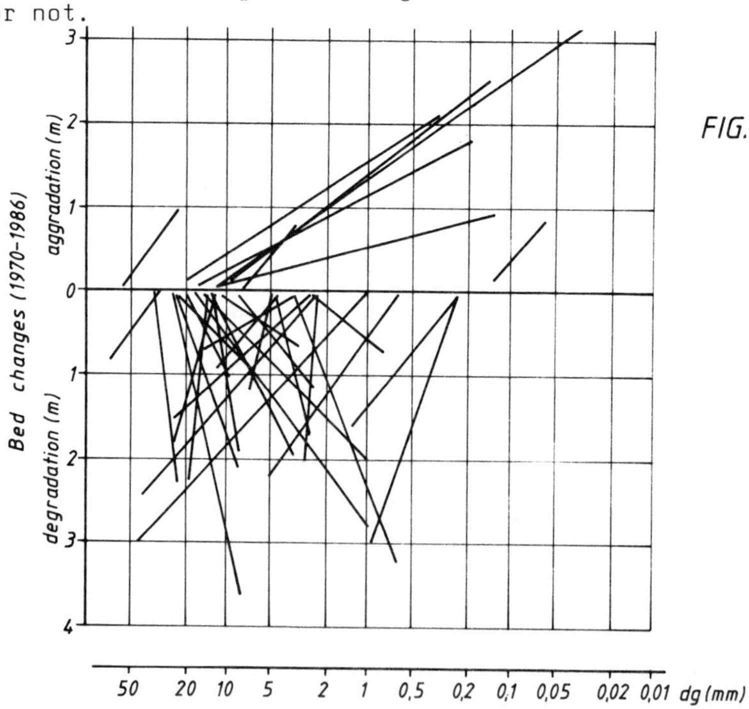


The answer is given in Fig 9. Fig 9 shows
- in logarithmic scale the local c h a n g e s of the mean grain--size d i a m e t e r of the bed/bottom samples taken in 1970 and

in 1986 from the area of actual dredging/training activities v e r s u s
- the simultaneous bed/channel a g g r a d a t i o n/d e g r a d a t i o n at the same spots from the cross-section surveys.

The working hypothesis has been that both processes are in time linear which naturally is not the case.

It is clearly shown that in the "shadows" of spur-dykes (groynes) and in similar places the expected aggrading tendency is in definite connection with the refinement of the bed material i.e. finer sediment has also settled. However, in spite of the expectations, in sites/holes/troughs of dredging no definite refinement can be confirmed. Note the same result of the hydraulic model tests in Section 2.

## 4. CONCLUSIONS

Hydraulic model tests have confirmed that within the river-channel with the given dimensions dredging pits must not be made shorter than the channel-width and wider than about half of it to avoid undesirable flow-disturbances and bed-erosion. However, the excavation can be performed deeper as before when constructing very mild upstream and downstream slopes for the pit the latter in a diffusor-like shape. The side-walls can be excavated vertically. Slopes concerning the given grain-size etc. conditions will develop by sliding (questions i and ii in Section 1).

The experiments have also confirmed that the well-known general bed-erosion, in our case induced by the head-drop at the upstream end of the dredged pit, will recede in the upstream direction to the next crossing even at reasonably mild water-surface slopes. Owing to the increased near-bottom-velocities in and around the pit no major filling up except of the slided wall-material will occur (question iii).

Extensive field surveys and samplings on a 60 km-long reach of the Danube river downstream of Budapest have confirmed that in excavated troughs with dimensions of 500 to 1000 x 100 to 150 x 2 to 4 m no major refilling occured for several years. Characteristic and overall refinement of the bed-material i.e. the sediment in the areas of excavations could not be proved (question iii). That is, from the points and needs of the water users and human exploitation the level-drops of low-flows are more severe consequences of the dredging activities than the intrusion itself into the channel-regime and sediment-balance.

The measured/calculated channel-balance of 15 years (including the excavated volumes) has shown a certain refilling process which conforms to the natural regime of the river-reach. Information available on the actual sediment-transport conditions are insufficient to confirm the channel-processes observed or to provide reasonable predictions of the consequences of further dredging activities (question iv).

It is worth noting that any more sophisticated numerical assesment of the situation is also hindered by the limited data availability and the complexeness of the system described.

Field investigations on a reach of 60 km upstream of Budapest including both branches of the river in Fig 1 have been performed in 1987. Further results confirming or disproving the findings of this paper may be presented on the Conference.

## 5. ACKNOWLEDGEMENTS

The skilful assistance of the staff of the River Training Experimental Station of "VITUKI" at the Rába river, Hungary and that of the survey team (head: C. Eng. O. Fiedler) from the Hydrographic Department of "VITUKI" is highly appreciated.

## 6. REFERENCES

(1) Kornis, S (Mrs): "Meander Travel in Alluvial Streams". In: Preprints of Int. Workshop on Alluvial River Problems. /Roorkee, India: March 18-20, 1980) Sarita Prakashan, New Delhi, pp 4/35-4/41.

(2) Laczay, I: "Bed-change processes in a sand-bearing river". In: Proc. IAHR Int. Symposium on River Mechanics (Bangkok, Thailand: January 9-12, 1973), Paper A-72.

(3) Laczay, I: "Sediment movement, local bed-interchanges and potential energy expenditure on sand-bearing rivers". In: Proc. IAHR XVI. Congress (Sao-Paolo, Brasil: 1975), Paper B-21, pp 168-172.

## 7. LIST OF FIGURES

International Conference on
RIVER REGIME
18-20 May, 1988

PAPER J1

# THE EFFECT OF BOAT TRAFFIC ON RIVER REGIME

P. N. GARRAD[1] and R. D. HEY[2],

[1] Forth River Purification Board, Colinton, Edinburgh.
[2] Environmental Sciences, University of East Anglia, Norwich, Norfolk.

SUMMARY : Since 1945 the width of Broadland rivers has increased dramatically. This is shown to be mainly due to wave attack by boat traffic aggravated by a decline in bankside vegetation. Management options to reduce bank erosion include curbs on boat speed and bank protection.

## 1.0 INTRODUCTION

### 1.1 Broadland

The Norfolk Broads are a series of interconnected lakes and rivers in Eastern England that discharge into the sea at Great Yarmouth (fig 1). These 200km of lock free waterways are particularly well suited to boat cruising. Indeed, Broadland became a popular sailing area at the turn of the century, and since the 1950's with the advent of fibre glass and the explosion of leisure demand in affluent post war Britain, the hire fleets have increased four fold and become predominatly motorised (fig 2). In 1981, over 12,000 boats were registered for use by the Navigation Authority, the Great Yarmouth Port and Haven Commissioners.

### 1.2 Problems of erosion

The problem of bank erosion on the Yare and Bure river systems in recent years has become serious enough for considerable lengths of the bank to be piled in order to prevent undercutting of the flood walls. On the lower reaches these walls form important tidal embankments protecting adjacent farm land, much of which is below the level of the river. Piling is not only expensive (£300 per metre), but also results in the loss of natural bank vegetation and natural habitat, as well as bringing a visually intrusive element into the waterside. Even where flood walls are not threatened, bank erosion seriously damages the ecology and landscape of the Broadland rivers, and increases suspended sediment concentrations. The latter has resulted in high rates of siltation in some Broads, and increases dredging requirements.

International Conference on River Regime, edited by W.R.White.

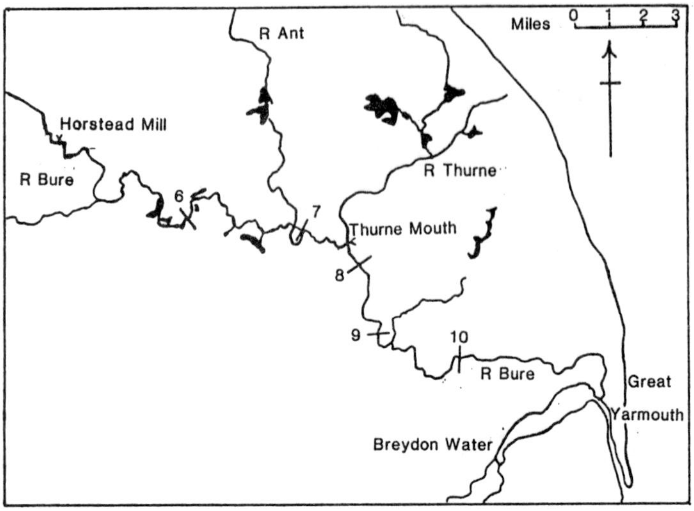


Fig 1. Hydrology of the Bure valley, indicating position of cross sections and field measurement sites.

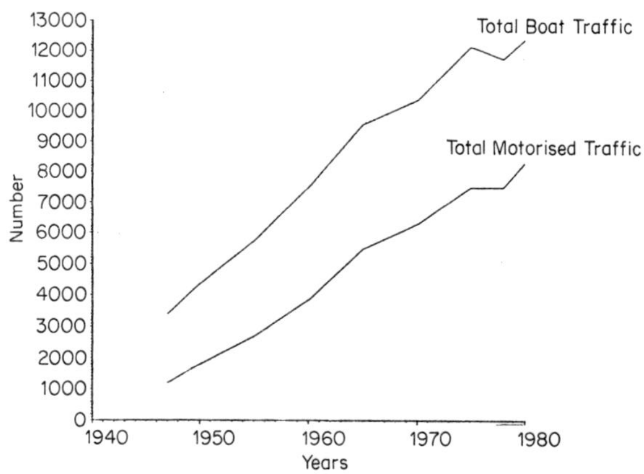


Fig 2. Pattern of boat registrations; Broadland waterways 1946-1980 (Source: ref 9).

## 2.0 EVIDENCE FOR BANK EROSION

### 2.1 Ordnance Survey maps

In any natural river system there will tend to be a certain amount of bank erosion. For rivers in dynamic equilibrium erosion tends to be balanced by an equal amount of accretion as meanders migrate across and down a flood plain, showing a long term stability. In Broadland the low flow velocities measured by Garrad (Ref 1) would suggest that the natural rate of channel migration would be minimal.

The increase in river width over historical times can be shown by comparing Ordnance Survey maps dated 1883, 1946, 1963 and 1976. This indicates that the lower reaches of the Bure has widened by as much as 16.0 metres in the period 1883 to 1976 (Fig 3), with markedly higher rates in recent years. For the period 1883-1946, the rate of bank retreat was of the order of 0.3 to 0.6 cm per year, whereas current rates range from 6.0 to 16.0 cm per year for the period 1946-1976, indicating channel width may be changing by up to 32 cm per year. The progressive increase in river width means that longer lengths of bank will need to be piled to prevent the channel encroaching within 5m of the flood wall, unless erosion can be halted or reversed.

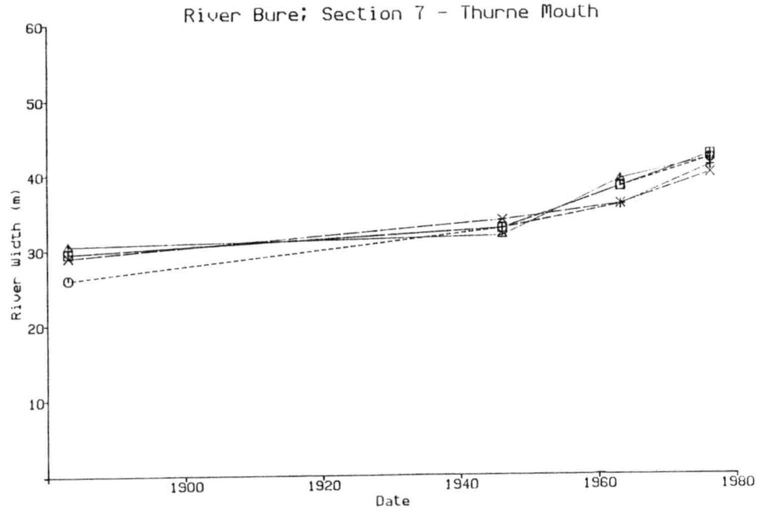


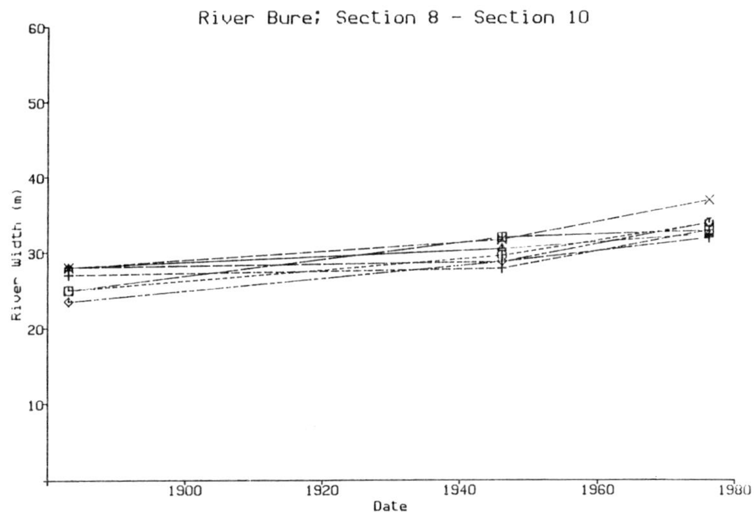


Fig 3. Change in river width at selected locations as taken from ordnance survey maps.

2.2 Aerial photographs

Measurement of river width can also be made using aerial photographs. The scale of such photographs can be accurately determined by comparing size of known objects on the ground, such as buildings etc. This gives a direct comparison between photographs of a different age. Estimates of river width from aerial photographs taken in 1946, 1962, 1970 and 1980 show how the river has increased at a rate of between 4 and 20 cm per year since 1970 at 5 sites upstream of section 7 (fig 4), and by as much as 3.5m at location (UB 1) since 1946.

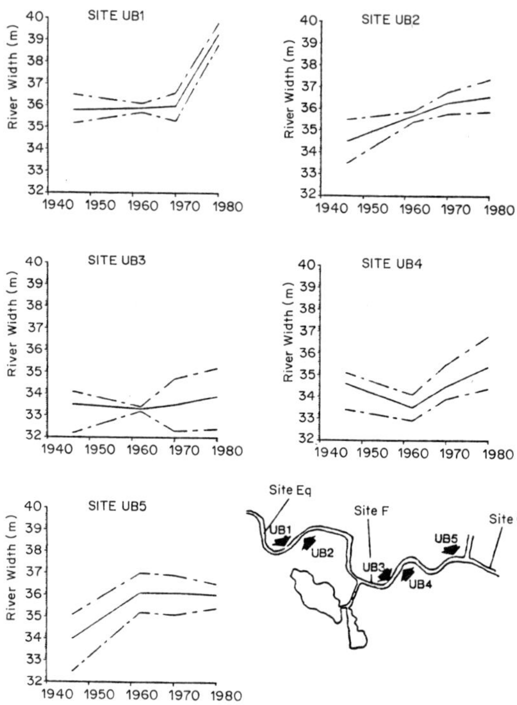


Fig 4. Change in river width as indicated by aerial photographs (Dashed lines indicate errors).

2.3 Models of sediment transport

Further estimates of changing river channel dimensions can be made by measuring the net loss of material from a reach of river. A model of suspended sediment transport for the middle reaches of the river Bure suggests a net deficit of 3480 tonnes of inorganic material per year between Horstead Mill and section 7 (Garrad, Ref 1). By calculating the area of river bank available for erosion, and accounting for known sources and sinks of material, a net erosion rate of between 1.9 and 5.3

cm is suggested, with a mean erosion rate of 3.7 cm per year. This assumes that all banks retreat at the same rate irrespective of vegetation type, and that parallel retreat occurs. No account was made of the amount of material lost to the river bed, since rate of dredging are too infrequent to allow a suitable estimate, and it is thus likely that the above figure is an underestimate of the true erosion rate.

## 2.4 Direct surveys of river cross sections

Sections surveyed by steel tape in 1981 (Garrad, Ref 2) were resurveyed in 1987 using the same technique. The results indicate a varying degrees of erosion, from 0.0 to 1.78 metres per year. In particular, some of the navigable sections have shown a marked increase - the upper entrance to Wroxham Broad has increased by 10.4 metres in this period. Some difficulty was encountered in resurveying some sections in that the datum points used in the 1981 survey (tree stumps, fence posts etc) had been eroded or removed by river bank erosion.

Sections surveyed by the Anglian Water Authority (Ref 3) in 1980, 1981, 1983 and 1985 (fig 5) also suggest changes in cross sectional shape at different points. Indeed, some of the 0.5m square concrete bench marks constructed well back from the river in 1976 have long disappeared as a result of bank erosion. These surveys indicate a large loss of material from the river banks themselves, whilst the beds have gained material. Section 6 on the river Bure has lost some 6.0 square metres of bank material in just under 5 years, whilst some sections further downstream have shown a net gain of material to the river bed or close to the banks.

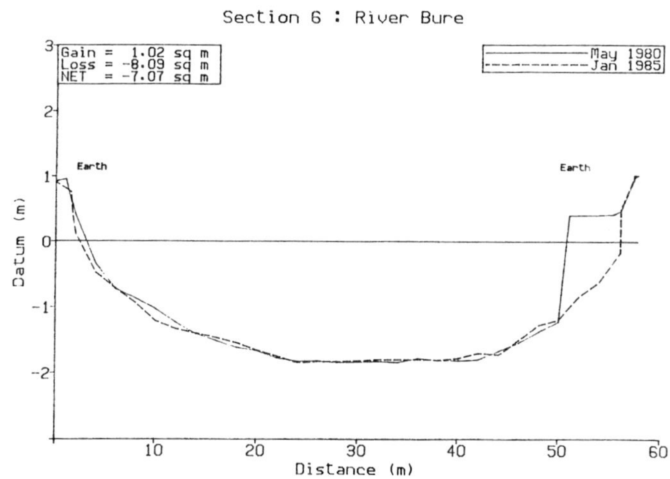

Section 7 : River Bure

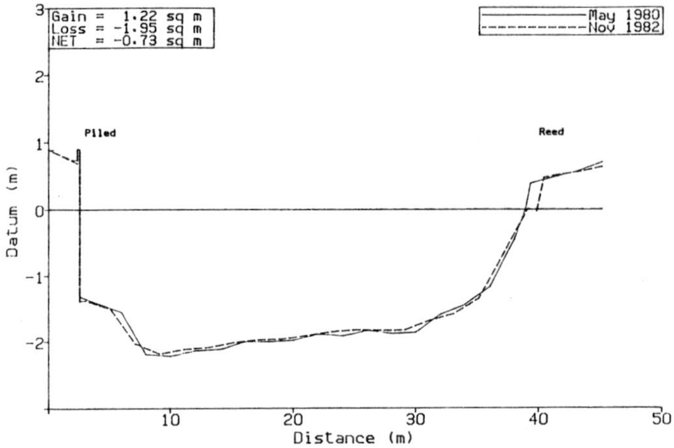


Section 8 : River Bure

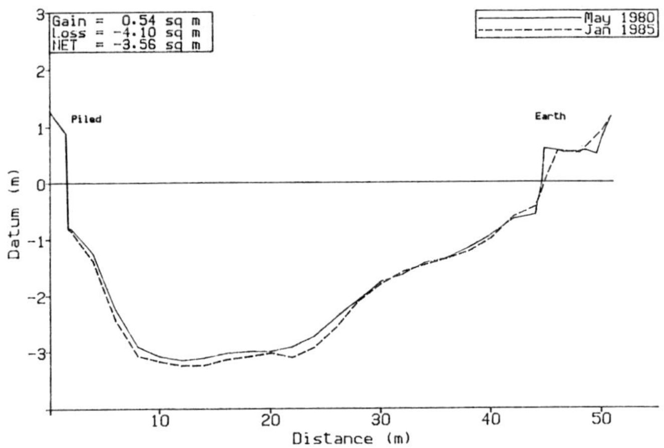


Section 9 : River Bure

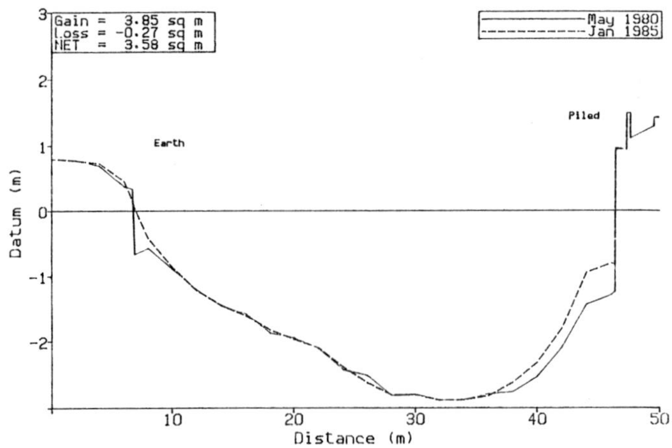

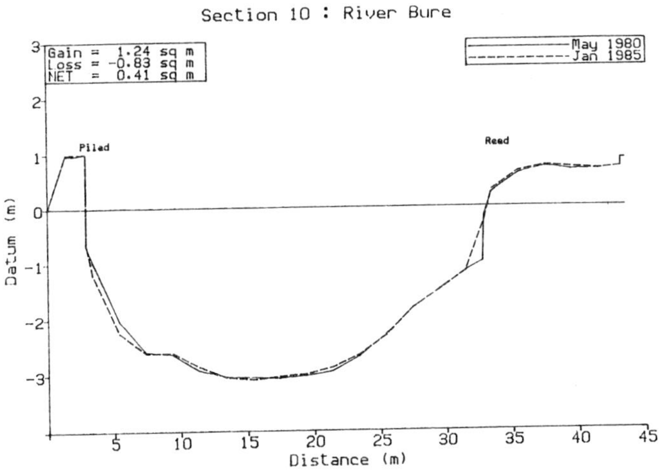


Fig 5. Change in river cross sectional shape over period May 1980 to January 1985, with associated bank types

## 3.0 BANK STABILITY AND THE CAUSES OF EROSION

The stability of a river bank is dependent on the erosional resistance of the bank material and the stresses acting upon it. Erosional resistance is related to soil type, vegetation and bank profile, although river water quality, dredging, trampling and animal activity can all influence the shear strength of the soil. The most significant of the applied stresses are those created by fluid flow, which may remove individual soil particles or larger aggregates. These may result from streamflow, tidal activity and wind or boat generated waves.

### 3.1 Resistance of bank material

#### 3.1.1 Soil Type

Soil profiles generally agree with the known sequence of depositional events in Broadland under repeated changes in sea level as examined by both Lambert and Pallis (Refs 4,5). The general trend is for silts near the seaward end, with an increasing amount of clay and peat in the lower part of the profile progressively landwards. In place the peat is quite thick (up to 1m). The middle reaches of the rivers have an organic clay - peat - clay sequence, with the peat getting progressivley thinner upstream. The upper reaches show a complex profile with inter-bedded sands and gravels.

Erosional resistance is primarly dependent on soil type, although its strength can be enhanced by the effects of vegetation or suitable bank grading. The shear strength of a soil type is dependent on the cohesion and the internal friction between its constituent particles. The soils in Broadland are mainly peats and clays; the peats have a high internal friction but low cohesion, and conversely clays have a low internal friction but a high cohesion. Since cohesion has less influence than internal friction on shear stength, it is easier to shear a block of clay than a block of peat, and it is often these weaker layers of clay or lenses of sands and gravels which are exploited first, and make certain banks more susceptible to erosion.

### 3.1.2 Vegetation

Three broad vegetational zones can be distinguished in Broadland; an area of grass banks around the mouths of the rivers, a large zone of reed fringed banks in the middle reaches, and an upper region of herbacious banks, carr and wooded areas.

In general, the greater the root depth and density of the root matter, the greater the bulk shear strength of the soil. Reeds and small trees, which are amongst the most common vegetation types in Broadland, provide both these requisites. Tall trees with large roots have only a small reinforcing effect, whilst grasses which have a very thick root mat only enhance the shear strength over a limited depth. Aquatic plants not only reinforce the strength of the bank through their root mat,but also help in reducing the fluid forces acting on the bank by locally increasing the frictional resistance to flow.

### 3.1.3 Bank Profile

All soils have a certain critical angle of rest, at or below which the bank is stable, whilst above this value it will tend to fail. For clays and silts with high water contents, the critical angle of rest is low, whilst for peats it is usually high. Typical bank profiles in the region can be divided into 4 classes; sloping, sloping with an undercut face, stepped and stepped with undercutting. Analysis of the soil sequences suggest that the undercutting is confined to the clay and silt horizons or to sand and gravel lenses, with the peat relatively resistant. In general, stable banks are composed predominatly of peat, whilst unstable banks have either no peat, or if present only in thin layers.

### 3.1.4 Indirect effects

Water quality can have an indirect effect on bank stability by its effect on aquatic plants. Aquatic vegetation can provide stability both through the root mat and by increasing the frictional resistance to fluid forces at the bank face. Osborne and Moss (Ref 6) indicated how water quality and turbidity cause preferential growth of algae over some higher plants by providing shade early in the growing season. The restricted growth of macrophytes can make the river banks more susceptible to erosion by reducing the natural protection. The causes of turbidity are thought to be a combination of phytoplankton blooms caused by over enrichment from sewage and nitrogen fertilizers (Moss, Ref 7), together with resuspension of bed sediments by boat traffic (Garrad and Hey, Ref 8).

Further erosion may be caused by the burrowing of water voles or coypu, whose tunnels may be enlarged by the action of waves, and by wetting and drying. These effects are restricted to the upper reaches of the bank around still water level, and are particularly important when the bank is exposed and not protected by a covering of vegetation or reed fringe. Loss of upper bank vegetation by grazing and by trampling by anglers can also have serious effects, as the cleared area has a degraded root mat. Creation of a vertical or steepened face beyond the critical angle by dredging may cause failure, and subsequent slumping.

## 3.2 Stresses acting on the bank

### 3.2.1 Stream flow and tidal activity

The most significant of the de-stabalizing factors are fluid forces. These not only cause erosion but allow transport of the removed material and thus allow further erosion to take place. Examination of the fluid forces in the upper and middle reaches of the River Bure show very low natural velocities. The maximum tidal velocity during spring tides at St Benets Abbey, measured in mid channel was 28 cm $s^{-1}$, whilst at the bank edge this may be reduced to less than 4 cm $s^{-1}$; sufficient to erode only clays and fine silts. This maximum velocity is reduced upstream as tidal influence declines (Garrad, Ref 2). During the passage of a boat however, local velocities may exceed 70 cm $s^{-1}$, (Payne and Hey, Ref 9) sufficient to erode and transport much of the river bed sediment, and damage the river banks, particularly when the boat waves break against the bank. Clearly, boats introduce high energy fluid forces into what is essentially a low energy environment.

### 3.2.2 Boat and wind waves.

The magnitude of fluid velocities within a wave, whether boat or wind induced, is dependent on its height, length, period, the depth of water in which the wave is travelling, and the position in the water column relative to the still water level. A single particle starting at rest on the surface, and subject to a wave form will follow a circular motion, with its path and time of travel determined by the wave height, and the wave period. The velocity of a particle can be expressed as ;

$$V = \frac{\pi}{T} H$$

where H is the wave height, T the period and V the velocity of the particle. Thus at the surface high amplitude short period waves create higher particle velocities than low amplitude long period waves. This effect diminishes lower in the water column; particle velocity decreases by a factor of two for every one ninth of the wave length below the surface.

The height and period of wind generated waves are determined by the wind strength and the fetch distance of open water over which the wind is acting. On bends or tree lined sections it is unlikely that stretches of open water will be long enough to allow the formation of significantly large waves for a sufficient period of time.

The height and period of waves causes by the passage of a boat are determined primarily by the boat speed, and to a lesser extent by hull design, proximity to the river bank and geometry of the waterway (Payne and Hey, Ref 9). There are three distinct elements associated with a boat's wake (Ref 10). First as a boat approaches, there is a slow rise in water level as water is pushed up in front of the vessel (Fig 6). Secondly, on passing, there is a rapid fall in water level exposing previously submerged portions of the bank, with strong reversed currents increasing the shear stress against the channel boundary. Finally there are a series of waves hitting the bank at an angle, and diminishing with height and angle at each wave until eventually only a gentle swell moving parallel to the bank remains. These may take the from of breaking waves. Waves generated at the bow of the vessel may interact with transverse stern waves producing interference peaks (Hey and Tovey, Ref 11).

The amount of water level rise and drawdown is related to the ratio of hull to channel cross sectional area; the blockage factor. Given present hull designs, this can be reduced by lowering boat speeds or maximising the distance between the boat and the bank. For most boats on the Bure, the blockage factor is low, and so displacement induced flows are negligable. The first of the diverging waves to hit the bank are the highest and have the shortest periods and wave lengths, and are therefore the most significant in terms of fluid velocities within the waveform. The relationship between boat speed and wave height for the UEA research boat "Envoy" (Fig 7) emphasises that boat induced waves are dependent on boat speed and its distance from the bank, up to a given point. After a given speed, these curves become asymtopic to a fixed wave height, and then decrease in height as planing speeds are reached (Schijf, Gates and Herbich, Refs 12,13).

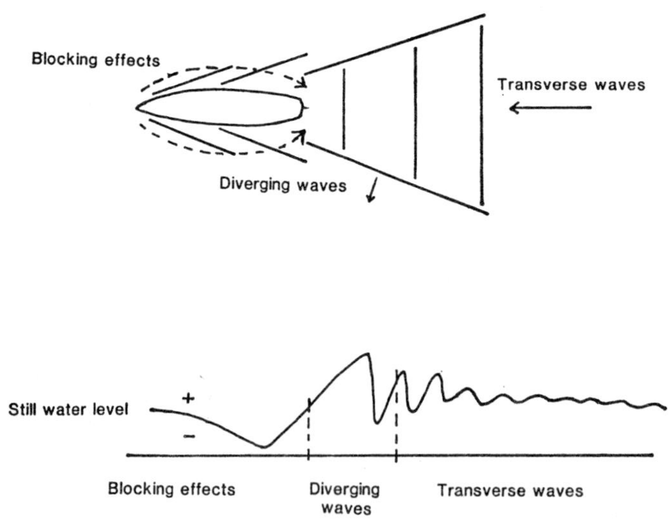


Fig 6. Types of wave caused by the passage of a boat, and subsequent variation in water level with time.

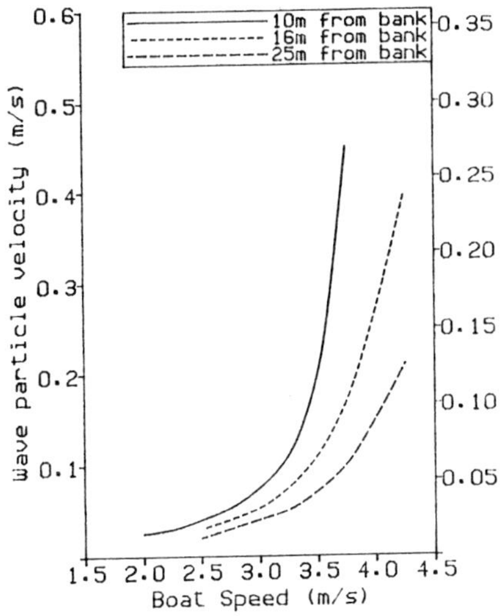


Fig 7. Relationship between wave height or particle velocity and boat speed.

### 3.2.3 Abrasion

In some parts of the river system, particularly the River Ant, physical abrasion of bank material by continued mooring on unprotected banks is a considerable problem.

## 4.0 MECHANISMS OF BANK FAILURE

### 4.1 Surface erosion on earth banks

If the shear stress exerted by the fluid forces acting on the bank are strong enough, they will remove individual particles. This removal will be most rapid in areas of high shear stress and locally weak banks. In most cases this results in exploitation of weak strata, generally below the rooting zone, resulting in undercutting. This will leave an undercut block which may fail en-masse once its stabilty passes beyond a critical limit. As shown above, the dominant fluid force operating on the middle and upper reaches of these river system are boat generated waves. These are dependent on wave characteristics and decrease in intensity with depth. Thus the maximum stress exerted on the bank by a wave is at the waterline. Banks above the waterline when subject to wave action, experience alternate wetting and drying which may result in accelerated erosion of the clays. Soils below the waterline experience a loss in stability during periods of drawdown due to the loss of bouyancy provided by the water. In the case of saturated overhanging blocks this can be sufficient to overcome stability and cause failure. This is the most common mode of failure in the middle reaches - undercutting occurs along a line of weakness, or thin weak strata such as a sandy lens, and failure occurs when the effect of gravity on the overhanging block overcomes the tensional strength of the bank material. This will often occur when the water level has dropped, and the block has its weight increased by previous saturation under waves or tidal action. Failure is more likley to occur when instability is induced by a degree of undercutting, although boat waves may provide the trigger by reducing lateral support during drawdown.

### 4.2 Reed fringed banks.

In reed fringed reaches the reeds form a floating shelf with a root mat up to 40cms thick. In an undisturbed state, the reed bed may grow out into the river with the limit of the overhang or 'hover' determined by the strength of the root mat or disrupting forces which may cause the hover to break. A hover can absorb small amplitude wind and boat waves, but once it reaches a critical size and is subject to larger waves, failure will occur. The mode of failure is generally for small sections to break off by cantilver failure after the generation of tensional cracks up to 100cm from the waterline. Periodically large hovers are displaced causing serious problems for navigation. The increase in boating activity in recent years has resulted in a greater frequency of waves and hence pressure on existing hovers. These can absorb wave energy, although when subject to high amplitude waves of short duration, together with drawdown effects, tensional cracks may develop at the head of the root mat and failure may occur. Once a hover has failed, then wave motion is no longer damped before it can reach the face, and if weaker strata are exposed, the process can start again. The critical controls for using reed hovers as bank protection are establishing a healthy reed growth with a deep root mat, and controlling wave energy below the threshold limit for failure.

### 4.3 Tree lined banks.

Mechanisms for failure in the tree lined reaches generally involve removal of the soil around the root mat by fluid forces as described above. Once this matrix is removed, the stability of the tree is jeopordised. The subsequent failure will depend on the rooting depth and hence size of tree, and can be aggravated by high winds. By keeping the trees at the river margins relatively young by active management, the roots can be kept small and hence bind the soil to a greater degree than trees with larger root systems. Additionally, those trees with dense water line foliage are better able to buffer wave energy, and hence protect the bank.

## 5.0 MANAGEMENT TO PREVENT FURTHER EROSION

The management required to reduce bank erosion must be directed towards either (i) reducing the forces acting on the river banks, or (ii) increasing the ability of the bank to withstand these forces. Of all the forces affecting bank stability, river velocities are very low and insufficient to cause erosion, whilst tidal velocities are only high enough to cause erosion in the lower reaches. In the upper and middle reaches of these rivers the only effective erosional forces are those produced by waves. Wind waves can under exceptional circumstances become appreciable erosive agents, but by far the most consistent agent is boat generated waves. The factors which control the bank's erosional resistance are its soil type, profile and vegetation cover. Only boat generated waves, the bank profile and its vegetation cover are controllable factors as far as practical management of the waterway is concerned.

### 5.1 Reducing forces acting on the river bank

To prevent erosion, or at least substantially reduce it, the following management options require consideration.

#### 5.1.1 Reduction of boat wash.

The critical factor is the shear stress generated by the wave when it hits the bank compared to the strength of the bank material. The reduction of wash by individual boats can be acheived by the following means.

(a) Speed reduction. The height of the wave created by a boat and its associated shear stress increase with speed. It should be possible to adjust speed limits to a level which generates waves at or below a level which the river bank can withstand. The setting of an approriate speed limit would then take account of the type of bank material, the shape of the bank profile, the prescence of protective vegetation and the width and depth of the channel. Given the wide range of situations and complications created by multiple boat passages, it is unlikely that a precise figure could ever be established. Since wave height increases exponentially with boat speed (fig 7) a reduction in boat speeds could cut wave heights to an acceptable level dramatically. Current information, would suggest that boat speeds should be kept below 5mph ($2.23\ m\ s^{-1}$) for most of the river systems.

An alternative would be to monitor bank erosion rates for current speed limits and relate these to bank material type. The biggest unknown is the extent to which speed limits are observed. The current speed limits, designed with regard to navigation and nuisance control, have been ineffective in reducing erosion. A complete review of speed limits to take account of the above factors is required.

(b) Demarkation of river banks prone to erosion. Since wave height decays with distance, (hence the relation with sailing line) controlling proximity to the bank could prove effective. However for most rivers, the demarkation of limits adjacent to the river banks would involve practical difficulties as well as being hazardous to navigation, given that many of the rivers are very narrow.

(c) Hull design. Experiments on different hull designs have shown little variation in wave height between many of the larger hire cruisers. Indeed, it is the smaller day boats, with the smallest period waves that potentially have the most damaging effects. Given the expected life span of boats used on the Broads, it is likely to be some time before an improved hull design, if available, would have significant impact.

(d) Number of boats. Controlling the number of boats on these waterways is impratical and would undoubtedly prove unpopular. The boat fleets provide a muti-million pound income for the region both directly to the boatowners, and indirectly for the subsidary industries associated with the tourist trade. It is doubtful if this would be an efficient way of reducing erosion unless boat numbers were reduced substantially.

## 5.2 Increase the erosional resistance of bank material

A second element in a management strategy to reduce erosion involves increasing the ability of the banks to withstand erosive forces. Introduction of suitable vegetation together with regrading bank profiles would help, particularly if reed can be re-established as this provides a sufficiently dense root mat and also acts as a barrier absorbing energy from the fluid forces as they approach the bank. Given a suitable sloping profile and clear water, reed can be encouraged to root to depths suitable for bank protection. Unfortunatly there are many places where there is no longer sufficient width of rand to permit such re-grading. Whilst theoretically the flood walls could be set further back to allow regrading, there would be major problems providing suitable foundations for the new wall because of the poor bearing capacity of the flood plain soils. Even where the width of rand is sufficient, the problems of reed die back could mean that re-establishment may only be temporary. Research on the causes of reed die-back, and on the conditions neccessary for re-establishemnat of reed is required. Even with the introdution of reed, there could still be a problem with undercutting below the root zone. Natural protection could be achieved by the introduction of macrophytes which root lower in the bank profile. This would require improvement in water quality by controlling turbidity through a reduction in boat movements, or by eliminating algal blooms particularly during the early part of the macrophyte growth cycle. Some form of submerged piling or geotextile revetments could also be used for toe protection (Bonham, Ref 14) should natural procedures fail.

## 6.0 CONCLUSIONS

Evidence from a variety of sources has shown how river widths are increasing at rate of between 1.9cm and 1.8 metres per year at various locations on the Broadland waterways. This is somewhat suprising given that these are naturally low energy environments. The main cause of bank erosion being boat induced waves.

In the short term, the most effective way of reducing river bank erosion would be by controlling boat wash. This can be achieved by reducing boat speeds to 5mph (2.23 m $s^{-1}$) or less in most regions. Additionally, suitable management to increase the resistance of the banks to erosion by regrading and introducing protective vegetation may also prove appropriate.

## NOMENCLATURE

V = velocity of particle in a wave
H = Wave height
T = Period of a wave

ACKNOWLEDGMENTS

The authors are grateful to the Anglian Water Authority, for kindly supplying data from their Broadland surveys. This work was supported by a grant from the Great Yarmouth Port and Haven Commissioners.

References

1. Garrad, P.N. 'Sediment flux and bank erosion in a tidal recreational waterway.' In: Proc. National Hydrology Symposium, British Hydrological Society, University of Hull, Sept 14-16 1987, Paper 24, 20pp.

2. Garrad, P.N., 'A Sediment budget of the River Bure, Norfolk.' Unpublished PhD presented to the University of East Anglia, Norwich, 1984.

3. Anglian Water Authority, Broadland survey data, private communication, 1987.

4. Lambert, J.M., Jennings, J.N., Smith, C.T., Green, C. and Hutchinson, J.N., 'The making of the Broads; a reconsideration in the light of new evidence.' Royal Geographical Memoirs, 3, 1960, John Murray Ltd, 153pp.

5. Pallis, M., 'The status of Fen and the origin of the Norfolk Broads.' R. Mackehoe and Co, 1961, Glasgow.

6. Osborne, P.L. and Moss, B., 'Paleolimnology and trends in the phosphorus and iron budgets of an old man made lake, Barton Broad, Norfolk.' Freshwater Biology, 7, 1977, pp213-233

7. Moss, B., 'Conservation problems in the Norfolk Broads and rivers of East Anglia. phytoplankton, boats and the causes of turbidity.' Biological Conservation, 12, 1977, pp95-114..

8. Garrad, P.N. and Hey, R.D., 'Boat traffic, sediment resuspension and turbidity in a Broadland river.' Journal of Hydrology, 1987 (In press).

9. Payne, S.J. and Hey R.D., 'River Management to reduce bank erosion; Yare and Bure river systems.' Broads Authority Research Series 4, Broads Authority, Norwich, 1982, 35pp.

10. British Transport Docks Board Research Station, 'Creation of wash ny pleasure craft', British Transport Docks Board Research Station report no R36, 1972.

11. Hey, R.D. and Tovey, N., 'Mechanics of bank erosion.' CIRIA.

12. Schijf, M.J.R., 'Ships in restricted channels', In: 17th International Navigation Congress, 1949, Brussells, Section 1 - communications, pp61-78.

13. Gates, E.T. and Herbich, J.B., 'The squat phenomenon and related effects of geometry', In: Proc. 25th Ann. Spec. Conf. 'Hydraulics in the Coastal Zone', at Texas A and M Univ.

14. Bonham, A.J. 'Bank protection using emergent plants against boat wash in rivers and canals.' Report No. IT 206, 1980, Hydraulics Research Station, Wallingford, OXON.

International Conference on
# RIVER REGIME
18-20 May, 1988

PAPER J2

## THE USE OF AN EMPIRICAL METHOD (EROSION POTENTIAL METHOD) FOR CALCULATING SEDIMENT PRODUCTION AND TRANSPORTATION IN UNSTUDIED OR TORRENTIAL STREAMS

Zoran Gavrilovic, dipl.eng.
Institute for the Development of Water Resources "Jaroslav Cerni"
Jaroslava Cernog 80, 11223 Beograd (Beli Potok), Jugoslavija

**SUMMARY:** *The Erosion Potential Method* can be used in estimates of erosion production in catchment areas and its transportation to a discharge point, or to an important river section. The method is based on identified situation and erosion interactions, topographic features, and general climatic characteristics. While it requires some field investigation, the method is incomparably faster than long-term observations.

## 1/ INTRODUCTION

The importance of an accurate estimate of erosion products transport is known to all those concerned with the load transport measurements. With modern equipment, the transport of suspended load can be reliably measured, whilst a reliable method for measurement of bed load movement has yet to be found solved. Bed load can only be reliably measured in geodetic surveys of flood sediments in lakes and impounding reservoirs, when it is too late to take protective measures or to give up construction. Transport and discharge of sediment load requires a measurement period not shorter than ten years - a long period for sometimes futile preliminary measurements at a site to be later abandoned due to excessive load hazard. As small impounding reservoirs are generally formed on small, mostly torrential streams, for which even the most elementary hydrometric data are not avialable, not to mention those of load transport, the problem of a preliminary estimate of erosion products becomes of primary importance. This is often the case in developing countries, where large storage reservoirs are constructed on rivers for which observation data are scarce or non-existent.

Systematic investigation of erosion intensity began in Yugoslavia forty years ago at the /***"Jaroslav Cerni"* Institute for the Development of Water Resources**/, which enabled the development of a ***method for quantitive classification of erosion (MQCE)***. It was found during these investigations that erosion intensity could be used in the estimates of load amounts carried by rivers all the way downstream, and investigations have been extended to include observations of load transportation to control sites. **(Ref. 1)**

The calculation method for production and transportation of eroded materials was developed from the investigation data, and was named ***"The Erosion Potential Method"* (EPM)**. The method of calculation procedure was defined more than twenty years ago and ever since it has been used in applied engineering in Yugoslavia and continously refined. **(Ref. 2)**.

The erosion potential method proved to be reliable when used in preliminary estimates of average annual amounts of load at potential or actual reservoir sites, regulated rivers or important structures and communities.

International Conference on River Regime, edited by W.R.White.

## 2/ DESCRIPTION OF THE EROSION POTENTIAL METHOD (EPM)

The name of the method itself indicates the cause-consequence relation of erosion intensity and load discharge in rivers. For correct use of the method, the identification of various forms of erosion must be mastered. For a better understanding of the method, however, a fundamental knowledge of the sources and forms of soil erosion is necessary..

Processes of erosion result from interactions at ***the foundation rock/soil level***, and trough ***topographic features, climatic factors, and land use***. Three important factors, of erosion development occur naturally *(foundation rock/soil level, topography, and climate)*, while land use is entirely man - dependent. **(Ref. 3)**.

The foundation rock/soil base is the level to which erosion progresses under the effect of climatic factors. The name itself indicates the surface soil level, composed of the foundation rock and top soil.

Topographic features are also important in the development of erosion processes, because the rain effect is much stronger on steep slopes than on gentle ones. The rain power effect is equal on any slope, but steep slopes allow greater wear and and transportation of rock particles and soil which are dislodged under rain drop impact.

It is well known that the factor most responsible for fast and intensive soil erosion is the human one, or rather the wrong and destructive mode of land use. Under conditions of wrong land use, the initial forms of surface erosion transform into deep ones. Deep erosion is caused by rills, rivulets, and sheet flow, and the percentage of its surface area is directly related to the destructability of the torrents.

In respect of the climatic agent, erosion occurs in two general forms, but the erosion potential method has been directed to wards research in to intensity of erosion by water. Mention should therefore be made of the general forms of erosion, typical of pluvial erosion: surface, deep and mixed erosions.

Surface erosion in its initial form is generally unnoticed until it begins to turn into deep processes, or mixed processes, which are the commonest.

As mentioned in the Introduction, the *erosion potential method* **(EPM)** is an advanced *method for quantitative classification of erosion* **(MQCE)**. The method for quantitative classification of erosion will be outlined, because the right use of this method is essential for estimates by the erosion potential method.

### 2/1. THE METHOD FOR QUANTITATIVE CLASSIFICATIONS OF EROSION (MQCE)

#### 2/1.1. DESCRIPTION OF THE RESEARCH METHOD USED

The study of erosion intensity is a very complex research problem, because a natural phenomenon is being investigated which has multiple dependence, only one of which is accessible for human intervention, others being incidental natural events.

Like elsewhere in the world, researchers of the ***"Jaroslav Cerni"*** **Institute for the Development of Water Resources** used the usual investigation procedure for multi-functional relations; they set up experimental stations for specific studies.

The aim has been accomplished by forming ***erosion plots*** on slopes at equal angles and on homogeneous foundation rock, but under diferent land use *(forest, crop field, orchard, pasture, etc.)*. The best shape of the erosion plot was also investigated, and the resulting rectangular shape adopted is the form, known as the ***golden rectangul format***. **(Fig.1.). (Ref. 4.).**

**Figure 1. Erosion plot**

Each erosion plot was a catchment area isolated from surrounding effects, and equiped with a totaliser for collecting run-off water and eroded material from the plot.

The entire area of experimental stations was covered with climatic instruments, with an increased density and higher accuracy for precipitation, air and soil temperature measurements.

The collected data were systematicaly processed and systematised, giving the first erosion intensity relations, as shown in **Fig.2.**

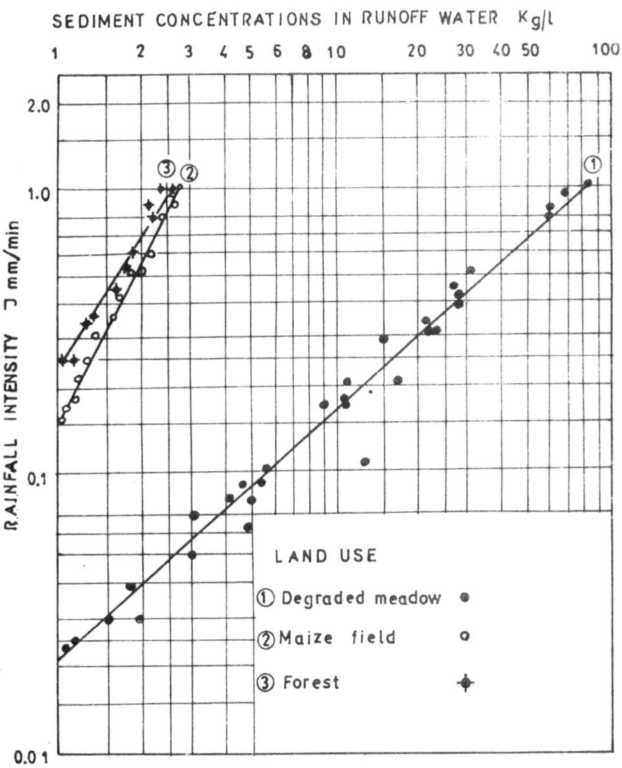


**Figure 2. The established erosion intensity relation to the land use**

Purely experimental results were insufficient for a conclusion to be drawn on the actual effect of one single factor, because the principal erosion agent, climate, could only be roughly selected. Repetition of the complete climatic picture and observation results was therefore not registered in a single case.

In order to acquire coefficients for correction of test results, the experimental stage of investigation was started. In the ***"Jaroslav Cerni"*** Institute Laboratory, special equipment was constructed for rain generation, with adjustable rainfall intensity. At the same time, lysimeters were installed for samples, which were slope - adjustible. **(Fig.3.).(Ref.1).**

Lysimetres accommodated soil samples from test plots, which were then attacked by a sequence of rains of controlled intensities and durations at different temperatures.

Various types of foundation rock/soil base were tested for various slopes and land uses. The collected data were processed to find the coefficients of erosion related to major factors. **(Ref. 1.)**.

## 2/1.2. RESULTS OF COMPLETED INVESTIGATIONS

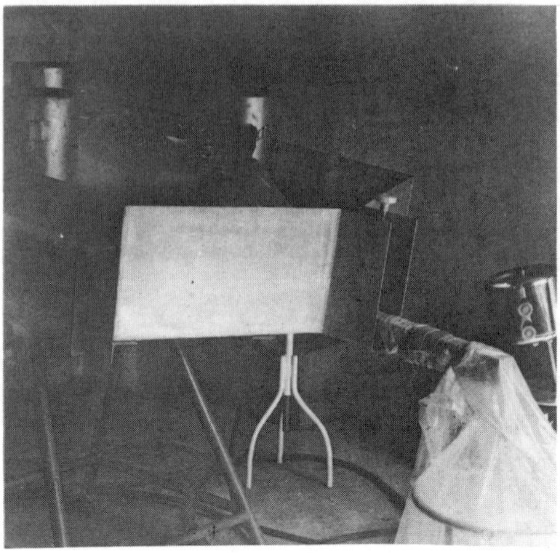

**Figure 3. Laboratory tests.**

The most important result of long research and experimental work is the method for defining the erosion coefficient, which in other methodologies was done by an visual estimate. The quantitative correlation of coefficients and erosion intensities was also defined. Because erosion is a phenomenon affecting the entire soil surface, it is classified into five groups, suitable for cartographic presentations of various erosion intensity incidences. A soil erosion map was prepared using the method for quantitative classifications of erosion, which is practical for use in projects unconcerned with erosion problems, but requiring an erosion situation picture ***(feasibility reports, project reports, etc.)***.

## 2/1.3. CALCULATION OF EROSION COEFFICIENT

Before calculating the coefficient of erosion for a land area or river basin, a geodetic map at a suitable scale should be prepared, together with aerial photoprints at the same scale showing the required details. The scale range used in practice was wide (from 1:5.000 to 1:500.000), with the accurracy corresponding to the map scale. **(Ref. 4.)**.

For a natural phenomenon such as erosion, a minimum of field investigation is necessary. The coefficients for further calculations are estimated in the field, and these are the coefficients of the observed erosion effects and the coefficients of land use. Other coefficients are determined in the office, or directly in the field (only for small areas).

## 2/1.4. COEFFICIENT OF THE OBSERVED EROSION PROCESS ( $\varphi$ ) AND ITS ESTIMATE

The coefficient of the observed erosion process **($\varphi$)** is a matter of visual estimate, which requires certain experience; nevertheless, the coefficient values in **Tab. 1.** have given satisfactory results. Values of the coefficient of observed erosion process **($\varphi$)** are graded into ten values, in a range from **1.0** to **0.1**. Description of the states of erosion for each coefficient value are recognizable for practical use. **(Ref. 5)**.

## 2/1.5. DETERMINING LAND USE COEFFICIENT ( Xa )

The idea for determining the coefficient of land use **(Xa)**, was taken from the American method developed by ***Browning-a***, and modified for specific use. Coefficient values are determined for various land uses, but those given in **Tab. 2.** are sufficient for common usage. **(Ref. 6.)**.

*Table* 1. COEFFICIENT VALUES FOR THE OBSERVED EROSION PROCESS ( $\varphi$ )

| No. | CONDITIONS OF INFLUENCE ON THE COEFFICIENT VALUE | MEAN VALUE |
|---|---|---|
| 1. | Basin or area fully attacked by gulleying and deep process of erosion | 1.0 |
| 2. | About 80% of area under rills and gullies | 0.9 |
| 3. | About 50% of area under rills and gullies | 0.8 |
| 4. | Entire area attacked by surface erosion: detritus and debris, few rills and gullies (deep erosion), and heavy karst erosion | 0.7 |
| 5. | Entire area attacked by erosion, but without visible deep effects (rills, gullies, rockfalls, etc) | 0.6 |
| 6. | 50% of area attacked by surface erosion, while the rest of the basin is unattacked | 0.5 |
| 7. | 20% of area attacked by surface erosion, and 80% unattacked | 0.3 |
| 8. | Land surface without visible erosion effect; minor rockfalls or slips in stream channels | 0.2 |
| 9. | Land surface without visible erosion effect; mostly crop fields | 0.15 |
| 10. | Land surface without visible erosion effect; mostly under woods or perennial crops (meadows, pastures etc.) | 0.1 |

*Table* 2. LAND USE COEFFICENT ( Xa )

| No | CONDITIONS OF INFLUENCE ON THE COEFFICIENT VALUE | MEAN VALUE |
|---|---|---|
| 1. | Denuded unarable land (badlands) | 1.0 |
| 2. | Fields ploughed up/down the hill | 0.9 |
| 3. | Orchards and vineyards without low vegetation | 0.7 |
| 4. | Field contour-farmed | 0.63 |
| 5. | Degraded forest and shrub on eroded soil | 0.6 |
| 6. | Dry mountain pastures | 0.6 |
| 7. | Meadows and similar perennial crops | 0.4 |
| 8. | Grass-grown and drained pastured | 0.3 |
| 9. | Good forest on steep slopes | 0.2 |
| 10. | Good forest on gentle slopes | 0.05 |

*Table* 3. COEFFICIENTS OF SOIL RESISTANCE TO EROSION ( Y )

| No. | FOUNDATION ROCK AND SOIL LEVEL | MEAN VALUE |
|---|---|---|
| 1. | Sand, gravel, and loose soil | 2.0 |
| 2. | Loess, tuff, saline soil, steppe soil and the like | 1.6 |
| 3. | Weathered limestone and marl | 1.2 |
| 4. | Serpentine, red sandstone, flysch deposits | 1.1 |
| 5. | Podzol, parapodzol, disintegrated schist: micaschist, gneiss, argillaceous schist, etc | 1.0 |
| 6. | Compact and schistose limestones; tera rossa and humose-silicate soils | 0.9 |
| 7. | Brown forest soil and mountain soils | 0.8 |
| 8. | Smonitsa, valley and back bog soils | 0.6 |
| 9. | Chernozem and alluvial deposits of good texture | 0.5 |
| 10. | Bare compact igneous rocks | 0.25 |

### 2/1.6. COEFFICIENT OF SOIL RESISTANCE TO EROSION ( Y )

As foundation rock/soil is the level eroded under atmospheric effects, it is important to know resistivities of various rock/soil types to erosion. **Table .3.** summarizes principal coefficients of soil resistance to erosion for commonest rocks and soils. **(Ref. 1.).**

### 2/1.7. EROSION COEFFICIENT CALCULATION FORM

The above mentioned coefficients form the basis for calculation of the erosion coefficients, but before using them, average land slope **(I)** schould be determined for each unit area *(basin or sub-basin)*. **(Ref. 1.).**

Units with common characteristics (coefficients of observed erosion process **(φ)** and land use **(Xa)**) should be separated on aerial photoprints. Coefficients of resistance to erosion **(Y)** should be determined for the separated units using geologic and soil maps. Coefficient of erosion **(Z)** is then calculated from the relation:

$$Z = Y * Xa * ( \quad + \; I^{1/2} )$$

The calculation will give a multitude of most diverse values, a different value being obtained for each area. Being impractical for mapping, the numerosity of data led to a classification of erosion process.

### 2/1.8. CLASSIFICATION OF EROSION

Coefficient of erosion is the most acceptable quantity for classification of erosion. Through comparisons with the situations in the field, a methodology has been developed for division of the erosion coeffecent values, into five categories. The division could have been more detailed, but five categories proved the optimum practical number for diagramatic representation of the state and areal incidence of erosion. For practical use and estimation each category was given a qualitative name and a respective coefficient of erosion value (Z). The graphical representation of areal incidence of erosion by categories is termed as ***"erosion map"***, and is a basis for application in other fields. Ranges of the erosion coefficient **(Z)** values are given in **Tab. 4.(Ref. 7).**

**Erosion category I** - ***Excessive erosion*** - *deep erosion process (gullies, rills, rockfalls, and the like)*
**Erosion category II** - ***Heavy erosion*** - *milder forms of excessive erosion*
**Erosion category III** - ***Moderate erosion***
**Erosion category IV** - ***Slight erosion***
**Erosion category V** - ***Very slight erosion***

**Table 4. RANGES OF THE EROSION COEFFICIENT (Z) VALUES**

| Erosion category | I | II | III | IV | V |
|---|---|---|---|---|---|
| **range of coeff. (Z) values** | Z > 1.0 | 0.71< Z <1.0 | 0.41<Z<0.7 | 0.20<Z<0.4 | Z< 0.19 |
| **Mean value of coefficient (Z)** | Z=1.25 | Z=0.85 | Z=0.55 | Z=0.30 | Z=0.10 |
| **Qualitative name of erosion category** | **EXCESSIVE** | **HEAVY** | **MODERATE** | **SLIGHT** | **VERY SLIGHT** |

A nomogram has been constructed for fast erosion classification in the field by categories to provide a rough estimate of erosion coefficient (Z) values **(Fig. 4)** The nomogram can be used where the surface slope (%) and other coefficients are determined from the Tables.

The use of the nomogram is shown in **Fig. 4.**, starting from point **"A"** *(surface slope in %)*, to point **"B"** *(values of erosion coefficient (Z) and erosion category)*. (Ref. 8.).

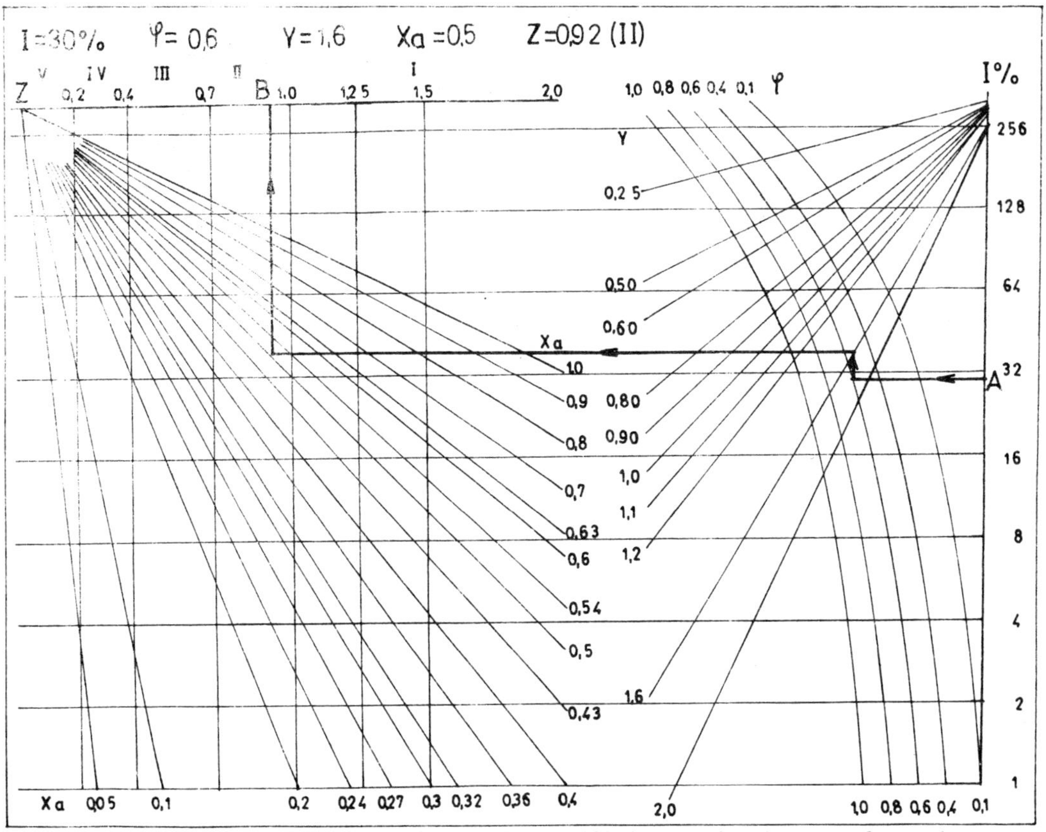


**Figure 4. Nomogram for determining the coefficient and category of erosion.**

The nomogram shown is duplicated on one sheet together with the coefficient tables, and this has become a handy method for fast field investigation.

## 2/1.9. ESTIMATE OF SEDIMENT PRODUCTION

Erosion categories in other methodologies are qualitative gradients of erosion intensity, which usualy do not give even an approximate quantitative value of sediment production, and are usable only in descriptive presentations.

The categories of erosion have, by the erosion quantitative classification method, both qualitative and quantitative representation expressed in the coefficient of erosion, which enables further calculation of sediment production from the relation **(Ref. 5.)**:

$W_{sp} = T * H * \quad * Z^{3/2}$ **m³/km²/year** where,

$W_{sp}$ is average annual specific production of sediments per **km²** in **m³/year**

T is temperature coefficient calculated from $T = ( (t/10) + 0.1)^{1/2}$

were **t** is mean annual air temperature in degrees Celsius (°C)

**H** is mean annual amount of precipitations in ***milimeters* (mm)*/year***

is **PI** number **(3,14...)**

**Z** is coefficient of erosion calculated by the described methodology.

Note that the dependence of the average annual production of sediment on the average amount of precipitation and temperature has been found. While such dependence seems rough and inaccurate, it approximates reality, because erosion is effected by nearly all rains and thermal variations, or, in short, the climate. It is the climate that is defined in terms of mean annual precipitations and temperatures, or the result of statistic processing of long-term precipitation and temperature data.

Average annual production of erosional sediments in the studied catchment area ($W_g$), is a multiplication product of the surface area (F) in **km²**, and the weighted averaged specific sediment production ($W_{sp}$).

$W_g = F * W_{sp}$ **m³/year**

## 2/2. CALCULATION OF SEDIMENT LOAD TRANSPORT AND DISCHARGE

The method of quantitative classification allows calculation of erosion-produced sediments, which, although very useful information, is not complete because not all of the produced sediment is discharged.

Eroded material is generally transported for a long time and fragmented on its way. Along the flow path, from the source to the discharge point, the trend of the grain size distribution curve is towards the fine grain-size. The variation in grain size distribution may be smaller or greater in relation to the mechanical strength of sediments, length of transportation, and level differnce of transportation route. Mechanical strenght of sediments to disintegration is discussed in the section on the coefficient of erosion, whereas the lenght of transportation route and level difference are elements of topographic features.

A correlation was noted in measured data between sediment production and discharge volumes, which has been ascribed to topographic feature effects. Discharged volumes are always lower than the produced ones. The difference in sediment volumes is due to a natural retention of sediments down the river.

### 2/2.1. CALCULATION OF SEDIMENT RETENTION COEFFICIENT

To invesigating the relationship between sediment retention coefficient and topography, various topographic features and their effects were examined. A satisfactory accuracy was given by the empirical formula for sediment retention coefficient **(Ru)** **(Ref. 2.)**:

$R_u = 4*(O * D)^{1/2} / ( L+10 )$

**where,** O is perimeter of the basin in **km**
L is lenght of the basin in **km**
D is mean level difference in the basin in **km**.

The mean basin level difference (D) is the difference between the mean altitude of the basin ($N_{sr}$) and the level of the discharge point or studied section ($N_u$).

$$D = N_{sr} - N_u \quad (km')$$

Each level difference is the basis for calculation of potential energy by Newton's law of gravitation. Assumptions on the dependance sediment load retention with main river bed gradient were checked during the investigation, but the results were not as expected. Only the mean level difference in the basin (D) gave satisfactory results; this is easy to explain.

Rainfalls, storm rains in particular, cause flow over the drainage area which carries worn material, accumulating it in tributary channels. The water-borne sediments gain in acceleration over the slopes in the basin, but its energy is proportional to the covered level difference; the mean value of this level difference being equivalent to the mean level difference of the basin (D).

## 2/2.2. CALCULATION OF SPECIFIC AND TOTAL SEDIMENT DISCHARGES

Sediment discharge or sediment transport calculated by this method are two identical values. The term ***"sediment discharge"*** is used for storage reservoirs where the sediments end their route, and the term ***"sediment transport"*** is used in calculations for any river section. Both terms imply average annual volume of sediment load which reaches or flows through a given river cross-section. The calculated sediment volume includes the sums of bed and suspended load volumes.

Specific mean annual discharge of eroded materials ($G_{sp}$) is the volume of the average annual sediment discharge from a square kilometre of drainage area, which is the multiplication product of specific mean annual sediment production ($W_{sp}$) and coefficient of sediment retention ($R_u$). **(Ref. 5.).**

$$G_{sp} = W_{sp} * R_u \quad (m^3/km^2/year)$$

Total mean annual discharge of eroded materials ($G_g$) is the multiplication product of total mean annual sediment production ($W_g$) and coefficient of sediment retention ($R_u$).

$$G_g = W_g * R_u \quad (m^3/year)$$

The erosion potential method does not account for the bed/suspended load volume ratio and their grain-size distribution curves. Where this information is necessary, a grain-size analysis must be carried out for bed and suspended loads and a budget account for suspended sediment discharge made which will give the required ratio.

## 3/ CORRELATION OF EROSION INTENSITY AND RESERVOIR SILTATION

The rate at which a reservoir fills up with sediments is related to the intensity of erosion, but the relation is multi-funtional rather than proportional or simple. **(Ref. 8.).**

The filling up of reservoir with sediments depends on four principal relations: erosion, topography, volume, and surface of the basin.

Intensity of erosion and topographic features in the basin control the specific reservoir siltation, expressed in annual rates **($m^3/km^3/year$)**.

The rate of specific siltation is the average volume of sediment expressed in $m^3$, annually discharged into the storage space from each $km^2$ of the drainage area.

The rate of specific siltation is only an indication of the drainage area potential, while the siltation value is determined by the multiplication specific siltation and the drainage area.

The above stated refer to the volume of sediment reaching a water storage area; the actual hazard of the reservoir filling up with eroded materials is something else.

If the specific rate of siltation is $G_{sp}$ **($m^3/km^2$/year)** and drainage area is **F ($km^2$)**, then

$$G_g = G_{sp} \times F \quad (m^3/\text{year})$$

where $G_g$ is the rate of average annual siltation in a reservoir. For an estimate of a reservoir section siltation risk, is the volume of the clogging material, reaching the reservoir within **fifty years** of its service age is important. This volume, $G_{50}$ **($m^3$)**, is:

$$G_{50} = G_g \times 50 \quad (m^3)$$

The volume of sediments which will reach a reservoir during fifty years is only the erosion potential expected to fill up the storage space. This volume is constant for a cross-section, and varies only with the variation of erosion intensity in the drainage area. It is independent of the storage volume variation.

Storage volume is cotrolled by natural and economic factors. Natural factors are storage capacity of water from the drainage area and potential size of the storage space. Economic factors define the storage volume in relation to water requirements and possible water use.

Water storage potential of the basin means the volume of water sufficient for normal operation and existance of the storage reservoir. The water storage potential varies from one basin to another, and from one region to another, but a rough average quantity can be given for specific water storage potential of a basin.

The specific water storage potential of a basin is estimated usin the available hydrologic methods. Where data are not available for a basin, the recommended quantity is **$V_{sp}$=120.000 $m^3/km^2$** ***(for Europe)***, which is the maximum storage volume that can be provided in a basin of one square kilometre in area. Consequently, water potential of a basin is:

$$V_p = V_{sp} \times F \quad (m^3)$$

Potential size of a storage space is controlled by topographic features of the river valley upstream of the proposed dam site. This volume can be smaller or larger than the volume of water potential from the basin.

Economic factors are also important to consider when determining the actual storage volume. The relocation of villages and infastructures within a future storage area usually reduce the storage much beyond the limitations imposed by topographic features. On the contrary, the state of development in the valley will allow an economical construction for a larger storage volume than the potential volume of water from the basin.

The very costly construction of a dam or impounding reservoir is mostly governed by economic considerations. Complete neglet of the erosion situation is not uncommon in such economic considerations, reasulting in the omission, or reduction, of costs for anti-erosional development in the basin. The reason for this attitude can only be explained by the fact that, for most of the reservoirs, the anti-erosional development costs, if included in the project cost, would make the project uneconomical. There are few projects or feasibility studies in which a reservoir siltation risk coefficient has been properly estimated, and consequently a reasonable estimate of erosion-control works.

This coeficient is not difficult to calculate, as will be explained. First to be calculated is the rate of potential water use from the basin (**$P_i$**) for the given storage volume. The rate for a given storage volume (**$V_a$**) and basin water potential (**$V_p$**) is:

$$P_i = V_a / V_p$$

The rate of water potential use from the basin can be higher than unity, but generally it is lower. The optimum quantity of the water potential use is unit or more, while quantities lower than unity are due to the above mentioned factors.

Because the erosion potential is unchanged, depending on the situation in, and size of, the drainage area, the coefficient of storage siltation risk is not difficult to determine. This coefficient (**$K_u$**), for average moderate erosion in the basin, is roughly proportional to the reciprocal of the rate of water potential use (**$P_i$**).

$$K_u = 1/P_i$$

If the coefficient of siltation risk is lower than unity, then the valley cross-section is suitable and economical, even including erosion-control works. Valley cross-sections, for which the coefficient is much higher than unity, are not suitable. For drainage areas heavily attacked by erosion processes, correction must be made for the increased sediment above that of moderate erosion. (**Ref. 8.**).

## 4/ A CASE EXAMPLE IN THE USE OF THE EROSION POTENTIAL METHOD

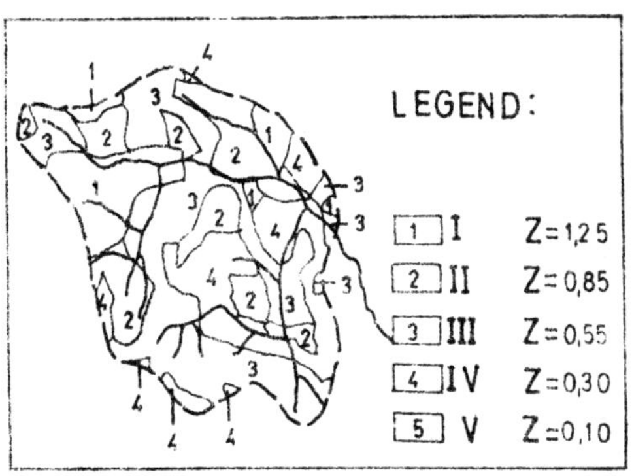


**Figure 5. Erosion Map**

Case examples can best illustrate the use of a method. A common case example has thus been selected; that being, calculation of erosion-produced material and its transport to a valley cross-section considered for a dam site, and expected siltation of the future storage reservoir.

The data base are a topographic map and erosi on mapon the same scale. The discussed case example is a reservoir storing water from a catchment area attacked by mixed processes of erosion ***(moderate erosion on average)***.

The erosion map of the catchment area is shown in **Fig. 5**. Topographic, climatic, and erosional characteristics found from maps and available data, are the following:

| | | | | |
|---|---|---|---|---|
| Catchment area surface: | $F=347.2$ km² | Areal incidence of erosion: | | |
| Catchment area length: | $L=27.2$ km | Category | I | 13.7 % |
| Catchment area perimeter: | $O=81.4$ km | Category | II | 22.3 % |
| Mean altitude: | $N_{sr}=958$ m | Category | III | 39.6 % |
| Mean level difference: | $D=0.470$ km | Category | IV | 24.4 % |
| Mean annual amount of precipitations: | 850 mm | Category | V | 0.0 % |
| Air temperature: | 9.7 °C | Coefficient of erosion for catchment area | | $Z=0.652$ |

Calculated production and discharge of sediments are:

| | Production of eroded material | Coefficient of retention | Sediment discharge |
|---|---|---|---|
| Specific: | $W_{sp}=$ 1454 m³/km²/year | $R_u=0.665$ | $G_{sp}=$ 967 m³/km²/year |
| Total: | $W_g=504829$ m³/year | $R_u=0.665$ | $G_g=335711$ m³/year |

Total sediment discharge at storage site for 50 years: $G_{50}=16,785.550$ m³

Potential water storage, found through hydrologic considerations, is $V_p=45,540.000$ m³. The site conditions allow for a maximum storage volume of $V_a=26,500.000$ m³. The rate of potential run-off water use is $P_i=0.58$. For this drainage area exposed to moderate erosion on average, the formula can be used for calculation of the expected siltation in the reservoir: $K_u=1.72$, indicating the feasibility of dam construction at a reasonable cost of erosion control works.

The importance of the above consideration is noted, as early as at storage site selection stage, with many large and small storage reservoirs which got filled with sediments within short periods. In instances where the preliminary estimate of the sediment production and discharge was made using the erosion potential method, significant futile investments in dam construction were avoided; if not, storage reservoirs would soon have become ineffective due to filling with erosion products.

## REFERENCES:


1. *Gavrilovic S.*:"Methodology for Classification of Erosion Processes and Mapping of Eroded Areas. Stage 1". Beograd, Institut "Jaroslav Cerni", 1966. (Izrada metodike za klasifikaciju erozionih procesa i kartiranje erozionih podrucja I faza).

2. *Gavrilovic S.*: "Proposed Clasification of Torrents for the Water Manangement Plan of Yugoslavia" Beograd, Institut "Jaroslav Cerni", 1968. (Predlog klasifikacije bujicnih tokova za izradu vodoprivredne osnove Jugoslavije).

3. *Kohnke H.* and *Bertrand R.A.*:"Soil Conservation". McGRAW-HILL, New York, 1959

4. *Gavrilovic S.*:"Contribution to the Methodology for Research in Erosion Plots" In: International Seminar: Erosion, Torrents, and Alluvial Deposits,Beograd 1970. (Doprinos metodici za naucna istrazivanja na erozionim parcelama).

5. *Gavrilovic S.*:"Engineering of Torrents and Erosion". Izgradnja, special isue, Beograd 1972. (Inzenjering o bujicnim tokovima i eroziji).

6. *Browing G.M.*:"A Method for Estimating Soil Management Requirements.Iowa, 1947

7. *Gavrilovic S.*:"Classification of Torrents in Grdelica Canyon and Quantitative Regime of Torrent-Borne Sediments" Beograd, 1957. (Klasifikacija bujicnih tokova Grdelicke klisure i kvantitavni rezim njihovih nanosa)

8. *Gavrilovic Z.*: "Storms and River Disasters - Erosion and Torrents"Conference on Disasters and Catastrophies (Budva YU, 1986). EROZIJA I BUJICNI TOKOVI).

International Conference on
RIVER REGIME
18-20 May, 1988

PAPER J3

# EARTHQUAKES AND RIVER REGIMES

Arun Bapat
Central Water & Power Research Station
Pune-411024 (India)

SUMMARY: Occurence of earthquake in river regime affects river in upper ridges, plains and delta regions. These effects are principally due to faulting, shaking, ground failures and tsunamis. The pre-seismic, co-seismic and post-seismic effects on rivers in India, New Zealand and Latin America are discussed.

1.0 Introduction: Rivers have been a perennial source of enrichment since the early times. Most of the ancient cultures have flourished in the vicinity of rivers and paradoxically some of these have been destroyed due to flood or earthquake. In modern times, industrial and technological progress has been achieved with the help of water which has helped in increasing agricultural yield and generating hydro-electric power. Because of the pervasive role of water in the affairs of man, development of hydrologic science has followed rather than led the applications, primarily water supply and hazard reduction under the leadership of civil and agricultural engineers. The cultivation of hydrology and its harmonious development has not occurred. It is therefore pertinent that efforts should be directed towards this objective. The global scale environmental problems in water management are increasing and have been constrained by deficiencies in basic hydrologic science. Problems of scaling, equilibrium, stability, teleconnections and space and time variability require a new attention to basic hydrologic research. The fundamental statistical-dynamical problem of hydrology remains unsolved. Understanding the dynamics at various space time scales in the presence of great heterogeneity is still a big problem. At present dynamical studies are confined only to flow of water. Earthquake occurrence in river regime is significant transient dynamic phenomenon which has not been considered seriously by researchers. The effect of earthquake on rivers though known to most of the scientists and engineers has not been seriously considered during the planning and development of river valley projects. The consideration of seismic force on body of dam has been limited by providing appropriate seismic co-efficient in design of dam. At some sites, studies are also undertaken for Reservoir Induced Seismicity (RIS). Both these studies account for dam and reservoir only and not the entire river regime, upstream or downstream of dam. However effects of earthquakes on water, flow mechanisms, sedimentation, river bed slope, change in courses of rivers etc. has not been considered.

International Conference on River Regime, edited by W.R.White.

It is proposed to discuss effects of earthquakes on river regimes as evidenced by several field reports on macroscopic post-seismic happenings.

2.0 Earthquakes: It is felt that a small paragraph about earthquakes would be useful to those who are not much conversant with this subject. These details have been taken from Richter (Ref 1). Earthquake constitutes a sub-surface geological phenomenon, along an established tectonic fault. Differential rates of movements on either sides of fault due to various dynamic forces, cause accumulation of stress energy over length of time. This mechanical energy is stored inside the rocks till it could withstand within its elastic limits. Further increase of stress could not be absorbed by the geological material and the rock yields to this stress. Sudden release of huge energy constitutes occurrence of earthquake. The energy thus released causes crushing of material in the sub-surface area, usually know as hypocenter. The point radially above hypocenter is known as epicenter, which is normally given by latitude, longitude and depth of focus which is the distance between epicenter and hypocenter in Km. The strength of earthquake is measured by Richter scale, which is logarithmic in nature and indicates the amount of energy released in ergs. Earthquakes of magnitude 5.0 and above are felt by all people in the epicentral area. Conventionally shocks of 5.0 - 6.0 magnitude earthquakes are known as mild, 6.1 - 7.0 as moderate and 7.1 and above as major earthquakes. The energies associated with magnitudes 5.0, 6.0, 7.0 and 8.0 are about $10^{19}$, $10^{20}$, $10^{22}$ and $10^{23}$ respectively (Unitergs). In the annals of seismic history, the largest magnitude recorded is 8.7 for Assam earthquake of 1897 and Chile earthquake of 1960. Part of energy released during earthquake travels in the form of vibrations and the graphical record on seismograph is known as seismogram. Records of medium and major earthquakes all over the world are freely available at the International Seismological Centre at Edinburgh (UK) and also at US Geological Survey. In addition every country has its own seismic centre.

2.1 Effects of Earthquakes: Occurrence of medium to major earthquake changes the geological scenario in the epicentral area. The destructive effects of earthquake are observed by modified Mercalli scale, usually known as Modified MM Intensity scale. It is always written in capital Roman numerals from I, II, III.... X, XI and XII. Normally intensity V is perceptible by human beings and XII, the highest intensity indicates total destruction. The intensity of an earthquake goes on decreasing with distance from epicenter and the highest intensity is known as highest isoseismal followed by contours of decreasing intensity.

2.2 Though occurrence of an earthquake is a transient phenomenon, which lasts for a few seconds, not more than 10-15 sec. under normal circumstances, the stress building activity is spread over several decades prior to the occurrence of the seismic event. The effects of earthquake on river regime are divided into three categories viz:

1. Pre-seismic effects
2. Co-seismic effects
3. Post-seismic effects

Usually Pre-seismic effects are seen on micro-scale and need special observations and are seen a few years prior to occurrence of earthquake. The Co-seismic effects are mostly on macro-scale and are observed for small duration, up to 100-200 hours which includes the main shock followed by several aftershocks which further contribute to the destabilization of the already weakened geological material. The post-seismic effects are on mega-scale and are extensive over river regime and could be seen for a few years. The release of large energy during earthquake causes effects on geo-environments. The destructive effects could be broadly classified in following four groups, which are based on seismic processes:

| | Type of effect | Evidenced by |
|---|---|---|
| a) | Faulting .. .. .. .. .. .. | Actual ground rupture |
| b) | Shaking .. .. .. .. .. .. | Vibrations of natural and man made objects |
| c) | Ground failure .. .. .. .. | Landslides, rockfalls, slips |
| d) | Tsunamis .. .. .. .. .. .. | High sea waves |

It is proposed to discuss above effects as manifested by different seismic processes in three different parts of river viz. upper ridges, plains and deltas from cases in India, New Zealand and Latin America.

3.0 Pre-seismic effects: The most significant pre-seismic effects as observed by seismologists is change in ground level, either sinking or upheaval of land (Ref 2). Ground tilt is a sure and reliable precursor of any earthquake. Prior to major earthquakes, level changes up to 4-8 meters have been observed. In the case of Alaska earthquake of 1964 of 8.25 magnitude, level changes of the order of 12 meters have been reported by Scott (Ref 3). There are reports of level changes prior to earthquake from several other seismically active countries of the world. The change in level causes change in river slope. If slope is reduced the velocity of river water reduces and causes additonal deposit of sediments and more spreading of water occurs during flood periods. If slope is increased the velocity of water is also increased and it causes excessive erosion. Such cases have been reported by hydrologists without knowing the seismic reason behind it.

4.0 Co-seismic effects: The most significant co-seismic effect is manifested in the form of massive displacement of geological materials, as a result of liquefaction in river regime encompassed by highest isoseismal. Due to liquefaction, several landslides occur and sometimes flow of river is temporarily blocked due to formation of natural dams. These dams give way after a few days and flash floods occur.

4.1 In Peru, the largest rockslide was caused by the 1970 earthquake of magnitude 7.7 when the western part of the summit of Mount Nevadas Huascaran collapsed (Ref 4). It broke away from shear, west peak of the mountain at an elevation of about 5500-6000 m. The volume of the displaced material has been estimated to be about 50-80 million $m^3$. The displaced mass took about 3-4 minutes for travel. The distance travelled was about 3 Km vertically and about 12 Km horizontally, which destroyed the 200-250 m high ridge. It blocked the Llanganuco river and caused the water level to rise and destroyed Santa river valley. In the deep glacial valleys of Cordillera Blanca, many lakes were formed behind morainal deposits that marked the retreating of ice front.

4.1.1 During the Peruvian earthquake, alluvial fans were dissected and these were filled up by recent sand and lagooned deposits. At Casma, differential compaction and lateral spreading probably produced by liquefaction of loose, saturated, sandy or granular materials occurred in beach deposits and produced many tension cracks. Because of differential compaction in some areas underlain by loose, saturated sand the areas were inundated by local groundwater rising above the ground surface.

4.2 In India the largest earthquake during this century occurred on 15th August 1950, with magnitude of about 8.6 in Assam province (Ref 5). The river Brahmaputra and its several tributaries were severely affected. River Subansiri which is one of the major tributaries, dried up on account of slip-dam across the river. The 40 m high dam burst on the night of 19th August 1950, causing widespread havoc in the Subansiri valley thereby inundating about 1000 $Km^2$ area with a population of about 50,000 people, which took toll of about 500 lives approximately.

4.3 In New Zealand, a landslide triggered by Inangahua earthquake, dammed the Buller river in 1968 (Ref 6). The landslide contained about $4.3 \times 10^6$ $m^3$ of material and formed a dam 12 m high that pounded the river for a length of 6 Km. When the dam failed suddenly 21 hours after the earthquake, approximately $1.5 \times 10^6$ $m^3$ of water was released causing flash flood.

5.0 Post-seismic effects: The major post-seismic effects are seen in the form of change in river course, heavy sedimentation, rise in bed level, change in granular and chemical composition of sediment, change in quality of water etc.

5.1 After the Peruvian earthquake, aerial photographs of the affected region showed strong evidence of post-seismic geomorphic activity along Santa river valley. A substantial amount of the mud flow and debris flow material resulted from bursting of natural morainal dammed glacial lakes and also from bursting of natural dams. The levels of rivers Llanganuco and Santa were raised by 2-3 meters. After the Assam earthquake the bed of river Brahmaputra and its several tributaries rose on an average by 3-4 meters. Several tributaries of Brahmaputra such as Subansiri, Dirgha, Kadam and Kakoi left their old channels at places and have cut new ones, thus causing heavy destruction of cultivated lands. Considerable changes took place in Brahmaputura due to movements of sand and fresh sediments which resulted in loss of navigability. Similar is the case of river Buller in New Zealand.

5.2 Seismosedimentation: Bapat (Ref 7) has defined the additional amount of sediments due to earthquake as:

"Geological materials which are crushed, fractured, loosened or displaced directly or indirectly as a result of earthquake induced processes such as landslides, landslips, slumps, mudflows, avalanches, liquefaction, rockfalls, rockslides etc. and which move under gravity in the catchment area of river are defined as Seismosediments and the process described as Seismosedimentation".

This amount of seismosediments has been quantified by the following mathematical relationship

$$Q = RPTF$$

where Q (Tons) is the amount of seismosediments during period T (Years), R (Tons/year) is average annual sedimentation rate and P is the probability of occurrence of an earthquake, F is the fractions of sediments. This relationship has been further modified to account for contributions by direct magnitude ranges of earthquakes as

$$Q = RT\ (P_1F_1 + P_2F_2 + P_3F_3)$$

where $P_1$, $P_2$ and $P_3$ are the probabilities of occurrences of earthquakes of three different magnitude ranges 5.5 - 6.4, 6.5 - 7.4 and $M \geqslant 7.5$ and $F_1$, $F_2$ and $F_3$ are the corresponding fractions of sediments due to earthquakes of three different magnitudes (Ref 8).

5.3 Tsunamis: The major earthquake related hazard to the coastal region of Peru is from destructive tsunamis. No such damage due to severe water waves was observed during the 1970 earthquake, presumably because the causative fault displacement was relatively deep seated. However, many large destructive tsunamis have accompanied previous major earthquakes along the Peruvian coast during historical period of the deltaic regions of rivers Rio Santa, Rio Nepena, Rio Casma and towns Chimbote and Pt. Casma were severely affected due to seismic waves. The 1883 Krakatoa volcanic explosion had generated tsunami waves which hit the eastern coast of India and also the New Zealand coast. In addition, several earthquakes in Latin America, Alaska and Japan have generated tsunamis which hit the Pacific area coastlines.

6.0 Discussion: In order to assess the likely effects of any potential earthquake/s on river regime, the following procedure could be used. These guidelines have been formed because data in this field are not available from different countries.

6.1 Examine from the available list of historical earthquakes whether there is any reasonable and finite probability of occurrence of medium to large earthquake during foreseeable future or the projected period of studies. This exercise is mostly probabilistic assessment of seismic risk. The probability of occurrence of earthquake, P could be determined from the available list of earthquakes and assuming statistical distribution of earthquake process to be Poissonian by standard relation

$$P = 1 - e^{-\lambda T}$$

where $\lambda$ is the mean annual frequency of earthquakes and T (years) is the period. If the statistical analysis indicates a finite ($P \geqslant 0.6$)

probability of an earthquake, examine the associated fault, river regime, geology and broadly locate the likely epicenter of the probable earthquake. The error associated in this location could be of the order of $\pm$ 50 Km. After assessing the likely location of epicenter, examine the area with the help of Band 4 or 7 satellite imageries for available period. If the area is really heading for any seismic activity, changes in surface drainage pattern would indicate it or there could be changes in river courses or small rivers may undergo termination. The change observed with the help of satellite imageries should be confirmed by ground checks. In addition, installation of micro-seismic net, measurement of other geological and seismological parameters would help in identifying the epicenter with an accuracy of $\pm$ 20 - 30 Km.

6.2 Work out the likely levels of accelerations due to predicted earthquake at different distances and its effect on local geology. Plot likely isoseismals for a few probable locations of epicenters. Examine this with reference to the river regime. Various likely effects could then be worked out. The effects discussed in this paper need not be taken as exhaustive, there could be some other effects depending upon local geo-environment. Allen (Ref 9) has given some useful relationship between earthquake magnitude frequency, epicentral distance and soft sediment deformations in sedimentary basins. Keefer (Ref 10) has studied data from 40 historical worldwide earthquakes to determine the characteristics, geologic environments and hazards of landslide caused by seismic events. He has further observed that maximum area likely to be affected by landslides in a seismic event increases from approximately 0 at magnitude 4.0 to about 500,000 $Km^2$ at highest magnitude. These relations could serve as useful guidelines.

7.0 Conclusions: It is observed that occurrence of medium to major earthquakes in the vicinity of river regime adversely affects the river regime in various ways. The effect can be as small as just a seiche in the river or as large as termination of the river itself. A realistic assessment of these likely effects would help in understanding the earthquake effects on river regime. Future river valley development projects could be planned in a better way so as to mitigate the earthquake effects.

References:


1. Richter, C.F.: "Elementary Seismology". San Francisco, W.F. Freeman and Co., 1958.

2. Rikitake, T.: "Earthquake Prediction". Amsterdam, Elsevier Publishing Co., 1977.

3. Scott, R.: "Soil mechanics and foundation engineering aspects of the Alaskan earthquake of March 1964". In: Proc. 3rd World Conf. on Earthquake Engineering (New Zealand 1965) Vol I, pp 157-172.

4. Cluff, Lloyd S.: "Peru earthquake of May 31, 1970, Engineering geology observations". Bull. Seism. Soc. Am. 60(5), pp 1561 - 1606, 1970.

5. Potdar, M.C.: "Preliminary report of the Assam earthquake, 15th August 1950". In: Geological Survey of India, Calcutta, India, 1950.

6. Adams, John: "Earthquakes, landslides and large dams in New Zealand". Bulletin of the New Zealand National Society for Earthquake Engineering, 14(2), June 1981, pp 93 - 95.

7. Bapat, Arun: "Generation of silt due to earthquakes". In: Proc. XIX Congress of International Association of Hydraulic Reserch (New Dehli, India), 1981, Vol. II, pp 43 - 48.

8. Bapat, Arun: "The phenomenon of seismosedimentation". In: Proc. 2nd International Workshop on Alluvial River Problems (Dept. of Civil Engineering, University of Roorkee, Roorkee, India, 1985) pp 47 - 52.

9. Allen, J.R.L.: "Earthquake magnitude-frequency, epicentral distance, and soft sediment deformation in sedimentary basins". Sedimentary Geology, 46, January 1986, pp 67-75.

10. Keefer, David K.: "Landslides caused by earthquakes". Bulletin Geological Society of America, 95(4), April 1984, pp 406 - 421.

International Conference on
**RIVER REGIME**
18-20 May, 1988

**PAPER J4**

# REGIME THEORY AND ENVIRONMENTAL CHANGE - IRRECONCILABLE CONCEPTS?

J. Lewin[1], M. G. Macklin[2] and M. D. Newson[2]
1. Department of Geography, University College of Wales, Aberystwyth, U.K.
2. Department of Geography, University of Newcastle upon Tyne, U.K.

SUMMARY: Regime theory as used in river engineering makes the assumption of dynamic equilibrium; however, both intrinsic and extrinsic factors in gravel-bed rivers lead to channel instability which undermines engineering designs/ structures. Using examples from Northern England and Wales a spectrum of geomorphological evidence and guidance is made available to engineers.

## 1. INTRODUCTION

The regime concept in river engineering makes both spatial and temporal assumptions about the adjustment of channels to river flow. The spatial assumptions are rational: the channel is adjusted to accommodate the dominant discharge. The temporal assumptions are seldom quoted, although Blench (Ref 1) is bold enough to suggest that the regime approach treats average river behaviour over 20 to 40 years. It is perhaps more useful, since this paper seeks to advance a geomorphological contribution to river engineering, to suggest that regime applies to "engineering time" (Refs 2 and 3). Engineering time (around a century or less) is that period over which an engineering project is designed to last. It is over just such a period that our data base of rates and directions of natural channel change is now established and it is becoming clear that there are systematic inconsistencies in the assumption of an equilibrium channel form, even a dynamic one, over such periods. River equilibrium states have been through a period of redefinition of late (e.g. Schumm (Ref 4); sources of instability considered here are both intrinsic and extrinsic.

Intrinsic influences, recently revealed by studies of sediment transport processes, derive from the intermittent movement of sediment seawards in river systems - on what Ferguson (Ref 5) calls a "jerky conveyor belt". This applies particularly to gravel bedload; the river systems considered here are gravel-bed systems. At one extreme of the recorded "jerky" behaviour one may cite the evidence of Church and Jones (Ref 6) that course sediment moves through channels as a sequence of large depositional zones or megaforms. At the other extreme (in both space and time) is the riffle-pool sequence. Both influence channel dynamics and morphological response.

The extrinsic sources of river instability over engineering timescales are perhaps more obvious, though may seldom be considered by practical engineering: changes of flood magnitude and frequency and changes of sediment supply or calibre.

The process of river channel adjustment to new environmental constraints is not immediate and may take several decades to complete. River channel change also has a spatial context, in that upstream changes may have effects which

International Conference on River Regime, edited by W.R.White.

work their way downstream with a temporal lag between cause and effect especially in the middle and lower reaches of river basins. Allen (Ref 7) has provided comprehensive theoretical and practical guidance on these aspects of fluvial sediment systems.

The result is that many river channels may be actually enlarging, contracting, aggrading or incising rather than in a state of equilibrium. To establish whether a river is adjusted or adjusting, and the probable nature and magnitude of future channel change is of particular importance for engineers managing river systems.

## 2. ENVIRONMENTAL CHANGE, INSTABILITY AND BRITISH RIVERS

It is perhaps especially important to demonstrate to engineers in Britain that patterns of river instability pose a systematic threat to design in some of our river systems. Vanoni (Ref 8) points to historic differences in attitude by British and American river engineers, the former espousing regime approaches and the latter approaching via sediment transport dynamics. Another national feature is the impressive stability of British lowland river channels. Ferguson (Ref 5) refers to the very large differences in stream power between upland and lowland rivers and Dury (Ref 9) reports that the Great Ouse passed a 1000-year flood without measurable channel change!

Newson (Ref 3) has compiled evidence from British geomorphologists working in Britain for channel instability in a variety of upland and "piedmont" locations. These may be classified as follows:

2.1 Response to, and recovery from, large floods. Detailed geomorphological compilations of "rare" flood effects on channels have appeared during the last decade. Impressive amongst these in terms of the potential danger of regime assumptions during recovery are the Howgill event of 1982 (Refs 10 and 11) and the Hurricane Charley event of 1986 (Ref 12).

2.2 Changes in flood magnitude and frequency. These may be influencing channel capacity through:

(i) Climatic change, i.e. increasing rainfall inputs (Ref 13),

(ii) Changes in land-use which increase (Ref 14) or decrease (Ref 15) flood magnitude,

(iii) Changes imposed progressively by river regulation from upstream dams (Ref 16),

(iv) Changes resulting from declining floodplain storage. Over long periods these may be caused by changes in agriculture and settlement. Over shorter periods channelization and embankment may increase the stream power and have a (relatively) sudden destabilizing effect on channels (Ref 17).

2.3 Changes resulting from sediment supply fluctuation

(i) Land-use is under an intense spotlight currently with increasing sediment yields recorded from afforestation (Refs 18, 19, 20, 21) and intensive agriculture.

(ii) Metal mining has, in the past, produced very marked increases in sediment loads in Britain with resulting channel metamorphosis (Ref 22),

(iii) Pressure on aggregate supplies has encouraged gravel extraction from rivers on an industrial scale in several river systems, promoting a morphological response (Ref 23).

(iv) The effects of river regulation on sediment supply for downstream channel adjustment (Ref 24).

Such compilations of individual reports establish little more than a warning that regime conditions are unlikely to apply at a variety of locations in British gravel-bed rivers. It is, therefore, important to move towards more detailed case studies to illustrate the nature of the observed channel instabilities, their location and duration.

## 3. CASE STUDIES OF THE TIMESCALES OF CHANNEL RESPONSE TO AND RECOVERY FROM ENVIRONMENTAL CHANGE

3.1 The South Tyne. The North Pennine area of northern England was the most extensive and productive base metal ore field in Great Britain, accounting for a third of the total dressed ore produced in the country between 1840 and 1940 (Ref 25). Extending over an area exceeding 2500 $km^2$ the waste products of metal mining influenced in a very significant way both the quantity and quality of sediment supplied to river systems in the region. Before large-scale mechanisation of the mines in the latter part of the eighteenth century waste appeared to have been transported alongside "natural" bedload and suspended sediment in a manner that did not disrupt, in any major way, pre-existing patterns of channel development and floodplain deposition. But by the mid-nineteenth century in many of the most intensively-mined parts of the orefield, notably in the headwaters of the Tyne, Tees and Wear rivers, dramatic channel and floodplain metamorphoses had taken place. These involved a change from meandering to braided channel patterns associated with valley floor aggradation that has been followed by incision and reversion to a single channel in the present century.

The timescale over which channels and their floodplains responded to and recovered from this sediment-supply-induced episode of river instability is well documented within the Rivers Nent and West Allen, the principal tributaries of the River South Tyne (Fig. 1). Most river systems in the Northern Pennines are characterised by valleys consisting of a series of steep, narrow, bedrock or drift-confined reaches where floodplains are poorly developed, that alternate with sections of lower gradient with wider valley floors and laterally extensive recent floodplain sediments. These alluvially-controlled sedimentation basins, over the period of topographic documentation ($\leq$ 200 years), have generally been characterised by rapid lateral channel migration and are invariably sites of current river instability. In this respect the Nent and West Allen floodplains, near Blagill and Blackett Bridge respectively, are generally representative of most river valleys in the Northern Pennines affected by metal mining. At both sites (Figs. 2, 3) relict, unvegetated gravel bars, typical of those found within a mobile braided river system, are presently being incised by a single-thread channel of much lower sinuosity.

Historic maps of the floodplain at Blagill (Fig. 4) date the transformation of the River Nent from a single, laterally-stable channel to an aggrading, gravel-dominated braided river to between 1840 and 1861. Channel and floodplain metamorphosis appears to have been initiated by an influx of a large quantity of bed load calibre sediment following a large flood in the mid 1840s (Ref 26). Coarse flood deposits of this age are also preserved in the West Allen valley at Blackett Bridge (Fig. 3), where a progressive reduction in floodplain width, active gravel area and total channel sinuosity associated with channel incision, reflects prolonged post-flood channel and floodplain readjustment dating from the mid-nineteenth century (Ref 27).

Although metal mining-sediment linkages can be clearly identified within the fine fraction (< 2000 μm) in the Nent and West Allen floodplains (both are grossly contaminated by lead and zinc), examination of coarser sediments involved in the valley flow aggradation at Blackett Bridge shows only a minor proportion to be directly the product of mining. Coarse mine waste, however, is abundant on the floodplain of the Nent at Blagill; spoil from nearby mines here provided a ready source of coarse sediment that may have indirectly prepared the site for metamorphosis but certainly directly perpetuated the transformation by continuing to supply waste of a bedload calibre to the river channel.

3.2 The Rheidol and the Ystwyth. The Rheidol and Ystwyth are two small and adjacent catchments (respectively 182 $km^2$ and 170 $km^2$) draining to the sea in mid-Wales: in many respects similar (Ref 24), they differ in that the Rheidol has been regulated since 1963 for hydro-electric power generation. Water from 34% of the catchment is impounded in Nant-y-Moch reservoir, and is then fed through pressure tunnels, secondary impoundments and generating stations, eventually being returned to the floodplain channel some 10 km from the river mouth. In this lower river section, discharge and sediment loads have been considerably modified. Flow extremes have been reduced, coarse material entrainment frequency has been lowered, and the supply of fine sediment in suspension has decreased. This may be appreciated by comparing the sediment discharge of the regulated Rheidol with the unregulated Ystwyth (Fig. 5).

There are continuing adjustments to channel morphology as a result. In the immediate vicinity of the major dam, there is a build-up of in-channel sediments from tributary inputs, these being no longer flushed away by the main river (Ref 28). In the lowest 10 km of the river, lateral erosion of the meandering river continues, but the supply of fine sediment is now inadequate to provide a fine bar-top and overbank sediment cover for the coarse lateral accretion channel deposits. The result is a progressive coarsening of flood-plain and channel bank sediments, and this process of adjustment to altered sediment supply will continue until the floodplain has been worked across by the river: a matter of many decades and probably centuries. Such a length of equilibrative period needs to be considered for other regulated rivers in Britain. Williams and Wolman report U.S. data on this problem (Ref 29).

Both Rheidol and Ystwyth have been grossly affected by metal mining activity. Valley floor development has seen phases of aggradation and incision since the last glaciation of the area, with a sequence of terraces being best appreciated on the Rheidol (Fig. 6) (taken from Ref 30). Thus the highest level, some 10 m above the present river in both valleys, can be related to glacial outwash; the lowest (T5, floodplain complex) involves aggradation and incision related to mining activity and its aftermath. This can be confirmed by reference to historical maps and documents. On the Ystwyth at Trawscoed, for example, (Ref 23), it appears that peak nineteenth-century mining activity was followed by floodplain aggradation of some metres totally, with braided river channels and the development of splays of gravel out across the pre-mining floodplain (Fig. 7). In this century, with the cessation of mining, the river has incised itself into and become confined within the nineteenth-century deposits, such that an abnormally deep channel has been created, with the floodplain hardly liable to flood from the main river at all. Judging from undermined bank protection works and from extended observations of the channel for nearly two decades, the river appears still to be incising in places.

3.3 The Tywi. Downstream of Llandovery, where the upper Tywi (232 $km^2$) and Bran (67 $km^2$) combine, the Tywi flows in a major 'piedmont' alluvial trough. Here the actively migrating river (Fig. 8) is affected by adjustment to environmental changes of three major kinds:

Channel patterns are transitional between braiding and meandering. Under natural conditions, major floods appear to generate major lobate gravel bedforms with avalanche fronts and a multichannel system. In intervening time periods (over periods of years) there is a reversion to a meandering habit, with bars becoming islands or attached to one side of the channel or another. Thus adjustment and recovery to large floods is a natural characteristic of the river at this point and notions of equilibrium must here include and encompass such a fluctuating behaviour pattern.

The river is now regulated for water supply purposes by Llyn Brianne (1972). This appears to have eliminated flow extremes and the channel pattern of 1971 (Fig. 8) now follows a single course with a decreased area of exposed and bare gravel alongside this channel. Lateral erosion continues (the rate of lateral reworking has been analysed using air photography and does not appear to have decreased) though the width of the high-flow channel has effectively decreased. Adjustment to altered discharge and sediment conditions will presumably continue for some decades.

Gravel extraction has led to localised channel responses. The convenience and safety of extracting shoals has led to a wider, braided channel pattern in the affected reaches. This effect is localised but another, incipient, result is the loosening of sediments in abandoned extraction sites; during the inevitable and seasonal avulsions of the main flow thread into these zones bedload is easily entrained and transported. The downstream effect is sedimentation of the pool-riffle sequence (which has ecological implications) and increased channel width (affecting flood protection and agriculture). Water authority policy at downstream sites has recently become one that will not provide local bank protection in the face of such demonstrable instability of channel plan.

## 4. CONCLUSIONS

Recent alluvial histories of the Rivers Nent and West Allen in the South Tyne basins thus show the extended timescales over which adjustment to new equilibrium conditions following environmental change may take place. Furthermore, many river systems in the Northern Pennines, or for that matter other metal mining areas of upland Britain (Ref 31), are still undergoing changes in channel morphology, readjusting to new rates of sediment supply, fifty years after the cessation of mining. One practical consequence of channel incision has been a significant reduction in floodwater storage in many Northern Pennine headwater streams resulting in increased flood damage in downstream middle and lower river reaches. It is clear from investigation of floodplains in north-east England that channel instability problems require appreciation of relict as well as the functional components of a river system particularly when instability cause and effect may be lagged over several decades.

The Rheidol and Ystwyth appear still to be adjusting to altered sediment supply. The morphology of these alluvial rivers can best be regarded as equilibrating rather than as having actually achieved a state of equilibrium, and it is clear that such processes have to be regarded as taking decades and centuries to achieve.

If engineers are invited to intervene to "cure" river instability of the types detailed above they clearly need to know the diagnostic signs. Table 1 is an attempt at a qualitative geomorphological guide, keyed to those forms of instability over engineering timescales we have listed in Section 2. Table 2 is a guide to documentary sources of guidance on river channel change. Stevens, Simons and Richardson (Ref 32 ) conclude that engineers need to accept that river form will change during the lifetime of a project, <u>if the form is</u>

not controlled. Structural controls on form are, however, nowadays unpopular economically and ecologically. Vanoni (Ref 8) sees the river engineer falling back from prediction to experience and intuition. Hey (Ref 33) would like to see a move forward to fully rational river mechanics. In some ways geomorphologists contribute between these extremes, perhaps erring towards "quantified intuition". As interest in geomorphology shifts from the observation of equilibrium states to the recognition of multiple stable and unstable equilibria (Ref 34), it is clear that regime assumptions need setting in a more dynamic framework. To abandon regime bespeaks a replacement; dialogue between engineering and geomorphology is desirable to derive this framework.

## 5. REFERENCES


(1) Blench, T: "Regime behaviour of canals and rivers". London, Butterworths, 1957, 138 pp.

(2) Hickin, E.J: "River channel changes: retrospect and prospect". Special Publication 6 of the International Association of Sedimentologists, 1983, pp 61-83.

(3) Newson, M. D: "River basin engineering - fluvial geomorphology". Journal of the Institution of Water Engineers and Scientists, 40 (4), 1986, pp 307-324.

(4) Schumm, S.A: "The fluvial system". New York, John Wiley and Sons, 1977, 338 pp.

(5) Ferguson, R.I: "Channel form and channel changes". In British Rivers, (ed. J. Lewin), London, George Allen and Unwin, 1981, pp 90-125.

(6) Church, M. and Jones, D: "Channel bars in gravel-bed rivers". In "Gravel-bed rivers", (eds. R. D. Hey, J. C. Bathurst and C. R. Thorne), Chichester, John Wiley and Sons, 1982, pp 291-338.

(7) Allen, J.R.L: "Reaction, relaxation and lag in natural sedimentary systems: general principles, examples and lessons". Earth-Science Reviews, 10, 1974, pp 263-342.

(8) Vanoni, V.A: "Fifty years of sedimentation". Journal of Hydraulic Engineering (ASCE), 110 (8), 1984, pp 1022-1057.

(9) Dury, G.H: "Magnitude-frequency analysis and channel morphology". In "Fluvial Geomorphology", (ed. M. Morrisawa), London, George Allen and Unwin, 1981, pp 91-121.

(10) Harvey, A.M: "Geomorphic effects of a 100-year storm in the Howgill Fells, Northwest England". Zeitschrift für Geomorphologie, 30 (1), 1986, pp 71-91.

(11) Wells, S.A. and Harvey, A.M: "Sedimentologic and geomorphic variations in storm-generated alluvial fans, Howgill Fells, Northwest England". Geological Society of America Bulletin, 98, 1987, 182-198.

(12) Macklin, M.A., and Newson, M.D: "Hurricane Charley: flooding in Swaledale and Teesdale". Seminar Paper, University of Newcastle, Department of Geography (in prep.)

(13) Higgs, G: "Environmental change and flood hazard in the upper Severn catchment". In "The hydrology, sedimentology and geomorphological implications of floods", (eds. P. Carling and K. J. Beven), Chichester, John Wiley and Sons (in press).

(14) Robinson, M: "Changes in catchment runoff following drainage and afforestation", Journal of Hydrology, 86, pp 71-84.

(15) Newson, M.D. and Robinson, M: "Effects of agricultural drainage on upland streamflow: case studies in mid-Wales". Journal of Environmental Management, 17, 1983, pp 333-348.

(16) Petts, G.E: "Morphological changes of river channels consequent upon headwater impoundment", Journal of the Institution of Water Engineers and Scientists, 34(4), 1980, 374-382.

(17) Leeks, G.J., Lewin, J. and Newson, M.D: "Channel change, fluvial geomorphology and river engineering: the case of the Afon Trannon, mid-Wales", Earth Surface Processes and Landforms (in press).

(18) Newson, M.D: "The erosion of drainage ditches and its effect on bed-load yields in mid-Wales", Earth Surface Processes, 5, 1980, pp 275-290.

(19) Robinson, M. and Blyth, K: "The effect of forestry drainage operations on upland sediment yields: a case study". Earth Surface Processes and Landforms, 7, 1982, 85-90.

(20) Burt, J.P., Donohoe, M.A. and Vann, A.R: "A comparison of suspended sediment yields from two small upland catchments following open ditching for forestry drainage". Zeitschrift für Geomorphologie, N.F. 51, 1984, 51-62.

(21) Stott, T.A., Ferguson, R.I., Johnson, R.C. and Newson, M.D: "Sediment budgets in forested and unforested basins in upland Scotland", International Association of Scientific Hydrology, Publication 159, 1986, 57-68.

(22) Lewin, J., Bradley, S.B. and Macklin, M.G: "Historical valley alluviation in mid-Wales". Geological Journal, 18, 1983, pp 331-350.

(23) Leeks, G.J., Newson, M.D. and Lewin, J: "River gravel extraction: implications for fisheries and river engineering on the Tyne". (Report to Welsh Water and paper in prep.).

(24) Grimshaw, D.L. and Lewin, J: "Reservoir effects on sediment yield". Journal of Hydrology, 47, 1980, pp 163-167.

(25) Dunham, K.C: "The Geology of the Northern Pennine Orefield 1". Memoirs of the Geological Survey of Great Britain, 1948.

(26) Macklin, M.G: Channel and floodplain metamorphosis in the River Nent, Cumberland. In "Quaternary river landforms and sediments in the Northern Pennines", (eds. M.G. Macklin and J. Rose), 1986, pp 19-33. British Geomorphological Research Group/Quaternary Research Association.

(27) Macklin, M.G. and Aspinall, R.J: Historic sedimentation in the River West Allen, Northumberland: a case study of channel change in an upland gravel-bed river in the Northern Pennines. In "Quaternary river landforms and sediments in the Northern Pennines", (eds. M.G. Macklin and J. Rose), 1986, pp 7-17. British Geomorphological Research Group/Quaternary Research Association.

(28) Petts, G.E: "Sedimentation within a regulated river". Earth Surface Processes and Landforms, 9, 1984, pp 125-134.

(29) Williams, G.P. and Wolman, M.G: "Downstream effects of dams in alluvial rivers". United States Geological Survey, Professional Paper, 1286, 1984, 83 pp.

(30) Macklin, M.G. and Lewin, J: "Terraced fills of Pleistocene and Holocene age in the Rheidol valley". Journal of Quaternary Science, 1, 1986, pp 21-34.

(31) Lewin, J. and Macklin, M.G: "Metal mining and floodplain sedimentation in Britain". In "International Geomorphology", (ed. V. Gardiner), 1986, John Wiley & Sons, pp 1009-1027.

(32) Stevens, M.A., Simons, D.B. and Richardson, E.V: "Non-equilibrium river form". Journal of the Hydraulic Division (ASCE), 101 (HY5), 1975, pp 557-566.

(33) Hey, R.D: "River mechanics", Journal of the Institution of Water Engineers and Scientists, 40 (2), 1986, pp 139-158.

(34) Thornes, J.B: "Fluvial processes". In "Process in geomorphology", (eds. C. Embleton and J. Thornes), London, Edward Arnold, 1979, pp 213-271.

TABLE 1

FIELD INDICATORS OF CHANNEL INSTABILITY

| Field evidence | Instability tendency and probable sequence |
|---|---|
| **Floodplain morphology** | |
| 1. Inactive multichannel system on floodplain | (a) (i), (iii)→(ii),(iv) adjustment timescale measured in decades<br>OR<br>(b) Possible channelisation |
| 2. Terraces on valley-floor alluvium | (a) (i)→(ii) adjustment timescale measured in decades<br>OR<br>(b) (ii)→tendency of British rivers to incise during the Holocene |
| **Channel morphology** | |
| 1. Undermined bank protection structures, bridge footings, exposed tree roots on bankside | (a) (ii)<br>OR<br>(b) Faulty design |
| 2. Buried structures, contracting bridge openings | (a) (i)<br>OR<br>(b) Over-designed channel |
| 3. Erosion of both channel banks, increase in river width | (a) (iii)<br>OR<br>(b) Under-designed channel |
| 4. Channel infilling and contraction | (a) (iv)<br>OR<br>(b) Over-designed channel |
| 5. All flood flows conveyed within channel | (a) (ii), (iii)<br>OR<br>(b) Designed channel enlargement works |
| **Floodplain sedimentary sequences** | |
| 1. Buried soils within alluvium | (a) (i)<br>OR<br>(b) Made ground |
| 2. Textural changes in vertical sequences of alluvium | |
| e.g. Thick coarse within-channel sediments overlying out-of-channel fines | (a) (i)<br>OR<br>(b) Made ground |
| e.g. 2. Fine sediment overlying terraced coarse within-channel deposits | (a) (ii)<br>AND/OR<br>(b) (iv) |

KEY

Instability tendencies

| | | | | | |
|---|---|---|---|---|---|
| Vertical | (i) | aggradation | Lateral | (iii) | enlarging |
| | (ii) | incision | | (iv) | contracting |

TABLE 2

DOCUMENTARY SOURCES FOR IDENTIFYING POSSIBLE CAUSES OF CHANNEL INSTABILITY

| | Documentary sources | Possible cause of channel instability |
|---|---|---|
| 1. | Water Authority records | Changes in flood and/or precipitation characteristics (magnitude, frequency, hydrograph parameters) |
| 2. | UK Land-use surveys, annual agricultural returns, Forestry Commission | Land-use change (afforestation, agricultural practice, land drainage) |
| 3. | Water Authority records | Channel engineering (channelisation, bank protection) |
| 4. | Water and planning authorities | Changes in sediment regime (gravel extraction, mining, industrial activity) |

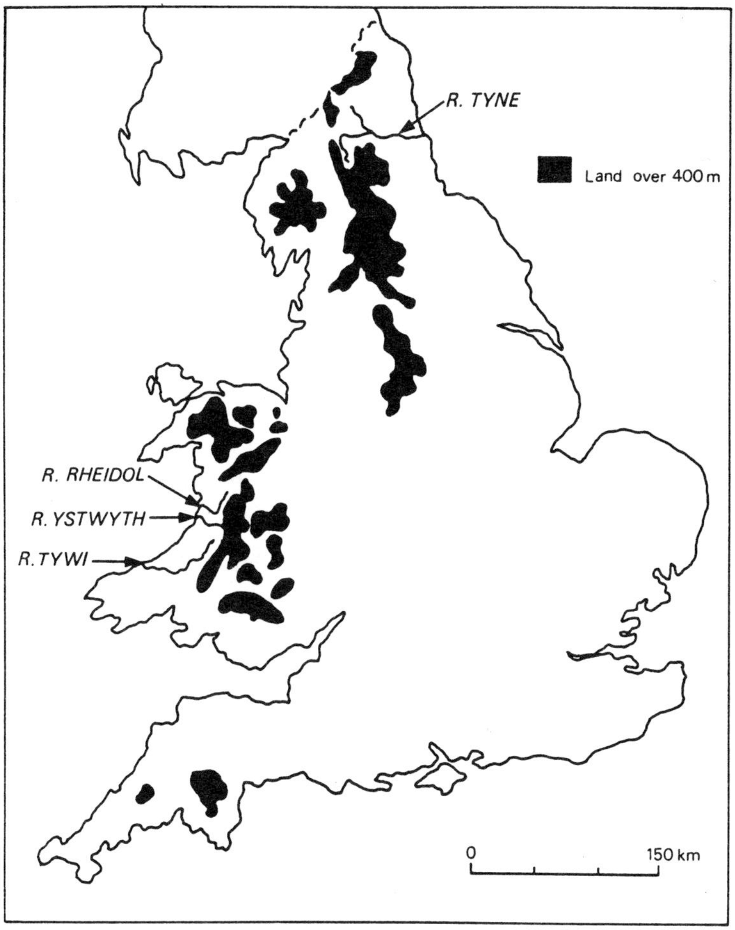


LEWIN, MACKLIN AND NEWSON FIGURE 1
Location of field sites England and Wales

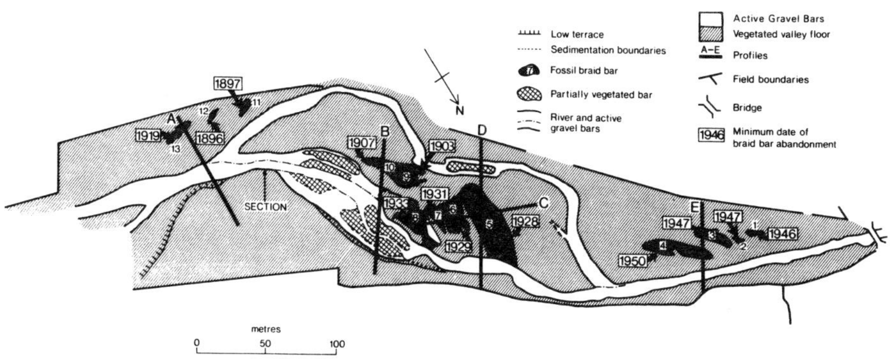


LEWIN, MACKLIN AND NEWSON FIGURE 2
River Nent and its changing channel plan near Blagill

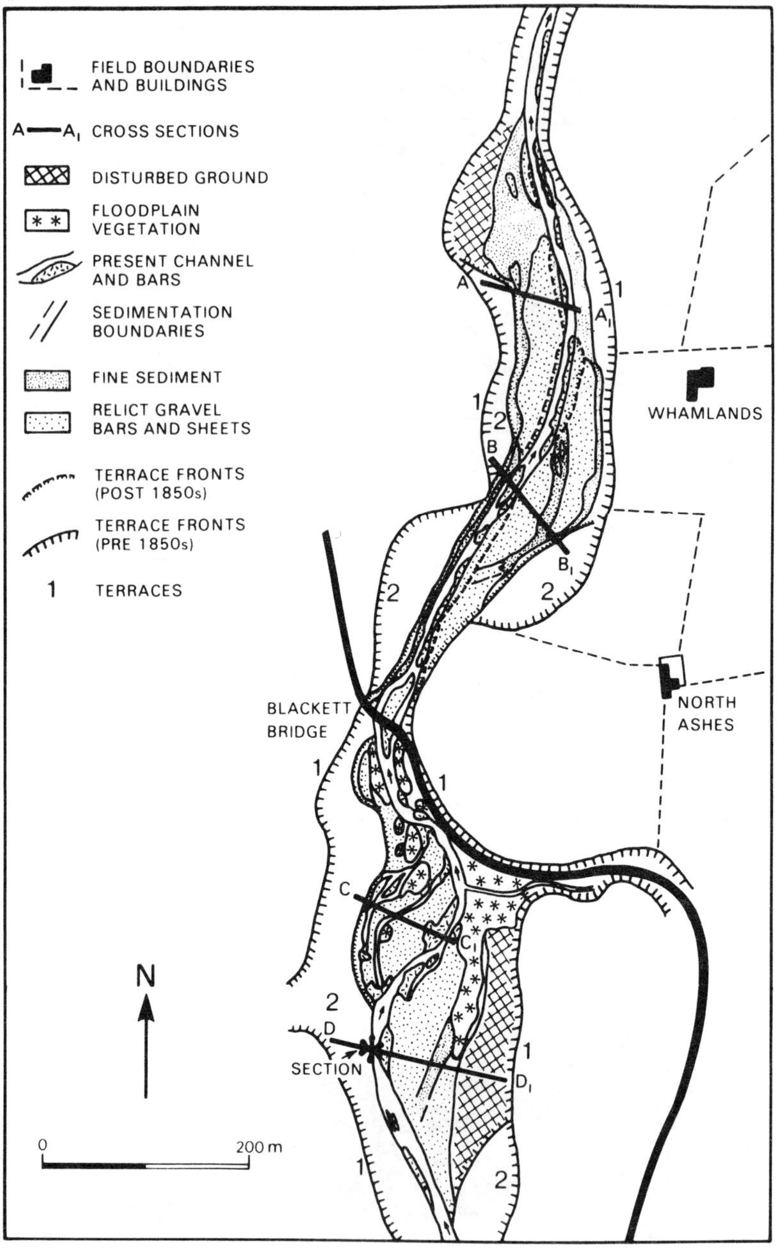


LEWIN, MACKLIN AND NEWSON FIGURE 3
Channel pattern on the River West Allen at Blackett Bridge

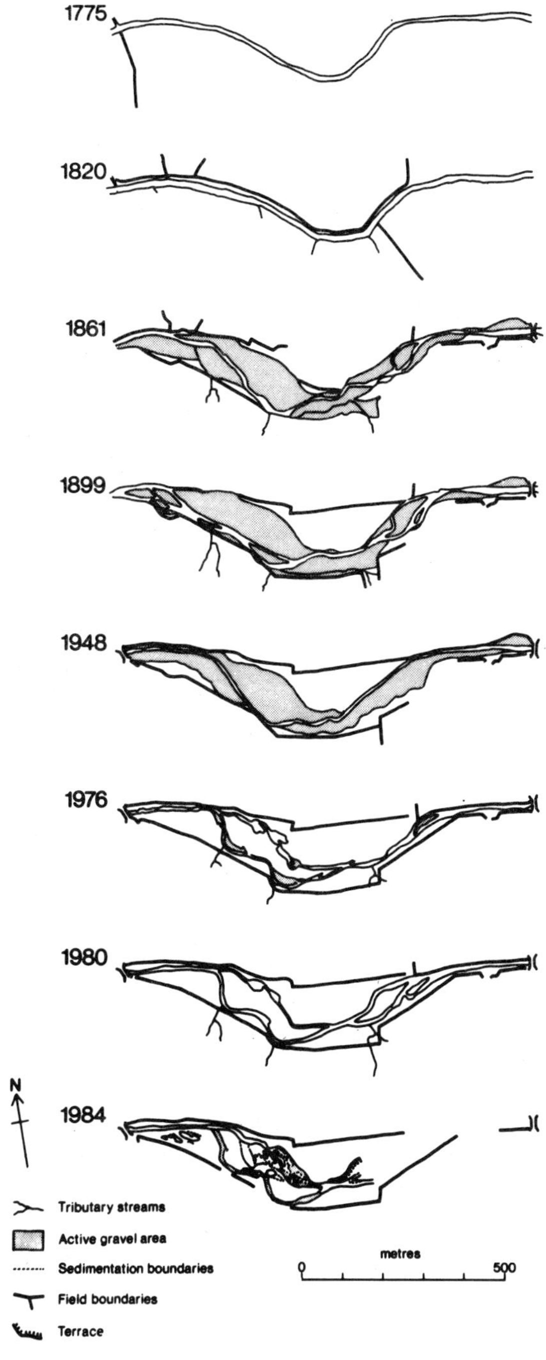


LEWIN, MACKLIN AND NEWSON FIGURE 4
Historic maps of the channel and floodplain of the West Allen at Blagill

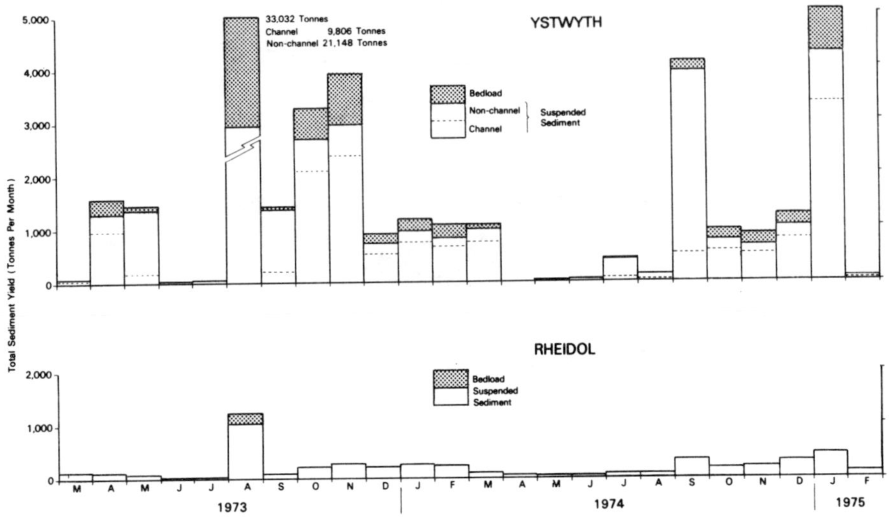


LEWIN, MACKLIN AND NEWSON FIGURE 5
Comparative sediment discharge of the regulated Rheidol and unregulated Ystwyth rivers, mid-Wales.

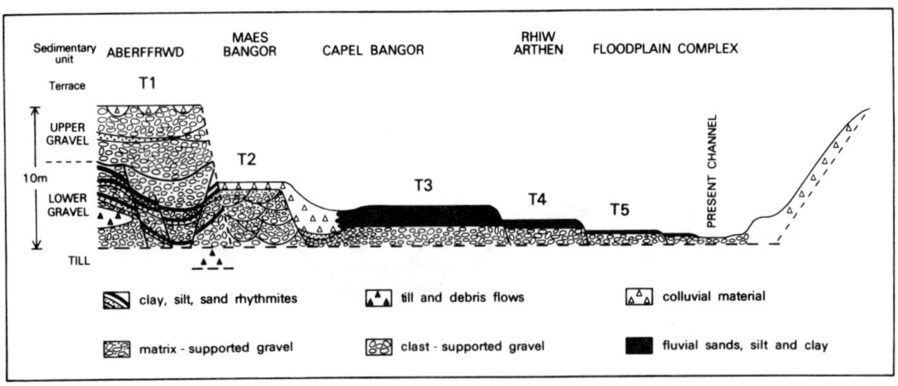


LEWIN, MACKLIN AND NEWSON FIGURE 6
Terrace sequences of the River Rheidol

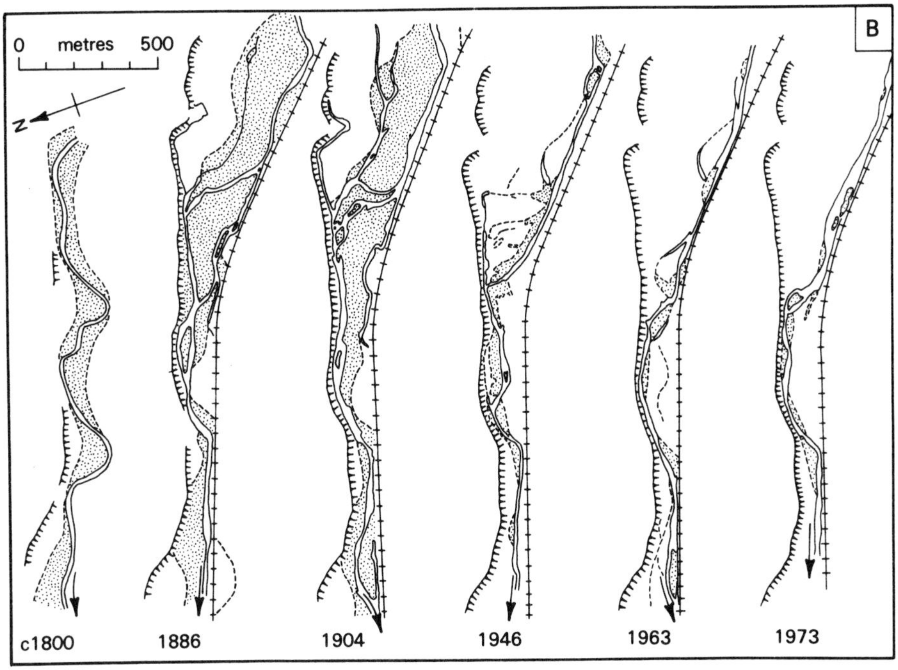


LEWIN, MACKLIN AND NEWSON FIGURE 7
Channel changes on the River Ystwyth during mining disturbance and recovery

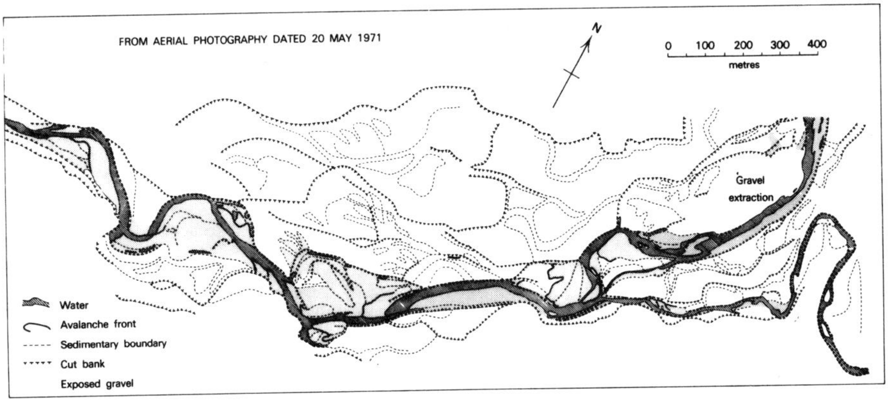


LEWIN, MACKLIN AND NEWSON FIGURE 8
Migrating channel of the River Tywi, downstream of Llandovery.